BioMEMS Devices

BioMEMS Devices

Edited by

Azrul Azlan Hamzah
Institute of Microengineering and Nanoelectronics (IMEN), Universiti Kebangsaan Malaysia, Bangi, Selangor, Malaysia

Jumril Yunas
Institute of Microengineering and Nanoelectronics (IMEN), Universiti Kebangsaan Malaysia, Bangi, Selangor, Malaysia

Anis Nurashikin Nordin
Department of Electrical and Computer Engineering, International Islamic University Malaysia, Kuala Lumpur, Malaysia

ELSEVIER

ACADEMIC PRESS
An imprint of Elsevier
elsevier.com/books-and-journals

Academic Press is an imprint of Elsevier

125 London Wall, London EC2Y 5AS, United Kingdom
50 Hampshire Street, 5th Floor, Cambridge, MA 02139, United States

ISBN: 978-0-443-24015-7

For Information on all Academic Press publications visit our website at https://www.elsevier.com/books-and-journals

Publisher: Mara Conner
Acquisitions Editor: Sonnini Yura
Editorial Project Manager: Sumit Yadav
Production Project Manager: Maria Bernard
Cover Designer: Mark Rogers

Typeset by Aptara, New Delhi, India

Contents

Contributors

Norazreen Abd Aziz
Faculty of Engineering and Built Environment, Universiti Kebangsaan Malaysia, Bangi, Selangor, Malaysia

Siti Fatimah Abd Rahman
School of Electrical and Electronic Engineering, Universiti Sains Malaysia, Engineering Campus, Nibong Tebal, Pulau Pinang, Malaysia

Nur Shahira Abdul Nasir
Centre of Liberal Studies, National University of Malaysia, Bangi, Selangor, Malaysia

Arifah Syahirah Abdul Rahman
Institute of Microengineering and Nanoelectronics (IMEN), Universiti Kebagsaan Malaysia, Bangi, Selangor, Malaysia

Anis Amirah Alim
Institute of Microengineering and Nanoelectronics (IMEN), Universiti Kebangsaan Malaysia, Bangi, Selangor, Malaysia

Mohd H.S. Alrashdan
Department of Electrical Engineering, College of Engineering, Al-Hussein Bin Talal University, Ma'an, Jordan

Hafeez Anwar
Department of Physics, University of Agriculture Faisalabad, Faisalabad, Pakistan

Abdul Basit
Harbin Institute of Technology, Nangang, Harbin, China

Muhamad Ramdzan Buyong
Institut Kejuruteraan Mikro dan Nanoelektronik, Universiti Kebangsaan Malaysia, Bangi, Selangor, Malaysia

Thikra S. Dhahi
Physiotherapy Department, Medical & Health Technology College, Southern Technical University, Basrah, Iraq

Celine Elie-Caille
Institut FEMTO-ST, UMR 6174 CNRS, Université de Franche-Comté, Besançon, France

Subash C.B. Gopinath
Department of Neonatology, Saveetha Medical College and Hospital, Saveetha Institute of Medical and Technical Sciences, Chennai, Tamil Nadu, India; Faculty of Chemical Engineering & Technology, Universiti Malaysia Perlis (UniMAP), Arau, Perlis, Malaysia; Institute of Nano Electronic Engineering, Universiti Malaysia Perlis (UniMAP), Kangar, Perlis, Malaysia

Mohd Nizar Hamidon
Electrical and Electronic Engineering Department, Faculty of Engineering, Universiti Putra Malaysia (UPM), UPM Serdang, Selangor, Malaysia

Azrul Azlan Hamzah
Institute of Microengineering and Nanoelectronics (IMEN), Universiti Kebangsaan Malaysia, Bangi, Selangor, Malaysia

Abdul Manaf Hashim
Malaysia-Japan International Institute of Technology, Universiti Teknologi Malaysia, Kuala Lumpur, Malaysia

Ahmad Ghadafi Ismail
Institute of Microengineering and Nanoelectronics (IMEN), Universiti Kebangsaan Malaysia, Bangi, Selangor, Malaysia

Aziah Ismail
Institute of Research in Molecular Medicine (INFORMM), Universiti Sains Malaysia, Health Campus, Kubang Kerian, Kelantan, Malaysia

Nur Farhani Ismail
Institute of Microengineering and Nanoelectronics (IMEN), Universiti Kebangsaan Malaysia, Bangi, Selangor, Malaysia

Mohd Hafizuddin Junaidi
Faculty of Electronic Engineering and Technology, Universiti Malaysia Perlis, Arau, Malaysia

Shama Kazmi
Monash Institute of Medical Engineering (MIME), Monash University, Melbourne, VIC, Australia

Mazlee Mazalan
Faculty of Electronic Engineering and Technology, Universiti Malaysia Perlis, Arau, Malaysia

Sawal Hamid Md Ali
Faculty of Engineering and Built Environment, Universiti Kebangsaan Malaysia, Bangi, Selangor, Malaysia

P. Susthitha Menon
Institute of Microengineering and Nanoelectronics (IMEN), Universiti Kebangsaan Malaysia, Bangi, Selangor, Malaysia

Umi Zulaikha Mohd Azmi
Nuclear Emerging Technology Center, Technical Support Division, Malaysian Nuclear Agency, Kajang, Selangor, Malaysia

Nur Nasyifa Mohd Maidin
Institute of Microengineering and Nanoelectronics (IMEN), Universiti Kebangsaan Malaysia, Bangi, Selangor, Malaysia

Ahmad Fairuzabadi Mohd Mansor
Department of Electrical and Computer Engineering, Kulliyyah of Engineering, International Islamic University Malaysia, Kuala Lumpur, Malaysia

Nashrul Fazli Mohd Nasir
Biomedical Electronic Engineering Program, Faculty of Electronic Engineering and Technology, Universiti Malaysia Perlis, Arau, Malaysia

Anas Mohd Noor
Biomedical Electronic Engineering Program, Faculty of Electronic Engineering and Technology, Universiti Malaysia Perlis, Arau, Malaysia

Budi Mulyanti
Study Program of Electrical Engineering, Universitas Pendidikan Indonesia, Bandung, Indonesia

Sh. Nadzirah
Institute of Nano Electronic Engineering, Universiti Malaysia Perlis (UniMAP), Kangar, Perlis, Malaysia; Institute of Microengineering and Nanoelectronics (IMEN), Universiti Kebangsaan Malaysia, Bangi, Selangor, Malaysia

Anis Nurashikin Nordin
Department of Electrical and Computer Engineering, International Islamic University Malaysia, Kuala Lumpur, Malaysia

Mehmet Ozsoz
Department of Biomedical Engineering, Near East University, Mersin, TRNC, Turkey

Fahreddin Palaz
Department of Surgery, College of Medicine, University of Florida, Gainesville, FL, United States

Roer Eka Pawinanto
Study Program of Industrial Automation Engineering Education and Robotics, Universitas Pendidikan Indonesia, Bandung, Indonesia

Ruslinda A Rahim
Lynas Rare Earths, Kuantan, Pahang, Malaysia

R. D. A. A. Rajapaksha
Department of Nano Science Technology, Faculty of Technology, Wayamba University of Sri Lanka, Kuliyapitiya, Sri Lanka

Santheraleka Ramanathan
Department of Chemical Engineering, Faculty of Engineering, University of Malaya, Kuala Lumpur, Malaysia; Institute of Nano Electronic Engineering, Universiti Malaysia Perlis (UniMAP), Kangar, Perlis, Malaysia

Muhammad Akmal Suhaimi
Institute of Microengineering and Nanoelectronics (IMEN), Universiti Kebangsaan Malaysia, Bangi, Selangor, Malaysia

Ioana Voiculescu
Mechanical Engineering Department, The City College of New York, The City University of New York, New York, NY, United States

Jumril Yunas
Institute of Microengineering and Nanoelectronics (IMEN), Universiti Kebangsaan Malaysia, Bangi, Selangor, Malaysia

Noraziah Mohamad Zin
Center for Diagnostic, Therapeutic and Investigative Studies, Faculty of Health Sciences, Universiti Kebangsaan Malaysia, Jalan Raja Muda Abdul Aziz, Kuala Lumpur, Malaysia

Arash Zulkarnain
Institut Kejuruteraan Mikro dan Nanoelektronik, Universiti Kebangsaan Malaysia, Bangi, Selangor, Malaysia

Preface

Biological microelectromechanical systems (BioMEMS) is an emerging field, stemming from the classic microelectromechanical systems (MEMS), where mechanical microelements and electronic circuits converge to create a function-specific system to serve specific purposes. An MEMS system can either sense, or actuate, or do both simultaneously. As its name suggests, BioMEMS is a category of MEMS that handles biological targets and can detect changes in its environment. Recently, BioMEMS has become a key technology in the biomedical field, as it can miniaturize large laboratory equipment into rapid and portable devices, enable highly precise treatments, and allow real-time monitoring of human health.

This book is intended for technologists, researchers, technicians, and students working in the field. On the sensing front, this book covers electrical, electrochemical, optical, micromechanical, and microelectromechanical sensors. This book then delves into mechanisms and devices allowing biomolecules and cells manipulation such as microneedles, micromixers, micropumps, and biomolecules manipulation and filtration using dielectrophoresis. These micromachines allow for precise drugs and targets control in an MEMS-based drug delivery system. As a progression from the basic BioMEMS building blocks, this book assesses point-of-care biosensors for diagnosing diseases, BioMEMS uses in medical implants, BioMEMS embedded in the human body, and the use of lab-on-chip for diagnostic purposes.

In the final part of this book, data transfer, data analysis, and management technology for biomedical devices are assessed. This book also discusses strategies to bring new BioMEMS devices from the laboratory to the market. Finally, this book summarizes emerging BioMEMS technologies and how these could be implemented to improve patients' or consumers' health and well-being now and in the future.

The authors would like to extend their heartfelt gratitude to their families, colleagues, and friends for their unwavering support and invaluable feedback during the publication journey of this book. This endeavor would not have been possible without their encouragement and guidance.

CHAPTER

Introduction to BioMEMS and microsystems

1

Azrul Azlan Hamzah[a], Anis Nurashikin Nordin[b] and Jumril Yunas[a]
[a]*Institute of Microengineering and Nanoelectronics (IMEN), Universiti Kebangsaan Malaysia, Bangi, Selangor, Malaysia,* [b]*Department of Electrical and Computer Engineering, International Islamic University Malaysia, Kuala Lumpur, Malaysia*

Bio microelectromechanical systems (BioMEMS) is a cutting-edge and quickly developing discipline that emerged from a combination of biology and microelectromechanical systems (MEMS). The way we approach biological research, therapies, and diagnostics has been completely transformed by these systems, allowing a combination of both microfabricated and biological components.

As a subdiscipline of MEMS, this emerging field integrates microscale transducers, microfluidic networks, and functionalized surfaces to interface with biological entities such as biomolecules and tissues. Advances in MEMS techniques such as soft lithography, silicon micromachining, polymer processing, and additive manufacturing have accelerated the translation of laboratory-scale bioanalytical concepts into robust microsystems. The miniaturization of these analytical systems produced distinct advantages such as reduced sample volumes, rapid response times, lower power consumption, and portability. These features are critical for low-cost, decentralized diagnostics and precision medicine . BioMEMS technologies, ranging from lab-on-a-chip (LOC) devices to implantable biosensors, are pushing the boundaries of traditional medical and biological tools, offering unprecedented precision, sensitivity, and portability.

This book provides a complete review of recent advances in BioMEMS, with a special emphasis on biosensing technologies and their applications in drug administration, diagnostics, and therapeutic interventions.

By leveraging on recent breakthroughs in materials science, advanced semiconductor microfabrication, and improved understanding of biological processes, researchers have developed sophisticated microscale devices capable of precise biological target detection, drug delivery, and controlled cellular manipulation that mimic complex biological systems. These techniques allow fabrication of high-aspect ratio structures to produce high surface-area-to-volume ratios and highly sensitive biosensors for rapid biochemical detection. This technique is very advantageous for biomedical devices as it allows the incorporation of microfluidics, sensors, and actuators into BioMEMS devices known as LOC systems.

Microsystem technology demonstrates its primary advantage, particularly in medical devices, in its ability to reduce the dimensions of equipment while increasing the performance. Miniaturization is crucial in the medical device industry because it enables less invasive procedures and faster recovery. By utilizing microsystems, we can improve the accuracy and precision of medical devices, resulting in more effective diagnosis and treatment of diseases in the future. LOC systems can perform various medical tests on a single chip. These devices are small enough to fit in a handheld phone or any

BioMEMS Devices. DOI: https://doi.org/10.1016/B978-0-443-24015-7.00001-2

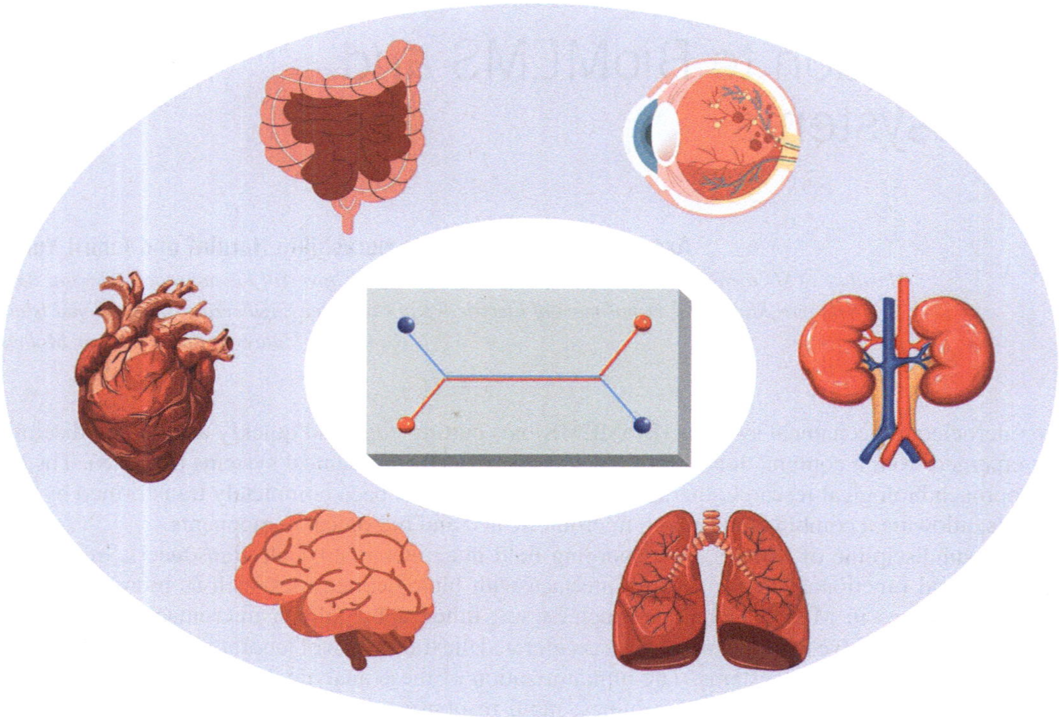

FIGURE 1.1

Organ-on-a-chip.

mobile instrumentation but can precisely analyze blood, urine, or saliva samples for a variety of medical conditions. LOC devices can help diagnose diseases early, leading to more effective treatments and better patient outcomes. These platforms are critical for investigating cellular behaviors, medication interactions, and disease causes in a controlled context, therefore improving the precision of biomedical research and diagnoses (Chircov & Grumezescu, 2022; Motreuil-Ragot et al., 2022).

Among BioMEMS technology's most important accomplishments is the "Organ-on-a-Chip (OoC)" platforms, which are biomimetic environments that replicate the physical and mechanical stresses found within human tissues. An example of the OoC system is shown in Fig. 1.1. With the use of these platforms, scientists can produce functioning models of human organs, such the liver, which are more precise than those made using conventional cell culture techniques in simulating in vivo circumstances. Three-dimensional tissues can be built from individual cells, and this bottom-up method makes it easier to examine the connections and functioning of organs in a controlled setting. These systems' capacity to recreate physiological circumstances has been further improved by the combination of microfluidics and biomimetics, making them indispensable instruments for disease modeling and drug testing.

Another important application of BioMEMS is in wearable health monitoring. These devices integrate miniaturized, low-power sensors with flexible and stretchable substrates to enable continuous, real-time measurement of physiological parameters in daily life. BioMEMS technologies allow a combination of electrochemical, optical, and micromechanical sensing elements to monitor biomarkers such as glucose, lactate, electrolytes, and cortisol, as well as physical signals including the heart rate, respiration, temperature, and motion. By replacing rigid silicon with breathable alternatives such as PDMS, soft BioMEMS devices allow development of biocompatible, epidermal electronics, and microfluidic sweat-analysis platforms. When combined with wireless communication modules and data analytics, wearable BioMEMS devices support remote health monitoring, early disease detection, and personalized medicine, reducing reliance on centralized clinical testing while promoting proactive healthcare management.

Other emerging areas that use BioMEMS are soft robotics and implantable sensors. Implantable devices are used to treat chronic conditions, such as heart disease, diabetes, and epilepsy. The integration of microsystem into the device can make these devices smaller, more durable, and more energy efficient, thereby improving patient comfort and reducing the risk of complications. In soft robotics, research is focused on new BioMEMS materials such as ionic polymer metal composites (IPMCs). IPMCs are smart soft, pliable polymers with embedded metal layers which can be used as sensitive biosensors that can detect minute changes in biological signals. IPMCs can be combined with hydrogels for increased sensitivity and responsiveness. An interesting application of IPMCs is as "artificial cilia" to detect tiny fluid flows in the body or as patches on the throat to monitor swallowing patterns in patients with dysphagia (Motreuil-Ragot et al., 2023).

Biosensing technologies within BioMEMS have also advanced significantly, notably in the creation of biosensors that can monitor biological parameters in real time. For example, piezoelectric flexural plate wave (FPW) sensors have been developed for point-of-care diagnostics, allowing for the quick detection of biomarkers in biological fluids. These sensors use acoustic wave propagation principles to achieve excellent sensitivity and specificity, making them ideal for disease diagnosis and monitoring applications (Dsa, 2018). Integration of nanotechnology with MEMS has resulted in the development of nanoscale sensors that can detect low concentrations of analytes, thus improving the capabilities of biosensing technologies (Pachkawade, 2023).

In drug delivery, BioMEMS technologies have opened the way for the development of sophisticated drug delivery systems (DDSs) capable of providing targeted and controlled release of treatments. Metal-organic frameworks (MOFs) have emerged as excellent drug delivery carriers due to their adjustable characteristics and biocompatibility. A recent study has shown that zirconium-based MOFs may encapsulate pharmaceuticals such as tetracycline hydrochloride, demonstrating their ability to increase the drug-loading capacity and stability (Hanafi et al., 2022). Bioactive chemicals can be integrated with MOFs to develop dual-DDSs that respond to specific biological stimuli, enhancing therapeutic effectiveness (Zhang et al., 2023).

BioMEMS can also be used to fabricate active delivery devices such as miniature piezoelectric actuators that can deliver medications to precise areas in the body, increasing therapeutic precision (Chen et al., 2021; Dsa, 2018). These piezoelectric MEMS devices may be programmed to respond to external stimuli, allowing for on-demand medication release, which is especially useful in controlling chronic illnesses where patient-specific dosage is critical (Ghazali et al., 2020; Walk et al., 2019).

Another BioMEMS device for medication delivery is microneedles, which are minimally invasive and can penetrate the skin to administer vaccinations and medicines with little discomfort (Algamili et al., 2021). Recent advances in microneedle design and manufacturing have focused on enhancing their mechanical strength and flow properties, providing effective drug administration even for strong medicines supplied in microdoses (Algamili et al., 2021). This device not only improves patient compliance, but also provides new options for delivering biologics and gene treatments.

BioMEMS applications can extend beyond medication delivery and diagnostics to include therapeutic applications. For example, biomicromotor tweezers have been proposed for noninvasive manipulation of biological cells, allowing for the delivery of therapeutic medicines to precise cellular regions (Karumuri et al., 2021). This technique has the potential to revolutionize cancer therapy by precisely targeting tumor cells, which can greatly enhance treatment outcomes.

BioMEMS is an emerging field with a wide range of exciting applications. The combination of biosensing technology with DDSs has the potential to alter the landscape of personalized medicine, allowing for more effective and focused therapeutic treatments. As research advances, the future of Bio-MEMS promises to provide novel answers to important healthcare concerns, eventually enhancing patient outcomes and quality of life.

References

Algamili, A., Haris, M., Khir, Md., Dennis, J., Ahmed, A., Omar, S., Ba Hashwan, S., & Junaid, M. (2021). A review of actuation and sensing mechanisms in MEMS-based sensor devices. *Nanoscale Research Letters, 16*. https://doi.org/10.1186/s11671-021-03481-7.

Chen, X., Zhuang, Y., Rampal, N., Hewitt, R., Divitini, G., O'Keefe, C. A., Liu, X., Whitaker, D. J., Wills, J. W., Jugdaohsingh, R., Powell, J. J., Yu, H., Grey, C. P., Scherman, O. A., & Fairen-Jimenez, D. (2021). Formulation of metal–organic framework-based drug carriers by controlled coordination of methoxy PEG phosphate: Boosting colloidal stability and redispersibility. *Journal of the American Chemical Society,, 143*(34), 13557–13572. https://doi.org/10.1021/jacs.1c03943.

Chircov, C., & Grumezescu, A. M. (2022). Microelectromechanical systems (MEMS) for biomedical applications. *Micromachines, 13*(2), 164. https://doi.org/10.3390/mi13020164.

Dsa, J. (2018). In vitro characterization of MEMS based piezoelctrically actuated drug delivery device for biomedical applications. *Biomedical Journal of Scientific & Technical Research, 11*. https://doi.org/10.26717/BJSTR.2018.11.002093.

Ghazali, F., Hasan, M., Rehman, T., Nafea, M., Mohamed Ali, M. S., & Takahata, K. (2020). MEMS actuators for biomedical applications: A review. *Journal of Micromechanics and Microengineering, 30*, 073001–073020. https://doi.org/10.1088/1361-6439/ab8832.

Hanafi, A. A, El-Kady, M., & Ayad, M. (2022). Insights into drug-loading capacity of bio-zirconium metal-organic frameworks nanoparticles for tetracycline hydrochloride. *Key Engineering Materials, 935*, 179–184. Retrieved from: https://doi.org/10.4028/p-5emz84 .

Karumuri, S., Mohammed, H., Guha, K., Kumar, P., Einsanwi, A., & Sravani, G. (2021). Design, simulation and analysis of micro electro-mechanical system microneedle for micropump in drug delivery systems. *IET Nanobiotechnology, 15*(5). https://doi.org/10.1049/nbt2.12013.

Motreuil Ragot, P., Hunt, A., Sacco, L. N., Sarro, P. M., & Mastrangeli, M. (2023). Manufacturing thin ionic polymer metal composite for sensing at the microscale. *Smart Materials and Structures, 32*(3), 035006. https://doi.org/10.1088/1361-665X/acb305.

Motreuil-Ragot, T., De Wagenaar, H., & Sarro, M. (2022). Ionic polymer metal composite-based microfluidic flow sensor for bio-MEMS applications. In *Proceedings of IEEE Sensors: 2022–October* https://doi.org/10.1109/SENSORS52175.2022.9967281.

Pachkawade, V. (2023). *N/MEMS biosensors: An introduction* (pp. 1–10). Springer. https://doi.org/10.1007/978-981-19-8714-4_1.

Walk, C., Wiemann, M., Goertz, M., Weidenmüller, J., Jupe, A., & Seidl, K. (2019). A piezoelectric flexural plate wave (FPW) bio-MEMS sensor with improved molecular mass detection for point-of-care diagnostics. *Current Directions in Biomedical Engineering, 5*(1), 265–268. https://doi.org/10.1515/cdbme-2019-0067.

Zhang, X., Zhao, Z., Xu, M., Yang, S., Xie, Y., Zhang, B., & Fan, J. (2023). A curcumin-based MOF embedded Ag NPs and modified with polydopamine for dual-drug delivery and dual biological window responsive synergetic cancer therapy. *Particle & Particle Systems Characterization, 40*(9), 2300047. https://doi.org/10.1002/ppsc.202300047.

CHAPTER

Electrical and electrochemical sensors

2

Umi Zulaikha Mohd Azmi[a], Siti Fatimah Abd Rahman[b], R. D. A. A. Rajapaksha[c], Thikra S. Dhahi[d] and Anis Amirah Alim[e]

[a]*Nuclear Emerging Technology Center, Technical Support Division, Malaysian Nuclear Agency, Kajang, Selangor, Malaysia,* [b]*School of Electrical and Electronic Engineering, Universiti Sains Malaysia, Engineering Campus, Nibong Tebal, Pulau Pinang, Malaysia,* [c]*Department of Nano Science Technology, Faculty of Technology, Wayamba University of Sri Lanka, Kuliyapitiya, Sri Lanka,* [d]*Physiotherapy Department, Medical & Health Technology College, Southern Technical University, Basrah, Iraq,* [e]*Institute of Microengineering and Nanoelectronics (IMEN), Universiti Kebangsaan Malaysia, Bangi, Selangor, Malaysia*

2.1 Electrochemical (static microstructures/microsystem)

Electroanalytical methods in analytical chemistry are grounded in the fundamental analytical signals of current, charge, and potential. This is due to the inherent nature of electrochemical biosensors, where reactions that occur involve the production or consumption of ions or electrons in the surrounding medium, leading to measurable alterations in the electrical signal. The analyte in the electrochemical media exhibits two distinct behaviors, namely bulk technique and interfacial technique. Bulk techniques involve the measurement of properties of the entire solution, while interfacial techniques focus on measuring the properties of the species located at the electrode-solution interface. An illustration of a bulk electrochemical approach involves the measurement of solution conductivity. In this context, the level of conductivity exhibits a direct relationship with the concentration of ions that are dissolved inside the medium. In contrast, the utilization of pH electrodes for the purpose of measuring the pH of a solution serves as an illustration of an interfacial electrochemical approach. The integration of nanostructures into electrodes and biosensors has led to a significant revolution in various biomedical sectors.

Electrochemical systems can be classified into two main types based on the number of active electrodes in the electrolyte solution: (i) bulk solution approaches, where electrodes are submerged in different electrochemical cells, and (ii) nano- and micro-volume-based systems, where electrodes are enveloped in a small volume, such as a solution droplet. Two-electrode systems are the simplest electrochemical systems and are suitable for static electrical response-based systems. The working electrode is sensitive to the analyte concentration, while the counter electrode provides a reference potential for the working electrode and completes the electrical circuit. In two-electrode systems, the counter electrode should ideally maintain a constant potential so that any changes in the overall system potential can be attributed solely to the working electrode. In dynamic systems, current flow can alter the concentration of species in the electrolyte, leading to changes in the potential at the counter electrode over time. One solution to this problem is to use three-electrode systems, which replace the counter

BioMEMS Devices. DOI: https://doi.org/10.1016/B978-0-443-24015-7.00014-0

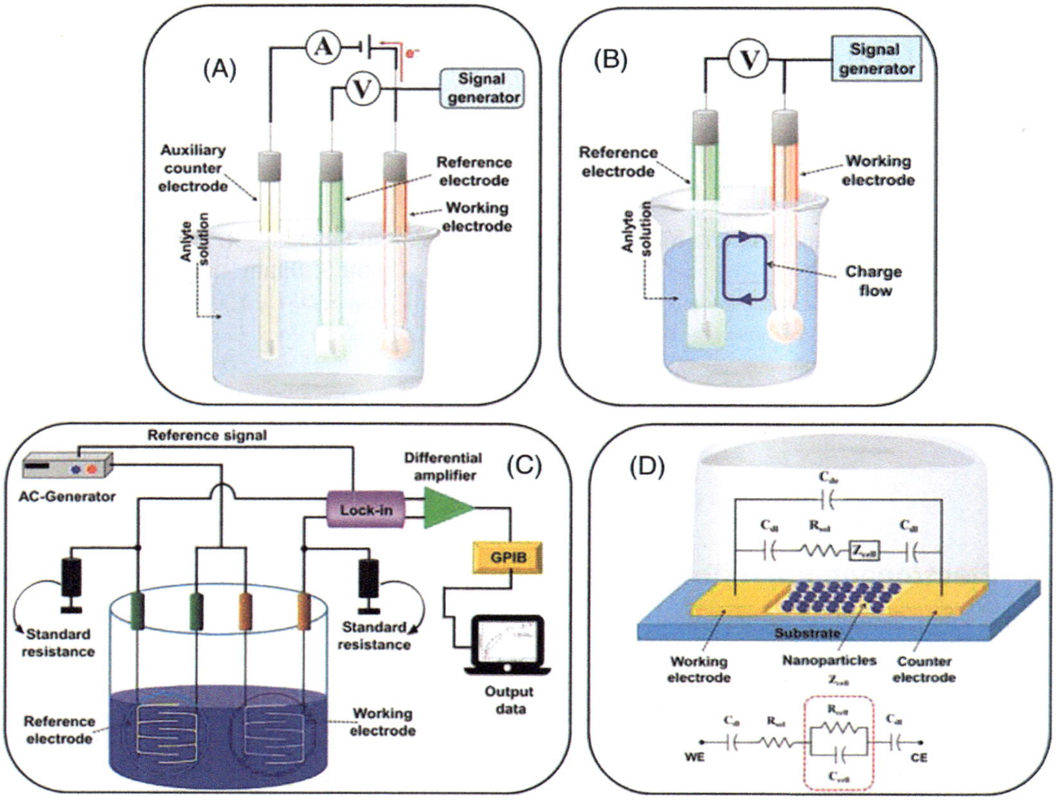

FIGURE 2.1

Schematic representations of (A) amperometric / voltammetric, (B) potentiometric, (C) conductometric, and (D) impedimetric-based analytical measurements.

electrode with two additional electrodes: a reference electrode and an auxiliary electrode. The reference electrode is used to ensure a constant voltage, while the auxiliary electrode provides a path for current flow. Both two-electrode and three-electrode systems can be used in three basic experimental designs: (i) static techniques, where the potential is measured under no current flow ($i = 0$); (ii) dynamic techniques where the potential is measured while controlling the current ($i \neq 0$); and (iii) dynamic techniques where the current is measured while controlling the potential. Although there are only three distinct categories of electrical signals, different analytical designs can be employed, as shown in Figs. 2.1 and 2.2.

The following subsection discusses the unique applications, utilization, and potential prospects of analytical systems that combine nanoparticle (NP) sensors, focusing on the aforementioned analytical signals. The measurement techniques used depend mainly on the applied electrical signal and the detection mechanism.

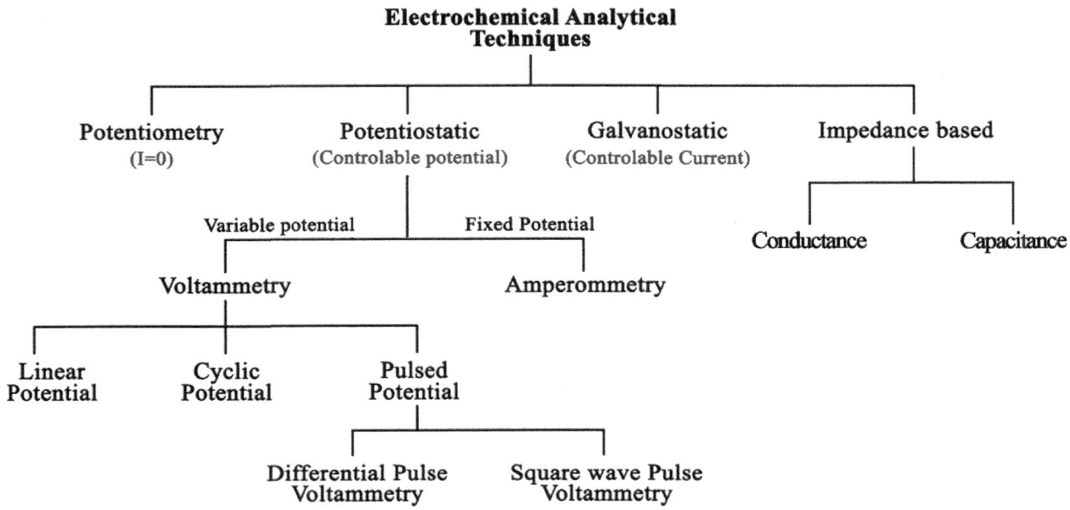

FIGURE 2.2

Electrochemical analytical techniques.

2.1.1 Potentiometry

Potentiometry-based electrochemical sensors are analytical tools designed to detect and quantify specific ions or molecules in a wide range of samples. These sensors rely on the measurement of electrical potential or voltage to provide valuable insights into the concentration of target analytes. Potentiometric sensors rely on a potential difference between an indicator electrode and a reference electrode in a two-electrode system (Fernando et al., 2013; Rajapaksha et al., 2021; Subramani et al., 2018; Wang et al., 2019). This potential difference is highly sensitive to the ion or molecule of interest, leading to a distinct potential response. The sensor's selectivity for the target analyte is a key feature, enabling precise measurements even in complex matrices like clinical samples, food products, or biological specimens. In potentiometric sensors, selective membranes and ion-selective thin film electrodes (ISEs) are commonly used to detect gasses (CO_2, NH_3), ions (K^+, I^-, F^-, Na^+, etc.), and pH (Rajapaksha et al., 2018; Yahaya et al., 2017). Integration with biosensors further expands the application of potentiometry, allowing the transduction of biological signals into electrical responses.

Bratov et al. have highlighted the versatility of potentiometric sensors, emphasizing their applicability in classifying and analyzing intricate samples, including clinical specimens, food products, and biological analytes (Bratov et al., 2010). Further, they highlighted that simple analysis techniques, low limit of detection (LOD), reproducibility, and rapidity are some merits of ISEs. Batool et al. expanded the horizon of potentiometric sensor applications by harnessing ISEs to transduce biological signals into electrical signals (Batool et al., 2019). Liang et al. have detected trace levels of triclosan in toothpaste by developing molecularly imprinted polymers (MIPs) in the form of nanobeads, embedded within ISEs with a polymeric membrane (Liang et al., 2013). Their achievement includes a remarkable LOD at 1.9×10^{-9} mol L^{-1} in a pH 10.1 environment.

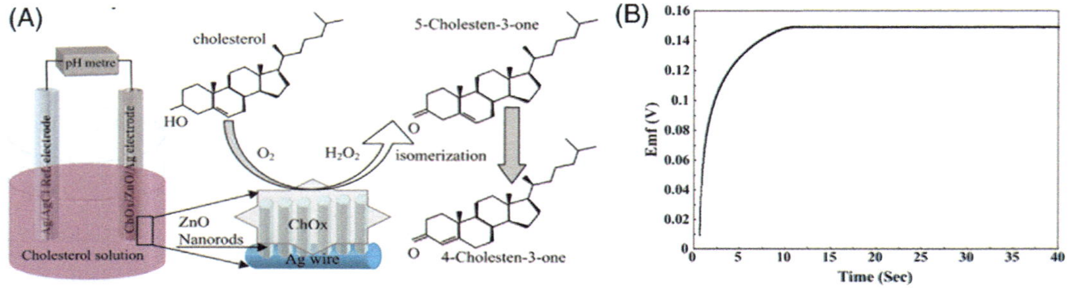

FIGURE 2.3

Potentiometric analytical techniques, (A) working bioelectrode comprising with ZnO nanorods for cholesterol sensing, (B) The response of output cholesterol concentration of ChOx/ZnO/Ag bioelectrode (1×10^{-4} M).

Reproduced with permission from Israr et al. (2010). Copyright © 2010 Elsevier B.V.

Another impressive development comes from Cevik and the research team, who crafted a potentiometric urea biosensor using Fe_3O_4 NPs (Çevik et al., 2013). These NPs are prized for their low toxicity, superb biocompatibility, and exceptional magnetic properties. Importantly, they enable high enzyme loading and enhance electro-catalytic functions through increased surface area. Israr et al. have developed a ZnO nanorod-based cholesterol biosensor by potentiometric detection (Fig. 2.3) (Israr et al., 2010). These nanorods, with diameters ranging from 125 to 250 nm and a length of approximately 1 μm, offer excellent electron communication, outstanding biocompatibility, and high ionicity. They adopted the ChOx/ZnO/Ag working bioelectrode with a standard Ag/AgCl reference electrode, resulting in a biosensor displaying a sensitivity of 35.2 mV per decade. Kharitonov's work adds another dimension to potentiometric sensing, with a focus on papaverine hydrochloride detection (Kharitonov, 2006). Utilizing four types of Keggin-type POMs, namely $H_3PMo_{12}O_{40}$, $H_3PW_{12}O_{40}$, $H_4SiMo_{12}O_{40}$, and $H_4SiW_{12}O_{40}$, Kharitonov successfully fabricated highly sensitive, rapid, and stable sensors for this application. These advancements underscore the versatility and potential of potentiometry in the electrochemical field.

2.1.2 Amperometric

Amperometric-based electrochemical sensors are powerful tools used to detect and quantify target analytes in various samples. These sensors operate by measuring the electric current generated as a result of oxidation or reduction reactions within biochemical systems, and they do so during a fixed time interval at a constant potential (Hayat et al., 2014). The detection mechanism of amperometric sensors relies on the direct proportionality between the generated current and the concentration of the electroactive species present in the sample, as reported by Rathee and colleagues (Rathee et al., 2016). These sensors offer numerous advantages, including high sensitivity, selectivity, cost-effectiveness, and easy integration into continuous analysis systems.

Thevenot et al. elucidated that amperometric biosensors employ carbon or metal electrodes, along with chemically modified electrodes (CME), to detect a range of substances, including oligonucleotides, alcohols, sugars, O_2, and, phenols (Thevenot et al., 2001). Ronkainen et al. emphasized the amperometric technique's popularity in biocatalytic and affinity sensors due to its low LOD and simplicity

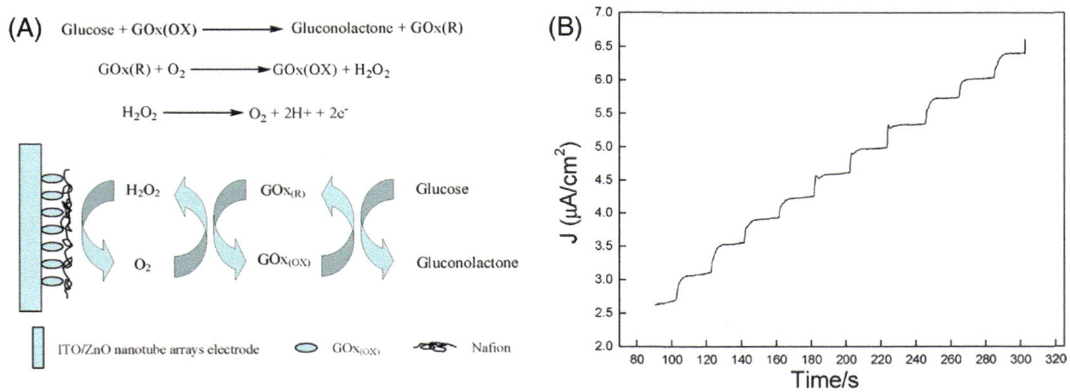

FIGURE 2.4

(A) The glucose-sensing mechanism in a modified electrode using an amperometric analytical method. (B) Glucose concentration of "GOx/ZnO nanotube array/ ITO electrodes", the typical amperometric response curve.

Reproduced with permission from Yang et al. (2009). Copyright © 2009, American Chemical Society.

(Ronkainen et al., 2010). The fixed potential inherent to amperometry results in a shallow charging current, minimizing noise and enhancing the LOD. Moreover, nanoparticles are commonly harnessed in amperometric biosensors, offering physicochemical properties, surface charge density, shape, and control over size. Rashidi and colleagues designed amperometric biosensors for cholesterol detection, enhancing sensor performance with Au NPs and graphene (Gr) combined with carbon nanotubes (CNTs) (Rashidi et al., 2018). The resulting biosensor demonstrated a sensitivity of 6.6 μA μM^{-1} cm^{-2} and an impressive 0.07 μM LOD, showcasing its selectivity, sensitivity, stability, repeatability, and reproducibility (Fig. 2.4, Yang et al., 2009). ZNT's characteristics, such as a high surface-to-bulk ratio, excellent biocompatibility, chemical stability, electrochemical activity, and high isoelectric point (9.5), contribute to the sensor's high sensitivity, with a LOD of 10 μM and a sensitivity of 30.85 μA cm^{-2} mM^{-1}.

2.1.3 Conductometric

This approach is widely employed in numerous electrochemical analyses, displaying considerable potential. In this context, the medium properties, such as the resistance or conductance, are quantified while observing electrochemical reactions taking place within the solution. The overall change in charge inside the electrolytic medium is directly proportional to the entire variation in ions or electrons. This method involves the generation of an electric field through the use of a sinusoidal voltage signal, namely an alternating current (AC). The purpose of this electric field is to mitigate the adverse consequences associated with double-layer charging, concentration polarization, and Faradaic efficiency. In addition to the negligible utility of the reference electrode, the capacity for miniaturization and cost-effective large-scale production are notable advantages of conductometric biosensors. Ambrosi et al. define conductance as the capacity to facilitate the movement of electrical charges along specific pathways within solid or ionic liquids (Ambrosi et al., 2008). The study findings indicated that the identification of resistance or conductance took place subsequent to the attraction of the biological analytes to the

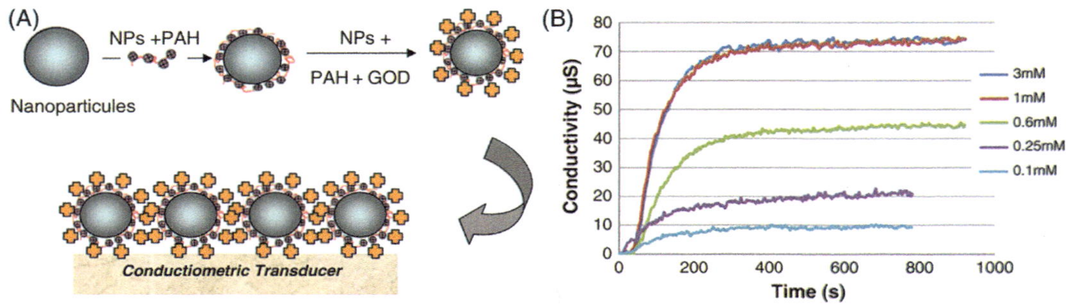

FIGURE 2.5

(A) Illustrates the schematic representation of the biosensor's operational principle for the detection of glucose. (B) The conductance response of the glucose sensor when magnetic nanoparticles are present in a 5 mM phosphate buffer solution with a pH of 7.3, for varying glucose concentrations.

Reproduced with permission from Nouira et al. (2013). Copyright © 2012 Elsevier B.V.

surface. The primary challenges associated with this approach are limited sensitivity and non-specificity. In order to address these challenges, various nanostructures play a crucial role as they have the ability to induce substantial alterations in resistance based on the binding components present on their surface (Ambrosi et al., 2008; Ronkainen et al., 2010).

Ansari and colleagues have developed a conductometric biosensor for glucose detection using nano-baskets made of SnO_2 (Ansari et al., 2008). The features of metal nanoparticles in biosensor applications include thermal stability, chemical inertness with biomolecules, low-temperature operation, and huge surface area, as emphasized by the researchers. They claimed that during the oxidation process, oxygen liberates the confined electron to the conduction band of SnO_2. Conversely, in the reduction process, the liberated electrons are returned to the conduction band of the material. The sensor exhibited a sensitivity range of 0 to 10 millimolar (mM). Nouira et al. have successfully developed a conductometric biosensor for glucose detection by utilizing both gold nanoparticles (GNPs) and magnetic nanoparticles in their fabrication process (Nouira et al., 2013). According to their report, it was said that the incorporation of GNPs would result in an improvement in conductivity (Fig. 2.5). The biosensor's sensitivity, when utilizing magnetic nanoparticles (NPs), was determined to be 70 μM/mM, with a corresponding LOD of 3 μM. In comparison, the biosensor's sensitivity with GNPs was found to be 45 μM/mM, with an LOD of 9 μM. Based on their study, it was seen that the stability of the sensor exhibited a notably high level. Furthermore, Kim et al. successfully produced a membrane strip immunological sensor for detection of human serum albumin (Kim et al., 2000). This sensor utilized nanoscale gold colloids with a size of 20 nm, which were modified with polyaniline. Gold colloids are found to have a crucial function in facilitating charge transfer, while the polymeric conducting layer serves as a conduit, facilitating the proximity of adjacent particles in order to augment the charge transfer mechanism.

2.1.4 Impedimetric

Electrochemical impedance spectroscopy, sometimes known as EIS, is an important analytical method that has a wide range of applications in many different areas of research. This method makes it possible

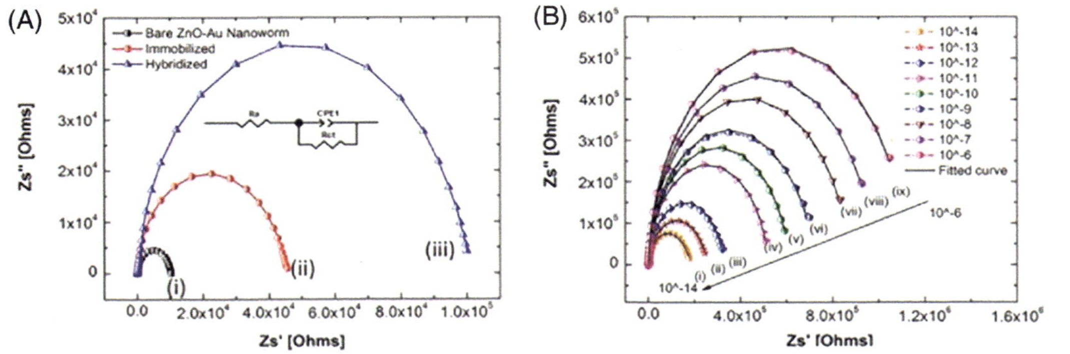

FIGURE 2.6

Impedance spectra for (A) functionalization, immobilization, and target hybridization (B) with the detection range of 0.1 mM to 1 pM.

to investigate the equilibrium state of resistive and capacitive properties of materials while they are present in an electrochemical medium. Impedance is measured inside the system by adjusting the frequency of the AC throughout a wide range in order to cover as much ground as possible. At higher frequencies, the Nyquist plot of EIS displays a segment in the shape of a semicircle. The diameter of this semicircle is directly proportional to the resistance of charge transfer (Rct) that is taking place at the electrode surface. This property has the potential to be utilized in the process of characterizing the interface properties of the electrode surface. In EIS, the Nyquist plot additionally incorporates a linear segment that is connected to the phenomenon of diffusion. Hence, the EIS technique can be employed for monitoring the alterations made to the electrode in order to produce the biosensor (Abd Rahman et al., 2021; Thevenot et al., 2001).

Perumal et al. successfully synthesized a hybrid nanostructure composed of Zinc Oxide-gold (ZnO/Au) using the aqueous hydrothermal technique. In this study, the authors employed the sputtering approach to introduce doping of Au-nanoparticles (AuNPs) onto the ZnO lattice during its growth process (Perumal et al., 2015). Impedance spectroscopy is a technique that offers enhance selectivity and achieves a sensitivity of femtomolar (10 fM) levels, as shown in Fig. 2.6. The researchers, led by Lin and colleagues, developed impedimetric biosensors using GNPs for the detection of the foodborne pathogen *Escherichia coli* O157:H7 (Lin et al., 2019). The utilization of GNPs in biosensors is primarily motivated by their distinctive electrical properties, ease of synthesis, exceptional biocompatibility, and their high surface-area-to-volume ratio, as previously elucidated. According to the findings that were provided in their research, the self-assembly of GNPs with a 1,6-hexanedithiol (1,6-HDT) modified gold electrode makes the process of transferring electrons from $Fe(CN)_6^{3-/4-}$ to the electrode easier. Moreover, it was suggested that the incorporation of graphene nanoparticles (GNPs) could potentially enhance the current density due to their inherent electroactive characteristics.

Jia et al. have successfully fabricated an impedimetric biosensor for the detection of salmonella using nanocomposites of reduced graphene oxide (rGO) and multiwall CNT (MWCNT) (Jia et al., 2016). The rGO-MWCNT nanocomposites exhibit a high level of electron transfer capability, which renders them effective as an amplification device for improving the LOD. In addition, MWCNTs enable a significant increase in surface area, hence enhancing the bonding capacity of probe aptamers. The utilization of rGO in this biosensor offers additional benefits such as exceptional chemical durability and high mechanical strength. MWCNT exhibit enhanced electrical conductivity, hence facilitating improved passage of electric charge between the analyte and electrode. Chowdhury and group created a label-free impedimetric biosensor utilization of Fe_3O_4 nanoparticles, which imparts magnetic characteristics to the system and serves as anchoring sites for graphene quantum dots (GQDs) for cancer cell detection (Chowdhury et al., 2018). The impedimetric response sensitivity of sensors designed for detecting two types of cancer cells, namely HeLa and MCF-7, is shown to be 16.7 and 13.1 times greater, respectively, compared to the sensitivity of the original sensor electrode.

2.1.5 Voltammetry

In voltammetry, the method quantifies the electrical current produced in relation to the applied potential and can be classified into three main types: cyclic voltammetry (CV), linear sweep voltammetry (LSV), and pulsed voltammetry, which encompasses differential pulse voltammetry and square wave pulse voltammetry (SWV). Linear sweep voltammetry and CV are widely employed techniques to identify specific molecular species in the sample by characterizing the oxidation and reduction processes. Additionally, these techniques offer valuable insights into chemical reactions that are triggered by electron transfer mechanisms. The utilization of voltammetry is of great significance in electrochemical analysis that involves nanoparticles, primarily due to its notable attributes such as high sensitivity, cost-effectiveness, and compatibility with diverse equipment (Rajapaksha et al., 2017).

Ambrosi et al. provide a comprehensive overview of nanoparticle analysis, categorizing the methodologies into two primary approaches: direct and indirect. The indirect methods encompass several techniques such as catalytic and voltammetric stripping analysis procedures (Ambrosi et al., 2008). In the field of direct voltammetric approaches, researchers such as Wang et al. have shown the utilization of nanoparticles without the need for preliminary dissolution. This was exemplified in their study where they developed a methodology based on GNPs for electrochemical DNA hybridization (Wang et al., 2002). Wang's research team has successfully achieved high sensitivity and low detection limits, with an amazing sensitivity value of 12.3 seconds per milliliter per nanogram (s mL/ng) and a detection limit of 150 per milliliter (pg mL^{-1}). In a study conducted by McKenzie and Marken, direct voltammetric detection was employed to analyze the properties of Fe_2O_3 nanoparticles. The characterization process involved the utilization of tin-doped indium oxide (ITO) electrodes (McKenzie & Marken, 2001).

The indirect method of voltammetric stripping analysis encompasses the electroanalytical approaches that facilitate the oxidative dissolution of nanoparticles into specific metals. This approach has notable advantages such as high sensitivity and low detection limits (Ambrosi et al., 2008). In indirect voltammetric techniques, the catalytic action of nanoparticles is a crucial factor that plays a role, particularly because it is utilized to amplify sensitivity. Lai et al. successfully developed an electrochemical biosensor strip utilizing carbon nanoparticles (CNP) for the detection of Japanese Encephalitis Virus (JEV) (Lai et al., 2017). The results obtained from the CV analysis demonstrate the prompt and efficient response exhibited by biosensors. A steady current was achieved within a duration of 10 minutes. In this study, it was observed that carbon nanoparticles (CNP) exhibit enhanced electrical conductivity when a significant quantity of Japanese encephalitis virus (JEV) antibodies is included.

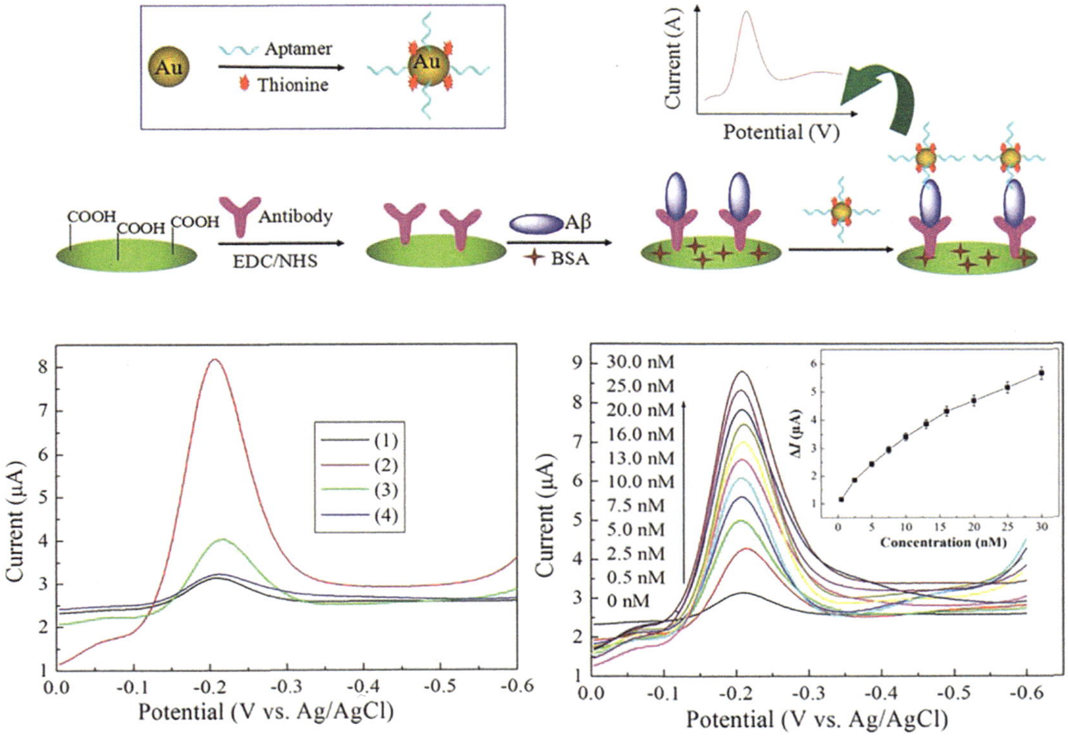

FIGURE 2.7

GNP label aptamer-based target electrochemical assays based on voltammetric technique.

Tran et al. employed CV and SWV techniques in their study to detect hydrogen peroxide (H_2O_2). Ag NPs were employed as an electrocatalyst in the reduction of H_2O_2 to produce water and dioxygen, as reported in reference (Tran et al., 2018). Papagianni et al. have successfully devised a sensor, known as PANI-POM (where POM represents XM_{12}, with X representing P or Si, and M representing Mo or W), for the purpose of detecting bromate (BrO_3^-) in drinking water and flour extract. This detection method employed CV (Papagianni et al., 2012). They found that PANI–PMo_{12} demonstrates exceptional catalytic properties in the reduction of bromate ions. In addition, the detection of *E. coli* utilizing a large surface volume (LSV) technique with titanium dioxide (TiO_2) nanoparticles (NPs) functionalized on an interdigitated electrode has been reported by Nadzirah et al. (Nadzirah et al., 2015).

Pulse wave voltammetry, a sophisticated voltammetry technique, involves the integration of differential pulse voltammetry (DPV) and SWV. Fig. 2.7 demonstrates the utilization of the DPV approach in conjunction with a GNP label aptamer-based electrochemical test for the detection of Alzheimer's disease (AD) (Zhou et al., 2016). The LOD of the biosensor was successfully obtained at a concentration of 100 pM. This achievement can be attributed to the signal amplification resulting from the high loading of thionine on the GNPs. Liao and co-worker have developed a very sensitive and efficient

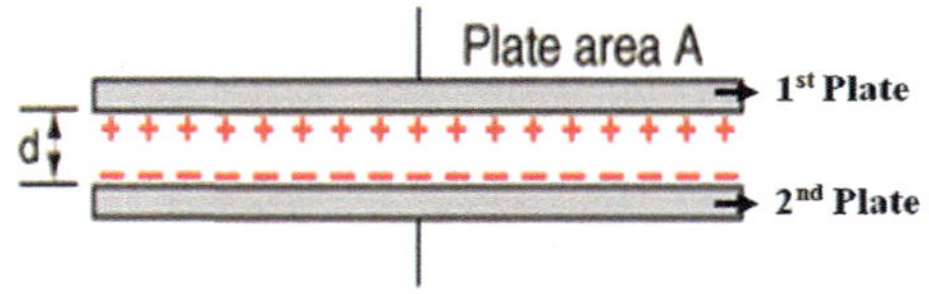

FIGURE 2.8

Schematic illustrates the capacitive of two plates or electrodes.

electrochemical DNA sensor for the purpose of detecting Escherichia coli (*E. coli*) (Liao & Ho, 2009). Square wave voltammetry was employed to detect the current, while Au nanoparticles were utilized to create a favorable surface for the immobilization of DNA probes. Rassas et al. created a glucose biosensor for glucose detection by doping GNP's in voltammetric measurements (Rassas et al., 2019). Additionally, it was discovered that the presence of GNPs led to an increase in the rate of electron transfer. The square wave voltammetry experiment yielded distinct findings indicating that the introduction of Au NPs through doping resulted in an observable increase in the intensity of the anodic peak. Additionally, a notable shift towards the positive side was observed in the peak potential value. The researchers emphasized the significance of the glucose sensor's performance, which was primarily attributed to the electrocatalytic characteristics of GNPs.

2.1.6 Capacitive (medium properties—permittivity)

A capacitive is a passive electronic component that stores energy in the form of an electrostatic field. In its simplest form, a capacitive consists of two conducting plates separated by an insulating material called the dielectric. The capacitive or capacitance is directly proportional to the surface areas of the plates and is inversely proportional to the separation between the plates. Capacitance also depends on the dielectric constant of the substance separating the plates, which can be represented in Equation (Curulli, 2021):

$$C = \varepsilon A/d$$

where $\varepsilon = \varepsilon o \times \varepsilon r$. Here, εo demonstrates the dielectric constant for air (~8.854×10^{-12} F/m), εr represents relative permittivity of the dielectric materials in the free space between the plates, *C* represents the capacitance value, *A* represents the surface area of the plate structures, and *d* demonstrates the size of the distance between the two plates or electrodes (Dhahi et al., 2011), as illustrated in Fig. 2.8. The standard unit of capacitance is the farad, F.

The experience of the past shows that throughout constant technology improvement, electronic devices have become more reliable, faster, more powerful, and less expensive, through the reduction in the dimension of circuits. These advantages drive the development of modern micro and nanoelectronics systems. The long-term goal of this development is to create nano and microstructures and circuits with superior sensitivities. Recently, there has been extensive studies and significant progress in the fabrication and testing of high-frequency MEMS resonators. In the case of parallel plate electrostatic transducers, extending the resonant frequency of micromechanical resonators to the GigaHertz range often entails shrinking the gap spacing between the electrodes and the resonator.

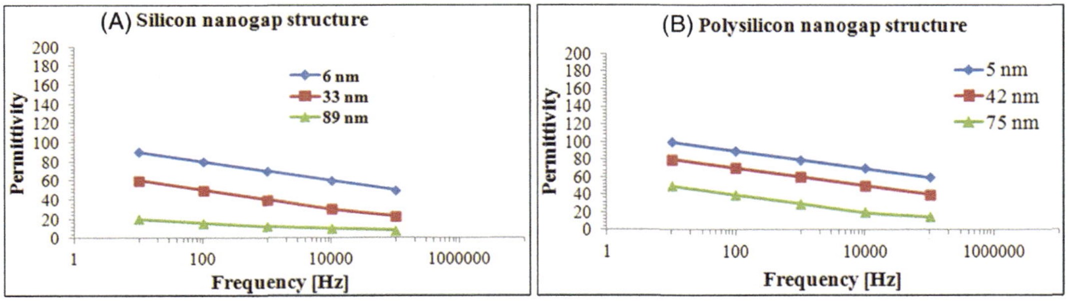

FIGURE 2.9

Permittivity vs frequency characterization with fabricated two plates of (A) silicon material and (B) polysilicon material.

Today, nanostructures have a new class of geometric properties with two electrodes (plates) separated with a desired distance structure to improve and develop new techniques for the fabrication method and high performance in the biosensor application. Nanostructure devices are interesting because the material between two electrodes, which are separated by few micro or nanometers can be examined by conducting various types of electrical measurements such as capacitance-frequency (Cf) and impedance. For example, the conduction of electricity or storage of charge gives rise to capacitive behavior which can be monitored by measuring Cf. For the sensitive measurements of electrical parameters, the electrodes should usually be made of metal so that by applying a small voltage across the electrodes one can measure the conduction of electrons through the nanogaps that sit between the electrodes.

Permittivity is the quality of the medium to respond to the changes in the reorientation of the dipoles in response to the changing electric field. At a low frequency signal, consider about the time interval available for reorientation of the dipole little carefully. While the low frequency means the time interval of change of polarity of the voltage signal applied is sufficient to allow the dipoles to reorient and therefore response is full. On the other hand, at high frequency, the change in signal polarity may be too short and therefore the dipoles do not get time to respond to the changing field. So, at high frequency, the dipoles are too low to respond to the changing electric field and thus the medium is not sensitive to changes, and this is reflected in low permittivity.

The permittivity reflects the index of a material's ability to transmit current without imposing a resistance (Saif & Poopalan, 2011). On the other hand, conductivity is a measure of a substance's capacity to allow the passage of current. Thus, the relation of permittivity and conductivity is proportional. Therefore, an increment of permittivity and conductivity with gap-size (distance) reduction suggests the fabricated nano-design structure can be used in sensing applications with the minimum supply of power. Nanogap-based devices with tightly confined electrodes have advantages for nano-sensing (Dhahi et al., 2011). The electrochemical response can be greatly improved by reducing the distance of the electrode gap, allowing for high sensitivity and rapid detection. As the surface interacts with the foreign materials, that is, pH and biomolecules, the surface charge across the gap induces electron excitation, which is caused by the double-layer phenomenon of gap electrochemistry.

In terms of the electrical characterization results, the permittivity trends are changed when changing the materials of the electrodes (two plates), as shown in Fig. 2.9. The results suggest that the permittivity

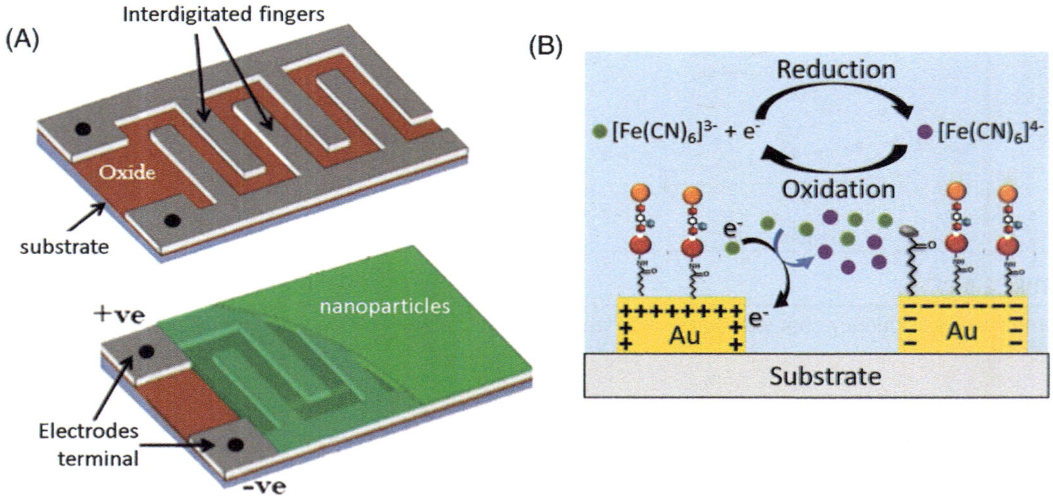

FIGURE 2.10

(A) Schematic illustration of interdigitated microelectrodes and (B) electrochemical-based detection using IDE sensors.

characterization is performed in terms of frequency using different semiconductor and doped semiconductor. The main trends give decreasing permittivity with increasing frequency. Moreover, in electrostatics, the permittivity plays an important role in determining the capacitance of a capacitor.

2.2 Electrode-based (static microstructures/microsystem)

2.2.1 Interdigitated electrodes-based

Considerable interest has been shown in using interdigitated electrode (IDE) arrays for sensing applications, as these arrays show promise for the realization of low-cost point-of-care devices, for example, cancer diagnosis, pH level detection, enzyme and heart disease, due to their ease of miniaturization, batch fabrication, and automation (Rajapaksha et al., 2018). IDE consists of two interlocking interdigitated fingers, with each section having a series of parallel comb-like electrode structures that are connected and separated by insulators. Increasing the active area with the use of IDE arrays and decreasing the distance between the finger electrodes helps to improve the sensitivity. However, smaller distances between the finger electrodes are known to require costly and time-consuming processes, for example, e-beam lithography technique. Alternatively, higher sensitivity can be achieved with deposition of nanoparticles between the finger electrodes of an IDE, which can shorten the distance between the fingers (Fig. 2.10(A)).

In the electrical aspect, IDE is an interesting device architecture, with the presence of small amplitude sinusoidal excitation signals. The high sensitivity levels required for many applications can be reached using a low-cost IDE with its submicron dimensions of electrode width and gap, thus eliminating the need for expensive conventional macro-electrodes technologies. Furthermore, for response measurement, they do not require a third reference electrode, thus making the system simpler for the realization of products. The application of interdigitated electrode arrays is not only limited to capacitive and resistive sensors but can also be extended to surface acoustic wave (SAW) and electrochemical impedimetric sensors. Based on the basic principle of SAW, the IDE transducer is designed to drive the piezoelectric wave generators and detectors used in SAW sensors (Abubakar et al., 2020). The interdigitated piezoelectric elements of the sensors are excellent candidates for efficiency analysis due to the lengthy, serpentine gap region between the two electrodes that provides a large contact area between the piezoelectric material and the current that drives the oscillations of the piezoelectric material. Thus, SAW-based IDE offers additional advantages of improved sensitivity, ease of fabrication process, rapid kinetics, increased signal-to-noise ratio, and a large electrode aspect ratio (Abubakar et al., 2020).

On the other hand, Rahman et al. demonstrated the electrochemical-based impedimetric method for interfacial investigation of the stepwise modification process by using interdigitated microelectrodes as detectors (Rahman et al., 2021). Measurements of the variations in charge transfer resistance (Rct) and total impedance (Z) versus frequency resulting from the lectin-glycan PSA binding were used to develop a detection strategy. The change in the Rct values corresponded to the charge transfer process between the $[Fe(CN)_6]^{3-/4-}$ redox species and the modified interfacial layer on the two electrode plates of the IDE, as depicted in Fig. 2.10(B). Depending on the electrode voltage, a faradaic current is produced by the species' charge transfer. As a result, the interfacial impedance is sensitive to the presence of compounds that are immobilized or adsorbed.

2.2.2 Transistor-based

A transducer is an analytical tool that provides an output quantity with a specific relationship to the input quantity. In electrically based techniques, a surface-based reaction would produce a detectable signal corresponding to changes in conductivity or resistivity of the test sample, which can be measured by the conductometric devices. The construction of these devices has been made possible by the rapid development of semiconductor technology and sensor integration with microelectronic devices, such as field-effect transistor (FET) devices based on transistor technology.

The conventional FET is a three-terminal device consisting of source, drain, and gate electrodes. The gate regulates the current flow between two electrodes (source and drain) in a semiconducting material. For example, in the case of a p-type substrate, applying a negative gate voltage results in the accumulation of the charge carriers (positive holes) in the channel region and thus resulting in an increase in conductivity, as shown in Fig. 2.11 (A). A special form of bioFET for sensing applications is illustrated in Fig. 2.11(B). In conversion of an FET into a transductor device, the metal gate electrode is replaced by a dielectric interface (e.g., Al_2O_3, SI_3N_4, SiO_2) with a selective chemistry. Upon analyte binding, the surface charge introduced at the interface is analogous to applying a new gate voltage across the interface, which in turn influences the density of charge carriers in the channel region, and hence conductivity. Therefore, if negatively charged target molecules, for example, bind to the surface of a p-type Si-FET, this acts as a negative gate voltage to the device and leads to an accumulation of positive hole carriers (Fig. 2.11(B)). Accordingly, the conductance level of the device changes in response to the binding events of the charged target molecules.

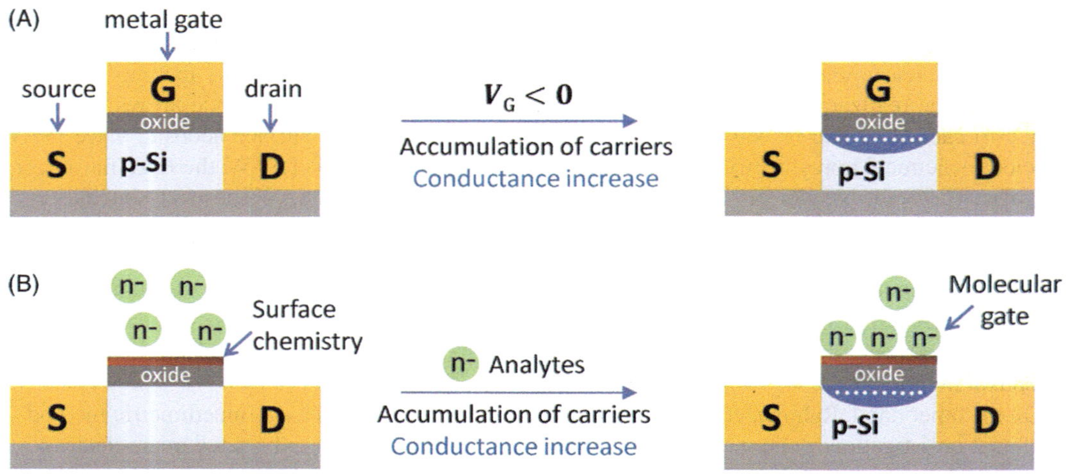

FIGURE 2.11

Schematic of (A) a typical FET device and (B) bioFET-based sensor device.

Recently, there is tremendous interest in the combination of nanostructures, such as nanowire with FET technology for biosensing application. The nanowire incorporated into FET devices was first introduced by Cui and co-worker in 2001. The usefulness of nanostructures-based FET biosensing offers downsized devices for higher performance and integration. These nanoscale structures overcome the sensitivity limitations of earlier planar FET sensors (Cui et al., 2001). The FET consists of a nanowire channel between the source and the drain terminals. The nanowire surface can be bio-functionalized so that a biomolecular binding event can create an electric field, similar to the control electric field applied to a conventional FET. The difference of the FET from the conventional channel is that the nanowire itself serves as a conductive channel. The binding of biomolecules to the surface of the NW leads to depletion or accumulation of carriers in the entire cross-section of the wire (Rahman et al., 2016), as compared to only the surface region of a planar device. SiNW-FET exhibits greater sensitivity than planar FET devices owing to reduced dimensionality and large surface area-to-volume ratio.

SiNW-FET exhibits greater sensitivity than planar FET devices owing to reduced dimensionality and a large surface area-to-volume ratio. While the reduced dimensionality of silicon nanowires significantly enhances FET sensitivity, further advancement in biosensing performance requires materials that combine high carrier mobility, structural simplicity, and full surface exposure. Graphene, a two-dimensional atomic crystal, fulfills these criteria and has therefore attracted substantial attention as a next-generation FET channel material.

Graphene-based field-effect transistors (GFETs) utilize single-layer or few-layer graphene as the conducting channel for ultrasensitive biosensing applications. Compared to one-dimensional nanowire systems, graphene provides a two-dimensional atomic framework that ensures full surface exposure of the conductive channel (Hamzah et al., 2017). This structural configuration allows direct electrostatic interaction between adsorbed biomolecules and the entire channel, leading to pronounced modulation of carrier concentration. In addition, graphene exhibits exceptionally high carrier mobility (up to 10^5 $cm^2/V \bullet s$) and ambipolar transport behavior, enabling continuous tuning between electron and hole

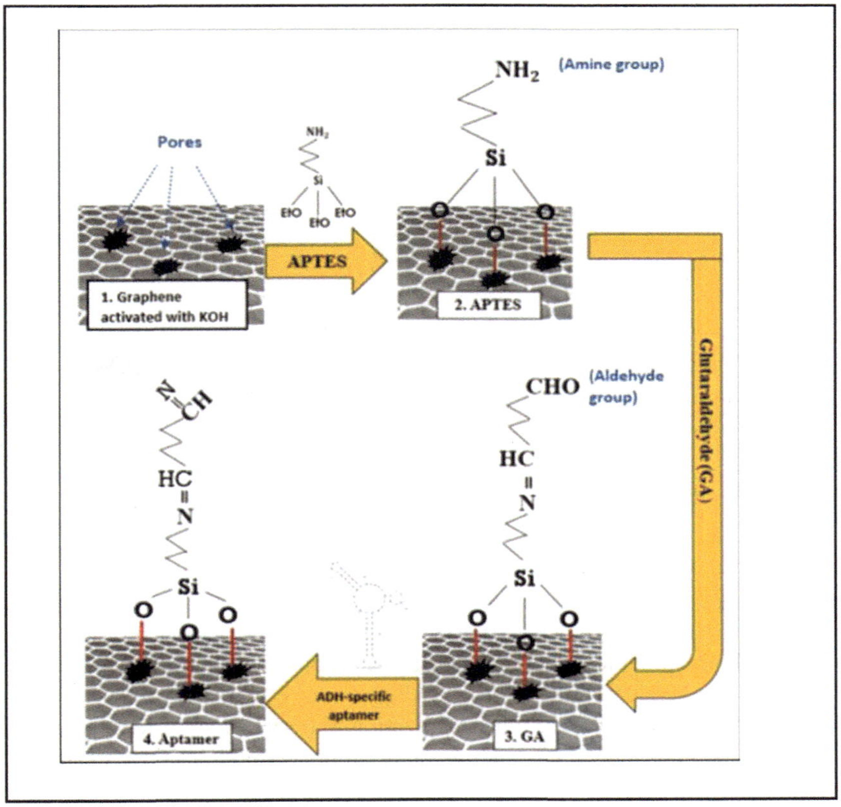

FIGURE 2.12

Schematic illustration of all stages in surfaces functionalization of graphene.

Reproduced under the terms of the CC-BY license from Selvarajan et al. (2020).

conduction via gate bias. The resulting shift in the Dirac point and transfer characteristics form the basis for highly sensitive detection mechanisms (Wang & Jia, 2018).

Graphene is commonly synthesized through chemical vapor deposition (CVD) on catalytic metal foils such as copper (Cu) or nickel (Ni), followed by transfer onto insulating substrates (Khalid et al., 2017). Various graphene transfer approaches, including directional transfer, conventional wet transfer, and electrolysis bubbling methods, are applied to form suspended or embedded graphene channels (Song et al., 2021). Although the graphene transfer method yields high-quality graphene suitable for proof-of-concept demonstrations, its limited scalability has led to the development of transfer-free direct growth of graphene on insulating substrates using metal catalysts, thereby minimizing contamination and structural damage commonly introduced by polymer-assisted transfer processes (Shan et al., 2022).

Most GFET devices employ graphene on silicon/silicon dioxide (Si/SiO_2) substrates, with metallic contacts such as gold (Au), copper (Cu), or silver (Ag) serving as source and drain electrodes (Khosravi Rad et al., 2024). Electrode fabrication in GFET devices is commonly performed through

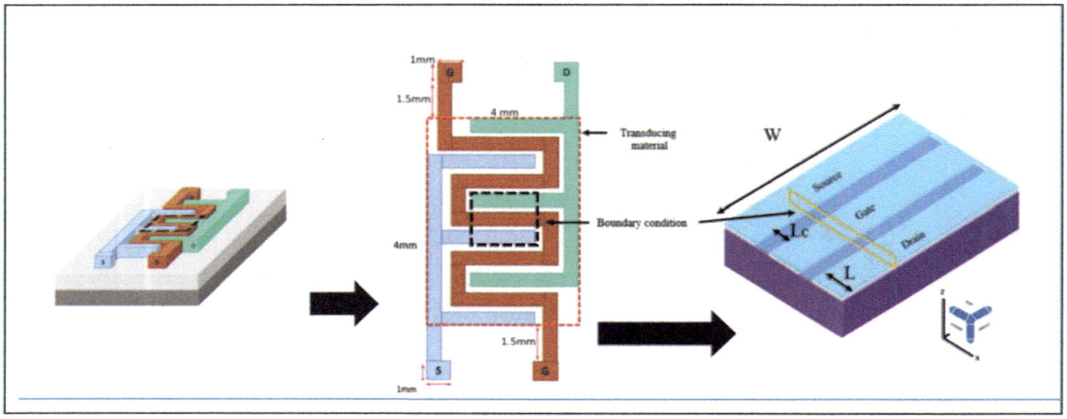

FIGURE 2.13

Pictorial representation of simulated design.

photolithography, metal deposition, and lift-off processes, which add fabrication complexity and potential contamination. Alternative approaches such as direct metal deposition using stainless-steel shadow masks could be employed to better preserve the intrinsic properties of the nanomaterials (Alim et al., 2024).

For biosensing applications, surface functionalization plays a critical role in achieving selectivity. Chemical activation methods Fig. 2.12, such as potassium hydroxide (KOH) treatment, are used to introduce reactive sites on graphene, followed by linker molecules such as APTES and glutaraldehyde (GA) to immobilize bioreceptors, enzymes, antibodies, or aptamers (Selvarajan et al., 2020). These functional layers enable electrostatic gating or charge transfer interactions upon analyte binding.

To evaluate and optimize these devices, three-dimensional simulations Fig. 2.13 based on the Finite Element Method (FEM) are conducted using COMSOL Multiphysics or Ansys Lumerical charge transport solvers (Singh et al., 2021). Additionally, statistical optimization techniques such as the Taguchi method are applied to systematically vary geometrical parameters, including channel length and electrode dimensions to maximize drain current (I_D) output and overall sensing performance (Alim et al., 2023). Such modelling approaches allow systematic optimization of geometrical and electrical parameters prior to experimental realization.

In addition to pure carbon structures, hybrid nanocomposites combining metal oxides with graphene derivatives have been investigated to further enhance sensitivity and selectivity. Titanium dioxide (TiO_2) integrated with graphene oxide (GO) or reduced graphene oxide (rGO) is a commonly reported system. In such approaches, TiO_2 in the anatase phase and GO are prepared using sol-gel techniques, followed by spin-coating onto substrates (Nadzirah et al., 2020). Thermal annealing under controlled argon/hydrogen flow reduces GO to rGO while simultaneously improving TiO_2 crystallinity and interfacial coupling (Roslan et al., 2024). The incorporation of TiO_2 enhances charge separation and surface reactivity, while rGO provides improved electrical conductivity, creating synergistic sensing effects.

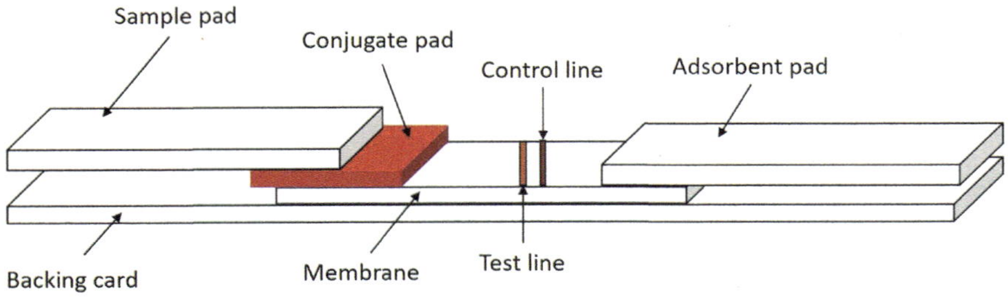

FIGURE 2.14

Typical layout of a lateral flow test strip.

2.2.3 Electrochemical immunochromatographic

Immunochromatography is a combination of chromatography (separation of the sample molecules and reagents based on migration on a solid support by capillary flow) and immunochemical reaction to detect the presence of target analyte in a sample. A typical lateral flow test format (Fig. 2.14) usually consists of a nitrocellulose or filter paper membrane that carries the liquid sample from the sample pad via the conjugate release pad (e.g., labelled antibodies with colloidal gold (commonly used)), along a strip. When an immunocomplex with the immobilized detection antibody is formed, a red-purple line is generated. Migration of the samples through the membrane is around 5–15 min, which is rapid to detect the presence of target analyte by naked eyes, qualitatively. This format has been widely commercialized, with COVID-19 rapid antigen tests and pregnancy tests serving as prominent examples familiar to consumers. However, this technique is lack of sensitivity and often give false negative results because it relies on visual readout or colorimetric screening, with detection limits typically in the range of 1-100 ng/mL depending on the target analyte (Ngom et al., 2010). The subjective nature of visual interpretation and the inability to detect low analyte concentrations have driven the development of more sensitive detection methods. Therefore, other transducers are required to combine with immunochromatography technique to improve sensitivity and enable quantitative measurements.

Recently, electrochemical techniques have been employed to combine with immunochromatographic technique, namely electrochemical immunochromatographic (EIC) technique because electrochemical detection offer attractive features such as high sensitivity, selectivity, low instrumentation cost, and the ability to miniaturize with improved analytical performance (Quesada-González & Merkoçi, 2015). This technique measures the electrical signal and quantitatively detects the analyte in the sample. As discussed in part 1, there are several electrochemical techniques that can be used as a signal measurement, but for EIC, the most popular techniques that are commonly used are voltammetry, amperometry, and impedimetric (Perju & Wongkaew, 2021). For instance, square wave voltammetry (SWV) has demonstrated excellent performance in detecting trichloropyridinol, a metabolite biomarker of exposure to organophosphorus insecticides, with a detection limit as low as 0.1 ng/mL and a wider dynamic range (Wang et al., 2011). More recently, Deenin et al., 2023 developed an electrochemical lateral-flow device for COVID-19 antigen detection that achieved 50-fold higher sensitivity than traditional antigen test kits, demonstrating the continued advancement of EIC technology (Deenin et al., 2023). The electrode materials commonly used in these systems include screen-printed electrodes

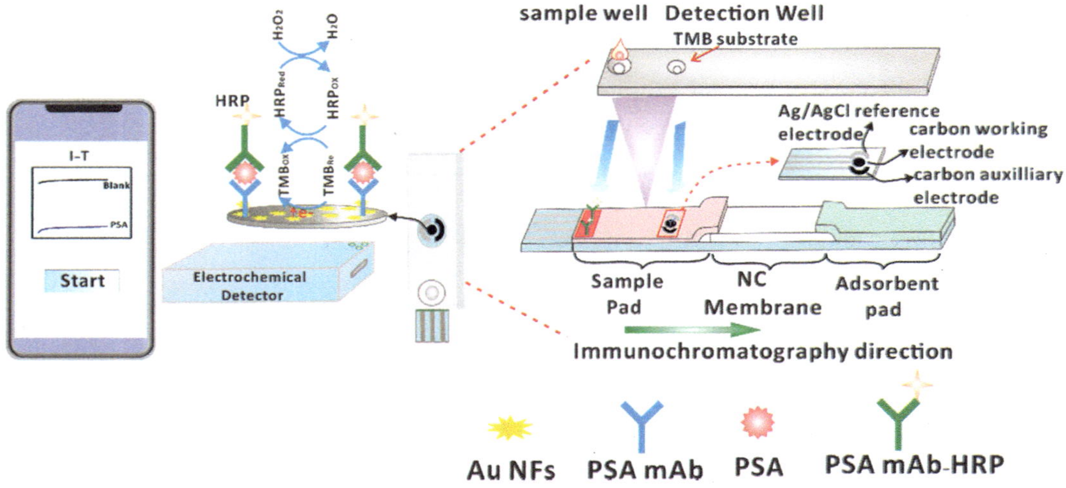

FIGURE 2.15

Schematic illustration of external electrodes mounted below the lateral flow strip.

Reproduced under the terms of the CC-BY license from Dou et al. (2022).

(SPEs) made from carbon, graphene, and CNTs, which offer excellent electrochemical properties, cost-effectiveness, and ease of mass production.

Three distinct approaches exist for incorporating electrochemical transducers into lateral flow immunochromatography (Perju & Wongkaew, 2021). The first method involves external electrodes mounted above, below, or at the end of the lateral flow strip. This approach has been demonstrated by several research groups, including the development of a portable biosensor utilizing gold nanoflower-modified screen-printed carbon electrodes positioned beneath the lateral flow strip for prostate-specific antigen detection (Fig. 2.15), achieving rapid 15 minute detection with clinical accuracy (Dou et al., 2022), and the work of Nandhakumar et al., 2023, who developed a redox cycling amplified electrochemical lateral-flow immunoassay with a screen-printed electrode positioned below the test line for decentralized sensitive insulin detection, achieving a detection limit of 12 pM in biological fluids like saliva and serum (Nandhakumar et al., 2023).

The external electrode configuration allows for independent optimization of both the immunochromatographic strip and the electrochemical sensor. The second approach, known as ex situ electrochemical detection, involves cutting the test line and conducting electrochemical measurements outside the test strip (Larpant et al., 2021; Nian et al., 2012). For example, Larpant and co-workers successfully developed an EIC using this approach for wireless detection of cotinine in urine samples by combining competitive immunochromatography with a near-field communication (NFC) tag (Fig. 2.16). The positive and negative samples could be discriminated at levels higher than 12 ng/mL, representing a significant improvement over visual detection at 200 ng/mL (Larpant et al., 2021). Another notable example is the work by Nian et al., 2012, who demonstrated ex situ detection of human chorionic gonadotropin by transferring the test zone to a screen-printed electrode surface, achieving a detection limit of 0.5 mIU/mL (Nian et al., 2012). This approach provides flexibility in electrode selection and measurement conditions.

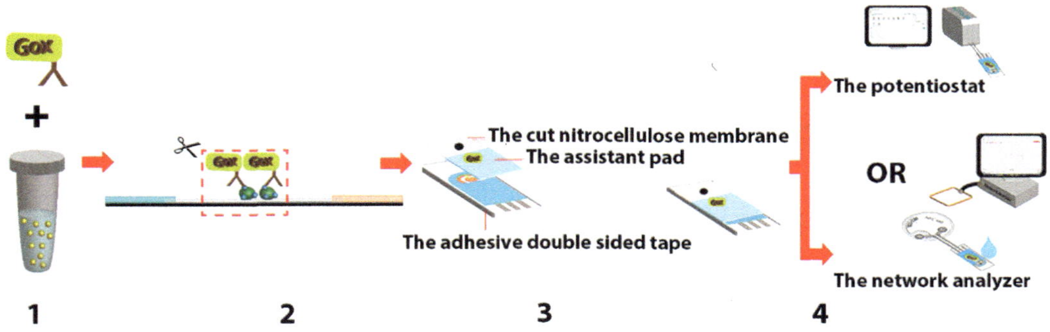

FIGURE 2.16

Schematic diagram of analytical procedure for cotinine determination by electrochemical LFA biosensor. The nitrocellulose membrane is cut and attached to the AgNP/HRP/AuNP-modified SPE.

The third strategy involves internal electrodes, where the electrodes are directly fabricated onto the lateral flow strip, or printing an electrode onto the nitrocellulose membrane, or embedding a complete 3-electrode system into the strip itself (Abarintos et al., 2025). A novel approach developed by Calucho et al., 2024 demonstrates complete integration of reduced graphene oxide electrodes into LFA strips using a CO_2 laser to simultaneously reduce graphene oxide and pattern nitrocellulose, creating leak-proof connection sites by exposing the backing material. The reduced graphene oxide is then transferred onto nitrocellulose using a roll-to-roll system with a wax printer mechanism, where applied pressure facilitates the transfer while maintaining fluidic properties, as illustrated in Fig. 2.17 (Calucho et al., 2024). This scalable fabrication method addresses limitations of traditional screen-printed electrode superposition approaches. Each of these strategies comes with its own set of advantages and disadvantages, which must be carefully considered based on the specific application requirements, target analyte, expected concentration range, and end-user capabilities.

For instance, placing electrodes above, below, or at the end of a lateral flow strip is a straightforward integration that combines existing lateral flow immunochromatographic technology with a well-established and reliable electrochemical transducer. However, ensuring intimate contact between the electrode surface and the lateral flow strip is crucial for achieving the desired detection sensitivity. The tightness or looseness of this contact can impact reproducibility between lateral flow strip devices. To address this issue, chip holders capable of specifying the exact location and ensuring consistent contact between the lateral flow strip and electrode should be developed. Several commercial and research prototypes have incorporated pressure-controlled clamps or alignment guides to standardize this contact interface. On the other hand, the strategy involving cutting the test line and performing ex situ electrochemical measurement leverages well-established technologies to yield reliable measurement results. This approach eliminates the need for intimate contact or specific electrode location concerns, as the extracted sample can be precisely positioned on the electrode surface. Nevertheless, poor reproducibility may arise due to improper handling by end users, particularly in point-of-care settings where trained personnel may not be available. The cutting process itself can introduce variability if not performed consistently. Therefore, the incorporation of pre-scored cutting line or integrated cutter on the lateral flow strip could be a good solution to minimize handling issues and ensure consistent

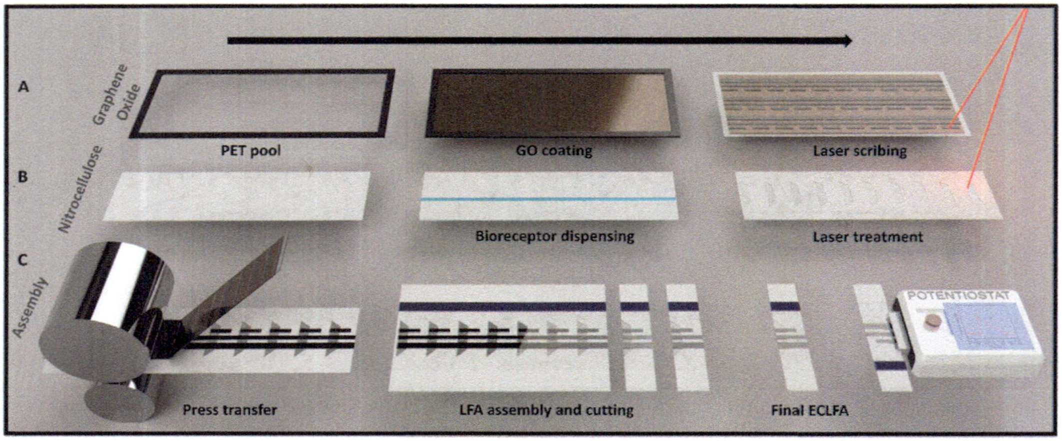

FIGURE 2.17

The fabrication steps of electrochemical LFA strips for internal electrode approach. (A) graphene oxide is dried in a frame and laser-reduced to form electrodes, (B) bioreceptors are applied to nitrocellulose and laser patterning creates connection sites by exposing the backing material, and (C) the reduced graphene oxide is transferred onto the membrane, assembled with other lateral flow components, cut into individual strips, and connected to a potentiostat for electrochemical detection.

sample transfer. Last but not least, creating electrodes directly on the lateral flow strips has a significant potential to produce a robust, fully integrated EIC device that eliminates handling steps and ensures optimal electrode-sample contact. Recent advances have shown promising developments in this area, with Calucho et al., 2024 demonstrating the use of reduced graphene oxide electrodes integrated with lateral flow assays for advanced point-of-care diagnostics, and Ying et al., 2024 developing an electrochemical lateral flow immunoassay with built-in electrodes for ultrasensitive and wireless detection of inflammatory biomarkers (Calucho et al., 2024; Ying et al., 2024). The main challenges include maintaining electrode conductivity and stability in the porous, hydrophilic environment of the lateral flow membrane, as well as ensuring that electrode fabrication does not interfere with capillary flow or biomolecule immobilization. Despite these challenges, this approach represents the most promising direction for developing truly miniaturized, user-friendly EIC devices, necessitating further research attention in the future. The continued development of EIC technology holds promise for expanding point-of-care diagnostics beyond simple qualitative detection to quantitative measurements with clinical-grade accuracy, making them essential tools for decentralized healthcare, environmental monitoring, and food safety applications.

2.2.4 Magnetic wave-based

For centuries, humans have been dedicated to advancing magnetic sensor technology. The magnetic compass, one of the earliest magnetic sensors, was used to navigate by sensing the Earth's magnetic field

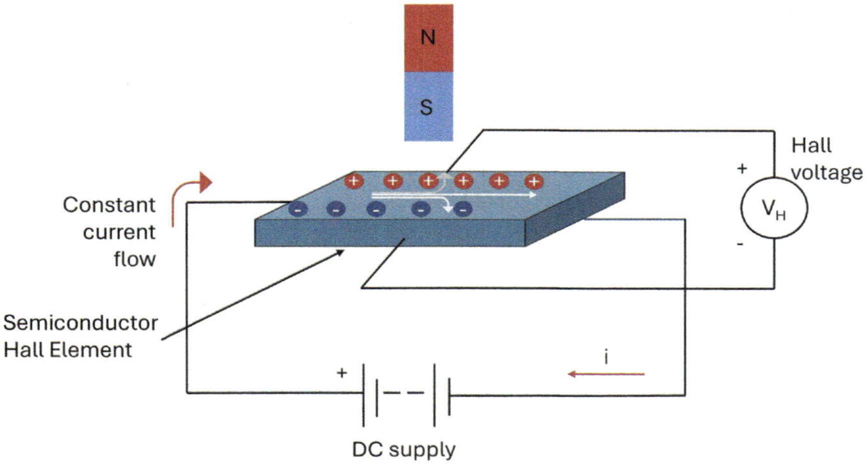

FIGURE 2.18

Schematic diagram of hall effect sensor and how the magnet works.

(Khan et al., 2021). Over time, advancements in technology led to the development of various types of magnetic sensors, including those used in magnetic storage devices like hard drives and magnetic tape. These sensors have played a crucial role in computing, providing a means for data storage and retrieval (Lenz, 1990). In more recent times, miniaturized solid-state magnetic sensors such as Hall Effect, giant magnetoresistance (GMR), tunnel magnetoresistance (TMR), anisotropic magnetoresistance (AMR), and giant magnetoimpedance (GMI) have become integral components in a wide range of applications, including biosensors and navigation systems (Khan et al., 2021). The evolution of these sensors has been driven by the need for higher sensitivity, smaller footprint, and better integration capabilities for point-of-care diagnostics and biomedical applications.

Planar microfabrication techniques are employed to produce these solid-state magnetic sensors, which deliver high sensitivity within a compact form factor. Since these magnetic sensors can be fabricated using CMOS-compatible processes, they enable simultaneous integration of sensing elements and electronic circuitry on a single chip, creating system-on-chip solutions that are particularly well-suited for Internet of Things implementations (Khan et al., 2021). The integration capability allows for on-chip signal processing, digital conversion, and wireless communication, enabling the development of portable, user-friendly diagnostic devices. Therefore, these sensors can be produced in very compact and low-cost systems that are suitable for mass production and deployment in resource-limited settings.

The Hall Effect sensor is a type of magnetic sensor that operates on the principle of the Hall Effect, which was discovered by Edwin Hall in 1879. Fig. 2.18 illustrates the schematic diagram of the Hall Effect sensor and it can be described as follows: when current flows through a conductor or semiconductor in the presence of an external magnetic field applied perpendicular to the current direction, the movement of charge carriers is affected by the Lorentz force, causing them to deflect. This deflection results in the accumulation of electrons on one side of the conductor, generating an additional electric field perpendicular to both the magnetic field and the current plane. This charge accumulation produces a measurable potential difference across the conductor, known as the Hall voltage (Fan et al., 2021). This type of sensor is one of the well-known magnetic sensors, exhibiting high magnetic moment sensitivity over a wide range of magnetic fields in both low and room temperature conditions. Moreover, they generally offer a highly linear response without being affected by magnetic saturation, which allows them to operate effectively even under large polarizing magnetic fields (>1.0 T) that can effectively magnetize MNPs (Žurauskienė, 2023). The compatibility of Hall sensors with standard CMOS fabrication processes had enabled their integration with on-chip electronics, allowing for compact, low-power biosensing systems suitable for point-of-care applications.

In biosensing applications, Hall Effect sensors detect the presence of magnetic nanoparticles (MNPs) or magnetic beads that are used as labels attached to target biomolecules. The detection principle involves two main approaches: direct magnetic field detection and magnetic relaxation detection. In direct detection, when magnetically labeled biomolecules bind to the sensor surface or pass over it, the magnetic field generated by the MNPs induces a Hall voltage that is proportional to the magnetic field strength and thus to the number of bound magnetic labels. The Hall voltage can be measured in real-time, enabling quantitative detection of the target analyte concentration. In magnetic relaxation-based detection, the magnetic dipole moments of superparamagnetic nanoparticles are first aligned by applying a pulsed magnetizing field, which is then rapidly switched off. The Hall sensor then detects the relaxation field of the MNPs as their magnetic moments decay through the Néel relaxation mechanism. This approach offers superior temporal resolution and can characterize the magnetic properties of different MNP types, enabling the differentiation of various magnetic labels in a single sample (Liu et al., 2011). Advanced Hall sensor designs have been successfully applied for detecting magnetically labeled biomolecules, including DNA detection using micro-Hall biosensors (Togawa et al., 2006), which enables molecular-level detection and opens possibilities for studying biomolecular interactions using magnetic techniques.

These sensors have been studied and investigated both numerically and experimentally for diverse medical and biotechnology applications. Hall Effect sensors have been successfully applied for detecting circulating tumor cells (CTCs) in cancer patients, demonstrating a superior performance with 96% detection rate in patient samples by measuring multiple tumor-specific surface proteins using magnetically labeled cells (Issadore et al., 2012). The integration of micro-Hall sensors into microfluidic platforms has enabled ultrasensitive enumeration of rare CTCs, with the ability to process small sample volumes and detect magnetically labeled cells with high sensitivity (Issadore et al., 2012). The integration of Hall Effect sensors with microfluidic platforms has also been demonstrated for enzyme-linked immunoassays (ELISA), where magnetic beads are used as labels to detect analytes in a sandwich ELISA format, with the complete assay performed on chip within approximately 2 hours (Bhalla et al., 2013). Beyond biomarker detection, Hall sensors have found applications in monitoring blood pulse wave velocity for cardiovascular diagnostics, characterizing soft biological materials for tissue engineering, and assisting in magnetic resonance imaging (MRI) for improved spatial resolution (Fan et al., 2021). The miniaturization integration of Hall sensors with CMOS technology has enabled the development of portable diagnostic platforms that can perform sensitive detection of biomarkers in clinical samples without requiring bulky instrumentation or magnetically shielded environments. Additionally, the integration of micro-current lines with Hall sensors has been demonstrated to manipulate magnetic beads through localized field gradients, allowing rapid accumulation of beads onto the sensor surface within seconds for real-time biomolecular detection (Sandhu et al., 2007). The advantages of Hall sensors include simple design and manufacturing technology, low cost, low power consumption,

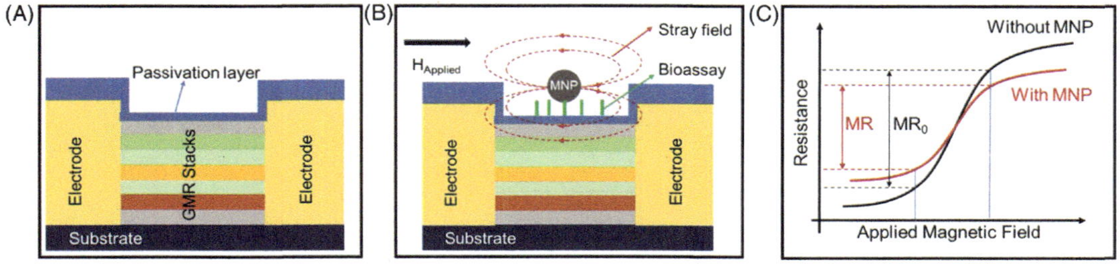

FIGURE 2.19

Schematic diagram of GMR biosensor with its functionalization and the R-H curve of the with and without conjugated MNPs on the sensor surface.

possibility of miniaturization, non-invasive measurement capability, and room temperature operation, making them particularly attractive for decentralized healthcare and resource-limited settings compared to other magnetic sensing systems, such as superconducting quantum interference devices (SQUIDs), that require complex cryogenic infrastructure.

Apart from Hall Effect sensors, GMR sensors have emerged as a powerful tool for ultrasensitive, multiplexed assays, and real-time sensing (Yao et al., 2022). GMR is based on quantum mechanical effect generated from various structures of magnetic thin film. The exhibiting phenomenon is known as spin-dependent scattering, where the changes of sensor's resistance are affected by the changes of magnetization due to the binding of MNPs to a biological element on a GMR sensor (Adem et al., 2020). Fig. 2.19 shows a typical GMR biosensor, which consists of GMR thin films, passivation layers, and electrodes (or electrical contact). The passivation layers serve to shield against degradation when exposed to biological fluids. Under an applied magnetic field, the MNPs that are in the vicinity of the GMR sensor surface will generate stray fields, which alters the MR of the sensor. There are several GMR thin films microstructures have been developed such as multilayer, spin-valve, and granular. The first GMR structure was found to be multilayer, where the structure consists of several alternating ferromagnetic (FM) layers and non-magnetic (NM) spacer layers, as shown in (Fig. 2.20(A)). Under zero applied magnetic field, the magnetization directions of every two adjacent FM layers are set to be antiparallelly aligned, which leads to a high resistance state of the system. If a large magnetic field is applied, the magnetizations in all FM layers will be aligned to the field direction, and this results in a low resistance state. However, despite these superior GMR ratios, multilayer structures demand large operating fields for magnetic switching and exhibit nonlinear field-response characteristics, which limits their practical utility in biosensing applications. In Fig. 2.20(B), the granular structure is composed of magnetic particles distributed throughout a nonmagnetic metallic host matrix, representing another microstructural approach. In this configuration, randomly oriented magnetic moments of the granules produce zero net magnetization and maximum resistance without applied fields. Strong magnetic fields align these particle moments, minimizing system resistance. However, granular structures show limited utility in biosensor applications due to weak GMR signals at low applied fields and poor sensitivity to small magnetic field changes. The spin-valve microstructure, in contrast, incorporates two FM layers separated by a NM spacer, plus an antiferromagnetic (AFM) pinning layer, as illustrated in

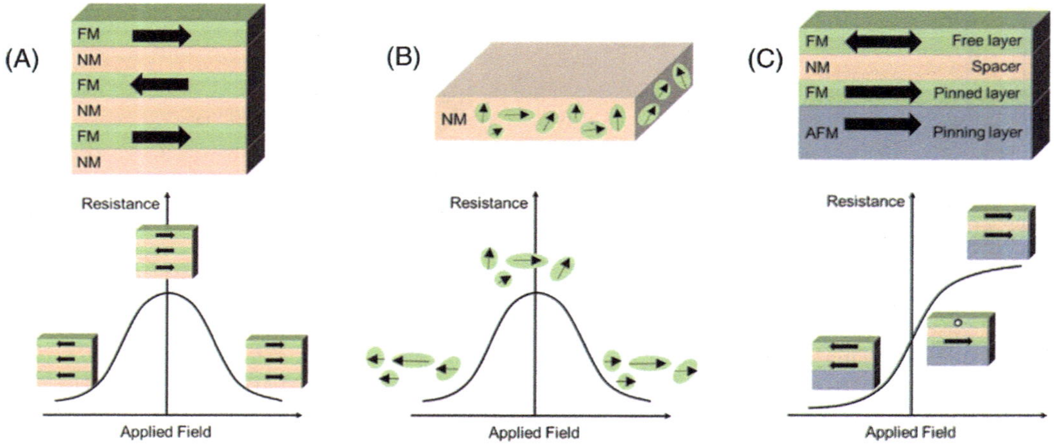

FIGURE 2.20

Schematic views of (A) GMR multilayer structure, (B) granular structure, (C) spin-valve structure, and their corresponding change of resistance in response to applied field curves.

Reproduced under the terms of the CC-BY license from Liang et al. (2022).

(Fig. 2.20(C)). This AFM component locks one FM layer's magnetization direction through exchange bias, creating a fixed reference layer, while the remaining FM free layer maintains rotational freedom under applied fields. Optimal linearity is typically achieved by setting the magnetization angle between the pinned and free layers at 90° when no field is present, allowing for maximum sensitivity to small field changes in both directions. As a result, spin-valve microstructure becomes an ideal for biosensing due to the combination of low saturation requirements and linear output.

Recent advances have shown remarkable progress in GMR biosensor performance. For instance, GMR biosensors have been demonstrated for real-time quantitative detection of protease activity, successfully detecting as low as 4 nM of cysteine protease papain (Adem et al., 2020). Additionally, Zurnansyah et al., 2024 reported a GMR chip-based sensor using green-synthesized Fe_3O_4/rGO nanocomposites as magnetic labels for bovine serum albumin (BSA) protein detection, achieving a sensitivity of 22.98 mV/(mg/mL) and an LOD of 0.3 mg/mL (Zurnansyah et al., 2024). Benefiting from the development of nanofabrication techniques, an individual GMR sensor is often in the size of the micrometer range, which enables hundreds to thousands of GMR sensors to be integrated on a single GMR chip.

Beyond Hall Effect and GMR sensors, other magnetoresistive technologies have shown promise in biosensing applications. TMR sensors operate based on the quantum mechanical tunneling effect in magnetic tunnel junctions (MTJs), which consist of two FM layers separated by a thin insulating barrier. When electrons tunnel through this insulating barrier, the tunneling probability depends on the relative magnetization alignment of the two FM layers. When the magnetizations are parallel, electrons tunnel more easily (low resistance state), whereas anti-parallel alignment results in higher resistance due to spin-dependent tunneling. This phenomenon produces much larger resistance changes compared to GMR sensors. TMR sensors offer even better sensitivity and resolution than GMR sensors due to their larger magnetoresistance ratio, properties that are very important for biosensor applications (Wibowo et al., 2024). This enhanced sensitivity enables the detection of single magnetic nanoparticles, opening possibilities for ultra-sensitive diagnostic applications such as detecting rare CTCs or quantifying very low concentrations of disease biomarkers. Jin et al., 2021 demonstrated a magnetic tunnel junction sensor with sensitivities over 18%/Oe, capable of detecting magnetic nanoparticle amounts below 500 ng and concentrations as low as 0.01 mg/mL (Jin et al., 2021). Ghemes et al., 2023 further validated this approach, measuring a maximum magnetoresistance ratio of 53% and a sensitivity of 1.24%/Oe (Ghemes et al., 2023). Wibowo et al., 2024 highlight the technology's promise for point-of-care diagnostics, particularly in detecting rare circulating tumor cells and low-concentration biomarkers (Wibowo et al., 2024). These studies collectively suggest TMR sensors could revolutionize early disease detection by providing unprecedented sensitivity in magnetic nanoparticle sensing.

AMR sensors are based on the anisotropic magnetoresistance effect, discovered in FM materials where the electrical resistance depends on the angle between the current direction and the magnetization direction within the material. In FM materials, the resistance is highest when the current flows parallel to the magnetization and lowest when perpendicular. This anisotropy arises from spin-orbit coupling effects in the FM material. When an external magnetic field is applied, it rotates the magnetization direction, causing a change in resistance that can be measured. Recently, this technology has been successfully adapted for high-stakes diagnostic applications, ranging from genome profiling of antimicrobial resistance genes and the real-time discrimination of single nucleotide polymorphisms to the quantification of cancer biomarkers and viral pathogens. A distinct frontier for AMR, diverging from GMR and TMR, lies in its mechanical ductility; this has enabled the fabrication of robust, flexible sensors on polymeric substrates for "electronic skin" and wearable health monitors that can withstand thousands of bending cycles (Khan et al., 2025; Wang et al., 2016). Vera et al., 2023 have successfully demonstrated a high-performance implantable AMR sensors based on $La_{0.67}Sr_{0.33}MnO_3$ (LSMO) thin films for biomedical applications, showing promise for in vivo monitoring and implantable diagnostic devices (Vera et al., 2023). Ultimately, while AMR may offer lower absolute sensitivity than TMR, its superior linearity in low fields, fabrication simplicity, and stability in integrated microfluidic "Lab-on-a-Chip" platform position it as a pragmatic, cost-effective engine for the next generation of decentralized point-of-care diagnostics and digital health ecosystem.,

GMI sensors exploit the giant magnetoimpedance effect, where the AC impedance of a soft FM conductor changes dramatically in response to an external DC magnetic field. The physical mechanism involves the skin effect: when an AC flows through magnetic material, it generates a circular magnetic field that affects the magnetic permeability of the material (Jimenez et al., 2022). The penetration depth (or skin depth) of the AC current depends on this permeability. When an external magnetic field applied, it changes the magnetic permeability of the material, which in turn alters the skin depth and consequently the effective impedance. This effect is most pronounced in soft magnetic materials such as amorphous ribbons or thin film multilayers with high magnetic permeability. GMI sensors can exhibit large impedance changes in response to external magnetic fields, providing high sensitivity to small magnetic field variations. These sensors have been explored for biosensing applications requiring rapid response times and ultra-sensitive detection. Recent advances in GMI sensor technology have shown the capability to detect liver cancer biomarker, alpha-fetoprotein (AFP) with a very low detection limit of 0.2 ng/mL, excellent specificity, and rapid response time (Zhu et al., 2019). More recently, GMI sensors have been applied for the detection and quantification of doxorubicin anticancer drugs labeled to Fe_3O_4 superparamagnetic nanoparticles, demonstrating their potential for the therapeutic drug monitoring and personalized cancer treatment (Kirat et al., 2024). Beyond cancer diagnostics, GMI sensors have been successfully applied for rapid pathogen identification, including detection of *E. coli* and SARS-CoV-2, highlighting their versatility for infectious disease diagnostics (Jimenez et al., 2022; Yang et al., 2016). A distinctive frontier for GMI technology lies in the development of magneto-LC-resonance (MLCR) sensors capable of differentiating magnetic nanoparticles by their coercivity, enabling multi-analyte detection and enhanced selectivity in complex biological samples (Hwang et al., 2025). Additionally, GMI sensors based on amorphous ribbons and thin-film multilayered structures have been developed for continuous flow detection of magnetic particles, with applications extending to biomedical diagnostics and tissue characterization (Jimenez et al., 2022). Recent innovations have also focused on flexible wearable GMI devices that utilize the sensor's high sensitivity for non-invasive, continuous monitoring of respiratory motion and vital signs, opening new possibilities for remote health monitoring and point-of-care applications. Ultimately, GMI biosensors offer superior sensitivity compared to traditional magnetoresistive platforms in certain applications, positioning them as a robust solution for both early-stage disease screening and real-time physiological health monitoring.

The integration of magnetic sensors with microfluidics has created powerful lab-on-a-chip platforms that combine sample preparation, analyte capture using magnetic beads, and sensitive magnetic detection in a single device. These integrated systems have been developed for various applications, including nucleic acid detection for infectious diseases, protein biomarker quantification for cancer diagnostics, and therapeutic drug monitoring. The use of magnetic nanoparticles as labels offers several advantages over conventional fluorescent or enzymatic labels, including stability, lack of photobleaching, minimal background signal in biological samples, and the ability to manipulate particles using external magnetic fields for enhanced assay performance. The development of green-synthesized magnetic nanocomposites, such as Fe_3O_4/rGO, provides environmentally friendly alternatives for magnetic labelling while maintaining excellent detection performance (Zurnansyah et al., 2024). Recent innovations have focused on developing smartphone-based magnetic sensing platforms that utilize the built-in magnetometer or integrate miniaturized magnetic sensors with mobile devices, enabling low-cost, portable diagnostic solutions accessible in remote and resource-limited settings. The continued advancement of magnetic sensor technology, coupled with innovations in nanomaterials, microfluidics, and signal processing algorithms, promises to expand the capabilities and applications of magnetic biosensors in clinical diagnostics, environmental monitoring, and food safety testing. The ability to perform real-time biomolecule detection without complex instrumentation positions magnetic biosensors as a key technology for decentralized healthcare and point-of-care diagnostics in the future.

2.2.5 Acoustic wave-based

Acoustic wave biosensors operate on the principle that the propagation characteristics of acoustic waves through a piezoelectric material are highly sensitive to changes in the physical properties at the material's surface or within its volume. When an alternating electrical signal is applied to electrodes on a piezoelectric substrate, the material undergoes periodic mechanical deformation, generating acoustic waves. These waves can propagate either through the bulk of the material (bulk acoustic waves, BAW) or along its surface (SAW), depending on the device geometry and electrode configuration.

The sensing mechanism relies on several physical phenomena. Mass loading is the primary detection mechanism: when target biomolecules bind to a recognition layer on the sensor surface, the added mass decreases the acoustic wave velocity, resulting in a measurable shift in resonant frequency or phase. This relationship is described by the Sauerbrey equation for mass-sensitive detection (Mujahid et al., 2019). Additionally, changes in viscoelastic properties, conductivity, and dielectric constant of the interfacial layer can also affect wave propagation. The sensitivity of acoustic sensors is fundamentally related to the operating frequency—higher frequencies yield greater mass sensitivity, as the fractional frequency change for a given mass loading is proportional to the square of the frequency. This principle drives the development of thin-film, high-frequency devices.

However, a significant challenge for acoustic wave biosensors is the operation in liquid environments, which is essential for biological sensing applications. When acoustic waves propagate into a liquid medium, substantial energy dissipation occurs due to viscous damping, dramatically reducing the quality factor (Q) of the resonator (Voiculescu & Nordin, 2012). A low Q-factor broadens the resonance peak and reduces measurement precision, thereby limiting sensitivity. This fundamental challenge has shaped the evolution of acoustic biosensor designs, leading to innovations in device structures and integration with microfluidic systems to minimize liquid loading effects while maintaining high sensitivity.

The Quartz Crystal Microbalance (QCM) represents the earliest and most established form of acoustic wave biosensor, utilizing thickness-shear mode bulk acoustic waves in AT-cut quartz crystals. In a QCM, the entire thickness of the quartz crystal (typically 100–300 μm) oscillates in a shear mode when driven by electrodes on opposite faces of the crystal. The resonant frequency, typically in the 5–20 MHz range, depends inversely on the thickness of the substrate (Mujahid et al., 2019). QCM devices established several key concepts for acoustic biosensors: the direct relationship between surface mass loading and frequency shift, the ability to perform label-free detection in real-time, and the potential for studying biomolecular interactions and binding kinetics (Lim et al., 2020).

Despite their pioneering role, conventional QCM devices face limitations for modern biosensing applications. Their relatively low operating frequencies limit mass sensitivity compared to thin-film alternatives. The requirement for bulk crystals makes miniaturization and array fabrication challenging, because reducing thickness beyond a certain point would make them mechanically unstable and fragile (Mujahid et al., 2019). Most significantly, energy trapping in liquids severely degrades performance (Zhao et al., 2024), requiring careful consideration of liquid cell design. Nevertheless, QCM technology laid the groundwork for understanding acoustic biosensing and remains valuable for applications requiring robustness and well-established protocols, particularly in research laboratories studying protein-protein interactions, antibody–antigen binding, and cellular adhesion phenomena.

Film bulk acoustic resonators (FBARs) are bulk acoustic devices that operate on the same fundamental principle as QCMs but employ a significantly different structural architecture that yields superior performance. An FBAR consists of a thin piezoelectric film (typically 0.5–3 μm thick) sandwiched between two metal electrodes, which enable it to operate at frequencies ranging from sub-GHz to approximately 10 GHs. When voltage is applied, a standing acoustic wave resonates within the film at a frequency inversely proportional to its thickness. The critical design feature is the acoustic isolation of the active piezoelectric region from the substrate, achieved either through an air cavity beneath the structure (conventional FBAR) or via acoustic Bragg reflectors (solidly mounted resonators (SMRs)), as illustrated in Fig. 2.21. In conventional air-cavity FBARs, the substrate beneath the bottom electrode is etched to create a suspended cavity (typically 100–300 μm in lateral dimensions), forming a free-standing piezoelectric membrane that confines acoustic energy within the active region. The SMR architecture achieves equivalent acoustic isolation through a different mechanism: acoustic Bragg

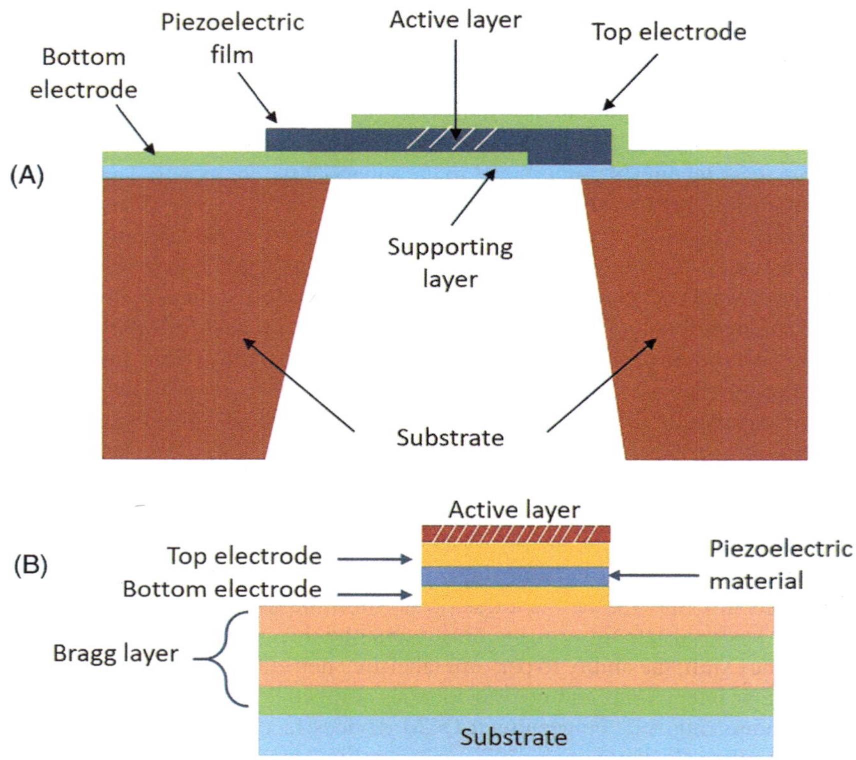

FIGURE 2.21

Schematic diagram of (A) conventional FBAR and (B) solidly mounted resonance (SMR).

reflectors positioned beneath the electrode stack reflect acoustic waves back into the piezoelectric film, preventing energy dissipation into the substrate. These reflectors consist of alternating layers of materials with contrasting acoustic impedances—commonly tungsten (W) and silicon dioxide (SiO_2) with an impedance ratio of approximately 4:1, or aluminum nitride (AlN) and SiO_2 with a ratio of about 3:1 (Lu et al., 2025; Yang et al., 2023). The large acoustic impedance mismatch at each interface causes a portion of the acoustic energy to be reflected at each interface of the multi-layered stack thus preventing energy leakage.

FBARs operate in two distinct resonance modes with fundamentally different liquid-phase performance characteristics: thickness longitudinal mode (TLM) and thickness shear mode (TSM). The mode depends on the piezoelectric crystal's c-axis orientation—perpendicular to the substrate for TLM and tilted approximately 20–35° for TSM (Fig. 2.18). This distinction is crucial for biosensing applications. TLM devices generate longitudinal waves that couple strongly to compressional waves in liquids, causing severe acoustic damping and dramatic Q-factor reduction, limiting their use to gas-phase sensing. Conversely, TSM devices produce shear waves that do not efficiently couple into liquids (which cannot support shear stress), maintaining high Q-values even when immersed in aqueous solutions (Chen et al., 2015). This property makes TSM FBARs the preferred configuration for biosensing and other liquid-phase detection applications. Additionally, liquid properties such as density and conductivity produce characteristic effects on resonance frequencies: increasing density lowers both series and parallel resonance frequencies through mass loading, while increasing conductivity produces divergent effects (decreasing series frequency but increasing parallel frequency), enabling discrimination between mass-loading and electrical property changes.

The selection of piezoelectric material critically determines FBAR performance, fabrication compatibility, and application suitability. Aluminum nitride (AlN) and zinc oxide (ZnO) are the two most extensively employed materials (Chen et al., 2015). AlN offers high acoustic velocity (enabling very high frequency operation), excellent chemical stability and biocompatibility, compatibility with CMOS fabrication for monolithic integration with electronics, and lower acoustic loss contributing to higher Q-factors (Lu et al., 2025). ZnO provides advantages, including smoother surface morphology, higher electromechanical coupling coefficient for more efficient energy conversion, and simpler synthesis at moderate temperatures, though it exhibits lower acoustic velocity and limited CMOS compatibility. Lead zirconate titanate (PZT) offers exceptional electromechanical coupling but suffers from higher acoustic attenuation, lower acoustic velocity, and biocompatibility concerns that limit practical biosensor applications. A significant challenge for both AlN and ZnO is sensitivity to temperature variations, with temperature coefficients of frequency (TCF) of approximately 25 ppm/°C and 40 ppm/°C, respectively. Temperature compensation strategies have been developed, including composite multi-layer structures (e.g., $Al/AlN/Al/SiO_2$) and passive compensation using micro-designed air-gap capacitors with bimetallic layers that exploit differential thermal expansion to counteract intrinsic material TCF, achieving near-zero temperature drift (Mujahid et al., 2019).

FBARs have been successfully demonstrated for biosensing in liquid media and gas detection applications, with their miniaturized design, array capability, MEMS fabrication compatibility, and straightforward operation making them highly suitable for integration with electronic and optoelectronic systems. The potential for monolithic integration with CMOS readout circuits enables compact, low-power sensing platforms appropriate for point-of-care diagnostics, environmental monitoring, and multiplexed biomarker detection. The combination of extreme sensitivity (picogram-level detection), high operating frequencies (enabling rapid measurements), robust performance in liquids (for TSM configurations), and system integration compatibility positions FBAR technology as a powerful and versatile platform for next-generation acoustic biosensors across clinical diagnostics, pharmaceutical development, and biotechnology applications.

SAW devices represent a complementary approach to acoustic biosensing, utilizing waves that propagate laterally along the surface of a piezoelectric substrate rather than through its thickness. The foundation of SAW technology dates to Lord Rayleigh's theoretical description of elastic wave propagation along a plane surface in 1885, though practical SAW devices did not emerge until the development of interdigital transducers (IDTs) in the 1960s. These IDT structures, consisting of periodic interdigitated metal finger electrodes resembling a comb pattern, efficiently convert electrical signals into mechanical surface waves and vice versa.

In principle, when an AC is applied, the vibrations will occur at the surface of the material and generate an acoustic wave. The changes in the mass or strain will change the resonance frequency, which will provide a quantifiable measurement correlated to the amount of mass or strain. The resonance frequency, amplitude, and propagation characteristics are determined by the design of the IDTs, the dimensions of the electrodes, the material properties, and the applied electrical current.

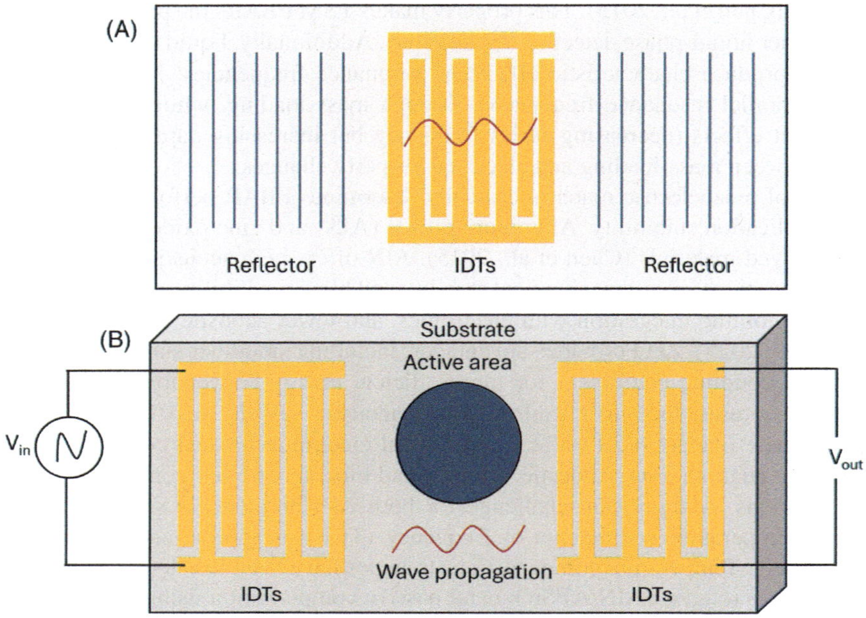

FIGURE 2.22

Schematic diagram of (A) one-port SAW resonator and (B) delay line SAW.

Common piezoelectric substrates for SAW devices include quartz, lithium niobate ($LiNbO_3$), and lithium tantalate ($LiTaO_3$), selected based on electromechanical coupling coefficient, temperature stability, and acoustic loss characteristics (Mujahid et al., 2019). The IDT electrodes are typically fabricated from chemically inert metals or alloys such as gold (Au), aluminum (Al), or multilayer structures like chromium/gold/chromium (Cr/Au/Cr) to ensure stability in biological environments and prevent corrosion. Standard photolithography techniques are employed to pattern these electrodes with spacing between gaps ranging from several micrometers to tens of micrometers, corresponding to operating frequencies from tens of MHz to over 1 GHz.

SAW devices can be configured in two primary architectures for biosensing: resonators and delay lines. In a SAW resonator configuration (one-port design), a single IDT is positioned centrally between two arrays of reflective gratings that trap acoustic energy within a cavity (Fig. 2.22(A)). The reflectors are typically composed of periodic metal strips or grooves that reflect acoustic waves back toward the IDT, creating a standing wave pattern at the resonant frequency. This configuration is characterized by a high quality factor and narrow resonance bandwidth, making it suitable for precise frequency measurements. Analyte detection is achieved by monitoring shifts in the resonant frequency; mass loading from biomolecular binding on the surface between the reflectors increases the effective surface density, decreasing the acoustic wave velocity and lowering the resonant frequency. Single-port SAW resonators are commonly incorporated into oscillator circuits for frequency-locked operation, providing continuous real-time monitoring of frequency changes with high resolution.

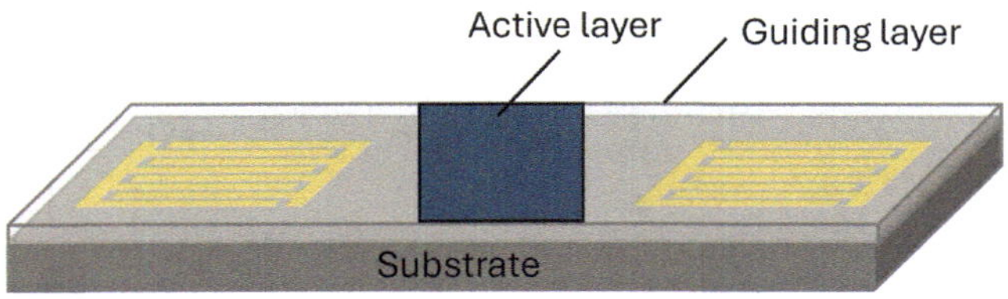

FIGURE 2.23

Schematic diagram of Love wave biosensor. It comprises of two interdigitated electrodes (IDTs) with a guiding layer such as PMMA. The functionalization of biosensor assay will be deposited on top of guiding layer or known as active layer.

In the SAW delay line configuration (two-port design), separate input (transmitting) and output (receiving) IDTs are positioned at opposite ends of the substrate with a sensing region between them (Fig. 2.22(B)). The input IDT converts an electrical RF burst into an acoustic wave that propagates across the substrate surface through the delay line region before being converted back to an electrical signal by the output IDT. The delay line area or active area is typically functionalized with a biorecognition layer (antibodies, aptamers, DNA, etc.) that selectively captures target analytes from a sample introduced to the surface. Biomolecular binding in this region causes changes in both the amplitude and phase of the transmitted signal due to mass loading and viscoelastic effects. By comparing the transmitted signal to a reference path, highly sensitive detection can be achieved. Two-port configurations offer advantages for differential measurements and are less sensitive to temperature fluctuations when reference channels are incorporated on the same substrate.

The liquid loading problem that affects FBAR devices is even more pronounced for conventional Rayleigh-mode SAW devices. The elliptical particle motion characteristic of Rayleigh waves possesses a significant vertical displacement component that couples efficiently to compressional waves in contacting liquids, causing severe acoustic damping and substantial quality factor degradation. To address the liquid damping challenge, researchers developed shear horizontal SAW (SH-SAW) devices that minimize energy dissipation during propagation through liquid media (Li & Bhethanabotla, 2019). Unlike Rayleigh waves with their elliptical motion, SH-SAW propagation involves purely horizontal particle displacement and parallel to the substrate surface but perpendicular to the wave propagation direction. This horizontal polarization is critical for liquid sensing applications, because liquids cannot support shear stress, shear horizontal waves couple much less efficiently to liquid media than vertically polarized Rayleigh waves. This weak coupling dramatically reduces energy loss and enables high-Q operation even when the sensor surface is in direct contact with aqueous solutions. The SH-SAW concept was further refined through the introduction of surface guiding layers, transforming the device into a Love wave sensor (Voiculescu & Nordin, 2012). Love waves are generated by depositing a thin guiding layer—typically materials such as silicon dioxide (SiO_2), polymethyl methacrylate (PMMA), or other polymers—directly onto the piezoelectric SAW substrate (Fig. 2.23). The essential requirement is that this guiding layer must have a lower shear acoustic velocity than the underlying substrate material. When this velocity contrast condition is satisfied, the guiding layer acts as an acoustic waveguide that traps shear horizontal wave energy near the interface, confining it within a shallow depth from the surface. This waveguiding effect produces two beneficial outcomes: it further reduces energy leakage

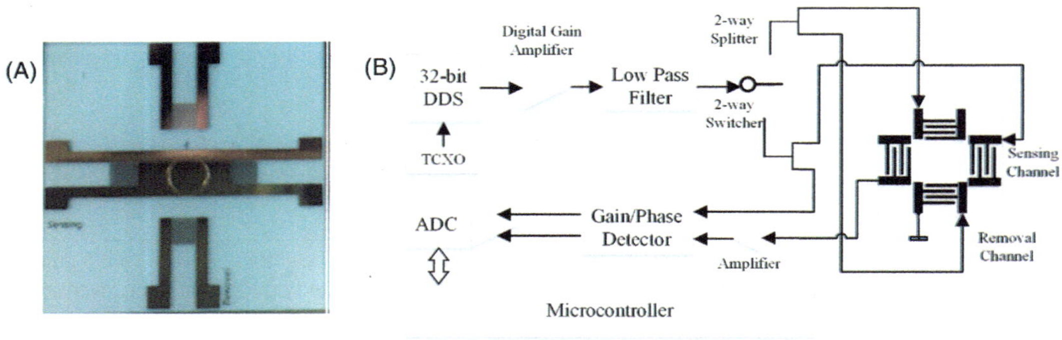

FIGURE 2.24

(A) Photograph of an orthogonal SAW and (B) Schematic of direct digital synthesizer (DDS) based orthogonal SAW system electronics

Reproduced under the terms of the CC-BY license from Li and Bhethanabotla (2019).

into the substrate bulk (improving quality factor), and it concentrates acoustic energy at the sensing surface where biomolecular binding occurs (enhancing mass sensitivity to surface perturbations). The combination of minimal liquid coupling (due to shear horizontal polarization) and enhanced surface energy concentration makes Love wave devices the preferred SAW configuration for liquid-phase biosensing applications.

More recently, advanced SAW platforms have integrated multiple wave modes to provide enhanced functionality beyond simple mass detection. Li and Bhethanabotla (2019) developed a dual-mode (orthogonal) SAW biosensor that simultaneously employs both Rayleigh wave and Love wave modes on a single device, as illustrated in Fig. 2.24 (Li & Bhethanabotla, 2019). This multimodal approach enables detection of target analytes (mouse anti-rabbit IgG in their demonstration) via the Love wave response while simultaneously utilizing the Rayleigh wave response to actively remove nonspecifically bound molecules from the sensor surface. The Rayleigh wave's vertical displacement component, which is detrimental for sensing in liquids due to damping, can be exploited beneficially to generate acoustic streaming and shear forces that dislodge weakly bound interfering molecules, effectively cleaning the sensor surface during measurement. This dual-mode configuration represents a sophisticated integration that addresses both sensitivity (through Love waves) and specificity (through Rayleigh wave-mediated surface cleaning) in a single platform, enhancing the reliability of biosensor measurements in complex biological samples.

2.3 Chemical (static microstructures/microsystem)

2.3.1 pH sensor—Ion analysis, electrolytes

Chemo-sensor technology is a combination of knowledge of chemicals with advances in microelectronics. Notably, field effect sensors can work based on the principle of detecting shifts in localized electric

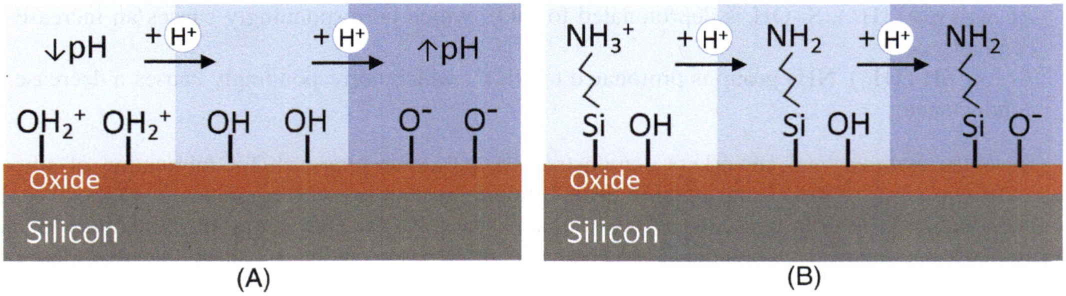

FIGURE 2.25

(A) At low pH the hydroxyl groups are protonated (OH_2^+), while at high pH the hydroxyl groups get deprotonated (O^-). (B) APTES modified silicon oxide surface for pH sensing.

potentials. This is due to chemical interactions at the device surface that cause a local charge accumulation or a change in the charge polarization. Bergveld and co-worker were the first to execute this theory by introducing the ion sensitive field effect transistor (ISFET) as a tool for measuring ion activities in aqueous solution of biological and chemical environments (van der Schoot & Bergveld, 1987). For this development, they proposed a theory that the ions in the aqueous solution would generate a charge interaction at the silicon oxide (SiO_2) interface, which led to a change in conductance. This hypothesis does not deviate far from the concept of semiconductor nanodevices that is being developed today.

In application of pH measurement, silicon nanowire (SiNW) sensors have the ability to detect ions in liquid solution that is attached to the NW surface. Change of pH value results in change of ion concentration that can be measured by ion-sensitive barrier on the top oxide. In such pH sensing, a layer of SiO_2, Si_3N_4 or Al_2O_3 acts as ion-sensitive barrier, since the hydroxyl group (OH) adsorbs charge to the top oxide, as depicted in Fig. 2.25(A). The reaction between silanol (Si-OH) groups of the SiO_2 and the hydrogen (H^+) ion in solution can be summarized as follows:

At high pH (↑H^+), Si-OH is deprotonated to SiO^- which correspondingly causes an increase in conductance.
At low pH (↓H^+), Si-OH is protonated to $SiOH_2^+$ which correspondingly causes a decrease in conductance.

Besides that, Cui et al. had demonstrated the modified SiO_2 surface with 3-aminopropyltriethoxysilane (APTES), which yields amino groups (NH_2) at the NW surface along with the naturally occurring silanol (Si-OH) groups of the oxide (Rahman et al., 2021). The covalent linking of NH_2 to Si-OH groups acts as a function of H^+ ion receptors that can undergo protonation or deprotonation reactions, as shown in Fig. 2.25(B). They found that the graph for modified SiO_2 with APTES shows a linear response between the conductance measurement and the pH value. This is due to the combined acid and base behavior of both NH_2 and Si-OH surface groups. Unlike unmodified SiNW, the reaction of hydrogen (H^+) ion for modified SiNW resulted in the change of NH_2 and Si-OH groups as described as follows:

At high pH (↑H^+), Si-OH is deprotonated to SiO^-, which correspondingly causes an increase in conductance.
At low pH (↓H^+), NH_2 group is protonated to NH_3^+, which correspondingly causes a decrease in conductance.

Recently, Rahman and coworkers demonstrated a top-down approach for fabrication of silicon nanowire-based pH sensors using electron beam lithography coupled with complementary metal oxide semiconductor (CMOS) technology (Abd Rahman et al., 2022). They show the sensitivity of the developed sensor in the detection of pH levels range of 4–10 based on the current-voltage relations. The functionalized silicon nanowire with APTES experienced protonation and deprotonation in response to the acidic solution ($NH_2 + H^+ \Longleftrightarrow NH_3^+$) and basic solution ($SiOH \Longleftrightarrow SiO^- + H^+$), respectively. The highest resistance change was observed for the detection of pH 4, followed by pH 7 and 10. This is due to the accumulation of hole carriers in the p-type Si nanowire, thus increasing the conduction for pH 10. This underlying concept can be illustrated in and Fig 2.25(B). Another reliable approach is the pH-to-time readout-based ISFET system, as presented by Jiang and group (Jiang et al., 2016). The capacitance (C) is pre-charged before detection and then discharged during pH sensing. The discharging time is dependent on the drain-source current of the transistor, which is further related to the pH level of the solution. In addition, the group further demonstrated the identification of bacterial levels in food resources by measuring pH changes in the culture medium (Jiang et al., 2018). They designed a pH-time-voltage conversion system, which is expected to further broaden the scope for ion sensing.

2.4 Chapter summary

This chapter provides a comprehensive overview of electrical and electrochemical sensing technologies organized into three major categories: electrochemical sensors, electrode-based sensors, and chemical sensors, all focusing on static microstructures and microsystems that transduce biochemical interactions into measurable electrical signals for biosensing, diagnostics, and analytical chemistry applications.

The chapter begins with fundamental electrochemical sensing techniques, examining six primary detection methods: potentiometry measuring potential differences for selective ion detection; amperometry measuring oxidation-reduction currents for high-sensitivity detection of electroactive species; conductometry detecting changes in electrical conductivity for label-free monitoring of ionic species and biomolecular binding; impedimetric sensing measuring complex impedance across frequency ranges to characterize electrode-electrolyte interfaces; voltammetry measuring current responses during potential sweeps to identify and quantify electroactive analytes; and capacitive sensing detecting changes in medium properties, particularly permittivity (dielectric constant). Each technique is contextualized within biosensor applications, highlighting specific advantages for detecting different analyte types across varying concentration ranges.

The electrode-based sensors section explores specialized electrode geometries and detection principles including interdigitated electrode (IDE) arrays with microscale spacing and deposition of nanoparticles between the finger electrodes enabling enhanced the sensor's sensitivity; transistor-based sensors such as bioFET that integrate sensing with semiconductor electronics for biosensing application; EIC systems combining lateral flow simplicity with electrochemical quantification through three integration strategies (external electrodes, ex situ detection, and internal electrodes), with recent developments demonstrating 50-fold sensitivity improvements for COVID-19 detection and sub-nanogram detection limits for insulin; magnetic wave-based sensors including Hall Effect (voltage generation under magnetic fields), GMR (spin-dependent scattering achieving femtomolar sensitivity), TMR (quantum tunneling with high magnetoresistance ratios), AMR (angle-dependent resistance for wearable devices), and GMI (impedance changes achieving high sensitivity) technologies utilizing magnetic nanoparticle labels for applications ranging from circulating tumor cell detection to pathogen identification and respiratory monitoring, with recent innovations including magneto-LC-resonance (MLCR) sensors for nanoparticle differentiation and temperature compensation for measurement stability; and acoustic wave-based sensors tracing evolution from 1880s piezoelectric discovery through QCM development to modern MEMS devices, examining FBARs with thin piezoelectric films (ZnO, AlN) operating at GHz frequencies achieving picogram detection through thickness shear mode for liquid-phase biosensing, alongside SAW devices using interdigitated transducers on piezoelectric substrates progressing from Rayleigh waves to shear horizontal SAW and Love waves (with guiding layers) to overcome liquid damping challenges.

The chemical sensors section addresses pH sensors and ion-selective electrodes operating through ISFET principles for measuring hydrogen ion activity with applications in biological and chemical environments, with MEMS integration enabling miniaturized solid-state designs overcoming traditional glass electrode limitations for continuous monitoring in point-of-care settings.

Throughout the chapter, emphasis is placed on practical advantages, including MEMS/CMOS fabrication compatibility enabling miniaturization and integration of sensing with electronics, array configurations for multiplexed detection, microfluidic integration for automated sample handling, and point-of-care suitability through portable low-cost implementations. These electrical and electrochemical sensing technologies represent powerful, versatile platforms addressing critical needs in clinical diagnostics, pharmaceutical development, environmental monitoring, and food safety testing, with continued innovation promising expanded capabilities for decentralized healthcare, personalized medicine, and real-time biosensing that improve patient outcomes and enable proactive health management.

References

Sandhu, A., Kumagai, Y., Lapicki, A., Sakamoto, S., Abe, M., & Handa, H. (2007). High efficiency hall effect micro-biosensor platform for detection of magnetically labeled biomolecules. *Biosensors and Bioelectronics, 22*(9–10), 2115–2120. https://doi.org/10.1016/j.bios.2006.09.021.

Selvarajan, R. S., Rahim, R. A., Majlis, B. Y., Gopinath, S. C. B., & Hamzah, A. A. (2020). Ultrasensitive and highly selective graphene-based field-effect transistor biosensor for anti-diuretic hormone detection. *Sensors (Switzerland), 20*(9), 2642. https://doi.org/10.3390/s20092642.

Shan, J., Fang, S., Wang, W., Zhao, W., Zhang, R., Liu, B., … Liu, Z. (2022). Copper acetate-facilitated transfer-free growth of high-quality graphene for hydrovoltaic generators. *National Science Review, 9*(7), nwab169. https://doi.org/10.1093/nsr/nwab169.

Singh, P., Abedini Sohi, P., & Kahrizi, M. (2021). Finite element modelling of bandgap engineered graphene FET with the application in sensing methanethiol biomarker. *Sensors, 21*(2), 580. https://doi.org/10.3390/s21020580.

Wang, Z., & Jia, Y. (2018). Graphene solution-gated field effect transistor DNA sensor fabricated by liquid exfoliation and double glutaraldehyde cross-linking. *Carbon, 130*(7), 758–767. https://doi.org/10.1016/j.carbon.2018.01.078.

Abarintos, V., Piper, A., & Merkoci, A. (2025). Electrochemical lateral flow assays: A new frontier for rapid and quantitative biosensing. *Current Opinion in Electrochemistry, 54*, 101750. https://doi.org/10.1016/j.coelec.2025.101750.

Abd Rahman, S. F., Arshad, M. K. M., Gopinath, S. C. B., Fathil, M. F. M., Sarry, F., & Ibau, C. (2021). Glycosylated biomarker sensors: Advancements in prostate cancer diagnosis. *Chemical Communications, 57*(76), 9640–9655. https://doi.org/10.1039/d1cc03080a.

Abd Rahman, S. F., Yusof, N. A., Md Arshad, M. K., Hashim, U., Md Nor, M. N., & Hamidon, M. N. (2022). Fabrication of silicon nanowire sensors for highly sensitive pH and DNA hybridization detection. *Nanomaterials, 12*(15), 2652. https://doi.org/10.3390/nano12152652.

Abubakar, S., Khalid, N., Abd Rahman, S. F., Tee, T. S., Hamidon, M. N., Talib, Z. A., … Paiman, S. (2020). Fabrication and characterization of nanostructured zinc oxide on printed microcontact electrode for piezoelectric applications. *Journal of Materials Research and Technology,, 9*(6), 15952–15961. https://doi.org/10.1016/j.jmrt.2020.11.038.

Adem, S., Jain, S., Sveiven, M., Zhou, X., O'Donoghue, A. J., & Hall, D. A. (2020). Giant magnetoresistive biosensors for real-time quantitative detection of protease activity. *Scientific Reports, 10*(1), 7941. https://doi.org/10.1038/s41598-020-62910-2.

Alim, A. A., Roslan, R., Nadzirah, Sh., Saidi, L. K., Menon, P. S., Aziah, I., … Hamzah, A. A. (2023). Geometrical characterisation of TiO2-rGO field-effect transistor as a platform for biosensing applications. *Micromachines, 14*(9), 1664. https://doi.org/10.3390/mi14091664.

Alim, A. A., Roslan, R., Nadzirah, S., Dee, C. F., Abdul Hamid, M. A., & Hamzah, A. A. (2024). Direct growth graphene via atmospheric pressure chemical vapour deposition. *International Journal of Nanoelectronics and Materials, 17*(June), 293–299. https://doi.org/10.58915/ijneam.v17iJune.871.

Ambrosi, A., Merkoçi, A., & de la Escosura-Muñiz, A. (2008). Electrochemical analysis with nanoparticle-based biosystems. *TrAC - Trends in Analytical Chemistry, 27*(7), 568–584. https://doi.org/10.1016/j.trac.2008.05.008.

Ansari, S. G., Ansari, Z. A., Wahab, R., Kim, Y.-S., Khang, G., & Shin, H.-S. (2008). Glucose sensor based on nano-baskets of tin oxide templated in porous alumina by plasma enhanced CVD. *Biosensors and Bioelectronics, 23*(12), 1838–1842. https://doi.org/10.1016/j.bios.2008.02.022.

Batool, R., Rhouati, A., Nawaz, M. H., Hayat, A., & Marty, J. L. (2019). A review of the construction of nano-hybrids for electrochemical biosensing of glucose. *Biosensors, 9*(1), 46. https://doi.org/10.3390/bios9010046.

Bhalla, N., Chung, D. W. Y., Chang, Y.-J., Uy, K. J. S., Ye, Y. Y., Chin, T.-Y., … Pijanowska, D. G. (2013). Microfluidic platform for enzyme-linked and magnetic particle-based immunoassay. *Micromachines, 4*(2), 257–271. https://doi.org/10.3390/mi4020257.

Bratov, A., Abramova, N., & Ipatov, A. (2010). Recent trends in potentiometric sensor arrays—A review. *Analytica Chimica Acta, 678*(2), 149–159. https://doi.org/10.1016/j.aca.2010.08.035.

Calucho, E., Álvarez-Diduk, R., Piper, A., Rossetti, M., Nevanen, T. K., & Merkoçi, A. (2024). Reduced graphene oxide electrodes meet lateral flow assays: A promising path to advanced point-of-care diagnostics. *Biosensors and Bioelectronics, 258*, 116315. https://doi.org/10.1016/j.bios.2024.116315.

Çevik, E., Şenel, M., & Baykal, A. (2013). Potentiometric urea biosensor based on poly(glycidylmethacrylate)-grafted iron oxide nanoparticles. *Current Applied Physics, 13*(1), 280–286. https://doi.org/10.1016/j.cap.2012.07.025.

Chen, Y.-C., Shih, W.-C., Chang, W.-T., Yang, C.-H., Kao, K.-S., & Cheng, C.-C. (2015). Biosensor for human IgE detection using shear-mode FBAR devices. *Nanoscale Research Letters, 10*(1), 69. https://doi.org/10.1186/s11671-015-0736-3.

Chowdhury, A. D., Ganganboina, A. B., Park, E. Y., & Doong, R.-A. (2018). Impedimetric biosensor for detection of cancer cells employing carbohydrate targeting ability of concanavalin A. *Biosensors and Bioelectronics, 122*, 95–103. https://doi.org/10.1016/j.bios.2018.08.039.

Cui, Y., Wei, Q., Park, H., & Lieber, C. M. (2001). Nanowire nanosensors for highly sensitive and selective detection of biological and chemical species. *Science, 293*(5533), 1289–1292. https://doi.org/10.1126/science.1062711.

Curulli, A. (2021). Electrochemical biosensors in food safety: Challenges and perspectives. *Molecules, 26*(10), 240. https://doi.org/10.3390/molecules26102940.

Deenin, W., Yakoh, A., Pimpitak, U., Pasomsub, E., Rengpipat, S., Crespo, G. A., & Chaiyo, S. (2023). Electrochemical lateral-flow device for rapid COVID-19 antigen-diagnostic testing. *Bioelectrochemistry, 152*, 108438. https://doi.org/10.1016/j.bioelechem.2023.108438.

Dhahi, Th. S., Hashim, U., Ali, M. E., Ahmed, N. M., & Nazwa, T. (2011). Fabrication of lateral polysilicon gap of less than 50 nm using conventional lithography. *Journal of Nanomaterials, 2011*, 1–8. https://doi.org/10.1155/2011/250350.

Dou, Y., Li, Z., Su, J., & Song, S. (2022). A portable biosensor based on Au nanoflower interface combined with electrochemical immunochromatography for POC detection of prostate-specific antigen. *Biosensors, 12*, 259. https://doi.org/10.3390/bios12050259.

Fan, H., Wang, J., Feng, Q., Hu, Q., Zuo, S., Nabaei, V., & Heidari, H. (2021). Detection techniques of biological and chemical hall sensors. *RSC Advances, 11*(13), 7257–7270. https://doi.org/10.1039/D0RA10027G.

Fernando, C. A. N., Liyanaarachchi, U. S., & Aarajapaksha, R. D. (2013). Explanation of the photocurrent quantum efficiency (Φ) enhancements through the CAN's model equation for the p-CuI sensitized methylviolet-C 18 LB films in the photoelectrochemical cells (PECs) and Cu/n-Cu 2O/M-C18/p-CuI solid-state photovoltaic cells. *Semiconductor Science and Technology, 28*(4). https://doi.org/10.1088/0268-1242/28/4/045017.

Ghemes, C., Dragos-Pinzaru, O.-G., Tibu, M., Lostun, M., Lupu, N., & Chiriac, H. (2023). Tunnel magnetoresistance-based sensor for biomedical application: Proof-of-concept. *Coatings, 13*, 227. https://doi.org/10.3390/coatings13020227.

Hamzah, A. A., Selvarajan, R. S., & Majlis, B. Y. (2017). Graphene for biomedical applications: A review. *Sains Malaysiana, 46*(7), 1125–1139. https://doi.org/10.17576/jsm-2017-4607-16.

Hayat, A., Catanante, G., & Marty, J. (2014). Current trends in nanomaterial-based amperometric biosensors. *Sensors, 14*(12), 23439–23461. https://doi.org/10.3390/s141223439.

Hwang, K. Y., Brown, D., Attanayake, S. B., Luu, D., Nguyen, M. D., Lee, T. R., & Phan, M.-H. (2025). Signal differentiation of moving magnetic nanoparticles for enhanced biodetection and diagnostics. *Biosensors, 15*(2), 116. https://doi.org/10.3390/bios15020116.

Israr, M. Q., Sadaf, J. R., Asif, M. H., Nur, O., Willander, M., & Danielsson, B. (2010). Potentiometric cholesterol biosensor based on ZnO nanorods chemically grown on Ag wire. *Thin Solid Films, 519*(3), 1106–1109. https://doi.org/10.1016/j.tsf.2010.08.052.

Issadore, D., Chung, J., Shao, H., Liong, M., Ghazani, A. A., Castro, C. M., … Lee, H. (2012). Ultrasensitive clinical enumeration of rare cells ex vivo using a micro-hall detector. *Science Translational Medicine, 4*(141), 141ra92. https://doi.org/10.1126/scitranslmed.3003747.

Jia, F., Duan, N., Wu, S., Dai, R., Wang, Z., & Li, X. (2016). Impedimetric *Salmonella aptasensor* using a glassy carbon electrode modified with an electrodeposited composite consisting of reduced graphene oxide and carbon nanotubes. *Microchimica Acta, 183*(1), 337–344. https://doi.org/10.1007/s00604-015-1649-7.

Jiang, Y., Liu, X., Dang, T. C., Huang, X., Feng, H., Zhang, Q., & Yu, H. (2018). A high-sensitivity potentiometric 65-nm CMOS ISFET sensor for rapid E. coli screening. *IEEE Transactions on Biomedical Circuits and Systems, 12*(2), 402–415. https://doi.org/10.1109/TBCAS.2018.2793861.

Jiang, Y., Liu, X., Dang, T. C., Yan, M., Yu, H., Huang, J.-C., … Chen, T.-T. (2016). A 512×576 65-nm CMOS ISFET sensor for food safety screening with 123.8 mV/pH sensitivity and 0.01 pH resolution. *Digest of Technical Papers - Symposium on VLSI Technology. 2016 IEEE Symposium on VLSI Technology*. Institute of Electrical and Electronics Engineers Inc https://doi.org/10.1109/VLSIT.2016.7573440.

Jimenez, V. O., Hwang, K. Y., Nguyen, D., Rahman, Y., Albrecht, C., Senator, B., … Phan, M.-H. (2022). Magnetoimpedance biosensors and real-time healthcare monitors: Progress, opportunities, and challenges. *Biosensors, 12*(7), 517. https://doi.org/10.3390/bios12070517.

Jin, Z., Koo, T. M., Kim, M. S., Al-Mahdawi, M., Oogane, M., Ando, Y., & Kim, Y. K. (2021). Highly-sensitive magnetic sensor for detecting magnetic nanoparticles based on magnetic tunnel junctions at a low static field. *AIP Advances, 11*, 015046. https://doi.org/10.1063/9.0000189.

Khalid, Aishah, Mohamed, Mohd Ambri, & Umar, Akrajas Ali (2017). Graphene growth at low temperatures using RF-plasma enhanced chemical vapour deposition. *Sains Malaysiana, 46*(7), 1111–1117. https://doi.org/10.17576/jsm-2017-4607-14.

Khan, M. A., Sun, J., Li, B., Przybysz, A., & Kosel, J. (2021). Magnetic sensors-A review and recent technologies. *Engineering Research Express, 3*(2). https://doi.org/10.1088/2631-8695/ac0838.

Khan, W. U., Alissa, M., Ma, H., Bhatti, U. A., Alghamdi, A., Alshehri, M. A., & Albelasi, A. (2025). Enhancing diagnostic reliability in non-invasive health Monitoring: An analytical framework for optimizing magnetic sensor-skin interactions in biomedical applications. *Materials Today Bio, 34*, 102259. https://doi.org/10.1016/j.mtbio.2025.102259.

Kharitonov, S. V. (2006). Electrochemical response characteristics and analytical application of papaverine ion-selective membrane electrodes. *Analytical Letters, 39*(2), 259–273. https://doi.org/10.1080/00032710500476912.

Khosravi Rad, Babak, Mehrfar, Amir Hossein, Sadeghi Neisiani, Zahra, Khaje, Mahdi, & Majd, Abdollah Eslami (2024). Effect of fabrication process on contact resistance and channel in graphene field effect transistors. *Scientific Reports, 14*, 9190. https://doi.org/10.1038/s41598-024-58360-9.

Kim, J.-H., Cho, J.-H., Cha, G. S., Lee, C.-W., Kim, H.-B., & Paek, S.-H. (2000). Conductimetric membrane strip immunosensor with polyaniline-bound gold colloids as signal generator. *Biosensors and Bioelectronics, 14*(12), 907–915. https://doi.org/10.1016/S0956-5663(99)00063-9.

Kirat, G., Erdoğan, A., & Aksan, M. A. (2024). GMI-based biosensor for the detection and quantification of doxorubicin anticancer drugs labeled to Fe3O4 superparamagnetic nanoparticles. *Sensors and Actuators A: Physical, 373*, 115400. https://doi.org/10.1016/j.sna.2024.115400.

Lai, H. C., Chin, S. F., Pang, S. C., Sum, M. S. H., & Perera, D. (2017). Carbon nanoparticles based electrochemical biosensor strip for detection of Japanese encephalitis virus. *Journal of Nanomaterials, 2017,*(1), 3615707‘. https://doi.org/10.1155/2017/3615707.

Larpant, N., Kalambate, P. K., Ruzgas, T., & Laiwattanapaisal, W. (2021). Paper-based competitive immunochromatography coupled with an enzyme-modified electrode to enable the wireless monitoring and electrochemical sensing of cotinine in urine. *Sensors, 21*(5), 1–16. https://doi.org/10.3390/s21051659.

Lenz, J. E. (1990). A review of magnetic sensors. *Proceedings of the IEEE, 78*(6), 973–989. https://doi.org/10.1109/5.56910.

Li, S., & Bhethanabotla, V. R. (2019). Design of a portable orthogonal surface acoustic wave sensor system for simultaneous sensing and removal of nonspecifically bound proteins. *Sensors, 19*(18), 3876. https://doi.org/10.3390/s19183876.

Liang, S., Sutham, P., Wu, K., Mallikarjunan, K., Wang, J.-P. (2022). Giant Magnetoresistance Biosensors for Food Safety Applications. *Sensors, 22*, 5663. https://doi.org/10.3390/s22155663.

Liang, R., Kou, L., Chen, Z., & Qin, W. (2013). Molecularly imprinted nanoparticles based potentiometric sensor with a nanomolar detection limit. *Sensors and Actuators, B: Chemical, 188*, 972–977. https://doi.org/10.1016/j.snb.2013.07.110.

Liao, W.-C., & Ho, J. A. A. (2009). Attomole DNA electrochemical sensor for the detection of escherichia coli O157. *Analytical Chemistry, 81*(7), 2470–2476. https://doi.org/10.1021/ac8020517.

Lim, H. J., Saha, T., Tey, B. T., Tan, W. S., & Ooi, C. W. (2020). Quartz crystal microbalance-based biosensors as rapid diagnostic devices for infectious diseases. *Biosensors and Bioelectronics, 168*, 112513. https://doi.org/10.1016/j.bios.2020.112513.

Lin, D., Pillai, R. G., Lee, W. E., & Jemere, A. B. (2019). An impedimetric biosensor for E. coli O157:H7 based on the use of self-assembled gold nanoparticles and protein G. *Microchimica Acta, 186*(3), 169. https://doi.org/10.1007/s00604-019-3282-3.

Liu, P., Skucha, K., Megens, M., & Boser, B. (2011). A CMOS hall-effect sensor for the characterization and detection of magnetic nanoparticles for biomedical applications. *IEEE Transactions on Magnetics, 47*(10), 3449–3451. https://doi.org/10.1109/TMAG.2011.2158600.

Lu, H., Hao, X., Yang, L., Hou, B., Zhang, M., Wu, M., … Ma, X. (2025). Recent advances in AlN-based acoustic wave resonators. *Micromachines, 16*, 205. https://doi.org/10.3390/mi16020205.

McKenzie, K. J., & Marken, F. (2001). Direct electrochemistry of nanoparticulate Fe2O3 in aqueous solution and adsorbed onto tin-doped indium oxide. *Pure and Applied Chemistry, 73*(12), 1885–1894. https://doi.org/10.1351/pac200173121885.

Mujahid, A., Afzal, A., & Dickert, F. L. (2019). An overview of high frequency acoustic sensors—QCMs, SAWs and FBARs—chemical and biochemical applications. *Sensors, 19*(20), 4395. https://doi.org/10.3390/s19204395.

Nadzirah, S., Azizah, N., Hashim, U, Gopinath, S. C. B., & Kashif, M. (2015). Titanium dioxide nanoparticle-based interdigitated electrodes: A novel current to voltage DNA biosensor recognizes E. coli O157:H7. *PLoS ONE, 10*(10), e0139766. https://doi.org/10.1371/journal.pone.0139766.

Nadzirah, S., Hashim, U., Gopinath, S. C. B., Parmin, N. A., Hamzah, A. A., Yu, H. W., & Dee, C. F. (2020). Titanium dioxide–mediated resistive nanobiosensor for E. coli O157:H7. *Microchimica Acta, 187*, 235. https://doi.org/10.1007/s00604-020-4214-y.

Nandhakumar, P., Martin, C. M. S., Arévalo, B., Ding, S., Lunker, M., Vargas, E., … Wang, J. (2023). Redox cycling amplified electrochemical lateral-flow immunoassay: Toward decentralized sensitive insulin detection. *ACS Sensors, 8*(10), 3892–3901. https://doi.org/10.1021/acssensors.3c01445.

Ngom, B., Guo, Y., Wang, X., & Bi, D. (2010). Development and application of lateral flow test strip technology for detection of infectious agents and chemical contaminants: A review. *Analytical and Bioanalytical Chemistry, 397*(3), 1113–1135. https://doi.org/10.1007/s00216-010-3661-4.

Nian, H., Wang, J., Wu, H., Lo, J.-G., Chiu, K.-H., Pounds, J. G., & Lin, Y. (2012). Electrochemical immunoassay of cotinine in serum based on nanoparticle probe and immunochromatographic strip. *Analytica Chimica Acta, 713*, 50–55. https://doi.org/10.1016/j.aca.2011.11.028.

Nouira, W., Maaref, A., Elaissari, H., Vocanson, F., Siadat, M., & Jaffrézic-Renault, N. (2013). Comparative study of conductometric glucose biosensor based on gold and on magnetic nanoparticles. *Materials Science and Engineering C, 33*(1), 298–303. https://doi.org/10.1016/j.msec.2012.08.043.

Papagianni, G. G., Stergiou, D. V., Armatas, G. S., Kanatzidis, M. G., & Prodromidis, M. I. (2012). Synthesis, characterization and performance of polyaniline-polyoxometalates (XM 12, X = P, Si and M = Mo, W) composites as electrocatalysts of bromates. *Sensors and Actuators, B: Chemical, 173*, 346–353. https://doi.org/10.1016/j.snb.2012.07.020.

Perju, A., & Wongkaew, N. (2021). integrating high-performing electrochemical transducers in lateral flow assay. *Analytical and Bioanalytical Chemistry, 413*(22), 5535–5549. https://doi.org/10.1007/s00216-021-03301-y.

Perumal, V., Hashim, U., Gopinath, S. C. B., Haarindraprasad, R., Foo, K. L., Balakrishnan, S. R., & Poopalan, P. (2015). 'Spotted nanoflowers': Gold-seeded zinc oxide nanohybrid for selective bio-capture. *Scientific Reports, 5*(1), 12231. https://doi.org/10.1038/srep12231.

Quesada-González, D., & Merkoçi, A. (2015). Nanoparticle-based lateral flow biosensors. *Biosensors and Bioelectronics, 73*, 47–63. https://doi.org/10.1016/j.bios.2015.05.050.

Rajapaksha, R. D. A. A., Hashim, U., Afnan Uda, M. N., & Fernando, C. A. N. (2018). High-performance electrical variable resistor sensor for E.coli O157:H7 detection. *Journal of Telecommunication, Electronic and Computer Engineering, 10*(1–17), 61–64. https://jtec.utem.edu.my/jtec/article/view/4167.

Rajapaksha, R. D. A. A., Hashim, U., Natasha, N. Z., Uda, M. N. A., Thivina, V., & Fernando, C. A. N. (2017). Gold nano-particle based Al interdigitated electrode electrical biosensor for specific ssDNA target detection. In *Proceedings of the 2017 IEEE Regional Symposium on Micro and Nanoelectronics, RSM 2017. 2017 IEEE Regional Symposium on Micro and Nanoelectronics (RSM)*. Institute of Electrical and Electronics Engineers Inc https://doi.org/10.1109/RSM.2017.8069167.

Rajapaksha, R. D. A. A., Fernando, C. A. N., Balakrishna, R. G., Kumar, V., & See, P. (2021). An insight in photocurrent generation mechanism on Cu2O quantum dot sensitized Cu/p-CuI photo-electrochemical cell and efficient H2 generation at Cu/p-CuI/Cu2O electrolyte interface. *Materials Science and Engineering: B, 270*, 115205. https://doi.org/10.1016/j.mseb.2021.115205.

Rahman, S. F. A., Arshad, M. K. M., Gopinath, S. C. B., Fathil, M. F. M., Sarry, F., & Nor, M. N. M. (2021). Impedimetric lectin biosensor for prostate cancer detection. In *2021 IEEE International Conference on Sensors and Nanotechnology (SENNANO)* (pp. 9–12). IEEE Xplore.

Rahman, S. F. A., Yusof, N. A., Hashim, U., Hushiarian, R., M. N. , M. Nuzaihan, Hamidon, M. N., ... Fathil, M. F. M. (2016). Enhanced sensing of dengue virus DNA detection using O2 plasma treated-silicon nanowire based electrical biosensor. *Analytica Chimica Acta, 942*, 74–85. https://doi.org/10.1016/j.aca.2016.09.009.

Rajapaksha, R. D. A. A., Yahaya, N. A. N., Uda, M. N. A., & Hashim, U. (2018). Development of portable electronic reader for picoampere detection for two-electrode based amperometric biosensor applications. *AIP Conference Proceedings, 2045*(1), 020021. https://doi.org/10.1063/1.5080834.

Rashidi, K., Mahmoudi, M., Mohammadi, G., Zangeneh, M. M., Korani, S., Goicoechea, H. C., ... Jalalvand, A. R. (2018). Simultaneous co-immobilization of three enzymes onto a modified glassy carbon electrode to fabricate a high-performance amperometric biosensor for determination of total cholesterol. *International Journal of Biological Macromolecules, 120*, 587–595. https://doi.org/10.1016/j.ijbiomac.2018.08.163.

Rassas, I., Braiek, M., Bonhomme, A., Bessueille, F., Raffin, G., Majdoub, H., & Jaffrezic-Renault, N. (2019). Highly sensitive voltammetric glucose biosensor based on glucose oxidase encapsulated in a chitosan/kappa-carrageenan/gold nanoparticle bionanocomposite. *Sensors, 19*(1), 154. https://doi.org/10.3390/s19010154.

Rathee, K., Dhull, V., Dhull, R., & Singh, S. (2016). Biosensors based on electrochemical lactate detection: A comprehensive review. *Biochemistry and Biophysics Reports, 5*, 35–54. https://doi.org/10.1016/j.bbrep.2015.11.010.

Ronkainen, N. J., Halsall, H. B., & Heineman, W. R. (2010). Electrochemical biosensors. *Chemical Society Reviews, 39*(5), 1747–1763. https://doi.org/10.1039/b714449k.

Roslan, R., Zin, N. M., Alim, A. A., Nadzirah, S., Zulkepli, N., Dee, C. F., ... Hamzah, A. A. (2024). Comparison between TiO2/rGO and TiO2 with various microelectrode gaps for biosensor applications. *International Journal of Nanoelectronics and Materials, 17*(Special Issue), 31–35. https://doi.org/10.58915/ijneam.v17iJune.832.

Saif, A. A., & Poopalan, P (2011). Effect of the film thickness on the impedance behavior of sol–gel Ba0.6Sr0.4TiO3 thin films. *Physica B: Condensed Matter, 406*(6–7), 1283–1288. https://doi.org/10.1016/j.physb.2011.01.017.

Tran, H. V., Huynh, C. D., Tran, H. V., & Piro, B. (2018). Cyclic voltammetry, square wave voltammetry, electrochemical impedance spectroscopy and colorimetric method for hydrogen peroxide detection based on chitosan/silver nanocomposite. *Arabian Journal of Chemistry, 11*(4), 453–459. https://doi.org/10.1016/j.arabjc.2016.08.007.

van der Schoot, B. H., & Bergveld, P. (1987). ISFET based enzyme sensors. *Biosensors, 3*(3), 161–186. https://doi.org/10.1016/0265-928X(87)80025-1.

Song, Yuqing, Zou, Wentao, Lu, Qi, Lin, Li, & Liu, Zhongfan (2021). Graphene transfer: Paving the road for applications of chemical vapor deposition graphene. *Small, 17*(48), 2007600. https://doi.org/10.1002/smll.202007600.

Subramani, I., Ayub, R. M., Gopinath, S. C. B., Arshad, M. K. M., Rajapaksha, R. D. A. A., & Ibau, C. (2018). Determination of functional head and tail groups of self-assembled monolayer (SAM) formed by 2-mercaptoacetate on aluminium oxide substrate. *International Journal of Nanoelectronics and Materials, 11*, 115–122.

Thevenot, Daniel R., Toth, Klara, Durst, Richard A., & Wilson, George S. (2001). Electrochemical biosensors: Recommended definitions and classification. *Biosensors and Bioelectronics, 16*(1–2), 121–131. https://doi.org/10.1016/S0956-5663(01)00115-4.

Togawa, K., Sanbonsugi, H., Sandhu, A., Abe, M., Narimatsu, H., Nishio, K., & Handa, H. (2006). Detection of magnetically labeled DNA using pseudomorphic AlGaAs/InGaAs/GaAs heterostructure micro-Hall biosensors. *Journal of Applied Physics, 99*(8), 08P103. https://doi.org/10.1063/1.2162041.

Vera, A., Martínez, I., Enger, L. G., Guillet, B., Guerrero, R., Diez, J. M., … Perna, P., et al. (2023). High-performance implantable sensors based on anisotropic magnetoresistive La0.67Sr0.33MnO3 for biomedical applications. *ACS Biomaterials Science & Engineering, 9*(2), 1020–1029. https://doi.org/10.1021/acsbiomaterials.2c01147.

Voiculescu, I., & Nordin, A. N. (2012). Acoustic wave based MEMS devices for biosensing applications. *Biosensors and Bioelectronics, 33*(1), 1–9. https://doi.org/10.1016/j.bios.2011.12.041.

Wang, J., Xu, D., & Polsky, R. (2002). Magnetically-induced solid-state electrochemical detection of DNA hybridization. *Journal of the American Chemical Society, 124*(16), 4208–4209. https://doi.org/10.1021/ja0255709.

Wang, L., Gopinath, S. C. B., Anbu, P., Rajapaksha, R. D. A. A., Velusamy, P., Pandian, K., … Lee, C.-G. (2019). Photovoltaic and antimicrobial potentials of electrodeposited copper nanoparticle. *Biochemical Engineering Journal, 142*, 97–104. https://doi.org/10.1016/j.bej.2018.10.009.

Wang, L., Lu, D., Wang, J., Du, D., Zou, Z., Wang, H., … Lin, Y. (2011). A novel immunochromatographic electrochemical biosensor for highly sensitive and selective detection of trichloropyridinol, a biomarker of exposure to chlorpyrifos. *Biosensors and Bioelectronics, 26*(6), 2835–2840. https://doi.org/10.1016/j.bios.2010.11.008.

Wang, Z., Wang, X., Li, M., Gao, Y., Hu, Z., Nan, T., … Sun, N.-X. (2016). Highly sensitive flexible magnetic sensor based on anisotropic magnetoresistance effect. *Advanced Materials, 28*(42), 9370–9377. https://doi.org/10.1002/adma.201602910.

Wibowo, N. A., Kurniawan, C., Kusumahastuti, D. K. A., Setiawan, A., & Suharyadi, E. (2024). Review—potential of tunneling magnetoresistance coupled to iron oxide nanoparticles as a novel transducer for biosensors-on-chip. *Journal of The Electrochemical Society, 171*(1), 017512. https://doi.org/10.1149/1945-7111/ad1f35.

Yahaya, N. A. N., Rajapaksha, R. D. A. A., Afnan Uda, M. N., & Hashim, U. (2017). Ultra-low current biosensor output detection using portable electronic reader. In *AIP Conference Proceedings*. American Institute of Physics Inc https://doi.org/10.1063/1.5002430.

Yang, K., She, G.-W., Wang, H., Ou, X.-M., Zhang, X.-H., Lee, C.-S., & Lee, S.-T. (2009). ZnO nanotube arrays as biosensors for glucose. *Journal of Physical Chemistry C, 113*(47), 20169–20172. https://doi.org/10.1021/jp901894j.

Yang, Y., Dejous, C., & Hallil, H. (2023). Trends and applications of surface and bulk acoustic wave devices: A review. *Micromachines, 14*(1), 43. https://doi.org/10.3390/mi14010043.

Yang, Z., Liu, Y., Chong, L., Xuecheng, S., & Zhou, Y. (2016). Ultrasensitive detection and quantification of E. coli O157:H7 using a giant magnetoimpedance sensor in an open-surface microfluidic cavity covered with an antibody-modified gold surface. *Microchimica Acta, 183*, 1831–1837. https://doi.org/10.1007/s00604-016-1818-3.

Yao, C., Ng, E., & Wang, S. X. (2022). An automated and mobile magnetoresistive biosensor system for early hepatocellular carcinoma diagnosis. *Biosensors and Bioelectronics, 202*, 113982. https://doi.org/10.1016/j.bios.2022.113982.

Ying, X., Fu, W., Zhu, L., Sun, T., Qi, M., Zhou, L., … Zhang, J. (2024). Electrochemical lateral flow immunoassay with built-in electrodes for ultrasensitive and wireless detection of inflammatory biomarkers. *Analytical Chemistry, 96*(26), 10630–10638. https://doi.org/10.1021/acs.analchem.4c01224.

Zhao, Y., Parlak, Z., Yu, W., French, D., Aquino, W., & Zauscher, S. (2024). Microfluidic QCM enables ultrahigh Q-factor: a new paradigm for in-liquid gravimetric sensing. *Microsystems & Nanoengineering, 10*, 116. https://doi.org/10.1038/s41378-024-00732-2.

Zhou, Y., Zhang, H., Liu, L., Li, C., Chang, Z., Zhu, X., … Xu, M. (2016). Fabrication of an antibody-aptamer sandwich assay for electrochemical evaluation of levels of β-amyloid oligomers. *Scientific Reports, 6*, 35186.

Zhu, Y., Zhang, Q., Li, X., Pan, H., Wang, J., & Zhao, Z. (2019). Detection of AFP with an ultra-sensitive giant magnetoimpedance biosensor. *Sensors and Actuators B: Chemical, 293*, 53–58. https://doi.org/10.1016/j.snb.2019.05.004.

Žurauskienė, N. (2023). Engineering of advanced materials for high magnetic field sensing: A review. *Sensors, 23*(6), 2939. https://doi.org/10.3390/s23062939.

Zurnansyah, Jayanti, P. D., Mahardhika, L. J., Kusumah, H. P., Ardiyanti, H., Wibowo, N. A., … Suharyadi, E. (2024). Real-time biomolecule detection using GMR chip-based sensor with green-synthesized Fe3O4/rGO nanocomposites as magnetic labels. *Sensors and Actuators A: Physical, 375*, 115493. https://doi.org/10.1016/j.sna.2024.115493.

CHAPTER

Optical sensors

3

P. Susthitha Menon[a], Mohd Nizar Hamidon[b] and Nur Shahira Abdul Nasir[c]

[a]*Institute of Microengineering and Nanoelectronics (IMEN), Universiti Kebangsaan Malaysia, Bangi, Selangor, Malaysia,* [b]*Electrical and Electronic Engineering Department, Faculty of Engineering, Universiti Putra Malaysia (UPM), UPM Serdang, Selangor, Malaysia,* [c]*Centre of Liberal Studies, National University of Malaysia, Bangi, Selangor, Malaysia*

3.1 Surface-plasmon resonance-based biosensors

Surface Plasmon Resonance (SPR)-based biosensors is an optical phenomenon that occurs when polarized light strikes a metal–dielectric interface at a specific angle under total internal reflection conditions, resulting in the resonant excitation of surface plasmons—coherent oscillations of free electrons at the metal surface. SPR is highly sensitive to changes in the local refractive index near the metal surface, which forms the basis for its application in optical sensing.

When p-polarized light (light with its electric field oscillating in the plane of incidence) passes through a high-refractive-index prism and strikes a thin metal film (typically gold or silver) deposited on the prism, part of the light's energy can be transferred to surface plasmons under the right conditions. This transfer of energy causes a sharp dip in the intensity of the reflected light, known as the SPR dip. The resonance condition depends critically on the refractive index of the medium immediately adjacent to the metal surface. Thus, any molecular binding events, such as the interaction between a target analyte and a capture ligand immobilized on the metal surface, will change the local refractive index and shift the SPR angle or resonance wavelength.

The SPR condition can be mathematically described by the momentum matching formula described in Equation 1 as shown in Fig. 3.1. where k_{sp} is the wave vector of the surface plasmons, λ is the wavelength of the incident light, and ε_m and ε_d are the dielectric constants of the metal and dielectric medium, respectively. The condition for resonance is that the parallel component of the incident light's wave vector matches k_{sp}.

The foundation of SPR as a distinct optical technique was laid in 1968, when Otto and Kretschmann independently demonstrated surface plasmon excitation using prism coupling methods. In the Otto configuration, a narrow air or dielectric gap separates the prism from the metal film, minimizing direct contact and enabling compatibility with high-refractive-index samples. Conversely, the Kretschmann configuration places a thin metal film directly onto a glass prism, offering a more practical setup for sensor integration. The Kretschmann arrangement is widely used in modern, compact, chip-based SPR systems. Here, incident light enters through the prism and, at a specific angle, excites surface plasmons at the metal–dielectric interface, resulting in a characteristic dip in reflected light intensity. When target molecules bind to ligands immobilized on the metal surface, they induce local changes in the

BioMEMS Devices. DOI: https://doi.org/10.1016/B978-0-443-24015-7.00007-3

$$K_{sp} = \frac{2\pi}{\lambda}\sqrt{\frac{\varepsilon_m \varepsilon_d}{\varepsilon_m + \varepsilon_d}}$$

FIGURE 3.1

Equation 1: SPR condition described by the momentum matching formula.

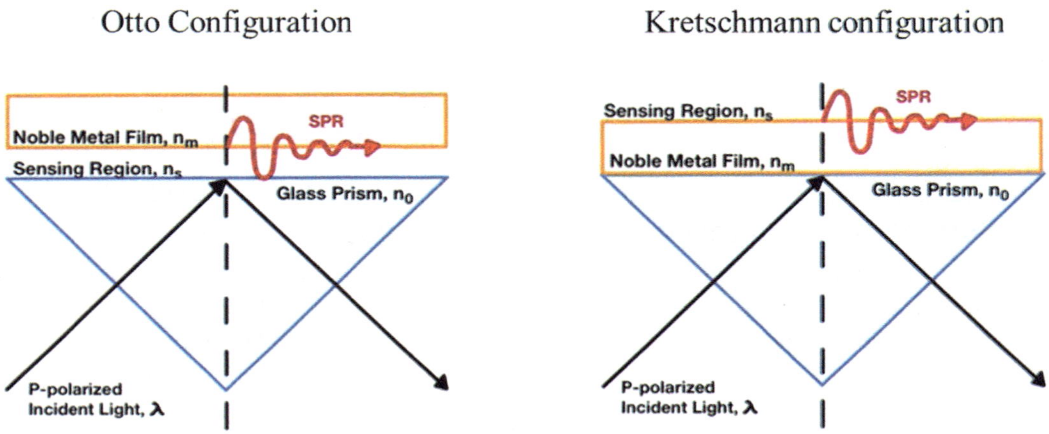

FIGURE 3.2

Otto and Krestchmann SPR configurations.

From Treebupachatsakul, T., Shinnakerdchoke, S., & Pechprasarn, S. (2021). Sensors, 21, *6164. https://doi.org/10.3390/s21186164.*

refractive index, shifting the resonance angle or wavelength. Monitoring these shifts in real-time allows researchers to extract kinetic data, such as association and dissociation rates, as well as equilibrium constants for molecular interactions such as antigen–antibody binding, DNA hybridization, and protein–protein interactions. Fig. 3.2 differentiates between the Otto and Krestchmann SPR configurations.

In recent years, significant efforts have been made to miniaturize SPR systems to create portable, cost-effective, and efficient sensors suitable for point-of-care (POC) diagnostics and field applications. Miniaturization is achieved through various strategies, including the use of microfluidics, fiber optics, and nanostructured materials. Fiberoptic SPR sensors, for instance, involve coating the core of an optical fiber with a thin metal layer, enabling remote sensing capabilities and flexible device designs. Microfluidic integration allows the handling of tiny sample volumes and precise control over the sensing environment, which is particularly useful for biological and chemical assays.

Moreover, the development of localized surface plasmon resonance (LSPR) sensors marks a key advancement in the miniaturization of SPR technology. Instead of using a continuous metal film, LSPR sensors utilize metallic nanoparticles, such as gold or silver nanospheres, nanorods, or nanoshells. When illuminated, these nanoparticles support localized plasmon oscillations that are confined to the nanoparticle surface. The resonance condition and the resulting optical properties are extremely sensitive to the local refractive index around each nanoparticle. Because LSPR sensors respond to binding events occurring directly on the nanoparticle surface, they can detect molecular interactions with high sensitivity using simpler optical setups—often eliminating the need for prisms or bulky optical components. Additionally, the strong electromagnetic field enhancement around the nanoparticles increases detection sensitivity, sometimes down to single-molecule levels.

The sensing mechanism after miniaturization becomes localized rather than propagative. In traditional SPR, the plasmon wave propagates along the metal–dielectric interface, and the interaction with the analyte is spread over a larger surface area. In contrast, LSPR focuses the plasmonic field around the nanoparticles, meaning that only molecules in the immediate vicinity of the nanoparticle affect the resonance condition. This confinement leads to faster response times and enables sensing in extremely small sample volumes. Furthermore, the tunability of nanoparticle size, shape, and material composition allows LSPR sensors to be optimized for different applications, covering a broad range of detection wavelengths from the visible to the near-infrared region.

3.1.1 Application of SPR-based sensor

SPR biosensors have emerged as a critical technology in the field of bioanalytics, allowing for real-time, label-free detection of biomolecular interactions. The SPR concept is based on the resonant oscillation of conduction electrons at the interface of a metal and a dielectric, which is sensitive to changes in refractive index produced by biomolecule binding events (Xia et al., 2019). Over the last decade, substantial breakthroughs in SPR technology have been reported, boosting its applicability in several fields, including medical diagnostics, environmental monitoring, and food safety (Nguyen et al., 2015; Nurrohman & Chiu, 2020).

Recent advances in SPR biosensors have greatly expanded their range of applications, notably in biomedical diagnostics. Over the past decade, numerous innovative researches have demonstrated the evolution and improved capabilities of SPR technology, emphasizing its promise for accurate biomolecular interaction analysis as well as its adaptability in detecting a wide range of biological targets.

Chiu et al. (2019) created a new high-affinity SPR aptasensor that uses carboxyl-graphene oxide to detect human chorionic gonadotropin (hCG) in clinical blood samples. To boost sensitivity, the design used aptamers immobilized on a graphene oxide surface, which increased optical and electrical conductivity. The study emphasized graphene oxide's higher SPR response, which was attributable to its large surface area and biocompatibility. With a detection limit of 0.1 ng/mL, the sensor showed great promise for clinical use. Because of its great sensitivity and specificity, the authors highlighted its potential as a powerful tool for early illness detection.

A study by Huang et al. used SPR biosensors to analyze the kinetics of interactions between graphene oxide sheets and proteins (Huang et al., 2013). In order to track binding events in real time and evaluate binding affinities and kinetics, the study used a label-free detection technique. Different protein concentrations were used in the process to assess the sensor's response, which was then examined using a Langmuir binding model. The results showed that robust protein adsorption was made possible by the special environment that the graphene oxide surface offered, which improved detection capabilities. According to the study's findings, the SPR biosensor is a useful instrument for biochemical research since it can efficiently track protein interactions.

In order to identify breast cancer antigens, Monteiro et al. created a microfluidic SPR biosensor (Monteiro et al., 2016). Multiple samples could be analyzed at once thanks to the methodology's integration of a plasmonic sensor and a microfluidic device. To increase the sensitivity of the SPR response, the sensor made use of nanohole arrays on gold sheets. According to the findings, breast cancer antigens could be detected by the microfluidic SPR biosensor at concentrations as low as 1 pg/mL. The benefits of this strategy, such as lower sample volume and higher throughput, were covered by the authors. They came to the conclusion that by offering quick and accurate detection techniques, the combination of microfluidics and SPR technology might completely transform cancer diagnostics.

3.2 Mass spectrometry

Mass spectrometry (MS) is a sophisticated analytical method that identifies and characterizes compounds based on their mass-to-charge ratios. It is essential in many domains, including biology, chemistry, and materials science, allowing researchers to analyze complicated biological samples with great sensitivity and precision. The approach works by ionizing molecules, separating the resultant ions in a mass analyzer, and detecting them to produce a spectrum that offers information on molecular composition, structure, and abundance.

MS has exceptional capabilities for detecting and analyzing biomolecules such as proteins, peptides, and nucleic acids in BioMEMS devices. Its combination with microfluidic devices expands its capabilities, allowing for quick and high-throughput studies in miniaturized configurations. This collaboration accelerates advances in proteomics, drug discovery, and diagnostics, highlighting MS's relevance as a foundational technique in current bioengineering and lab-on-chip applications.

3.2.1 Matrix-assisted laser desorption ionization-time of flight

Matrix-assisted laser desorption ionization-time of flight (MALDI-TOF) is a powerful analytical tool for detecting and characterizing biomolecules including proteins, peptides, and polymers. MALDI-TOF uses a laser to ionize materials contained in a matrix, resulting in charged molecules that are separated by mass-to-charge ratio using a time of flight analyzer, as shown in Fig. 3.3. MALDI-TOF is frequently used in proteomics, microbiology, and diagnostics because to its precision, rapidity, and ease of sample preparation. Its combination with BioMEMS devices and optical sensors paves the way for miniaturized, high-throughput systems that will improve biomolecule detection and analysis in a wide range of biomedical applications.

MALDI-TOF MS has developed as a critical tool in clinical microbiology, notably for the fast detection of harmful microbes. Recent investigations have demonstrated its usefulness and adaptability in a variety of applications. Furniss et al. used the MALDI Biotyper Sirius system to identify colistin resistance in *Escherichia coli*. The approach entailed analyzing lipid A structures, which revealed that MALDI-TOF may efficiently distinguish resistant strains, assisting in the development of novel diagnostics for Gram-negative bacteria (Furniss et al., 2019).

Weis et al. (2022) used machine learning techniques to predict antibiotic resistance directly from clinical MALDI-TOF mass spectra. The researchers used a collection of mass spectra from *Klebsiella pneumoniae* and *Staphylococcus aureus* isolates, with an emphasis on identifying spectral properties that correspond with resistance profiles. The approach includes preprocessing the mass spectra to

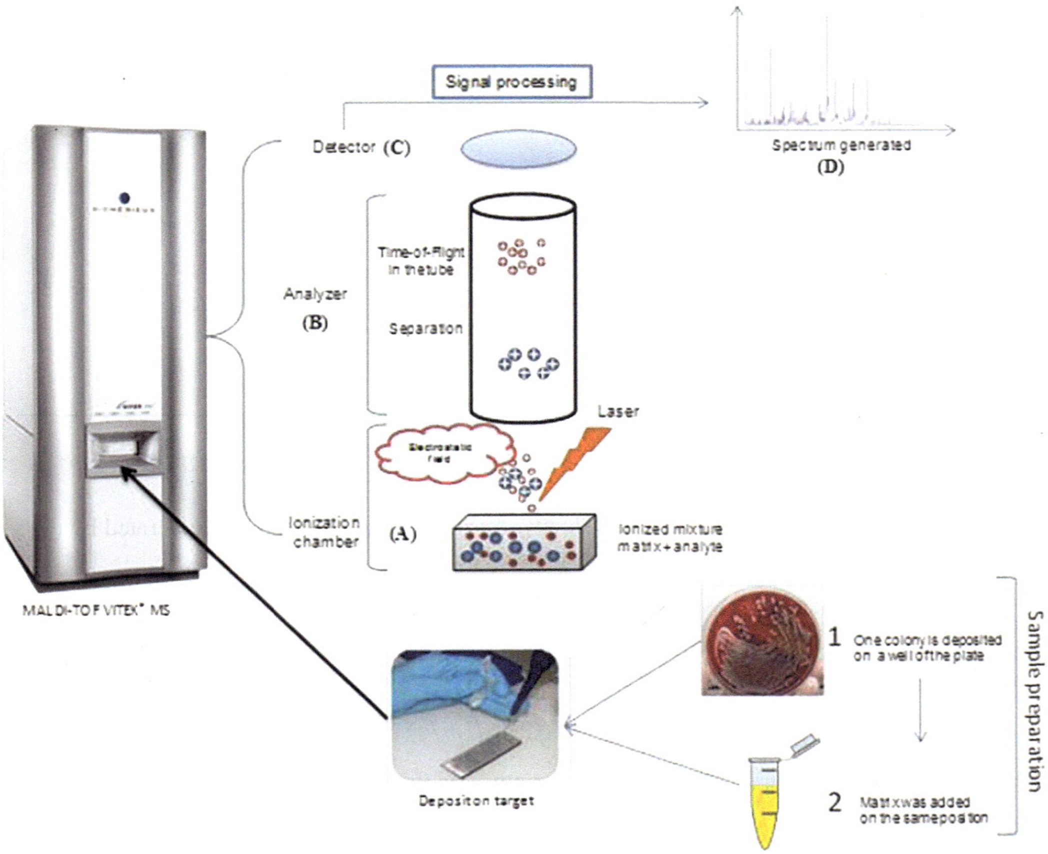

FIGURE 3.3

Operating principle and workflow of MALDI-TOF MS.

From Lo, C., Fall, B., Samb-Ba, B., Flaudrops, C., Faye, N., Mediannikov, O., Sokhna, C., Wade, B., Raoult, D., & Fenollar, F. (2017). African Journal of Microbiology Research, 11, *1360–1370. https://doi.org/10.5897/AJMR2016.8181.*

improve signal quality, followed by the use of several machine learning classifiers such as support vector machines and random forests. The results indicated that the machine learning models achieved high accuracy in classifying resistance profiles, with notable sensitivity and specificity metrics. Specifically, the models demonstrated the ability to differentiate between resistant and susceptible strains effectively, showcasing the potential for rapid resistance screening in clinical microbiology.

In MALDI-TOF MS specifically, ionization is achieved using a laser beam, and mass analysis is based on measuring the time it takes for ions to travel a fixed distance—lighter ions reach the detector faster than heavier ones. The basic equation that governs TOF MS is derived from kinetic energy:

$$KE = 1/2(mv)^2 = qV$$

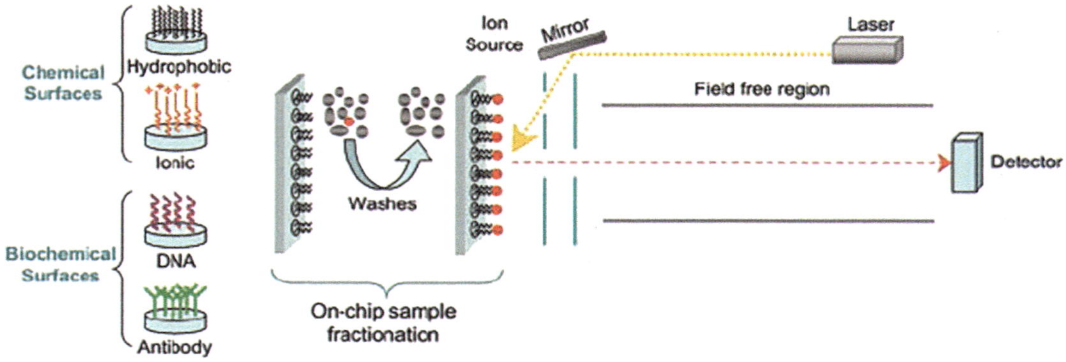

FIGURE 3.4

Surface-enhanced laser desorption/ionization (SELDI) mechanism.

From Cristea, I., Gaskell, S., & Whetton, A. (2004). Blood, 103, *3624–3634. https://doi.org/10.1182/blood-2003-09-3295.*

where m is the mass of the ion and v is its velocity. Assuming that all ions are accelerated by the same potential V, their kinetic energy can also be expressed as:

$$KE = zV$$

where z is the ion's charge and V is the accelerating voltage. Combining these formulas gives the relationship:

$$t = d/\sqrt{2V}\sqrt{(m/z)}$$

where t is the time of flight, d is the flight path length, V is the accelerating voltage, m is the mass, and z is the charge. This fundamental formula shows how the mass-to-charge ratio can be determined by measuring the flight time of ions. In MALDI-TOF MS specifically, ionization is achieved using a laser beam, and mass analysis is based on measuring the time it takes for ions to travel a fixed distance—lighter ions reach the detector faster than heavier ones.

3.2.2 Surface-enhanced laser desorption/ionization time-of-flight mass spectrometry

Surface-enhanced laser desorption and ionisation time-of-flight mass spectrometry (SELDI-TOF) as shown in Fig. 3.4 is a cutting-edge analytical technology that uses surface chemistry and laser desorption/ionization to identify biomolecules in a sensitive and specific manner. SELDI-TOF uses a specialized chip with surface changes that capture target molecules based on their biological features, such as hydrophobicity, charge, or affinity for certain ligands. Following laser desorption/ionization, the trapped molecules are ionized and analyzed with a TOF mass spectrometer, which produces a mass spectrum for exact identification and quantification.

In a research by Han et al. (2020) investigated hemorrhagic cerebral infarction, using SELDI-TOF MS to analyze blood samples from patients and healthy controls. The research involved statistical analysis to detect differentially expressed protein peaks. The results identified particular protein markers linked to hemorrhagic cerebral infarction, highlighting the technique's potential for early identification and monitoring of neurological disorders. The authors concluded that SELDI-TOF MS might be an effective diagnostic tool in clinical settings.

Zhou and Long (2022) looked at the use of SELDI-TOF MS to monitor serum proteins in patients with acute respiratory distress syndrome. To examine the reliability of the approach, the researchers used repeatability testing and differential protein peak detection. The results showed a high coefficient of variation of 13.64% across many chips, indicating that the measurements are very reproducible. The authors concluded that SELDI-TOF MS is a good tool for dynamic monitoring of serum proteins in ARDS, with the potential to dramatically improve patient management and treatment methods in clinical settings. This study emphasizes the utility of SELDI-TOF MS in giving vital insights into the metabolic alterations associated with ARDS.

3.3 Raman scattering-based optical biosensors

Raman scattering-based optical biosensors have emerged as a game-changing method in the field of bioMEMS devices, taking use of light–matter interaction principles to provide very sensitive and label-free biomolecule detection. The Raman effect explains the inelastic scattering of photons that interact with molecular vibrations, leaving a specific spectral "fingerprint" for each chemical species. Raman spectroscopy has become an indispensable tool in biological diagnostics, biochemical research, and environmental monitoring due to its capacity to identify and quantify molecular species without intrusive methods.

Raman scattering-based optical sensors are powerful analytical tools that utilize inelastic scattering of light to detect and analyze molecular compositions with high specificity. Raman scattering refers to the phenomenon where incident photons interact with the vibrational modes of molecules, leading to a shift in the energy (and thus wavelength) of the scattered light. This shift provides a molecular fingerprint unique to the chemical structure of the sample, enabling highly selective identification and quantification. Because of their non-invasive, label-free nature and high sensitivity to molecular changes, Raman-based sensors are widely used in fields such as biomedical diagnostics, environmental monitoring, food safety, and chemical process control.

The core principle behind Raman scattering is the inelastic interaction between photons and molecules. When monochromatic light (typically from a laser) is incident upon a material, most photons are elastically scattered, a process known as Rayleigh scattering, in which the scattered photons retain the same energy and wavelength as the incident light. However, a small fraction of the photons—approximately 1 in 10 million—undergo Raman scattering. In Raman scattering, photons exchange energy with the vibrational modes of the molecules. If the photon loses energy to the molecule, the scattered photon shifts to a lower energy (longer wavelength), called Stokes scattering. Conversely, if the photon gains energy from the molecule, it is shifted to a higher energy (shorter wavelength), called anti-stokes scattering.

3.3.1 Principles of Raman scattering

The energy shift (ΔE) corresponds exactly to the energy of a vibrational mode of the molecule and can be expressed as:

$$\Delta E = h\nu_{vib}$$

where h is Planck's constant and ν_{vib} is the vibrational frequency of the molecule. The Raman shift, often reported in wavenumbers (cm^{-1}), is given by:

$$\Delta\nu = (\nu_0) - (\nu_s)$$

where $\nu 0$ is the wavenumber of the incident light and νs is the wavenumber of the scattered light.

The resulting Raman spectrum a plot of intensity versus Raman shift serves as a molecular fingerprint for the sample under investigation.

The basic mechanism of a Raman sensor involves directing a focused laser beam onto the sample, collecting the scattered light, and analyzing it with a spectrometer. The Raman scattered light is separated from the much stronger Rayleigh scattered light using optical filters such as notch filters or edge filters. The filtered Raman light is then dispersed by a diffraction grating in the spectrometer and detected by a sensitive device like a charge-coupled device (CCD) camera. Each peak in the resulting Raman spectrum corresponds to a specific vibrational mode of the molecules in the sample, allowing for detailed chemical characterization.

Wei et al. conducted a critical work that focused on improving the quantitative capabilities of surface-enhanced Raman spectroscopy (SERS) by using SPR effects. The researchers used a confocal arrangement to limit the optical sampling volume, allowing them to maximize both elastic and inelastic scattering signals from analytes near plasmonic "hot spots" generated by gold nanoparticles (AuNPs). The enhancement factor attained was considerable, resulting in enhanced detection limits for several analytes, proving that the maximum field enhancement factor, indicated as |f_max|, is critical for SERS signal amplification. This study suggested that combining SPR with SERS might result in more sensitive biosensing platforms, which are critical for early illness diagnosis (Wei et al., 2018).

Another significant research by Hossain focusses on the label-free detection of microorganisms using an SERS method. The researchers used gold nanorods (AuNRs) decorated on a flake-like 3D graphene-based device to boost the SERS signal. The research includes analyzing the electrostatic interactions between bacterial samples and AuNRs, which greatly increased detection sensitivity. The study found that the SERS approach can identify bacteria at concentrations as low as 10^3 CFU/mL, indicating its potential for quick microbial diagnostics. This study concluded that integrating plasmonic nanostructures with SERS-enhanced biosensing capacities in complicated biological samples (Hossain et al., 2023). Fig. 3.5 shows the schematic depiction of a 3D graphene-AuNR-mediated device used as a label-free bacteria sensing platform, as well as an analysis of the integrated area in SERS enhancement with varied AuNR densities on the device.

Furthermore, Beeram et al. conducted a detailed assessment of current advancements in SERS-based plasmonic sensors for disease diagnostics and biomolecule detection. The assessment included a variety of developments in SERS technology, including the use of machine learning approaches to increase detection accuracy and speed. SERS, according to the authors, has become a prominent technique for applications in biology and medicine due to its nondestructive and label-free nature. They concluded that continued research and development in SERS technology, particularly in the context of machine learning integration, might considerably boost the performance of optical biosensors for clinical diagnosis (Beeram et al., 2023).

In summary, these investigations demonstrate the many uses and methodology of SERS-based optical biosensors, highlighting its promise for early illness detection and microbiological diagnostics. Advances in nanostructured materials and their integration with machine learning techniques have

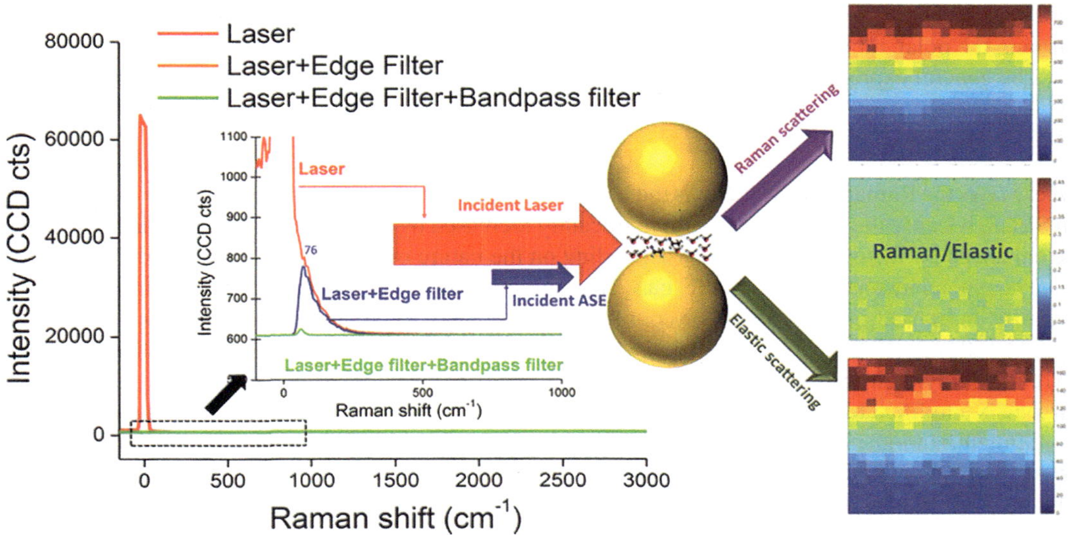

FIGURE 3.5

Schematic depiction of a 3D graphene-AuNR-mediated device used as a label-free bacteria sensing platform, as well as an analysis of the integrated area in SERS enhancement with varied AuNR densities on the device.

From Hossain, M.I., Nanda, S.S., Cho, S., Lee, B., Kim, B.J., Choi, J.-Y., & Yi, D.K. (2023). Gold nanorod density-dependent label-free bacteria sensing on a flake-like 3D graphene-based device by SERS. Biosensors, 13 *(11), 962. https://doi.org/10.3390/bios13110962.*

increased the capabilities of these biosensors, opening the path for their future usage in clinical and environmental monitoring applications.

3.4 Surface-enhanced Raman spectroscopy

Surface-enhanced Raman spectroscopy (SERS) is a powerful analytical technique that significantly amplifies the Raman scattering signals of molecules adsorbed onto rough metal surfaces or nanostructures, enabling highly sensitive molecular detection, often down to single-molecule levels. SERS builds upon the principles of conventional Raman scattering but introduces metallic nanostructures, such as gold or silver nanoparticles, to enhance the weak Raman signals through electromagnetic and chemical enhancement mechanisms. The extreme sensitivity, high specificity, and capability for miniaturization make SERS-based optical sensors ideal for applications in chemical sensing, biomedical diagnostics, food safety, and environmental monitoring.

The fundamental principle behind SERS is based on two primary enhancement mechanisms: electromagnetic (EM) enhancement and chemical (charge transfer) enhancement. Electromagnetic enhancement is the dominant mechanism and arises from the excitation of LSPRs in the metal nanostructures when illuminated by light. The LSPRs generate intense local electromagnetic fields at the metal surface, significantly boosting the Raman signal of nearby molecules. Chemical enhancement, although weaker, involves charge transfer between the adsorbed molecules and the metal surface, altering the polarizability of the molecules and thus enhancing the Raman signal.

The general expression for Raman scattering intensity is:

$$I_R \propto |E_{loc}|^4 \alpha^2$$

where I_{R1} is the Raman intensity, E_{loc} is the local electric field, and α\alpha is the molecular polarizability. In SERS, because the local field E_{loc} is significantly increased by LSPR effects, the Raman intensity can be enhanced by factors of 10^6 to 10^{14} depending on the nature of the metal substrate and the analyte.

The typical mechanism of a SERS-based optical sensor involves preparing an SERS-active substrate (usually composed of gold, silver, or copper nanostructures), applying the sample onto this substrate, and then illuminating the sample with a laser. The enhanced Raman scattered light is collected, filtered to remove Rayleigh scattering, and analyzed using a spectrometer. The unique Raman fingerprint of the analyte is amplified by the nanostructures, allowing for detection even at extremely low concentrations.

Miniaturization in SERS-based sensors is achieved by engineering nanoscale structures such as nanoparticles, nanopillars, nano-stars, or nano-islands on substrates. Advances in nanofabrication techniques, such as electron-beam lithography, nanoimprinting, and self-assembly methods, have enabled the mass production of reproducible and highly active SERS substrates. Integration with microfluidic systems further enhances miniaturization by allowing small volumes of liquid samples to be precisely manipulated and analyzed in real time.

After miniaturization, the sensing mechanism retains its core principles but becomes more efficient and versatile. Miniaturized SERS platforms use microfabricated arrays or colloidal suspensions of nanoparticles to create highly active and reproducible sensing surfaces. The miniaturized systems often use lower laser powers, portable spectrometers, and handheld Raman analyzers. By integrating microfluidics, sample delivery and signal acquisition can be automated and scaled down, making real-time, on-site analysis possible. Additionally, combining SERS with machine learning algorithms enables rapid data interpretation, even from complex mixtures.

Real-world applications of SERS-based optical sensors are abundant and impactful. In biomedical diagnostics, SERS sensors are used for early detection of diseases such as cancer by identifying biomarkers in blood, urine, or saliva samples. For example, SERS-based assays can detect circulating tumor DNA (ctDNA) or exosomes, providing noninvasive liquid biopsy solutions. In food safety, SERS sensors are employed to detect pesticide residues, foodborne pathogens, and contaminants in agricultural products. A notable example is the detection of melamine adulteration in milk using portable SERS devices. In environmental monitoring, SERS sensors are utilized to detect trace levels of pollutants, heavy metals, or toxins in water and soil samples. Furthermore, forensic applications have adopted SERS for the detection of drugs, explosives, and biological fluids at crime scenes, offering high sensitivity and specificity with minimal sample preparation. Fig. 3.6 shows as example of SERs optical sensor.

3.5 Chemiluminescence-based optical sensors

Chemiluminescence-based optical sensors rely on the emission of light produced by a chemical reaction without the need for external illumination. This spontaneous emission of light occurs when the reaction releases energy in the form of photons as reactants transition from a high-energy excited state to a

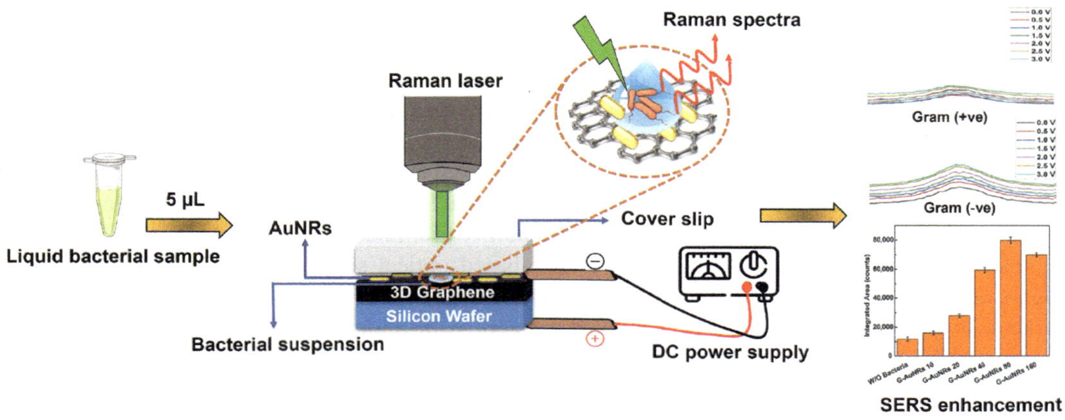

FIGURE 3.6

Example of SERS optical sensor.

From Suddala, K.C., Walter, N.G., & Burke-Aguero, D.H. (2014). In Riboswitch discovery, structure and function (Vol. 549, pp. 343–373). Academic Press. https://doi.org/10.1016/B978-0-12-801122-5.00015-5.

lower-energy ground state. Chemiluminescence sensors are prized for their extremely low detection limits, wide dynamic range, and high signal-to-noise ratios, making them valuable tools in clinical diagnostics, environmental monitoring, pharmaceutical testing, and forensic science.

The principle of chemiluminescence involves a chemical reaction in which one or more products are generated in an electronically excited state. When the excited product relaxes to its ground state, it emits a photon. The basic formula governing chemiluminescence intensity is:

$$I_{C}L \propto k[Reactant]$$

where I_{CL} is the chemiluminescence intensity, k is a rate constant dependent on reaction kinetics, and [Reactant] is the concentration of the reactant or analyte. The efficiency of photon production, known as quantum yield, determines the brightness of the chemiluminescence.

Typical chemiluminescent reactions used in sensing include luminol oxidation (commonly used in forensic blood detection), acridinium ester reactions, and enzymatic reactions such as those catalyzed by horseradish peroxidase (HRP) in the presence of hydrogen peroxide. The general mechanism of a chemiluminescence sensor involves introducing the sample into a reaction chamber containing reagents that react specifically with the target analyte. The chemical reaction emits light, which is then measured by a photodetector such as a photomultiplier tube (PMT), photodiode, or CCD camera. The intensity of the emitted light is directly proportional to the concentration of the analyte.

Miniaturization of chemiluminescence-based sensors has led to the development of microfluidic-based systems, lab-on-a-chip devices, and portable platforms capable of real-time analysis with minimal reagent consumption. Advances in microfabrication allow the integration of reaction chambers, fluidic channels, and optical detection elements into a single compact device. Portable chemiluminescence detectors may include miniaturized PMTs, fiber optics, or even CMOS cameras for light detection. The use of microfluidics allows for automated sample and reagent handling, precise control over reaction conditions, and reduced background noise.

After miniaturization, the sensing mechanism in chemiluminescence-based systems becomes faster, more efficient, and more sensitive. Microfluidic platforms ensure rapid mixing of reactants, leading to consistent and reproducible chemiluminescence signals. Miniaturized systems also benefit from reduced reagent and sample volumes, which is particularly important for high-throughput or POC applications. Advances in nanomaterials, such as incorporating nanoparticles or quantum dots, have further enhanced the chemiluminescence efficiency, stability, and tunability of miniaturized sensors.

Real-world applications of chemiluminescence-based optical sensors are widespread. In clinical diagnostics, chemiluminescent immunoassays (CLIAs) are widely used to detect hormones, cardiac markers, infectious disease agents, and tumor markers with high sensitivity. Examples include the Siemens ADVIA Centaur and Roche Elecsys analyzers used in hospitals for routine blood testing. In environmental monitoring, portable chemiluminescence detectors are employed to measure atmospheric pollutants such as nitrogen oxides (NOx) and sulfur dioxide (SO_2), offering real-time air quality assessments. In food safety, chemiluminescence assays detect bacterial contamination, toxins, and adulterants in food products. Forensic science uses luminol-based chemiluminescence to reveal trace amounts of blood at crime scenes, even after attempts to clean the evidence.

One notable example is the handheld **InnovaPrep ChemDetect** device, a miniaturized chemiluminescence platform designed for rapid biothreat agent detection in the field. Another emerging trend is the integration of chemiluminescence sensing into wearable devices for non-invasive health monitoring, opening new frontiers in personalized medicine.

3.6 Fluorescence-based biosensors

Fluorescence-based optical sensors are among the most widely used types of sensors in analytical science, biotechnology, environmental monitoring, and medical diagnostics. They operate on the principle of fluorescence, a process where certain molecules (fluorophores) absorb light at a specific excitation wavelength and re-emit light at a longer, lower-energy wavelength. The intensity, wavelength shift, lifetime, or polarization of the emitted fluorescence can be precisely measured and correlated with the concentration or presence of an analyte of interest. The exceptional sensitivity, specificity, and versatility of fluorescence make it a cornerstone technique for modern optical sensing technologies.

The fundamental principle of fluorescence involves the electronic excitation of a molecule from its ground state (S_0) to an excited singlet state (S_1) upon absorbing a photon. After a very short relaxation period (typically nanoseconds), the molecule returns to the ground state by emitting a photon of lower energy (longer wavelength) compared to the absorbed photon. The energy difference between the absorbed and emitted light is called the Stokes shift. The basic steps are often depicted in a Jablonski diagram. The fluorescence emission intensity I_f is directly related to the concentration of the fluorophore according to:

$$I_f = \phi_f \cdot \epsilon \cdot I_0 \cdot |C|$$

where

I_f = fluorescence intensity,
Φ_f = fluorescence quantum yield (efficiency of fluorescence),

ϵ = molar extinction coefficient (absorption efficiency),
I_0 = intensity of excitation light,
$|C|$ = concentration of fluorophore.

Thus, measuring the emitted fluorescence allows the quantitative determination of the target molecule.

In a typical fluorescence-based optical sensor, the mechanism involves exciting the sample with a light source (usually a laser diode, LED, or xenon lamp) at a specific wavelength that matches the absorption peak of the fluorophore. After excitation, the emitted fluorescence is collected at an angle (typically 90°) relative to the excitation path to minimize background noise. Optical filters or monochromators are used to isolate the emission wavelength, and a sensitive detector, such as a PMT or a CCD, records the fluorescence intensity. The measured signal is directly or indirectly correlated to the analyte concentration depending on the sensor design.

Miniaturization of fluorescence-based optical sensors has been a major trend, motivated by the need for portable, cost-effective, POC diagnostic devices and environmental field monitoring systems. Advances in microfabrication, nanotechnology, and microfluidics have enabled the development of compact fluorescence sensing platforms. Strategies for miniaturization include the use of micro-LEDs or laser diodes as excitation sources, integration of micro-optics such as miniaturized lenses and filters, microfluidic chips for sample handling, and miniaturized photodetectors like CMOS-based sensors. Lab-on-a-chip (LoC) devices are a prime example where fluorescence detection modules are integrated into a palm-sized system capable of analysing very small volumes (nano- to microliters) with high precision.

After miniaturization, the basic sensing mechanism remains the same—absorption and re-emission of photons—but becomes more efficient and user-friendly. Miniaturized fluorescence sensors typically use fiber optics to deliver excitation light and collect emission from very small sensing regions. The reduced sample volume and minimized optical path lengths improve the signal-to-noise ratio and reduce the overall analysis time. Moreover, miniaturized systems often integrate automatic sample handling (e.g., capillary flow in microfluidic chips) and can multiplex several analyses simultaneously. New materials, such as quantum dots (QDs) and carbon dots, offer tunable emission wavelengths and enhanced stability, further improving miniaturized sensor performance. Fluorescence-based biosensors have emerged as a key component of modern optical biosensing, providing excellent sensitivity, specificity, and adaptability in detecting a wide range of biomolecules. These biosensors give quantitative and qualitative insights into complicated biological and chemical systems by using the distinct features of fluorescence, including intensity, lifetime, and wavelength changes. Their applications are wide-ranging, encompassing clinical diagnostics, environmental monitoring, drug development, and food safety. Fluorescence-based biosensors work on the idea that an interaction between a target analyte and a fluorescent probe produces a detectable optical signal. Fluorescent dyes, quantum dots, and genetically encoded fluorescent proteins are often used as probes, each with unique benefits in terms of brightness, stability, and tunability. These sensors can be constructed to work via techniques, such as Förster resonance energy transfer (FRET), fluorescence quenching, or signal amplification, allowing for accurate monitoring of biological interactions. Integration with bioMEMS technology has greatly improved the capabilities of fluorescence-based biosensors. Microfluidic systems, for example, provide precise control of sample handling, resulting in high-throughput and multiplexed analysis. Furthermore, miniaturized optical components and sophisticated detection systems, such as CMOS-based photodetectors, have enabled the development of portable and POC fluorescence biosensors.

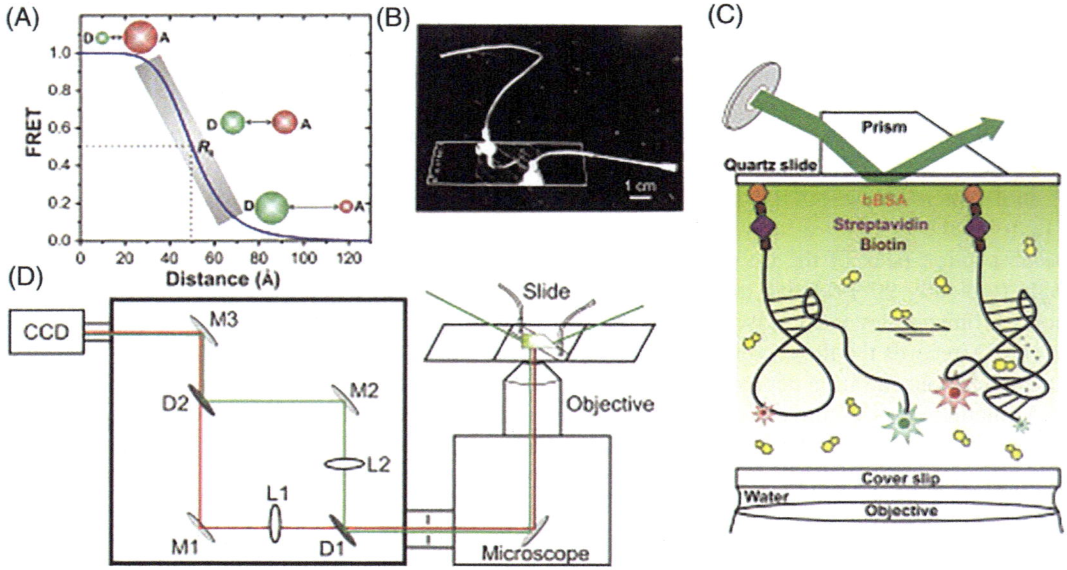

FIGURE 3.7

smFRET principles and TIRF microscopy setup.

From Suddala, K. C., Walter, N. G., & Burke-Aguero, D. H. (2014). In Riboswitch discovery, structure and function (Vol. 549, pp. 343–373). Academic Press. https://doi.org/10.1016/B978-0-12-801122-5.00015-5.

Habif et al.'s initial paper proposed a unique triple-modality biosensor known as CASPAM, which was built for multiplexed imaging of caspase network activity in live cells. The research used fluorescence correlation spectroscopy (FCS) to assess the stoichiometry of biosensors in the biological environment. The researchers were able to track caspase dynamics in real time using several fluorescent markers. The results showed that the biosensor could efficiently distinguish between different caspase activity, providing information on apoptotic signaling pathways. The scientists stated that CASPAM is a big step forward in the study of caspase networks, allowing researchers to visualize complicated biological processes with great temporal resolution (Habif et al., 2021).

3.6.1 Fluorescence resonance energy transfer-based optical sensors

Fluorescence resonance energy transfer (FRET) is a powerful and highly sensitive spectroscopic technique used extensively in optical sensing for detecting molecular interactions, conformational changes, and proximity relationships at the nanoscale Fig. 3.7. FRET-based optical sensors exploit the nonradiative transfer of energy from a donor fluorophore to an acceptor fluorophore when they are in close proximity (typically within 1–10 nm). This distance-dependent energy transfer enables precise detection of biological events, environmental changes, and chemical reactions at the molecular level, offering extremely high spatial resolution.

The theory of FRET is based on dipole–dipole coupling between two chromophores. When a donor molecule in an excited electronic state comes close enough to an acceptor molecule with overlapping absorption spectra, the donor can transfer its excitation energy directly to the acceptor without emitting a photon. This results in quenching of the donor's fluorescence and simultaneous excitation and emission of fluorescence from the acceptor. FRET efficiency is extremely sensitive to the distance between the donor and acceptor, making it a "molecular ruler" capable of detecting changes in the nanometer range.

The fundamental equation that governs FRET efficiency EE is:

$$E = 1/\left(1 + \left(r/R_0\right)^6\right)$$

where

E = energy transfer efficiency,
r = distance between donor and acceptor,
R_0 = Förster radius (the distance at which the energy transfer efficiency is 50%).

The Förster radius R_0 depends on the donor–acceptor pair properties and is calculated using:

$$R_0^6 = \left(9000(In10)K^2\phi D^J\right)/\left(128\pi^5 n^4 N_A\right)$$

where:

K^2= orientation factor between donor and acceptor dipoles,
ΦD = quantum yield of the donor,
J = spectral overlap integral,
n= refractive index of the medium,
N_A= Avogadro's number.

Thus, FRET efficiency is extremely sensitive to nanometer-scale changes in distance, making it ideal for probing molecular interactions.

The general mechanism of a FRET-based optical sensor involves labelling two biomolecules (or two regions of the same molecule) with a donor and an acceptor fluorophore. Upon excitation of the donor fluorophore at its characteristic wavelength, if the acceptor is within a close range, energy is transferred nonradiatively from the donor to the acceptor, leading to fluorescence emission from the acceptor instead of the donor. By measuring the changes in the fluorescence intensities of the donor and acceptor, the occurrence and magnitude of molecular events can be inferred.

Miniaturization of FRET-based optical sensing systems has become a crucial advancement, particularly for applications in live-cell imaging, portable biosensors, and POC diagnostics. Technologies such as microfluidic devices, nanoparticle-based FRET systems, and integrated fiberoptic probes have enabled the development of compact FRET sensor platforms. Quantum dots (QDs) and nanostructured materials have also been utilized as donors and acceptors, providing enhanced photostability, brightness, and tunable emission wavelengths. Microfabrication methods have allowed the construction of miniaturized chip-based FRET sensors that require minimal sample volume and enable high-throughput analysis.

Real-world applications of FRET-based optical sensors are numerous and impactful. In biomedical research, FRET is widely used for studying protein-protein interactions, enzyme activities, and conformational changes of biomolecules in live cells. For example, genetically encoded FRET biosensors have been developed to monitor intracellular calcium levels, cyclic AMP (cAMP) concentrations, and kinase activities with high temporal and spatial resolution. The **chameleon** family of FRET-based calcium sensors is a notable example, allowing real-time visualization of calcium dynamics in living cells.

In clinical diagnostics, FRET-based assays are used for detecting specific nucleic acid sequences, biomarkers, and pathogens. An example includes the **Cepheid GeneXpert system**, which uses FRET-based molecular beacons to detect tuberculosis (TB) DNA with high sensitivity and specificity. In environmental monitoring, FRET sensors have been employed to detect heavy metals like lead (Pb^{2+}) and mercury (Hg^{2+}) in water samples through binding-induced changes in donor–acceptor distances.

Another notable application is in drug discovery and screening, where high-throughput FRET assays are utilized to identify inhibitors of protein-protein interactions critical in disease processes. Pharmaceutical companies commonly use FRET-based technologies to accelerate the discovery of therapeutic compounds.

Moreover, wearable biosensors incorporating FRET have been developed for continuous monitoring of metabolites like glucose or lactate in sweat, presenting a non-invasive and user-friendly approach to health monitoring.

3.7 Bimolecular fluorescence complementation-based optical sensor

Bimolecular fluorescence complementation (BiFC) is an advanced and powerful technique used for optical sensing that relies on the reconstitution of fluorescence when two nonfluorescent fragments of a split fluorescent protein come together through the interaction of two proteins of interest. This method enables the visualization and monitoring of protein–protein interactions (PPIs) and complex molecular assemblies in living cells with high specificity and spatial resolution. BiFC-based sensors offer an intuitive and direct way to study dynamic biological processes, making them valuable tools in biomedical research, drug discovery, and diagnostics.

The theory behind BiFC is based on the principle that a fluorescent protein, such as green fluorescent protein (GFP) or its variants (e.g., YFP, CFP, mCherry), can be genetically split into two nonfluorescent fragments, typically at a site that allows each half to remain structurally stable but nonfunctional on its own. When these fragments are brought into close proximity by the interaction of two fused proteins, they refold and reassemble into a functional fluorescent complex. The formation of fluorescence indicates that the two proteins of interest are interacting within the cellular environment. This complementation is irreversible under normal conditions, providing a permanent mark of interaction events. Fig. 3.8 shows the schematic of BiFC optical sensing.

The main principles involved in BiFC include protein engineering, fluorescence reconstitution, and spatial proximity-induced molecular folding. Each fluorescent fragment is genetically fused to a protein or domain of interest. Upon expression in cells, if the target proteins interact, the two fragments come into close contact, enabling the reconstitution of the fluorescent structure and subsequent emission of fluorescence under appropriate excitation.

The efficiency of BiFC can be described in a simplified kinetic model where the formation rate of the fluorescent complex depends on the interaction rate (konk_{on}) and the dissociation rate (koffk_{off}) of the two binding partners:

$$F(t) \propto (1 - e(- k_o\, n[P_1][P_2]t))$$

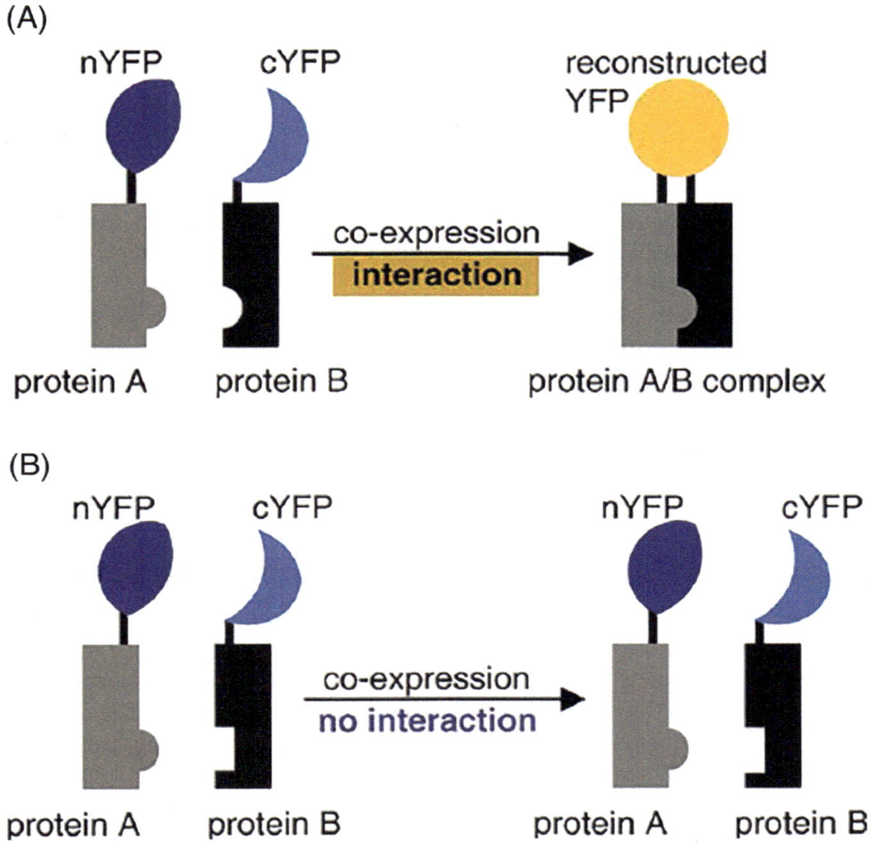

FIGURE 3.8

Schematic showing the BiFC optical sensing.

From Citovsky, V., Lee, L.-Y., Vyas, S., Glick, E., Chen, M.-H., Vainstein, A., Gafni, Y., Gelvin, S.B., & Tzfira, T. (2006). Journal of Molecular Biology, 362 *(5), 1120–1131. https://doi.org/10.1016/j.jmb.2006.08.017.*

where:

F(t) = fluorescence intensity at time, *t*,
$[P_1]$ and $[P_2]$ = concentrations of the two interacting proteins,
K_{on} = association rate constant.

Since BiFC complex formation is generally considered irreversible on a biological timescale, dissociation is minimal compared to other interaction assays like FRET.

The mechanism of the BiFC method typically starts with the genetic construction of two fusion proteins: each containing a non-fluorescent fragment of a fluorescent protein linked to a protein of interest. Upon co-expression in a living system (cells, tissues, organisms), if the two proteins interact, their associated fragments complement each other to form a functional fluorescent protein. The fluorescence is then excited at a specific wavelength, and the emitted light is detected by fluorescence microscopy, flow cytometry, or other optical detection systems.

Miniaturization of BiFC-based optical sensing has been an important development, particularly for high-throughput screening (HTS), POC diagnostics, and single-cell analysis. The miniaturization process involves integrating BiFC assays into microfluidic platforms, lab-on-a-chip devices, and nanostructured scaffolds. Microfluidic systems allow the manipulation of small volumes (picoliters to microliters) of cells or reagents, enabling rapid and parallel detection of interaction events. Advances in nanotechnology have also enabled the immobilization of BiFC pairs onto nanoparticles or nanowells, creating highly sensitive and multiplexed biosensors.

After miniaturization, the basic sensing mechanism remains the same—fluorescent protein reconstitution triggered by protein-protein interactions—but detection becomes faster, more sensitive, and feasible on smaller, portable devices. Miniaturized systems often incorporate micro-patterned arrays, where thousands of BiFC interactions can be monitored simultaneously using automated fluorescence detection. Moreover, improvements in optics, such as miniaturized CCD or CMOS sensors, allow the development of handheld BiFC readers for real-time analysis in clinical or field settings.

3.7.1 Applications of bimolecular fluorescence complementation based

BiFC has found widespread application across various fields of biological research and biotechnology such biomedical research, clinical diagnostics, drug discovery, and environmental monitoring.

BiFC has been instrumental in mapping complex protein interaction networks inside live cells. For example, it has been used to study signaling pathways like the MAPK/ERK cascade, where the interaction of kinases with their substrates can be visualized directly in living cells. Another notable application is the use of BiFC sensors to detect the interaction between viral proteins and host factors, crucial for understanding viral infection mechanisms such as in HIV or SARS-CoV-2 research.

On top of that, BiFC-based biosensors have been explored for detecting disease biomarkers. For instance, cancer diagnostic tools using BiFC can identify oncogenic protein interactions, such as the interaction between MDM2 and p53, which is relevant to tumor progression and therapeutic targeting. The irreversible fluorescence complementation characteristic of BiFC makes it a robust system for detecting transient or weak interactions that are difficult to capture by other methods.

BiFC is also employed in high-throughput screening platforms to identify inhibitors of protein–protein interactions. By expressing BiFC constructs in microtiter plates and applying chemical libraries, researchers can quickly screen for compounds that disrupt specific interactions, indicated by the loss of fluorescence.

Emerging applications of BiFC include environmental monitoring. Genetically engineered bacterial biosensors using BiFC have been developed to detect environmental pollutants by expressing BiFC pairs that interact only in the presence of specific toxins or heavy metals, leading to a visible fluorescent signal.

In summary, BiFC is a versatile and powerful technique for studying protein–protein interactions in living cells. Its applications span across various fields, including biomedical research, clinical diagnostics, drug discovery, and environmental monitoring, highlighting its significance in advancing our understanding of cellular processes and facilitating the development of novel therapeutic strategies.

3.8 Single fluorescent protein (FP)-based optical sensor

Single fluorescent protein (Single FP)-based optical sensors are a class of biosensors that rely on a single, genetically encoded fluorescent protein that undergoes changes in its fluorescence properties (such as intensity, wavelength, or lifetime) in response to specific biochemical or biophysical changes in the environment. Unlike techniques that require two separate fluorophores (such as FRET or BiFC), single FP-based sensors provide a simpler and highly efficient platform for real-time monitoring of dynamic cellular processes with minimal perturbation to the system being studied.

The theory behind single FP-based sensors revolves around the intrinsic ability of fluorescent proteins to change their optical properties when fused with specific sensing domains. These sensing domains are typically derived from naturally occurring receptor proteins, enzymes, or binding proteins that undergo conformational changes upon interaction with a target analyte (such as ions, metabolites, or signaling molecules). The conformational change affects the chromophore environment inside the fluorescent protein, leading to measurable shifts in fluorescence intensity, excitation/emission wavelengths, or even photo-switching behavior.

The fundamental principle of single FP-based sensing is that a single chromophore acts both as the reporter and the sensing unit. When the local environment (such as pH, ion concentration, or redox state) around the chromophore changes, it alters the protein's optical characteristics. There is no need for energy transfer between different molecules, simplifying the system and enhancing robustness.

One of the key parameters for single FP sensors is the fluorescence intensity II, which can be described by the general equation:

$$I = \phi \times \epsilon \times P$$

where

I = fluorescence intensity,
Φ = quantum yield of the fluorophore (fraction of photons emitted per photons absorbed),
ϵ = molar extinction coefficient (how strongly the fluorophore absorbs light at a given wavelength),
P = incident light power.

In single FP-based sensors, either Φ or ϵ (or both) can change when the sensing domain interacts with the target, leading to a detectable fluorescence signal change.

The mechanism of single FP-based sensors generally involves genetically fusing the fluorescent protein with a sensing domain that recognizes the specific analyte. Upon binding of the analyte, the sensing domain undergoes a conformational or chemical change that modifies the chromophore environment inside the protein barrel structure. This structural rearrangement either enhances, quenches, or shifts the fluorescence properties, allowing real-time optical detection of the target molecule without the need for external dyes or multiple labelling steps.

An important class of single FP-based sensors includes pH sensors (like pHluorin), calcium sensors (such as GCaMP), and redox sensors (like roGFP). For example, GCaMP sensors consist of a GFP variant fused to calmodulin and an M13 peptide; upon binding calcium, the calmodulin undergoes a conformational change that leads to a dramatic increase in GFP fluorescence.

Genetically encoded fluorescent protein-based biosensors as shown in Fig. 3.9 use different strategies to detect events inside cells. In translocation-based biosensors, the sensor moves to a specific cell

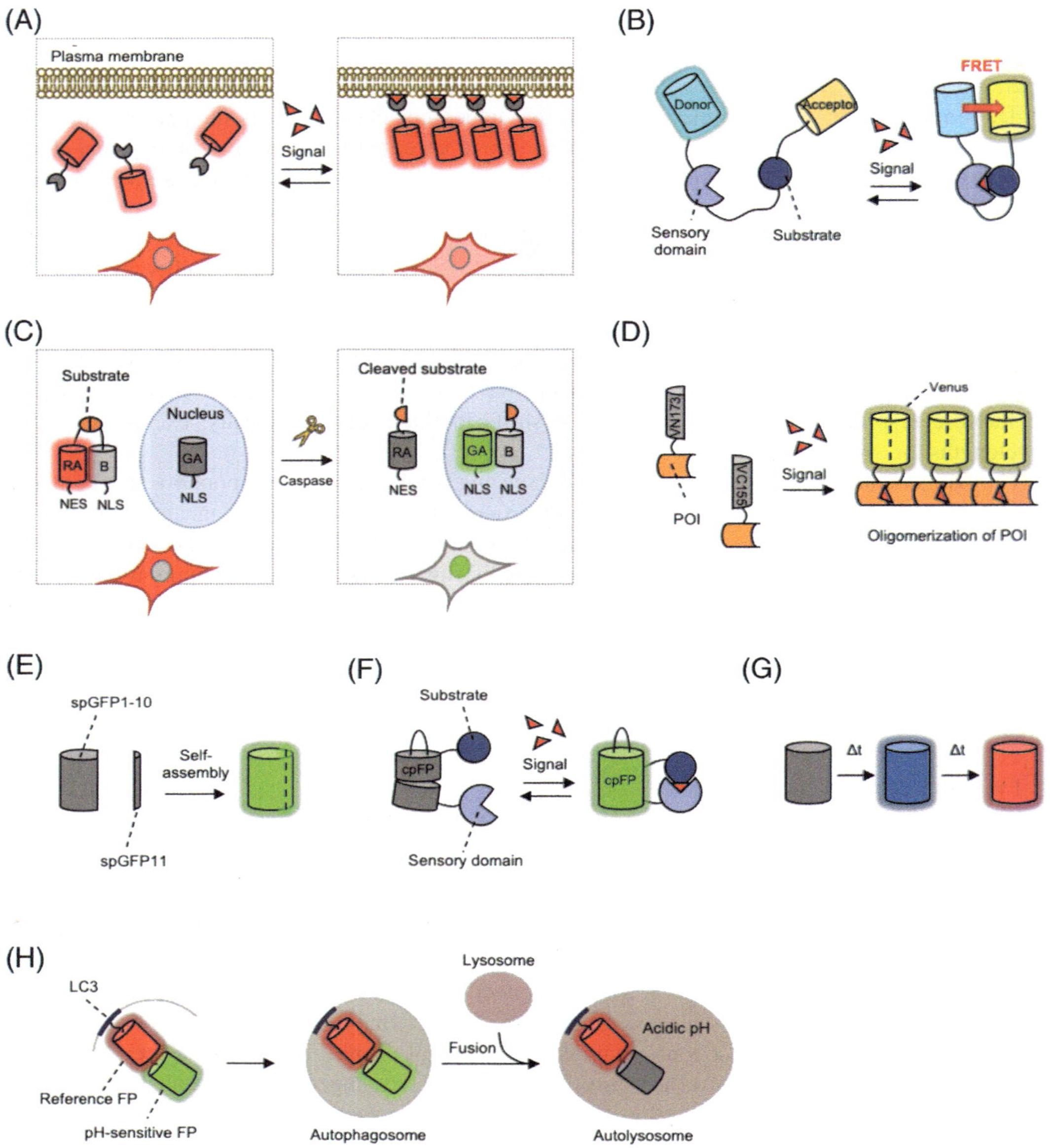

FIGURE 3.9

Fluorescent protein-based sensing methods used in genetically encoded biosensors.

From Kim, H., Ju, J., Lee, H. N., Chun, H., & Seong, J. (2021). Genetically encoded biosensors based on fluorescent proteins. Sensors, 21 *(3), 795. https://doi.org/10.3390/s21030795.*

location, such as the plasma membrane, when a target molecule is produced, making the signal visible. FRET-based biosensors work by changing shape in response to a signal, which brings two fluorescent proteins closer together and increases fluorescence. In ddFP-based biosensors, enzyme activity (such as caspase-3) cuts the sensor, causing changes in both fluorescence color and location inside the cell. BiFC-based biosensors produce fluorescence when two proteins interact, allowing split fluorescent protein fragments to reassemble and emit light. Similarly, split fluorescent proteins such as GFP1-10 and GFP11 can self-assemble to restore fluorescence. In cpFP-based biosensors, a signal-induced shape change increases the brightness of the fluorescent protein. Fluorescent timers change color over time, helping track the age of cellular events. Finally, autophagy flux sensors use pH-sensitive fluorescent proteins that lose intensity as autophagic vesicles become acidic, while a reference fluorescent signal remains unchanged.

Miniaturization of single FP-based optical sensors is critical for their application in high-throughput screening, implantable biosensors, portable diagnostic devices, and live-animal imaging. Advances in microfabrication, optoelectronics, and nanotechnology have enabled the integration of Single FP-based sensors onto microfluidic chips, nanoarrays, and fiberoptic probes. Because single FP sensors are genetically encoded, they can also be directly expressed in living cells, tissues, or organisms without the need for invasive implantation of external devices, offering true miniaturization at the molecular scale.

Miniaturized single FP sensors often incorporate integrated excitation sources (like miniaturized LEDs or lasers) and compact detectors (such as photodiodes or CMOS cameras). This integration reduces the overall size of the sensing system and enables in vivo monitoring of biochemical processes with minimal disruption. For instance, wearable biosensors using single FP constructs have been developed to monitor metabolites like lactate and glucose noninvasively.

After miniaturization, the sensing mechanism remains fundamentally unchanged: the fluorescent protein responds to specific biochemical stimuli by altering its fluorescence, which is then detected by a compact optical system. However, miniaturization improves signal collection efficiency, reduces noise, and enhances portability, making it possible to deploy single FP-based sensors in clinical, environmental, and remote settings.

3.8.1 Applications of fluorescent protein-based optical sensors

Single fluorescent protein (FP)-based optical sensors have found extensive and growing applications across various fields, offering real-time, noninvasive monitoring capabilities. Examples of its application are in the area of neuroscience, clinical research, environmental monitoring, and plant biology

In the realm of neuroscience, GCaMP sensors have transformed calcium imaging, enabling researchers to observe neural activity in living animals with high temporal and spatial resolution. Among the most advanced variants, GCaMP6 allows for the visualization of single action potentials in neurons. These sensors are utilized in fiber photometry systems, where they emit fluorescence in response to calcium binding, facilitating the recording of calcium transients in specific cell types.

In clinical settings, single FP-based sensors have been developed for continuous monitoring of blood chemistry in diabetic and critically ill patients. For instance, genetically engineered skin cells expressing a glucose-sensitive fluorescent protein have been proposed as a noninvasive glucose monitoring system.

Environmental monitoring also benefits from single FP-based sensors. Microorganisms engineered to express fluorescent proteins responsive to toxic metals or pollutants can act as living biosensors, providing an eco-friendly and cost-effective method for environmental surveillance.

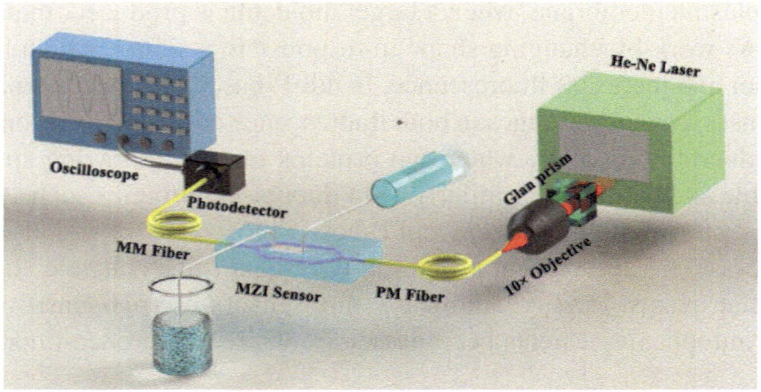

FIGURE 3.10

Schematic illustration of the MZI sensor measurement setup. PM fiber denotes polarization-maintaining fiber, and MM fiber refers to multimode fiber.

From Zhao, C., Xu, L., & Liu, L. (2021). Sensors, 21. *https://doi.org/10.3390/s21196600.*

Moreover, single FP-based optical sensors have been used in plant biology to track hormone signaling and stress responses in real time, providing insights into plant growth and adaptation mechanisms.

These advancements underscore the versatility and expanding utility of single FP-based optical sensors in diverse scientific domains.

3.9 Mach–Zehnder interferometer-based optical sensor

The Mach–Zehnder interferometer (MZI) is a pivotal optical device renowned for its sensitivity and versatility across various scientific and engineering disciplines. It operates on the principle of dividing a coherent light beam into two separate paths, introducing a phase shift in one path, and then recombining the beams to produce an interference pattern. This interference pattern is highly sensitive to phase differences, making the MZI an excellent tool for precise measurements in applications ranging from quantum mechanics to environmental sensing.

The MZI consists of two beam splitters and two mirrors arranged in a configuration that splits an incoming light beam into two paths Fig. 3.10. A phase shift can be introduced into one of the paths, and upon recombination at the second beam splitter, the resulting interference pattern depends on the phase difference between the two paths. The intensity II at the output ports can be expressed as:

$$I = I_0\left(1/2 \pm 1/2\cos\left(\Delta\phi\right)\right)$$

where I_0 is the initial intensity, and $\Delta\phi$ is the phase difference between the two paths. This equation illustrates how the interference pattern varies with the phase shift, enabling precise measurements of optical path differences.

The detection mechanism of the MZI relies on analyzing the interference pattern produced when the two light beams are recombined. Changes in the phase difference, due to variations in temperature, refractive index, or the presence of a sample, alter the interference pattern. By monitoring these changes, the MZI can detect minute variations in the environment or the properties of the sample under investigation.

Advancements in photonic integration have led to the miniaturization of the MZI. Integrated Mach–Zehnder interferometers (iMZIs) are fabricated on semiconductor substrates, utilizing materials like silicon photonics. This miniaturization allows for the development of compact and cost-effective sensors suitable for various applications, including portable devices and wearable technologies. The integration of iMZIs with microfluidic systems further enhances their utility in lab-on-a-chip devices, enabling real-time monitoring and analysis in biomedical diagnostics.

3.9.1 Applications of Mach–Zehnder interferometer

The MZI's sensitivity to phase shifts has led to its widespread application in numerous fields. In quantum mechanics, it serves as a fundamental tool for experiments exploring wave-particle duality and quantum interference. In telecommunications, MZIs are employed in optical modulators for controlling light signals in fiberoptic networks. Environmental monitoring utilizes MZIs to detect pollutants and measure atmospheric conditions with high precision. In biomedical diagnostics, MZIs are integrated into biosensors for detecting specific biomolecules, offering label-free and real-time analysis. Moreover, the development of portable and miniaturized MZI-based sensors has expanded their application to wearable health monitoring devices, providing continuous assessment of physiological parameters.

3.10 Absorption-based optical sensor

Absorption-based optical sensors are a category of sensing devices that rely on the change in the absorption of light by a material in response to the presence of a specific analyte or environmental condition. Unlike fluorescence-based sensors that detect emitted light, absorption-based systems measure the attenuation of light as it passes through or interacts with the sample. These changes in light absorption correlate with the concentration or presence of target molecules, making absorption-based sensors powerful tools for a wide range of applications, including chemical analysis, environmental monitoring, and biomedical diagnostics.

The theory behind absorption-based sensors is grounded in **the Beer-Lambert Law**, which quantitatively relates the absorption of light to the properties of the material through which the light travels. According to the Beer-Lambert Law:

$$A = \varepsilon \times c \times l$$

where

A = absorbance (no units),
ε = molar absorptivity or extinction coefficient (L $mol^{-1}cm^{-1}$)
c = concentration of the absorbing species (mol L^{-1}),
l = path length that the light travels through the material (cm).

Absorption occurs when photons are absorbed by molecules, promoting electrons from a ground state to an excited electronic state. The specific wavelengths at which absorption occurs depend on the

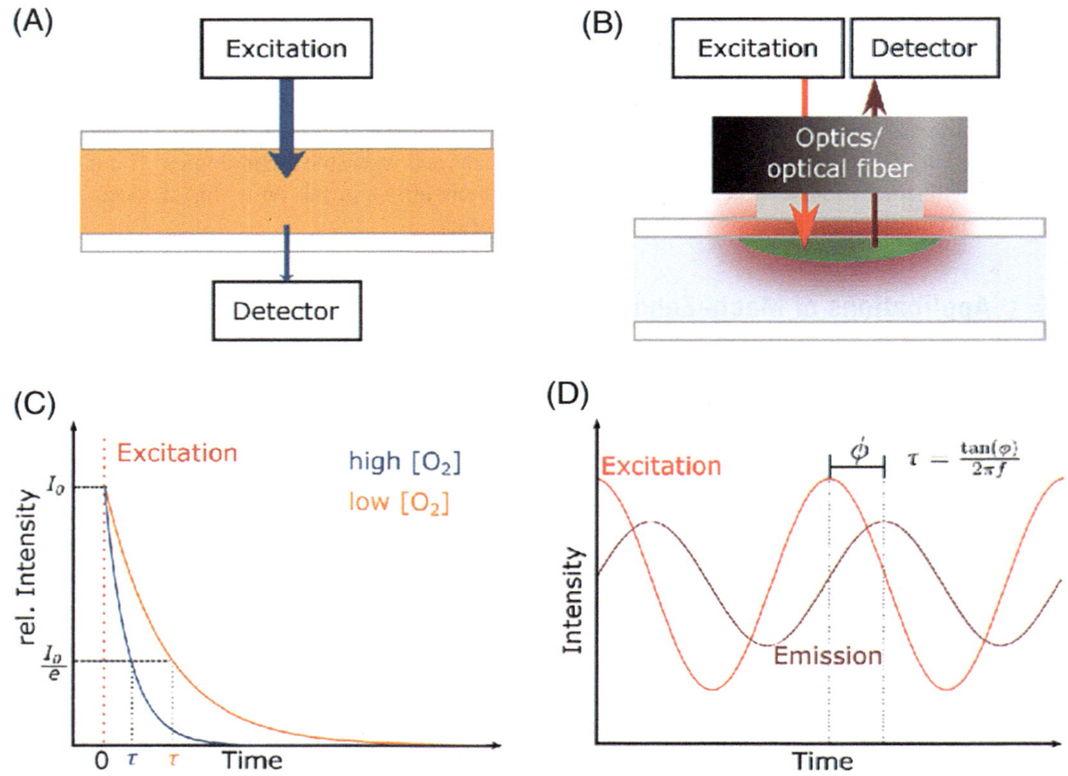

FIGURE 3.11

Absorption and luminescence optical sensor schematics.

From Fuchs, S., Johansson, S., Tjell, A., Werr, G., Mayr, T., & Tenje, M. (2021). ACS Biomaterials Science & Engineering, *XXXX. https://doi.org/10.1021/acsbiomaterials.0c01110.*

electronic structure of the molecules involved. This property allows different substances to be detected and quantified based on their unique absorption spectra.

One of the key principles involved in absorption-based sensing is the specific interaction between the sensing material and the target analyte, which leads to a measurable change in the absorption characteristics as shown in Fig. 3.11. This interaction can be either direct—such as a molecule absorbing light at a specific wavelength—or indirect, where an analyte induces a chemical or structural change in the sensing material that modifies its absorption properties.

A significant advancement in absorption-based sensors involves the use of photocatalysts, particularly semiconductor photocatalysts like titanium dioxide (TiO_2), zinc oxide (ZnO), and various metal–organic frameworks (MOFs). Photocatalysts can absorb light and generate electron–hole pairs, which participate in redox reactions. When the photocatalyst interacts with target analytes, the absorption profile changes due to alterations in the electronic properties or band structure, allowing for optical sensing.

The mechanism of absorption-based optical sensing typically follows these steps:

1. Light of a specific wavelength is directed onto the sensor.
2. The material absorbs light according to its inherent absorption characteristics.
3. In the presence of the analyte, the absorption behavior changes—either in magnitude, wavelength shift, or intensity.
4. The change is detected by measuring the difference in transmitted light compared to a reference, and this difference is correlated with the concentration of the analyte.

In systems utilizing photocatalysts, the sensing mechanism may involve photocatalytic reactions that either degrade the target molecules or produce a measurable intermediate that alters absorption. For example, the photocatalytic degradation of organic dyes like methylene blue or rhodamine B under UV light, monitored by absorbance decrease, forms the basis of several environmental sensors.

Miniaturization of absorption-based optical sensors has been a major focus to enhance sensitivity, portability, and integration into complex systems. Advances in nanotechnology have enabled the creation of nanostructured photocatalysts with high surface area-to-volume ratios, significantly boosting interaction with target molecules and improving sensor responsiveness. Additionally, integrating absorption sensors into microfluidic platforms allows for the analysis of very small sample volumes, crucial for applications like POC diagnostics or environmental monitoring.

For miniaturized devices, thin-film photocatalysts deposited on transparent substrates (e.g., glass, quartz) are commonly used. Microfabricated optical paths, such as waveguides or integrated microchannels, ensure precise control over light-matter interaction lengths. In these devices, absorbance changes can be detected using miniaturized photodetectors like CMOS sensors or photodiodes, combined with compact light sources such as LEDs or laser diodes.

After miniaturization, the basic sensing mechanism remains absorption measurement, but enhancements are achieved through better light management (using integrated optics), faster response times (due to reduced diffusion paths), and lower detection limits (thanks to higher sensitivity of nanomaterials and focused light beams). Importantly, miniaturization also allows for multiplexed sensing, where multiple analytes are detected simultaneously by spatially resolving different sensing zones on a chip.

3.10.1 Applications of absorption-based optical sensors

Absorption-based optical sensors have found diverse applications across environmental monitoring, biomedical diagnostics, food safety, and wearable health technologies.

Photocatalytic titanium dioxide (TiO_2) films are utilized in absorption-based sensors to detect air pollutants such as nitrogen oxides (NOx) and volatile organic compounds (VOCs) (Ma et al., 2025). These sensors operate by monitoring changes in the absorption spectrum as pollutants are oxidized on the photocatalyst surface.

Colorimetric biosensors employ enzymatic reactions that produce color changes detectable by absorbance (Yang & Ge, 2023). For instance, glucose monitoring systems based on glucose oxidase reactions produce hydrogen peroxide, which reacts with a chromogenic substrate to yield a color change proportional to glucose concentration, easily measured using absorption principles.

In the food industry, absorption-based sensors are employed to detect spoilage indicators such as ammonia or biogenic amines. Sensors designed with nanostructured photocatalytic coatings can indicate contamination or degradation of food products by visible color changes, providing rapid and noninvasive testing.

Moreover, wearable absorption-based devices are emerging, such as sweat sensors that monitor electrolyte balance by detecting absorption changes in microfluidic patches adhered to the skin. These devices integrate optical readout with mobile applications for real-time health monitoring.

These advancements demonstrate the versatility and expanding utility of absorption-based optical sensors in various scientific and medical fields.

3.11 Colorimetric-based optical sensor

Colorimetric-based optical sensors are a class of sensors that detect and quantify the presence of an analyte based on changes in color, which can be observed visually or measured with simple optical instruments. These sensors rely on the principle that certain chemical or biological interactions lead to a color change due to alterations in the optical absorption properties of the system. Colorimetric sensors are widely valued for their simplicity, low cost, and ease of interpretation, often requiring no sophisticated equipment for readout, making them ideal for POC diagnostics, environmental monitoring, food safety testing, and other applications.

The theory behind colorimetric sensing is primarily based on the **Beer-Lambert Law**, which describes how the intensity of light decreases as it passes through an absorbing medium (Mayerhöfer et al., 2020).

In a colorimetric assay, the presence or concentration of a specific analyte alters the chemical environment, causing a shift in the absorption spectrum. This results in a visible color change or an intensity variation that can be quantified by measuring absorbance at specific wavelengths.

Several key principles are involved in colorimetric sensing such as selective binding, chrogenic reaction, aggregation/dispersion mechanisms, and optical density change. Specific binding between a receptor (such as an antibody, aptamer, or chemical group) and the target analyte initiates the color change. The reaction between the analyte and a chromogenic substrate leads to the generation of colored products. In some cases, nanoparticles like AuNPs are used, where their aggregation state influences color due to plasmonic effects. The final detectable signal arises from the change in optical density due to the aforementioned processes.

The mechanism of a typical colorimetric method follows these basic steps:

1. The sensing material, often containing a chromophore or nanoparticle, is exposed to the sample.
2. If the target analyte is present, it binds to the sensing molecule or catalyzes a chemical reaction.
3. This interaction induces a color change visible to the naked eye or measurable with a simple spectrophotometer.
4. Quantification is achieved by correlating the degree of color change to the concentration of the analyte using calibration curves.

A prominent application of colorimetric-based sensing is the **lateral flow assay (LFA)**, a paper-based platform widely known for applications such as pregnancy tests and rapid infectious disease diagnostics (e.g., COVID-19 antigen tests). In LFA, a sample (such as blood, saliva, or urine) flows laterally across a membrane strip via capillary action. Specific binding interactions between the analyte

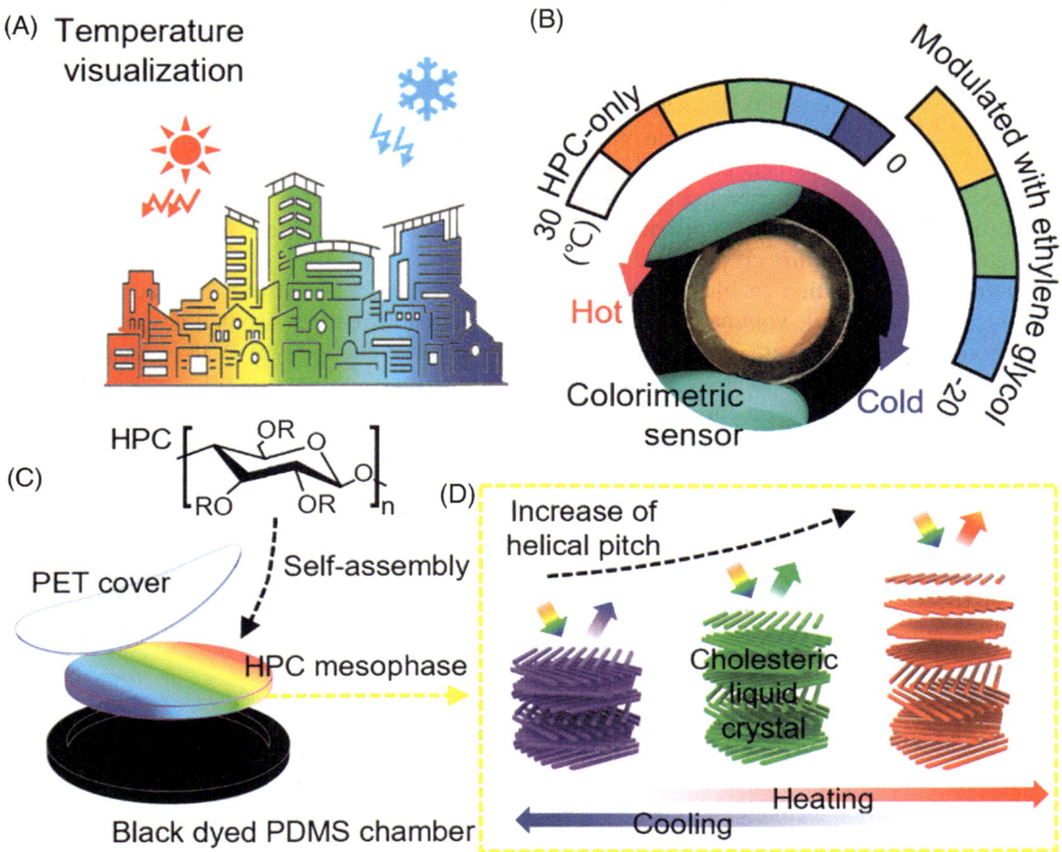

FIGURE 3.12

An HPC-based colorimetric sensor and a modified HPC sensor prepared using a simple design with a polymer chamber and a polyethylene terephthalate (PET) film.

From Yi, H., Lee, S.-H., Kim, D., Jeong, H.E., & Jeong, C. (2022). Colorimetric sensor based on hydroxypropyl cellulose for wide temperature sensing range. Sensors, *22 (3), 886. https://doi.org/10.3390/s22030886.*

and labelled detection molecules (often conjugated with colored nanoparticles like AuNPs) lead to the formation of visible colored lines at predefined test and control zones on the strip.

Another application of colorimetric-based sensing is HPC-based colorimetric temperature sensor, as shown in Fig. 3.12. Panel (a) shows how the sensor can visually display temperature changes using different colors, making it useful for temperature mapping in real environments. Panel (b) shows an actual image of the sensor and the range of temperatures it can detect, where the color changes from blue (cold) to red (hot). Panel (c) illustrates the materials used to build the sensor, including hydroxypropyl cellulose (HPC) that self-assembles into a structured layer, placed inside a black PDMS chamber and covered with a PET film to improve visibility. Panel (d) explains how the sensor works: when the temperature changes, the helical structure of the cholesteric liquid crystal made from HPC expands or contracts, which changes the reflected color. Heating increases the helical pitch and shifts the color toward red, while cooling reduces the pitch and shifts the color toward blue.

Miniaturization of colorimetric sensors has been a key focus to enhance portability, sensitivity, and multiplexing capabilities. In recent years, microfluidic paper-based analytical devices (μPADs) have emerged, where channels patterned on paper direct fluid flow and allow for multiple tests on a single small strip. Nano-engineering approaches have also been developed where nanomaterials with enhanced optical properties (e.g., plasmonic nanoparticles, photonic crystals) are embedded into the sensing platforms to amplify colorimetric signals.

Microfluidic integration enables the precise control of reagent mixing, reaction times, and sample handling, reducing sample volumes to microliters or less. Portable readers—sometimes integrated with smartphones—can capture images of the colorimetric response and analyze RGB values to provide quantitative results. Miniaturization thus not only preserves the simplicity of colorimetric assays but also introduces digital quantification, improving sensitivity and enabling real-time remote diagnostics.

After miniaturization, the basic sensing mechanism remains the detection of color changes resulting from analyte interaction. However, enhancements include:

Increased surface-area-to-volume ratio, improving reaction kinetics.

Reduced diffusion distances, leading to faster responses.

Higher sensitivity due to more efficient light–matter interaction and signal amplification using nanomaterials.

Multiplexing capabilities, allowing simultaneous detection of multiple analytes on a single device.

3.11.1 Applications of colorimetric-based optical sensors

Several real-world applications demonstrate the effectiveness of colorimetric-based optical sensors. Colorimetric detection technologies have become integral to various diagnostic applications, offering rapid, user-friendly, and cost-effective solutions across healthcare, environmental monitoring, food safety, and wearable health devices. Fig. 3.13 shows an example of a colorimetric based sensor which is a home pregnancy test

In the home pregnancy test, the hormone hCG in urine binds to antibodies conjugated to AuNPs, resulting in a visible line. This principle is employed in lateral flow immunoassays (LFIAs), which are widely used for POC diagnostics due to their simplicity and reliability.

Similarly, COVID-19 rapid antigen tests utilize lateral flow platforms where SARS-CoV-2 antigens are captured by specific antibodies, producing a colored line upon successful detection (He et al., 2023). These tests are particularly useful for identifying individuals at or near peak infection.

In environmental monitoring, colorimetric strips are employed for the detection of heavy metals like lead or mercury in water. These devices change color in response to specific ions, providing a fast and easy-to-use method for field testing without the need for laboratory equipment.

In food safety, colorimetric assays are used to detect pathogens such as *Salmonella* or *Listeria*. Some sensors are even capable of detecting pesticide residues on produce surfaces by a visible color change when in contact with specific contaminants.

Moreover, advanced developments have led to wearable colorimetric sensors, such as color-changing tattoos or skin patches that monitor biomarkers such as glucose, lactate, or pH through sweat analysis.

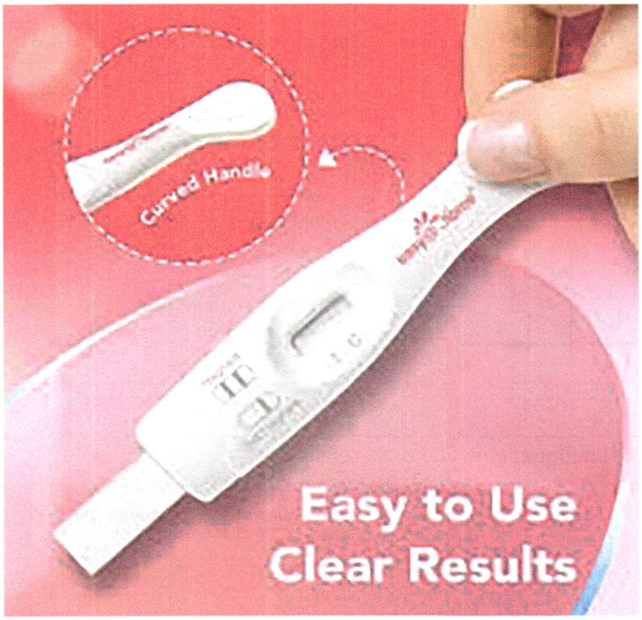

FIGURE 3.13

Home pregnancy test colorimetric optical sensor.

These wearable sensors integrate colorimetric reagents into flexible materials, providing continuous, noninvasive monitoring solutions.

These applications highlight the versatility and expanding utility of colorimetric detection technologies in diverse scientific and medical fields.

LFA-based biosensors have transformed POC diagnostics by providing a simple, fast, and inexpensive platform for detecting a wide spectrum of analytes (Vealan et al., 2023). These biosensors use capillary-driven fluid flow and particular biomolecular interactions to detect target molecules in minutes, making them excellent for use in clinical diagnostics, environmental monitoring, and food safety. The extensive application of LFAs, such as home pregnancy tests and COVID-19 antigen kits, demonstrates their versatility and accessibility. LFA-based biosensors function by transporting a liquid sample over a nitrocellulose strip that has been embedded with specialized reagents, such as capture probes and detection labels. As the sample passes over the strip, target analytes bind to the reagents, resulting in a visual signal, usually in the form of colored lines. Recent improvements in LFA technology have included optical biosensing components including fluorescence, chemiluminescence, and surface-enhanced Raman scattering (SERS) to improve detection sensitivity and specificity.

Wang et al.'s initial work used multiple cross displacement amplification (MCDA) with a nanoparticle-based lateral flow biosensor to detect *Mycoplasma pneumoniae* (Wang et al., 2019). The process involves amplification of target DNA with MCDA, followed by detection with a LFA. The biosensor was designed to produce findings in 30 minutes, making it ideal for POC testing. When

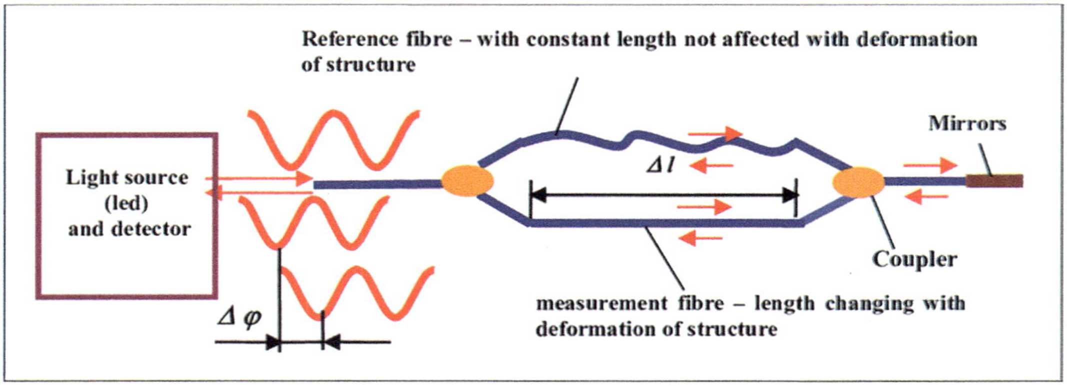

FIGURE 3.14

Fundamental principle of interferometric fiberoptic sensing.

From Mavar, K., Sekulić, D., & Wurth, D. (2007). Monitoring of Krk Bridge columns with interferometric fibre optic sensors. In Symposium Dubrovnik 2007: Concrete Structures – Simulators of Development, Proceedings (pp. 709–716). Zagreb, Croatia: SECON HDGK.

compared to clinical samples, the results showed a sensitivity of 95% and specificity of 100%. The authors concluded that the MCDA-LFB test is a potential diagnostic tool for quick identification of *Mycoplasma pneumoniae*, especially in resource-constrained environments.

Amalina et al.'s second study focused on building a nucleic acid-based lateral flow biosensor to detect *Salmonella typh*i and *Salmonella paratyphi* in stool samples (Amalina et al., 2021). The process required employing PCR to amplify particular DNA sequences from the pathogens, followed by detection using a lateral flow strip containing AuNPs. The results demonstrated that the biosensor could identify the target pathogens with 90% sensitivity and 95% specificity, exceeding standard approaches like agarose gel electrophoresis. The study found that the proposed biosensor provides a quick and sensitive way to identify *Salmonella* in clinical samples, which is critical for prompt diagnosis and treatment.

These findings demonstrate the adaptability and efficacy of LFA-based biosensors in a variety of applications, including pathogen detection and nucleic acid amplification. The combination of amplification methods and LFA technology increases sensitivity and specificity, making these biosensors useful for POC diagnostics.

3.12 Interferometric-based optical sensor

Interferometric-based optical sensors are powerful sensing devices that utilize the interference of light waves to detect extremely small changes in environmental conditions or molecular interactions. The fundamental concept behind interferometry is the superposition of two or more coherent light waves, which leads to constructive or destructive interference depending on their relative phase difference Fig. 3.14. Any variation in the phase, caused by changes in the optical path length, is thus translated into a measurable change in the intensity of the resulting light pattern. This sensitivity makes interferometric optical sensors invaluable tools across diverse fields such as biosensing, environmental monitoring, structural health diagnostics, and medical imaging.

The theory underlying interferometric sensors is based on the behavior of coherent light. When two coherent light beams combine, the phase difference between them governs the resulting intensity pattern. Mathematically, the phase difference ($\Delta\phi$) is expressed as

$$\Delta\phi = 2\pi/\lambda \times \Delta L$$

where λ represents the wavelength of light and ΔL denotes the difference in optical path lengths between the sensing and reference arms. The resulting intensity II of the interference pattern can be described by

$$I = I_0[1 + \cos(\Delta\phi)]$$

where I_0 is the maximum intensity. Small changes in the environment that affect the optical path length, such as variations in refractive index, pressure, temperature, or mechanical strain, can therefore be detected by monitoring shifts in the interference pattern.

The operation of interferometric optical sensors relies on several key principles. Coherence is essential; light sources, typically lasers, must possess a long coherence length to maintain a stable interference pattern. Phase sensitivity is critical, as the technique exploits even minute variations in phase to detect changes in the measured parameter. Additionally, multipath propagation, where the light is split into a sensing and a reference path, is a fundamental concept, ensuring that only changes induced in the sensing arm contribute to the interference signal. Finally, interference detection, either through fringe shifts or intensity modulation, provides the readout of the sensor.

In a typical interferometric sensor mechanism, a coherent light beam is divided into two paths. One beam travels through a sensing arm exposed to the target environment, while the other follows a reference path shielded from perturbations. After traversing their respective paths, the beams are recombined, and any difference in optical path length results in an interference pattern. Environmental changes, such as binding events in biosensing or mechanical deformation in structural monitoring, alter the refractive index or the physical length of the sensing arm, thereby shifting the interference fringes or changing the intensity pattern. By precisely measuring these changes, interferometric sensors can detect even extremely small perturbations, down to the level of nanometers or refractive index changes of 10^{-7} refractive index unit (RIU) or lower.

The miniaturization of interferometric sensors has significantly expanded their applicability and usability. Advances in microfabrication, including photolithography, microelectromechanical systems (MEMS) technology, and silicon photonics, have enabled the development of highly compact and integrated interferometric devices. Miniaturized versions, such as micro-Mach-Zehnder interferometers, Fabry-Pérot cavities on a chip, and fiberoptic interferometers, have emerged, offering the benefits of reduced size, enhanced portability, lower sample volume requirements, and potential for mass production. Silicon-on-insulator (SOI) platforms have been particularly important, allowing for the integration of multiple interferometric sensors onto a single microchip with high precision and reproducibility.

After miniaturization, the basic sensing mechanism remains rooted in interference-based detection; however, several enhancements emerge. Miniaturization allows for stronger confinement of the optical field, which increases the interaction between light and the analyte, thereby improving sensitivity.

Devices also exhibit faster response times due to reduced diffusion distances and can support multiplexed detection schemes, wherein many sensing elements operate simultaneously on a single platform. Furthermore, integrating miniaturized interferometers with electronics enables real-time signal processing, remote monitoring, and improved environmental stability through encapsulation and robust device architectures. Miniaturized fiberoptic interferometers, for example, can be deployed for real-time strain or temperature monitoring inside harsh environments such as oil wells or aerospace structures.

3.12.1 Applications of interferometric-based optical sensors

Interferometric-based optical sensors are widely utilized across various fields, offering high sensitivity and label-free detection capabilities. They are used in biosensing, environmental monitoring, structural health monitoring, medical imaging, and wearable health monitoring.

In biosensing, integrated Mach-Zehnder and Fabry-Pérot interferometers enable label-free detection of biomolecules such as proteins, DNA, and pathogens. For example, silicon photonic biosensors can detect cancer biomarkers in serum at attomolar concentrations without the need for fluorescent or enzymatic labels. Additionally, the interferometric reflectance imaging sensor (IRIS) platform allows high-throughput multiplexing of protein–protein, protein–DNA, and DNA–DNA interactions without fluorescent labels, quantifying biomolecular mass accumulated on the surface.

Fiberoptic interferometric sensors are extensively used to measure temperature, humidity, and gas concentrations, providing reliable data even in extreme conditions such deep-sea exploration or high-temperature industrial processes. These sensors can detect environmental pollutants, including toxins and pathogens, ensuring environmental safety and public health.

Fiberoptic Fabry–Pérot interferometers are embedded into structures such as bridges, dams, or aircraft to continuously monitor strain, vibrations, and deformations. This enables early detection of structural weaknesses, preventing catastrophic failures.

Interferometric sensors are integral to techniques such as optical coherence tomography (OCT), a noninvasive imaging method that uses low-coherence interferometry to obtain high-resolution cross-sectional images of biological tissues. OCT has revolutionized ophthalmology, dermatology, and cardiology, offering clinicians real-time insights into tissue microstructures with micrometer resolution.

Wearable interferometric sensors, integrated into textiles or flexible substrates, are being developed to monitor physiological parameters such as blood pressure, respiration rate, and body movements continuously. These devices integrate optical readout with mobile applications for real-time health monitoring.

These advancements demonstrate the versatility and expanding utility of interferometric-based optical sensors in various scientific and medical fields.

3.13 Chapter summary

Optical sensors form a cornerstone of BioMEMS technology by enabling highly sensitive, rapid, and often label-free detection of biological and chemical events. SPR sensors detect biomolecular binding through refractive index changes, supporting real-time analysis in microfluidic environments. MS approaches such as MALDI-TOF and SELDI-TOF complement BioMEMS platforms by enabling rapid molecular identification and profiling. Raman and SERS provide vibrational fingerprints with single-molecule sensitivity using nanostructured surfaces. Fluorescence-based sensors—including FRET,

BiFC, and genetically encoded fluorescent proteins—allow visualization of conformational dynamics, protein interactions, and intracellular signaling with high spatial and temporal resolution. Interferometric optical sensors, particularly MZIs, offer exceptional sensitivity to refractive index changes, enabling precise monitoring of biomolecular interactions on photonic chips. Absorption-based and colorimetric sensors provide simple, low-cost detection methods suitable for portable diagnostics and POC devices, often relying on enzymatic reactions or dye–analyte interactions to generate measurable optical changes. Collectively, these optical sensing modalities greatly enhance the analytical capabilities of BioMEMS devices, enabling applications in medical diagnostics, environmental monitoring, drug discovery, and personalized healthcare through miniaturized, integrated, and highly sensitive biosensing platforms.

References

Amalina, Z. N., Khalid, M. F., Rahman, S. F., Ahmad, M. N., Ahmad Najib, M., Ismail, A., & Aziah, I. (2021). Nucleic acid-based lateral flow biosensor for salmonella typhi and salmonella paratyphi: A detection in stool samples of suspected carriers. *Diagnostics, 11*(4). doi:10.3390/diagnostics11040700.

Beeram, R., Vepa, K. R., & Soma, V. R. (2023). Recent trends in SERS-based plasmonic sensors for diagnostics, biomolecules detection, and machine learning techniques. *Biosensors*. doi:10.3390/bios13030328.

Chiu, N.-F., Kuo, C.-T., & Chen, C.-Y. (2019). High-affinity carboxyl-graphene oxide-based SPR aptasensor for the detection of hCG protein in clinical serum samples. *International Journal of Nanomedicine, 14*, 4833–4847. doi:10.2147/IJN.S208292.

Furniss, R. C. D., Dortet, L., Bolland, W., Drews, O., Sparbier, K., Bonnin, R. A., Filloux, A., Kostrzewa, M., Mavridou, D. A., & Larrouy-Maumus, G. (2019). Detection of colistin resistance in escherichia coli by use of the MALDI biotyper sirius Mass spectrometry system. *Journal of Clinical Microbiology, 57*(12). doi:10.1128/jcm.01427-19.

Habif, M., Corbat, A. A., Silberberg, M., Silberberg, M., & Grecco, H. E. (2021). CASPAM: A triple-modality biosensor for multiplexed imaging of caspase network activity. *ACS Sensors, 6*(7), 2642–2653. doi:10.1021/acssensors.1c00554.

Han, Z., Wen, L., & Feng, L. (2020). Screening of serum protein biomarkers in hemorrhagic cerebral infarction by surface-enhanced laser desorption/ionization time-of-flight mass spectrometry (SELDI-TOF-MS) technology. *Annals of Translational Medicine, 8*(18), 1186. doi:10.21037/atm-20-6071.

He, J., Zhu, S., Zhou, J., Jiang, W., Yin, L., Su, L., Zhang, X., Chen, Q., & Li, X. (2023). Rapid detection of SARS-CoV-2: The gradual boom of lateral flow immunoassay. *Frontiers in Bioengineering and Biotechnology, 10*, 1090281. doi:10.3389/fbioe.2022.1090281.

Hossain, M. I., Nanda, S. S., Cho, S., Lee, B, Kim, B. J., Choi, J. Y., & Yi, D. K. (2023). Gold nanorod density-dependent label-free bacteria sensing on a flake-like 3D graphene-based device by SERS. *Biosensors*. doi:10.3390/bios13110962.

Huang, T.-Y., Chiu, N.-F., & Lai, H.-C. (2013). Kinetic analysis of graphene oxide sheet and protein interactions using surface plasmon resonance biosensors. In *2013 Conference on Lasers and Electro-Optics Pacific Rim. Kyoto: Optica Publishing Group* https://opg.optica.org/abstract.cfm?URI=CLEOPR-2013-ThF3_6.

Ma, H., Li, Y., Wang, C., Li, Y., & Zhang, X. (2025). TiO2-based photocatalysts for removal of low-concentration nox contamination. *Catalysts*. doi:10.3390/catal15020103.

Mayerhöfer, T., Pahlow, S., & Popp, J. (2020). The bouguer-beer-lambert law: Shining light on the obscure. *ChemPhysChem, 21*. doi:10.1002/cphc.202000464.

Monteiro, J. P., de Oliveira, J. H., Radovanovic, E., Brolo, A. G., & Girotto, E. M. (2016). Microfluidic plasmonic biosensor for breast cancer antigen detection. *Plasmonics, 11*(1), 45–51. doi:10.1007/s11468-015-0016-1.

Nguyen, H. H., Park, J., Kang, S., & Kim, M. (2015). Surface plasmon resonance: A versatile technique for biosensor applications. *Sensors*. doi:10.3390/s150510481.

Nurrohman, D. T., & Chiu, N.-F. (2020). Surface plasmon resonance biosensor performance analysis on 2D material based on graphene and transition metal dichalcogenides. *ECS Journal of Solid State Science and Technology, 9*(11). doi:10.1149/2162-8777/abb419.

Vealan, K., Joseph, N., Alimat, S., Karumbati, A. S., & Thilakavathy, K. (2023). Lateral flow assay: A promising rapid point-of-care testing tool for infections and non-communicable diseases. *Asian Biomedicine, 17*, 250–266. doi:10.2478/abm-2023-0068.

Wang, Y., Wang, Y., Quan, S., Jiao, W., Li, J., Sun, L., Wang, Y., Qi, X., Wang, X., & Shen, A. (2019). Establishment and application of a multiple cross displacement amplification coupled with nanoparticle-based lateral flow biosensor assay for detection of Mycoplasma pneumoniae. *Frontiers in cellular and infection microbiology, 9*, 325. doi:10.3389/fcimb.2019.00325.

Wei, H., Leng, W., Song, J., Willner, M. R., Marr, L. C., Zhou, W., & Vikesland, P. J. (2018). Improved quantitative SERS enabled by surface plasmon enhanced elastic light scattering. *Analytical Chemistry American Chemical Society, 90*(5), 3227–3237. doi:10.1021/acs.analchem.7b04667.

Weis, C., Cuénod, A., Rieck, B., Dubuis, O., Graf, S., Lang, C., Oberle, M., Brackmann, M., Søgaard, K. K., Osthoff, M., Borgwardt, K., & Egli, A. (2022). Direct antimicrobial resistance prediction from clinical MALDI-TOF mass spectra using machine learning. *Nature Medicine, 28*(1), 164–174. doi:10.1038/s41591-021-01619-9.

Xia, G., Zhou, C., Jin, S., Huang, C., Xing, J., & Liu, Z. (2019). Sensitivity enhancement of two-dimensional materials based on genetic optimization in surface plasmon resonance. *Sensors*. doi:10.3390/s19051198.

Yang, F., & Ge, L. (2023). Colorimetric sensors: Methods and applications. *Sensors, 23*, 9887. doi:10.3390/s23249887.

Zhou, Z., & Long, Y. (2022). Dynamic monitoring of serum protein in acute respiratory distress syndrome based on artificial neural network. *Computational and Mathematical Methods in Medicine, 2022*, 3542942. doi:10.1155/2022/3542942.

CHAPTER

Micromechanical and microelectromechanical sensors

4

Ahmad Ghadafi Ismail[a], Nur Farhani Ismail[a], Hafeez Anwar[b], Anis Amirah Alim[a], Nur Nasyifa Mohd Maidin[a] and Abdul Basit[c]

[a]*Institute of Microengineering and Nanoelectronics (IMEN), Universiti Kebangsaan Malaysia, Bangi, Selangor, Malaysia,* [b]*Department of Physics, University of Agriculture Faisalabad, Faisalabad, Pakistan,* [c]*Harbin Institute of Technology, University in Harbin, Nangang, China*

4.1 Introduction

Advances in micromechanical sensors have revolutionized the way biological, environmental, and medical signals are detected and interpreted. These sensors, operating at the intersection of physics, engineering, and biotechnology, provide unparalleled sensitivity and precision by translating minute mechanical changes into measurable electrical signals. Their ability to detect molecular interactions, environmental fluctuations, and physiological parameters in real time has positioned them as indispensable tools in modern diagnostics, monitoring, and research.

Among the diverse classes of micromechanical sensors, mass-sensitive biosensors such as quartz crystal microbalance (QCM) and surface acoustic wave (SAW) devices have emerged as frontrunners. By exploiting frequency shifts caused by mass variations, these sensors enable label-free detection of biomolecules, pathogens, and contaminants with remarkable accuracy. Their applications span biomedical diagnostics, food safety, and environmental monitoring, underscoring their versatility and impact.

Beyond QCM and SAW, other sensor technologies include cantilever-based, magnetic, piezoelectric, heat-transfer, sound-based (MEMS microphones), ultrasonic, and pressure biosensors that extend the functional landscape of micromechanical sensing. Each class brings unique detection principles and application domains, collectively shaping a multidisciplinary toolkit for healthcare, biotechnology, and industrial innovation.

This chapter provides a comprehensive exploration of these advanced micromechanical sensors, highlighting their operating principles, technological advancements, and transformative applications.

Chapter Overview

To guide the reader through this diverse field, the chapter is organized into the following sections:

- **Section 4.2: Mass-Sensitive Biosensors (QCM and SAW)**
- Explores quartz crystal microbalance and SAW sensors in detail.
- Discusses their mechanisms of mass detection, fabrication techniques, and applications in biomedical diagnostics, food safety, and environmental monitoring.
- Highlights case studies demonstrating real-time, label-free detection capabilities.
- **Section 4.3: Cantilever-Based Biosensors**

BioMEMS Devices. DOI: https://doi.org/10.1016/B978-0-443-24015-7.00012-7

- Examines the use of nano-scale mechanical deflections for protein detection, drug discovery, and disease diagnosis.
- Reviews innovations in microfabrication and integration with optical/electrical readouts.
- **Section 4.4: Magnetic Biosensors**
- Focuses on sensors utilizing magnetic nanoparticles for ultrasensitive detection.
- Explains their advantages in reducing noise and enabling detection at nano-level concentrations.
- Considers applications in point-of-care diagnostics and molecular imaging.
- **Section 4.5: Piezoelectric Biosensors**
- Discusses piezoelectric materials and their role in ultrasound imaging and noninvasive health monitoring.
- Highlights their integration into wearable devices and medical imaging systems.
- **Section 4.6: Heat-Transfer Sensors**
- Explores sensors based on thermal conductivity and heat-transfer mechanisms.
- Reviews their innovative applications in wearable technologies, therapeutic monitoring, and metabolic analysis.
- **Section 4.7 and 4.8: Sound-Based and Ultrasonic**
- Surveys MEMS microphones, ultrasonic sensors, and pressure biosensors.
- Explains how acoustic signals are harnessed for healthcare diagnostics, medical imaging, etc.
- **Section 4.9: Pressure Biosensors**
- Explains types of pressure biosensors.
- Explores the applications in medical such as respiratory monitoring, bio-functional versatility etc.

4.2 Mass sensitive based biosensor: Quartz crystal microbalance and surface acoustic wave

In recent years, molecularly imprinted polymers (MIPs) have enabled selective extractions that rival immunoaffinity-based separations and offer clear advantages over real antibodies for sensor applications. MIPs are intrinsically stable, robust, and capable of operating in extreme environments, making them simple to fabricate. The current cost of developing biosensors for proteins is high due to the use of pricy antibodies. Nevertheless, novel MIP-based sensor strategies are being developed as MIPs continue to demonstrate potential as viable alternatives to natural receptors. Among various types of physical transducers (electrochemical, piezoelectric, etc.), mass-sensitive devices, including SAW, surface plasmon resonance (SPR), and QCM have demonstrated increasing popularity in sensing applications.

In general, a successful biosensor consists of two primary components: a biological receptor or sensor element, and a transducer (Morales & Halpern, 2018). A signal processing unit, typically equipped with a display or printer, usually operates alongside the biosensor, as shown in Fig. 4.1 (Tetyana et al., 2021). The most common classification of biosensors is based on the type of transduction element used in the sensor, such as mass-sensitive elements. Mass-sensitive sensors are devices that detect small mass changes by converting them into quantifiable signals, usually through variations in frequency or resonance. These mass-sensitive sensors play a crucial role in applications where minute mass variations need to be monitored, such as in biochemical analysis, environmental monitoring, and

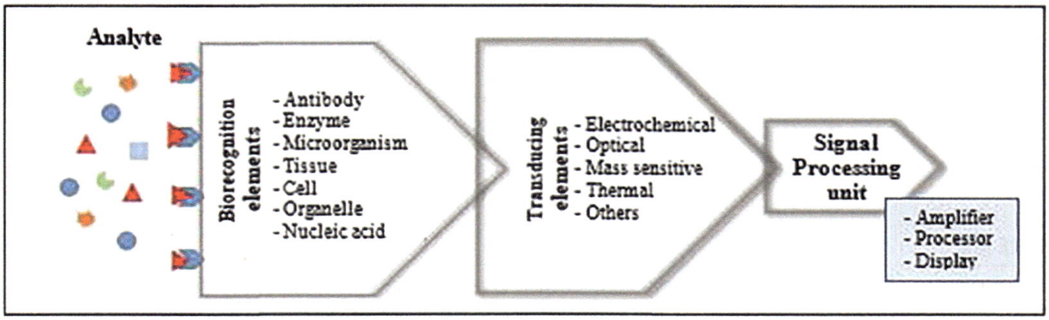

FIGURE 4.1

Schematic of biosensor components. Schematic illustration of biosensors on their transducing elements and biological recognition elements.

industrial processes. This chapter focuses on mass-sensitive biosensors, particularly QCM and SAW technologies. These types of biosensors have become essential tools in industries that require precise mass detection, including biomedical diagnostics, environmental monitoring, and industrial process control (Chen & Wang, 2020). Additionally, this chapter covers current issues, advantages, drawbacks, fabrication methods, operational mechanisms, and applications of these biosensors.

4.2.1 Quartz crystal microbalance biosensor

QCM based biosensors are classified under piezoelectric sensors and offer a sensitive alternative to traditional analytical methods by enabling real-time monitoring of mass changes from molecular binding events (Pohanka, 2021). QCM is one of the numerous acoustic transducers that is widely used and popular for sensor applications. It has been extensively utilized in chemical and biosensing fields due to its high sensitivity, robustness, compact size, and ease of integration with electronic measurement system (Mujahid, Afzal, & Dickert, 2019). However, to achieve effective selectivity and sensitivity, it is crucial to apply a sensing film to the QCM. Without coating materials, desired sensitivity and selectivity cannot be achieved. QCM biosensors are typically coated with a thin metallic layer that functions as an electrode on both surfaces of the crystal. When an alternating current is applied between the two electrodes, an acoustic wave propagates across the device and causes it to oscillate at a designated base frequency. The resonance frequency is contingent upon the physical parameters of the crystal and the medium. The alteration in the resonance frequency correlates with the mass deposited on the crystal as described by the Sauerbrey equation:

$$\Delta f = -\frac{2f_0^2}{A\sqrt{\rho_q \mu_q}}\Delta m$$

where Δf is the variation in resonance frequency, f_0 is the resonant frequency of the crystal, A is the active area of the sensors situated between the electrodes, ρ_q is the density of quartz, and μ_q is to the shear modulus of the quartz (Yılmaz et al., 2021).

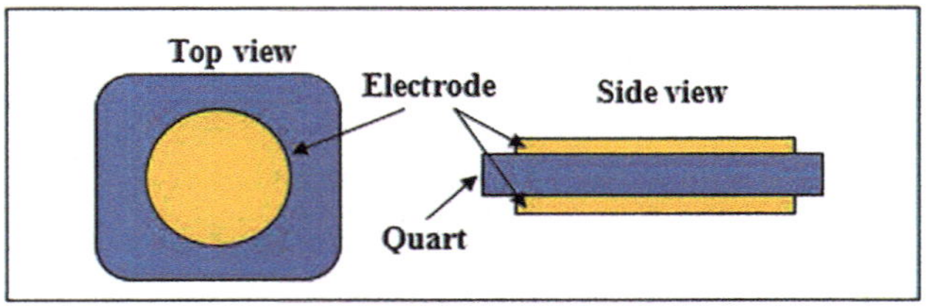

FIGURE 4.2

Basic Structure of a quartz crystal microbalance (QCM). This figure illustrates the key components of a quartz crystal microbalance (QCM).

4.2.1(a) *Principle mechanism of quartz crystal microbalance*

The working principle of a QCM biosensor relies on the thickness-shear mode of oscillation in a quartz crystal (Na Songkhla & Nakamoto, 2021). When a voltage is applied, the quartz crystal vibrates at its natural frequency due to its piezoelectric properties (Pohanka, 2021). This frequency of oscillation is highly sensitive to changes in mass on the crystal's surface, occurring when molecules or thin films adsorb onto it. The QCM operates by generating a mechanical oscillation with displacements perpendicular to the thickness of the crystal. When mass is deposited on the surface, it changes the mass loading, which in turn alters the oscillation frequency. This relationship is governed by the Sauerbrey equation by measuring this frequency shift, even very small changes in mass can be detected with high sensitivity, making QCM a valuable tool for applications requiring precise mass measurement at the molecular level (Alanazi et al., 2023). The details of the mechanism are explained in next section.

4.2.1(a)(i) Basic structure of quartz crystal microbalance

The physical structure of a QCM consists of a pair of electrodes attached to a piezoelectric crystal, typically made of quartz (SiO_2) (Zhao et al., 2022) as shown in Fig. 4.2. When an alternating electric field with a frequency close to the crystal's resonant frequency is applied to the electrodes, the crystal vibrates with intensity and stability, characterized by a high Q factor. The electrodes play a crucial role in energy trapping, contributing to the high Q factor that makes QCM an exceptionally stable resonator for various electronic applications, including main oscillators and timing circuits (Bonyadi et al., 2023). The QCM's characteristics are influenced by the crystal's vibration mode and the orientation of the quartz cut relative to its crystallographic axes (Zhao et al., 2022). The most commonly used cut, the AT-cut, is known for its temperature stability around room temperature (25°C) and operates in a thickness-shear mode, where the quartz displacement is perpendicular to its thickness. The AT-cut crystal is angled at approximately 35.15° to the crystal's z-axis, producing a frequency-temperature curve with an inflection point at 25°C. This provides a low temperature coefficient near room temperature. The fundamental frequency for AT-cut crystals typically ranges between 1 and 30 MHz. In addition to its fundamental frequency, QCM can also function in overtone modes, which are widely utilized in various applications. One of QCM's significant advantages is its high Q factor, which can reach exceptionally high levels, making it highly valuable for precision electronic uses.

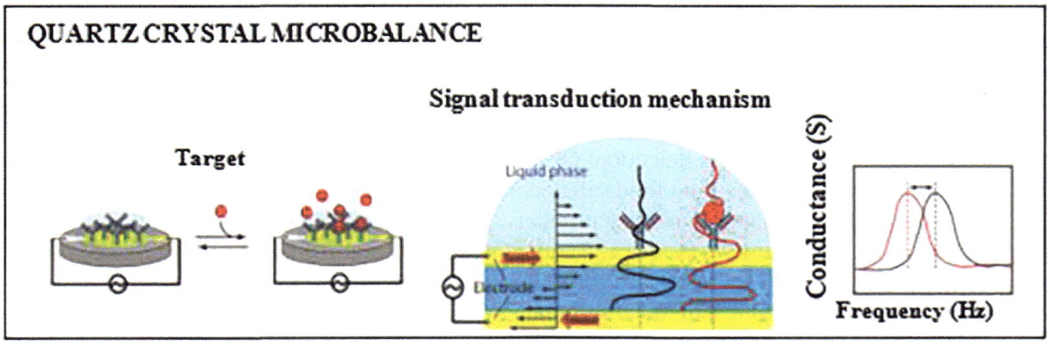

FIGURE 4.3

Working principle of the Quartz Crystal Microbalance (QCM) based on piezoelectric material. This figure illustrated the working principle of the Quartz Crystal Microbalance (QCM) using a piezoelectric material, such as quartz.

4.2.1(a)(ii) Piezoelectric effect of quartz crystal microbalance

The QCM sensor operates based on the piezoelectric effect (Pandya & Singh, 2022; Yan et al., 2012), as shown in Fig. 4.3. In certain crystals, mechanical strain leads to electrical polarization; when these materials are deformed, the displacement of atoms generates electrical dipoles. In crystals without inversion symmetry, the sum of these dipoles results in a net moment, creating the direct piezoelectric effect. Conversely, when an electric field is applied, the reverse piezoelectric effect occurs: the crystal experiences internal mechanical strain. This reverse effect underpins the operation of QCM. By applying an alternating voltage to the crystal's electrodes, the quartz crystal deforms and begins oscillating at its natural resonant frequency due to its piezoelectric properties (Zafirah et al., 2021).

4.2.1(b) ***Applications of quartz crystal microbalance as biosensor***

4.2.1(b)(i) In food industry

QCM biosensors have a wide range of applications in the food industry due to their sensitivity to mass changes, enabling real-time monitoring of food quality, safety, and other crucial parameters (Bonyadi et al., 2024). In food quality and safety monitoring, QCM biosensors detect foodborne pathogens such as *E. coli*, *Salmonella*, and *Listeria* by identifying specific biomolecules on food surfaces, thereby enhancing food safety and reducing the risk of foodborne illnesses (Ozalp et al., 2015). They are also used to detect spoilage indicators, such as volatile organic compounds (VOCs) released by decaying foods, which is particularly valuable for monitoring the freshness of perishable items such as meat, fish, and dairy (P J et al., 2021).

In allergen detection, QCM biosensors can monitor allergenic proteins from sources such as peanuts and gluten in food products, ensuring consumer safety by preventing accidental exposure to allergens. For quality control in food processing, these biosensors assess parameters such as moisture content, texture, and viscosity, which are essential for maintaining consistency in products such as dough, detecting spoilage in liquids, and optimizing textures in processed foods (Ye et al., 2019). QCM sensors are also valuable for monitoring aromatic compounds in foods and beverages, including coffee, tea, and alcoholic drinks, which supports quality control over flavors and aromas. Additionally, in packaging integrity and leak detection, QCM biosensors can detect gases like carbon dioxide and oxygen within modified atmosphere packaging (MAP) to extend shelf life, and they are used in tamper-evident packaging to detect breaches that compromise product safety (Lozano et al., 2018).

For contaminant and adulterant detection, QCM biosensors identify chemical contaminants such as pesticides and heavy metals, ensuring food safety compliance, and can detect adulteration, such as non-dairy fats in olive oil or melamine in dairy products, protecting consumers from fraudulent products. Finally, in smart packaging and shelf-life prediction, QCM sensors embedded in packaging can monitor biochemical changes to predict shelf life, offering consumers real-time freshness information, and can also track temperature and humidity to maintain optimal storage conditions from production to consumption (Kryuk et al., 2023)(Kryuk et al. 2023).

4.2.1(b)(ii) In environment

QCM biosensors have numerous applications in environmental monitoring due to their sensitivity and ability to provide real-time data on pollutants and ecological changes. In air quality monitoring, QCM biosensors detect airborne pollutants, including vVOCs, particulate matter, and gases such as nitrogen dioxide (NO_2), and sulfur dioxide (SO_2), which is crucial for tracking pollution levels and protecting public health (P J et al., 2021). In water quality analysis, QCM biosensors detect contaminants such as heavy metals, pesticides, and industrial chemicals in drinking water, wastewater and natural water bodies, offering essential insights for preserving aquatic ecosystems and ensuring safe human consumption. These biosensors are also effective in identifying biological contaminants, such as bacteria, viruses, and algae blooms in water and soil, which helps prevent waterborne diseases (Bragazzi et al., 2015).

Additionally, QCM sensors are valuable in detecting soil pollutants, including pesticides, herbicides, and heavy metals, aiding sustainable agricultural practices and environmental health by preventing toxic substance buildup (Justino, 2017). Furthermore, QCM biosensors can monitor environmental hormones and endocrine-disrupting chemicals (EDCs), such as phthalates and BPA, often found in water and soil due to industrial waste, supporting efforts to regulate pollution sources and protect ecological balance. QCM technology is also highly effective in detecting heavy metals like lead, arsenic, and cadmium in both soil and water at trace levels, enabling early intervention to prevent ecosystem degradation (Della Ventura et al., 2021).

Also, QCM biosensors support ecological monitoring and biodiversity assessment by tracking biochemical markers related to species presence and environmental stressors, which aids in biodiversity conservation and environmental research. These versatile applications make QCM biosensors invaluable for detecting and managing pollutants in air, water, and soil, ultimately safeguarding ecosystems and promoting human health.

4.2.1(b)(iii) In medical

QCM biosensors offer numerous applications in the medical field due to their high sensitivity and real-time analytical capabilities, playing a crucial role in diagnostics, monitoring, and research. In disease diagnosis, QCM biosensors detect biomarkers associated with conditions such as cancer, cardiovascular disease, and infections, enabling early detection and precise diagnosis. They are also highly effective in pathogen detection, identifying specific bacteria, viruses, and fungi in clinical samples to facilitate timely diagnosis and outbreak control. QCM biosensors play a key role in drug development and screening by analyzing drug–target interactions to assess efficacy and binding kinetics, expediting the identification of viable drug candidates (Akgonullu et al., 2022). Additionally, QCM technology supports antibody–antigen interaction studies, aiding the development of diagnostics and immunotherapies by characterizing binding properties crucial for targeted therapies.

In detecting hormones (e.g., insulin and cortisol) and small molecules (e.g., glucose, urea), QCM biosensors provide real-time monitoring for metabolic disorders, allowing personalized management of conditions such as diabetes and thyroid diseases. Also valuable for monitoring biomolecular interactions among proteins and nucleic acids, advancing biomedical research into cellular processes and disease mechanisms. QCM biosensors' adaptability makes them suitable for point-of-care testing, providing rapid, on-site diagnostics for biomarkers and pathogens in remote settings (Alanazi et al., 2023). Moreover, in therapeutic drug monitoring, QCM biosensors track drug levels in patients' blood to ensure safe and effective dosing, particularly for medications with narrow therapeutic ranges. These versatile applications underscore the role of QCM biosensors in improving diagnostics, advancing medical research, and personalizing patient care.

4.2.2 Surface acoustic wave biosensor

In the last 30 years, SAW devices have expanded beyond traditional uses in communications and signal processing. They are now widely applied across various fields, serving as identification tags, chemical and biosensors, and sensors for measuring physical quantities. SAW sensors are used in both consumer and professional systems, as they are passive (requiring no power supply) and can be accessed wirelessly, making them ideal for remote monitoring in harsh environments (Mandal et al., 2022). Operating within a frequency range of 10 MHz to several GHz, SAW sensors are characterized by ruggedness, compactness, stability, high sensitivity, low cost, fast real-time response, extremely small size, and light weight (Mujahid, 2017). Additionally, their fabrication is compatible with CMOS- and MEMS-integrated circuit technologies (Agostini & Cecchini, 2021).

SAW sensors are valuable for identifying moving objects and parts (ID tags) and for wireless measurement of temperature, pressure, stress, strain, torque, acceleration, tire-road friction, and humidity. They are especially well-suited for measuring pressure in car and truck tires, as well as monitoring tire-road friction. These features provide advantages over technologies such as capacitive and piezoresistive sensors, which require power and lack wireless capabilities. For example, a SAW pressure sensor weighing less than 1 g can be integrated into a car tire, where it measures pressure with a resolution of 5033.17 Pa. This setup enables drivers to monitor tire pressure from the cabin, enhancing safety, fuel efficiency, and tire lifespan. Such technology is particularly beneficial in the emerging run-flat (zero-pressure) tire market (Bruckner & Bardong, 2019).

SAWs were first discovered by Lord Rayleigh in 1885, and they are often referred to as Rayleigh waves (Ersoy & Waqar, 2023). A SAW is a type of mechanical wave that travels along a solid material's surface. Rayleigh demonstrated that SAWs could explain a previously mysterious component of seismic signals caused by earthquakes. With a typical acoustic wave velocity of about 3000 m/s, SAWs move much slower than electromagnetic waves. All acoustic wave devices and sensors rely on piezoelectric materials, such as quartz, to generate acoustic waves. Pierre and Paul-Jacques Curie discovered piezoelectricity in 1880, and Wilhelm Hankel named it in 1881. This phenomenon involves

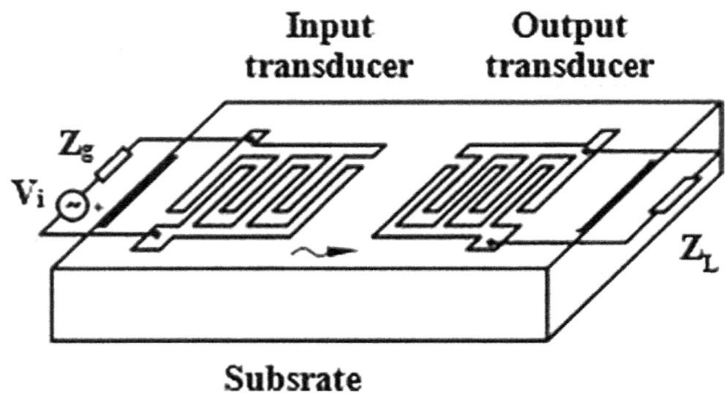

FIGURE 4.4

Basic structure of a Surface Acoustic Wave (SAW) device. This figure illustrates the basic structure of a Surface Acoustic Wave (SAW) device.

producing electric charges through mechanical stress, a process that is reversible: applying an electric field to a piezoelectric material induces mechanical stress. In piezoelectric acoustic wave sensors, an oscillating electric field generates a mechanical wave that propagates through the substrate, which is then converted back to an electric field for measurement.

The first SAW device was created in 1965 by White and Voltmer, who discovered how to launch a SAW in a piezoelectric substrate using an electrical signal. By 1970, SAW devices were being developed for radar pulse compression, oscillators, and bandpass filters in TVs and radios (2024). The 1980s saw a surge in SAW filter demand due to the rise of mobile radio and cellular telephones, leading to the annual production of approximately 3 billion SAW filters. In SAW biosensors, interdigital transducers (IDTs) generate and detect acoustic waves on the surface of a piezoelectric substrate (Xue et al., 2024). The acoustic energy in SAW devices is highly localized at the sensor surface, typically within one acoustic wavelength, making the effect independent of substrate thickness. One key advantage of SAW devices is their ability to operate across a wide frequency range, from MHz to GHz, allowing them to be used as resonators and filters in electronics and communication systems. These high-frequency devices offer improved sensitivity, enabling the detection of subtle surface changes, such as variations in accumulation, conductivity, and viscosity.

The development of SAW biosensors began in 1979 with Rayleigh-type SAWs, which were initially used for organic gas detection. Over time, SAW biosensors were adapted for liquid-phase sensing using horizontal shear SAWs (SH-SAWs). SH-SAW devices are highly effective for real-time liquid detection, making them ideal for biosensing applications. Today, SAW biosensors are increasingly utilized in medical diagnostics, particularly for detecting biomarkers linked to diseases such as cancer, chronic conditions, and infectious diseases. Their high sensitivity, versatility, and rapid response make them essential tools in biosensing and disease detection (Deng et al., 2023).

4.2.2(a) *Principle mechanism of the surface acoustic wave*

The fundamental process of SAW (Fig. 4.4) technology involves creating and propagating acoustic waves along the surface of a material, usually a piezoelectric crystal such as quartz. This process is widely used in sensors and electronic devices because of its sensitivity to surface conditions

(Anbalagan et al., 2024). SAWs are essentially electromechanical waves generated on the surface of a piezoelectric substrate. These waves travel along the crystal surface, guided by interdigitated electrodes (IDTs), which use electromechanical coupling between an input electrical signal and the piezoelectric substrate.

The design of IDTs can vary significantly based on the intended application and specific user requirements. IDTs are carefully designed to minimize insertion loss and enhance frequency responses. Certain parameters, such as electrode width, spacing, aperture, and the number of fingers, greatly influence the frequency bandwidth and electrical impedance of the SAW device. For SAW sensors, these IDTs act as transducers responsible for converting electrical energy to mechanical energy and vice versa. Each SAW device operates by propagating acoustic waves near the surface of a piezoelectric material, which confines most of the wave energy within a depth equal to one wavelength.

A basic SAW device includes two IDTs on a piezoelectric substrate, such as quartz. The input IDT generates the surface wave, while the output IDT receives it. The IDTs consist of interleaved electrodes made from a metal film on the piezoelectric substrate. The electrode width usually matches the gap between electrodes to maximize the conversion between electrical and mechanical signals. In industry, the smallest achievable electrode width is about 0.3 μm, allowing frequencies up to 3 GHz (Xue et al., 2024).

Common substrate crystals include quartz, lithium niobate, lithium tantalate, zinc oxide, and bismuth germanium oxide. These materials vary in piezoelectric coupling coefficients and temperature sensitivity (Lam, Wang, & Wang, 2010). ST quartz is often used for devices requiring high temperature stability. When a sinusoidal voltage of frequency, f is applied to the input IDT, it creates an electric field that causes a strain pattern with a periodicity of 2d, where dis the distance between electrode centers. If 2_d matches the surface wave wavelength, the surface wave propagates in both directions from the transducer. This wave induces a corresponding electric field at the output IDT, generating voltage at the impedance Z_L. The wave velocity, determined by the substrate material, typically ranges from 1500 to 4800 m/s (Bruckner & Bardong, 2019), which is much slower than electromagnetic waves. This slower velocity allows for the creation of small delay lines with significant delay.

SAW devices can function as either delay lines or resonators (Brukner & Bardong, 2019) (Fig. 4.5). Delay lines have uniformly spaced electrodes, providing a constant phase delay within a particular frequency range. Resonators, however, use IDTs to convert electrical signals to mechanical signals, but rely on wave reflections from metal strips or shallow grooves to create the desired amplitude and phase characteristics (Smagin et al., 2022). In both cases, the center or synchronous frequency of a SAW device is the frequency at which the output voltage is strongest, with output decaying as the frequency shifts away from this center frequency.

In summary, the flexibility of IDT geometry enables a wide variety of SAW devices. The key parameters in IDT design, such as electrode spacing and configuration, influence both the amplitude and phase characteristics of SAW devices, making them adaptable for various sensing and signal processing applications.

4.2.2(b) *Application of the surface acoustic wave*

4.2.2(b)(i) In food industry

SAW sensors are widely used in the food industry due to their high sensitivity, rapid response, and ability to detect small changes in mass and surface properties. Key applications include food quality

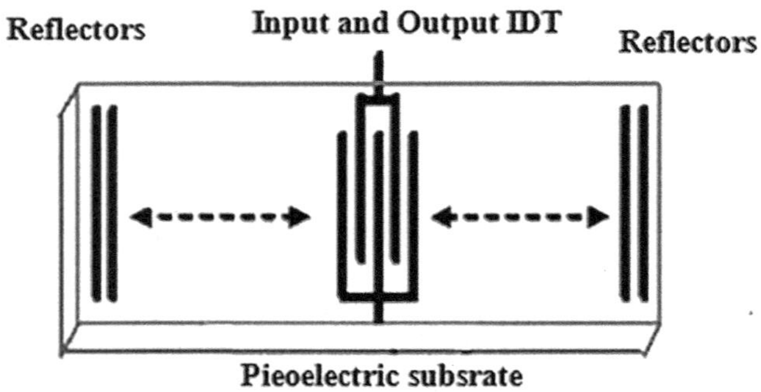

FIGURE 4.5

One-Port Surface Acoustic Wave (SAW) resonator. This illustration depicts a One-Port Surface Acoustic Wave (SAW) resonator, where a single input/output port is used for signal transmission.

and safety monitoring, quality control in food processing, packaging integrity, contaminant detection, and smart packaging solutions (Bruckner & Bardong, 2019).

SAW sensors play a vital role in enhancing food quality and safety by detecting pathogens, spoilage indicators, and allergens. For pathogen detection, SAW sensors can identify harmful bacteria, such as *E. coli, Salmonella*, and *Listeria*, by recognizing specific biomolecules on food surfaces. This capability is essential for ensuring food safety across various products. Additionally, SAW sensors can monitor food freshness in real time by detecting VOCs emitted by decaying foods, making them especially useful for perishable items such as meat, fish, and dairy. Beyond pathogens and spoilage indicators, SAW technology is also effective in allergen detection. By identifying allergenic proteins such as peanuts and gluten in food products, these sensors help protect consumers with allergies, further contributing to food safety and quality assurance (Ali et al., 2020).

In food processing, SAW sensors are crucial for maintaining quality through precise monitoring of moisture, texture, and flavor. By assessing viscoelastic properties, such as moisture content, texture, and viscosity, SAW sensors help ensure consistency in products such as dough, detect spoilage in liquids, and optimize texture in various processed foods. SAW technology is also instrumental in flavor and aroma analysis, as it enables the detection and quantification of aromatic compounds in foods and beverages. This capability is essential for maintaining the desired quality in products like coffee, tea, and alcoholic drinks (Ma et al., 2023).

SAW sensors enhance packaging integrity by detecting gas leaks and tampering. In MAP, SAW sensors can monitor gases such as CO_2 and oxygen, which helps extend the shelf life of packaged products by ensuring that optimal gas levels are maintained. Additionally, these sensors play a vital role in tamper-evident packaging by identifying breaches or leaks, which safeguards the integrity SAW technology is highly effective in detecting contaminants and adulterants in food. SAW sensors identify harmful chemical contaminants, such as pesticides, heavy metals, and other residues, ensuring that food products comply with safety standards. They can also detect food adulteration by recognizing unique molecular signatures. For example, they can identify melamine in dairy products or detect non-dairy fats in olive oil, which is essential for food authenticity and safety. and safety of products during transport and storage (Ma et al., 2023).

SAW sensors are integral to smart packaging solutions that offer real-time shelf-life monitoring. Embedded in packaging, these sensors track biochemical changes to predict shelf life, providing consumers with up-to-date information on food freshness. Additionally, SAW sensors monitor temperature and humidity, helping maintain optimal storage conditions. This ensures that food products retain their quality from production through to consumption (Mustafa, 2018).

4.2.2(b)(ii) In environment

SAW biosensors are increasingly valuable in environmental monitoring due to their high sensitivity, rapid response times, and ability to detect a wide range of biological and chemical pollutants (Justino, 2017). In detecting environmental pollutants, SAW biosensors can identify heavy metals, such as lead, mercury, and cadmium, in water sources by targeting specific ions or molecules, which is crucial for maintaining water quality and protecting public health. They also detect pesticide residues in soil and water, allowing for effective monitoring of agricultural runoff and preventing contamination of natural ecosystems. SAW biosensors are effective in identifying toxic gases like carbon monoxide, nitrogen dioxide, and sulfur dioxide in the air, which is essential for air quality monitoring and reducing exposure to hazardous pollutants (Hashemi Goradel et al., 2018).

For pathogen and microorganism detection, SAW biosensors can identify waterborne pathogens such as *E. coli, Salmonella*, and *Legionella*, which helps prevent waterborne diseases. These sensors also monitor biotoxins from harmful microorganisms, like algae blooms in freshwater and marine environments, which is vital for protecting both ecological and public health. SAW technology also supports environmental health and safety monitoring by detecting hazardous chemicals released by industrial processes. This enables real-time wastewater and emission monitoring to ensure compliance with environmental regulations. Additionally, SAW sensors help assess soil health by identifying specific microbial activity and contaminants, aiding in soil remediation and sustainable land management (Kumar et al., 2018).

In terms of volatile organic compound (VOC) detection, SAW biosensors monitor air quality by identifying VOCs such as benzene, toluene, and formaldehyde, which are common pollutants from industrial emissions, vehicle exhaust, and household products. This monitoring is essential for maintaining indoor and outdoor air quality, as well as providing an early warning system for environmental hazards near waste sites. SAW biosensors also contribute to climate and ecosystem monitoring by tracking changes in environmental factors such as temperature, humidity, and pH, which signal shifts in biodiversity or potential climate impacts. They detect greenhouse gases, such as methane and carbon dioxide, supporting efforts to mitigate global warming.

In agriculture and forestry, SAW biosensors monitor soil nutrients and fertilizer levels, helping to optimize agricultural practices and minimize environmental impacts from excessive fertilizer use. They also identify plant pathogens affecting crops and forests, enabling early interventions to prevent crop loss and forest damage. These diverse applications underscore the vital role SAW biosensors play in safeguarding the environment, public health, and sustainable practices (Mandal & Banerjee, 2022).

4.2.2(b)(iii) In medical

SAW biosensors have various medical applications due to their high sensitivity, rapid response, and ability to detect multiple biological molecules. For disease biomarker detection, SAW biosensors can identify cancer biomarkers such as prostate-specific antigen (PSA) and carcinoembryonic antigen (CEA). Detecting these biomarkers early is critical for cancer diagnosis and treatment. They are also useful for finding cardiovascular disease markers like troponin, which indicates heart attacks. Additionally, SAW sensors detect viral and bacterial biomarkers for infectious diseases such as HIV, hepatitis, and influenza, supporting early diagnosis and effective monitoring (2021).

SAW biosensors are effective in pathogen detection. They identify bacterial pathogens such as *Escherichia coli, Salmonella*, and *Staphylococcus aureus*, common sources of healthcare-associated infections. This is essential for quick infection management in clinical settings. SAW biosensors can also detect viral particles, like SARS-CoV-2, the virus causing COVID-19, through specific antigen recognition. This supports real-time virus detection, which is crucial for outbreak control (Castillo-Henríquez et al., 2020).

In managing chronic conditions, SAW biosensors help monitor glucose levels for diabetes patients, providing a less invasive option than traditional blood testing. They can also detect specific metabolites and proteins, enabling real-time monitoring of kidney and liver function. This helps with managing chronic kidney disease and liver cirrhosis. In drug development, SAW biosensors screen interactions between drug candidates and biological targets, aiding in the discovery of effective compounds. SAW sensors also monitor therapeutic drug levels in blood or plasma, particularly for medications with narrow therapeutic ranges, such as antibiotics and immune suppressants (Kukkar et al., 2022).

SAW biosensors play a role in hormone and protein monitoring. They can detect hormones such as cortisol and thyroid hormones, which supports the management of endocrine disorders such as hyperthyroidism. They also help monitor protein levels in autoimmune and neurodegenerative diseases, aiding in disease management. For point-of-care diagnostics, SAW biosensors are ideal for portable devices that allow quick testing at a patient's bedside or in remote locations. They can also be embedded in wearable monitors to track vital signs and biochemical markers continuously, supporting personalized healthcare (2021).

SAW biosensors are valuable in immunoassays and molecular diagnostics. They are sensitive to antigen–antibody interactions, making them useful for detecting antibodies or antigens in diseases, allergies, and autoimmune disorders. In genetic testing, SAW biosensors detect specific DNA or RNA sequences, enabling early diagnosis of genetic disorders and infectious diseases by recognizing unique genetic markers. These applications highlight the role of SAW biosensors in advancing medical diagnostics and personalized treatment (Sharma et al., 2021).

4.2.3 Conclusion

Mass-based biosensors are powerful platforms for monitoring molecular interactions, and they are often designed as label-free sensors. This means that the biosensor directly detects the interaction between the sensor surface and the analyte, without the need for specific reagents or complex sample processing. One of the most commonly used materials in these sensors is quartz, which possesses piezoelectric properties. Quartz crystals generate an electrical charge when mechanically deformed or compressed, due to the alignment of dipoles on the crystal surface. This piezoelectric effect makes quartz-based biosensors highly mass-sensitive, capable of detecting minute changes in mass on the sensor's surface, often down to the nanogram level. The QCM has become one of the most widely used mass-sensitive biosensor platforms, particularly in biosensing applications. These sensors are extremely sensitive to changes in mass on the crystal surface and are particularly effective for detecting target analytes with high precision. This sensitivity can be further enhanced by incorporating specialized

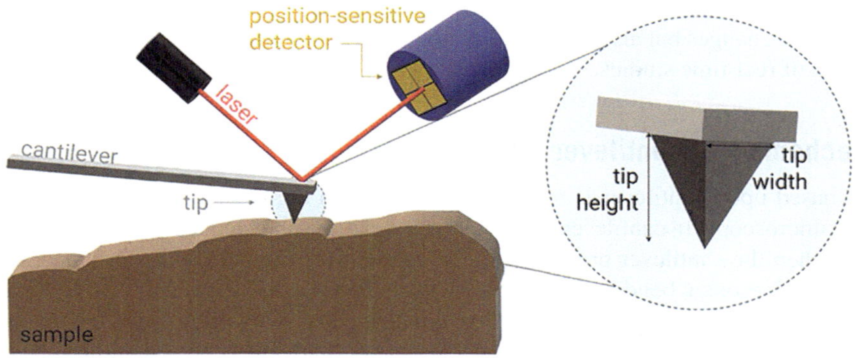

FIGURE 4.6

Cantilever's working sketch. This figure illustrates the working principle of a cantilever biosensor.

nanomaterials, which improve the sensor's performance and detection capabilities. The advancement of smart biosensors has led to increased detection sensitivity, which is critical for monitoring both known and emerging threats. Mass-sensitive biosensors are now used for the real-time detection and continuous monitoring of a variety of molecules, making them invaluable tools in various fields, including environmental monitoring, food safety, and medical diagnostics. The development of portable, reliable biosensors that can simultaneously and continuously monitor multiple analytes has gained significant attention, particularly in the wake of the COVID-19 pandemic. This shift toward more versatile, easy-to-use biosensors is a key area of research aimed at improving rapid detection capabilities and addressing global health challenges.

4.3 Cantilever biosensors

Biosensors usually operate due to the interaction of a biomolecular material with a mechanical instrument and that results in a mechanical motion of order of nano level, the changes occurring during the motion are minute and can be observed in real time measurements. Cantilevers can be categorized as the "mechanical instrument" responsible for demonstrating motion in contrast with interacting with the biomolecular materials thus employing biosensing tendencies. A cantilever is a flexible beam sort of structure that has one open end having a calibrated weight and one end fixed to a specific point, during its functionality, the cantilever beam subjects to bending and then restoration due to elastic restoring forces (Fig. 4.6). This anchored assembly may bend and restore back entirely depending upon the shear force it experiences while interacting with material under observation, provided that the material may be biological as well thus, giving applicability in terms of biosensing. Cantilevers can be incorporated into various beneficial assemblies as compared to other biosensing equipment, such as biochips. These sensors can detect surface stress which may be induced due to the deflection caused by molecular arrangements and can label the data in real time environment. As compared to other infrastructures, cantilever sensors are uniquely change sensitive to functionalized surfaces and that provides an observable mechanical variation. This assembly is quite capable of protein detections, the discovery of drugs and diagnosis applications (Kambhampati & Fritz, 2004). Not only these devices measure sensitive changes but also can be calibrated to shift their sensitivity within a range of 10^{-15} and 10^{-18} in terms of real-time studies.

4.3.1 Mechanism of cantilever biosensors

Biosensing based upon cantilever is mostly resembled with the mechanism employed in working of atomic force microscopy. In cantilever biosensors, the beam is subjected to interact with a sample placed underneath, when the cantilever probe (tip) interacts with the biomolecular substance, there occurs a bending trend and resonant bending as a response, this effect is caused by the force the beam experiences while interaction with the material. Although the bending capabilities entirely depend upon factors such as the "spring constant" of beam material and resonance frequency of the cantilever's up-and-down motion, these factors may be tailored via tempering with the cantilevers shape and the substance with which it is made of, (Eqs. 4.1) and (4.2) are relations for frequency "f_o" (resonant) and spring constant "k," respectively.

$$\text{fo} = \frac{1}{2\pi}\sqrt{L/m} \tag{4.1}$$

$$k = \frac{\text{E}wh3}{4L4} \tag{4.2}$$

where L = length of cantilever beam

m = mass of cantilever beam
h = thickness of beam
w = beam width
E = Young's Modulus

These parameters are responsible for effecting the operation of cantilever in their mode of static and dynamic operation, but in other modes of operation, there are some factors such as heat, stress, and pressure, which also account for the operational limits of cantilever biosensing. Cantilever's surface is functionalized with application-specific materials and then it is subjected to interact with a controlled motion by tweaking resonance frequency, the most common operation modes of cantilever sensing are divided into two categories as following.

4.3.1(a) *Dynamic mode of operation*

This mode of operation is not complicated in terms of controlled functionalization, the cantilever surface is functionalized on both sides, depending upon this arrangement there occurs a shift in resonant frequency caused by an increment in cantilever's mass, which eventually renders spring constant value as well. The variation in spring constant can lead to observation of even highly sensitive motion changes, the system assembly can be characterized via the energy dissipated and accumulated during vibratory motion of cantilever, although the energy measurement values can be used to carry out the working quality of sensor in this specific mode of operation. This mode is also capable to withstand various variation in quality whenever it is operated in liquids and air, but due to complexity involved in order to enhance the capability and performance of the cantilever assembly, dynamic mode is mostly not preferred over static mode of operation.

4.3.1(b) Static mode of operation

Cantilever operating in static mode has a specific mechanism involved, in this interaction mode the bending of cantilever is caused by the interaction between a specific surface point of cantilever and molecular sample's surface. As a consequence, there occurs a change in stress at surface, thus the cantilever deforms and is entirely dependent upon the forces induced during bio-reactive process occurring in molecular material. The forces produce due to two major responses, they are divided into two types. (1) When the surface of beam interacts with the forces generated by the interaction of neighboring molecules of biological sample, the response is denoted as non-lineal response and (2) whereas when cantilever interacts with forces caused by the bonds of sample, the phenomenon is quoted as lineal response. Nonlinear responses cause abrupt changes, which depend upon plethora of factors such as electrostatic interactions, Van der Waal forces, and surface absorbed impurities or molecules (Hagan et al., 2002). Static mode is advantageous to biological sensing as it only requires one portion of cantilever to be biofunctionalized, but at the same time, it is a task to be carried out with extreme complexity. This mode is unique in a sense that it is entirely dependent upon the surface stress changes caused at a specific portion of beam lever.

4.3.2 Fabrication of cantilever biosensors

The fabrication of cantilever judges a lot of factors regarding its performance, although there are some important parameters which are to be calibrated and keep in account for a balanced and stable sensing capabilities of cantilever-based sensing. The material that is to be incorporated for the fabrication of cantilever beam holds an important role as it judges the flexibility, stress, and the mechanical stress to be withstand by the beam apparatus. Usually polymeric materials, silicon and their nitrides are employed for beam fabrication, the choice of material is entirely subjective to the application at hand and thus to be chosen by researchers for corresponding desired task. While fabrication the size also plays a vital role, cantilever's resonant frequency is entirely dependent upon factors such as thickness, Young's modulus, length, and width as explained in Eq. 4.1. Resonant frequency judges the performance of cantilever sensing, supported by the fact that; high resonant frequency relates to high level of accuracy and sensitive motions (Boisen & Thundat, 2009). Usually, cantilever's can be fabricated via various techniques such as pattering via photolithography, dry etching using KOH. Kirstein et al. (2005) utilized a unique CMOS approach to synthesize a silicon-based cantilever for biosensing applications. DNA detection and various protein-based studies were carried out as well.

4.3.3 Detection mechanism using cantilever biosensors

Biological sample sensing can be generalized in an all-in-one (AIO) mechanism, a cantilever is usually placed in a clean environment followed by the deposition of a specific material, the specific material is labeled as "receptor material," its primary function is to formulate a binding tendency between cantilever surface and the surface of biological sample. This phenomenon is called "functionalization," and it varies with respect to mode of operation (static and dynamic). When the cantilever is brought close to

Table 4.1 Detection mechanisms for cantilever readouts.

Method	Accuracy	Feature	Advantage
Interferometry	0.1 Å	In liquid vicinity, sensitive to vibrations.	Can detect displacement variation even below Å range.
Capacitive	Average	Vibration sensitive in liquids.	Provides low noise in output.
Torsional	High	Exhibits and tunable quality factor.	Can detect highly sensitive changes.
Optical Beam Deflection	nm range	Complicated assembly dependent.	Can detect up to sub-nano level variations.
Piezoresistive	Almost 1 nm	Quite a simple assembly.	Can be easily connected to electronic devices.
Optical Diffraction	High	Quite simple setup.	It is precise than many other techniques.
MOSFET	High	Way more complicated than Piezoresistive assembly.	Provides uninterrupted output with less noise.

the surface of sample under observation. There occurs molecular interaction caused by the forces acting in between cantilever tip and sample, which leads to variation in energy values presented at cantilever's surface. The binding of receptor with sample is responsible to induce bending in cantilever's beam, which leads to a resonant frequency to become available. The resonance frequency allows the user to observe sensitive motions caused due to interaction among beam and sample. The sample placed under the action of cantilever can be moved via a piezoelectric crystal based assembly or other types of transducers for similar purpose, whereas a laser diode is present to focus a laser on cantilever, which directly measures the position of cantilever and thus produces a computational sketch, various detection techniques have been written as following (see Table 4.1) (Basu et al., 2020).

4.3.4 Applications of cantilever biosensors

Biosensing based upon mechanical cantilever apparatuses has been remarkable in plethora of applications. Wu et al. (2001) utilized a sensor based upon cantilever arrangement in order to analyze prostate-specific antigen (PSA) for biological studies. The instrument employed had been marked as a crucial prostate cancer detection, the apparatus successfully detected the traces of targeted disease via binding with PSA and producing a deflection pattern with high level sensitivity and accuracy. The sensory equipment was reported to detect an accurate value of 0.2 ng mL^{-1} despite facing interfering proteins through the sample batch, the result produced proven to be efficient, economical as compared to costly enzyme linked immunosorbent assay (ELISA) tests, thus making it an efficient study for biosensing applications. Okahata et al. (1998) employed quartz crystal-based microbalance assembly to study the kinetic pattern of hybridization of DNA. The sensor used a frequency input of 27 MHz for cantilever to analyze oligonucleotides in order to study the involved DNA binding. The biosensing studied a 0.6 $ngcm^2$ variation in deflection for each 1 MHz frequency input, showcasing a sensitive nature. Parameters such as ionic strength, mismatches of sequences, and temperature had been observed for their influence on kinetic of DNA hybridization, the proposed study offered a clear and reliable result as compared to typical complicated surface plasmon resonant techniques. Li et al. (2019) researched the efficacy of their modified microcantilever design to detect Mucin-1 (a biomarker most primarily used to indicate breast cancers in female). Mucin-1 traces were linked via employment of thiolated aptamers as receptor for surface functionalization of cantilever, a measurement of stress at surface had been recording in contrast with deflection pattern. An excellent study of 0.9 nM detection of Mucin-1 had been recorded, whereas the apparatus successfully detected 213 cells per mililitter for Michigan Cancer Foundation-7 (a cancer line for breast). The efficacy of instrument has been proven unique and precise based upon the accuracy achieved for the detection of cells thus, cancer-less cells were successfully indicated from cancerous ones. Etayash et al. (2015) functionalized the cantilever tip using peptides (18-4) in order to strongly bind with cells of breast cancer. Circulating cells of tumor had been successfully leveraged via their technique by the measurement of deflection pattern produced due to cell-surface adhesion. An accurate study of 50 to 100 cancer cells per mL was achieved resulting in efficient implementation of the instrument thus a cantilever based biosensing had successfully been reported. Fritz et al. (2000) reported accurate applicability of cantilever-based biosensors by functionalizing the surface via proteins and DNA samples. The stress-induced motion pattern resulted in detection of ligands and protein samples with quite an accuracy.

4.4 Magnetic biosensors

Nano level materials show extraordinary capabilities as compared to their bulk counterpart; researchers have been harnessing their unique characteristics in their favor. A material at bulk scale does not show any unique properties; to support this argument, gold is reactive at nano level but unreactive at bulk scale, this unique variation is due to availability of corner atom sites. Nano materials also show a class, which is magnetic in nature, labeled as magnetic nanoparticles (MNPs). Biosensing makes full use of magnetic behavior of these MNPs, a magnetic biosensor is class of micromechanical sensors, which uses magnetic field variations as observation tool for the understanding of interaction with a biological sample. Magnetic biosensors are unique in a way that their performance cannot be hindered via the intrusion of optical activities of biological materials and thus this leads to an output data with reduced noise as well.

4.4.1 Principle of magnetic biosensors

The working of a magnetic biological sensors can be explained in terms of assembly, generally the biological sample (analyte), which is to be studied is first made bonded with MNPs followed by the detection of magnetic field proposed by those MNPs. When the MNPs bond with analyte, an external magnetic field is applied to the system which causes variations to be produced, whereas a close by sensor is capable of detecting any variation produced in magneto-resistance. This term is directly in relation with the amount of effective MNPs bound to analyte, a noticed variation by the sensor gets analyzed and converted to a real time readable output signal.

A schematic diagram of a magnetic sensor has been shown in Fig. 4.7 At the base, there is a suitable substrate, and it is such that it does not provide any disturbance to the operation of sensor by unwantedly getting magnetized, typically silicon and similar semiconducting materials are employed

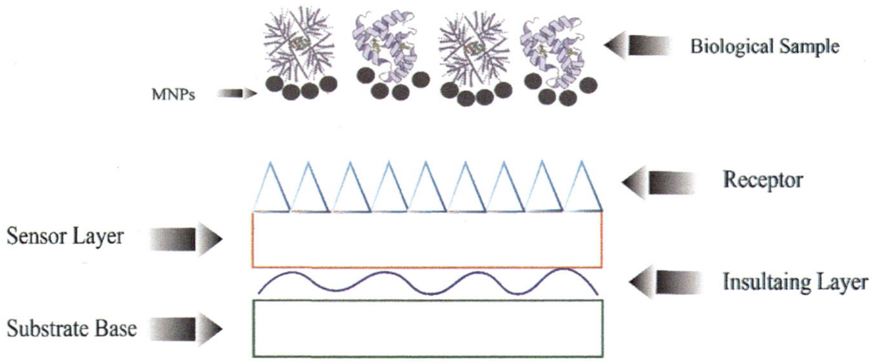

FIGURE 4.7

Schematic of magnetic biosensor. This figure illustrates the schematic for a working principle of a magnetic biosensor.

for base substrate, the substrate layer is also connected with contacts and directly wired to computational assembly for an output signal. Above the substrate there is deposited a material, which acts as insulation, this specific platform is pivotal for magneto-resistive variation and is often connected with magneto-resistive sensor, which is by far the most important part of the whole infrastructure. The assembly is then coated with suitable biological material atoms also quoted "biological receptor", in order to functionalize the sensor's surface and enable capture of under observation biological samples. The sample to be studied is prepared in advanced by incorporating MNPs over its surface which later allows the sensor to be fully operational. An apparatus capable of providing external magnetic field is used to magnetize the MNPs after they had successfully been bonded with the analyte (Buckner et al., 2016).

4.4.2 Detection mechanism for magnetic biosensing

The detection mechanism to obtain an output signal capable of being interpreted quite depends upon transduction phenomenon and type employed. Whenever a magnetic biosensor is subjected to study biological material a transducer is a core component, which converts that crucial interaction occurring in between the receptor and sample after it had been bonded with MNPs, into an electric signal (in some cases optical as well). The interaction is often named as "molecular event," during this the transducer translates all molecular interaction into a sort of signal pulse by analyzing the chemical activity. Types of transducers have been reported and their mechanisms have been explained as well:

- **Optical**: While an optical transducer is used in biosensing assembly, the MNPs are used to enhance the faint signal of luminescence (absorbance) obtained when binding of these MNPs occur with biological sample under study. The transducer detects this change in wavelength and reads out an output signal relating to nature of analyte.
- **Piezoelectric:** When MNPs bind with analyte, there can be observed a change in mass caused due to this specific binding and thus varies the resonant frequency of sample molecules, which is thus recorded by the transducer and an output file is obtained depending upon the shift in frequency.

Table 4.2 Transducer principle.

Sr #	Principle of Transduction	Biosensing Limit	MNPs Used	Reference
1	Electrochemical	1) 800 nm 2) 0.13 μm 3) No limit	1) $Fe_3O_4@SiO_2$/MWCNTs 2) Core–shell $Fe_3O_4@SiO_2$/ MWCNTs 3) Fe_3O_4 anchored on rGO	Arvand and Hassannezhad, (2014); Prakash et al. (2012)
2	Magnetic Field	1) 592.8 pm Oe^{-1} 2) 83 fM 3) 1.3×10^6 cells	1) NPs of Fe_3O_4 2) Cubic FeCo NPs 3) Carboxyl functionalized Fe_3O_4 NPs	Zu et al. (2014); Srinivasan et al. (2011); Hathaway et al. (2011)
3	Optical	0.65 ng mL^{-1}	$Fe_3O_4@Au$ NPs	Liang et al. (2012)
4	Piezoelectric	0.00128 HA units	Fe_3O_4 nanobeads	Li et al. (2011)

- **Magnetic field based**: Accumulation of MNPs can take place on the surface of analytes and that leads to a net variation in magnetic field which is detected via the transducer.
- **Electrochemical based**: Analyte atoms can experience electron transfers when MNPs are bound to them and that leads to an electrochemical variation, which can be observed to record an output signal to be interpreted.

These types of transducers each hold their uses in specific conditions and their sensitivity depends upon controllable conditions such as pressure and temperature. Different approaches of these transducer principles have been recorded and formulated into Table 4.2, which gives understanding of their effective efficiency of sensor as well.

4.4.3 Advantages of magnetic biosensors

MNPs play a vital role for the efficacy of magnetic biosensing applications, the strategy employed, and selection of transducers alongside careful tweaks can lead plethora of advantages of magnetic structural biosensors, it is summarized in Table 4.3.

4.4.4 Magnetic biosensing applications

A variety of studies have been carried out to support the applications of biosensing achieved via magnetic approach of materials. Ravalli and Marrazza (2015) synthesized an electrochemical sensor based upon MNPs in order to diagnose cancer related observations (Ravalli & Marrazza, 2015). Iron salts have been co-precipitated to fabricate magnetic material for the surface functionalization of sensor. The receptor successfully bound with the antibodies of sample under observation, the synthesized material showcased efficient biomarking tendencies for cancer cells. The proposed study was based upon measurement of voltage variation corresponding to electrochemical counterpart of magnetic sensor assembly. PSA detection limit of 0.5 pg per mL had been successfully recorder showcasing a sensitive efficacy of

Table 4.3 Features of magnetic biosensing.

Sr #	Feature	Description
1)	SNR	SNR refers to as "Noise to signal" ratio, magnetic structure accounts for a reduced background noise while interacting with complicated assembly of biological samples.
2)	LOD	Low limit of detection (LOD) is achieved in biosensing based upon magnetics. Magnetic field allows the analyte to concentrate strongly at sensor's surface and thus a sensitivity of even atto-molar units maybe achieved.
3)	Real time output	Magnetic sensing allows the real time monitoring of output signals and they can be instantly interpreted as well.
4)	No modification needed	Unlike other biomarkers and chips, magnetic sensing is independent of any modification needed to either enhance or operate the apparatus. A pre-set assembly will suffice.
5)	Harmless	The technique employed is quite nondestructive in nature as it entirely depends upon the magnetic interactions.
6)	Multitasking Tendency	The apparatus can be used to study multiple and yet complicated arrays of samples.

instrument. Akbari Nakhjavani et al. (2018) reported a highly sensitive electrochemical biosensor based upon magnetic bead layer for the detection of CA 15-3 breast cancer cells (Akbari Nakhjavani et al., 2018). An electrode based upon gold particles had been employed with streptavidin to cease biotinylated antibodies of monoclonal CA 15-3. Magnetic beads were used to signal amplification of electrochemical measurements which were carried out during the interaction phase. The sensor demonstrated a detection accurate limit of 15×10^{-6} U per mL thus, recording high sensitivity despite being in a human based sample under observation. The research produced a rapid method to diagnose and detect breast cells of cancer. Cai et al. (2021) constructed a useful magnetic sensor for biological studies and diagnosis, lateral flow immunoassay (LIFA) incorporated on gold MNPs was employed for PSA (Cai et al., 2021). A magnetic detector was used to study the quantitative analysis of cell detection, time interval of 15 min was noted. The method used recorded a detection value of 0.17 ng per mL which resulted in further detection of PSA cell concentration. The proposed LIFA approach resulted in economical and highly accurate cancer cell detection as well thus early indication maybe noted to prevent serious disasters. Koo et al. (2021) explored a study involving a magnetic sensor coupled with electrochemical behavior for variety of cancerous biomarkers for early detection (Koo et al., 2021). A novel nano material-based sensor for biological capabilities was fabricated, MNPs were loaded to enhance the detection sensitivity of the sensor, the sensor operated on the principle of separation of biomarkers from a complicated blood sample (Ghosh, 2020). A detection limit of femto to pico molar had been recorded, which resulted in real-time sensitive cancer detection studies.

4.5 Piezoelectric biosensors

Piezoelectric-based sensors are devices that utilize the piezoelectric effect to convert mechanical energy into electrical signals. These sensors convert mechanical signals such as pressure, vibration, force, or

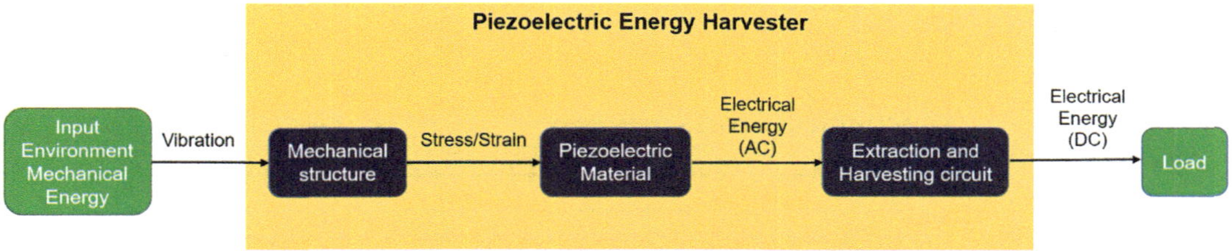

FIGURE 4.8

Structure of a vibration energy harvesting system from source to end user. This figure illustrates the structure of a vibration energy harvesting system, showing the flow of energy from the source to the end user.

acceleration into electrical signals, enabling precise measurements at the micro-scale. Their small size, high sensitivity, and low-power consumption make them ideal for many modern applications (Zhu et al., 2021).

Piezoelectric micromechanical sensors operate through a three-step process that converts mechanical energy into electrical signals. First, a mechanical input, such as vibration, pressure, or strain, is applied to the piezoelectric material, causing it to deform. This deformation triggers the piezoelectric effect, where the material generates electrical charges on its surface in response to the mechanical stress. Finally, the resulting voltage is measured and interpreted, providing a quantifiable electrical signal that corresponds to the magnitude and nature of the mechanical stimulus. This process enables precise monitoring and measurement in various applications. Fig. 4.8 shows the structure of the vibration energy harvesting system from the source to the end user.

4.5.1 Theory and mechanism of piezoelectric biosensors

The piezoelectric effect was initially identified in 1880 by French physicists Pierre and Jacques Curie. They explored the characteristics of different crystals, such as quartz and Rochelle salt, and showed that mechanical pressure could produce electrical charges on their surfaces. The term "piezoelectric" comes from the Greek word "piezein," meaning "to press" or "to squeeze," a term introduced in the late 19th century by Wilhelm Gottlieb Hankel (Mason, 1981). There are two distinct piezoelectric effects, namely, the direct effect and inverse effect. In direct piezoelectric effect, when mechanical pressure or strain is applied to a piezoelectric material, positive and negative charges move within the crystal structure. This generates a net electric dipole moment, causing a separation of charge. As a result, surface charges accumulate on the opposite sides of the material, producing a measurable voltage. The inverse piezoelectric effect occurs when an electric field is applied to a piezoelectric material, causing it to deform mechanically (Jain et al., 2015). Schematic diagram of direct and converse effect of piezoelectric materials is as shown in Fig. 4.9.

The constitutive Eqs. (4.3) and (4.4) for both the effects are given below:

$$\text{Direct Effect}: \ S_i = S_{ij}^{E} T_j + d_{im} E_{m.} \tag{4.3}$$

$$\text{Inverse Effect}: \ D_k = d_{jk} T_j + \varepsilon_{km}^{T} E_{m.} \tag{4.4}$$

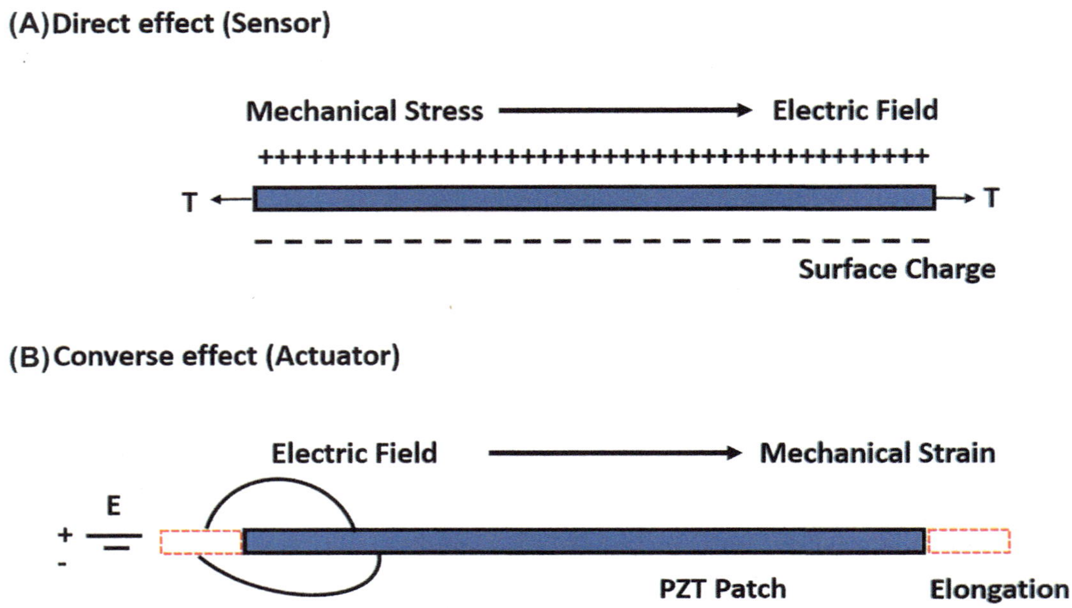

FIGURE 4.9

Direct and converse effect of piezoelectric materials. This figure illustrates the direct and converse effects of piezoelectric materials.

In this context, T represents stress, d is the piezoelectric constant, S denotes strain, D stands for electric displacement, and E indicates the electric field. S^Erefers to mechanical compliance at constant stress T. The indices I, j, and k correspond to different directions within the material's coordinate system, similar to the Cartesian axes (x, y, z), with values ranging from 1 to 3. Rotational motion around these three axes is indicated by the subscript "m," which takes values from 1 to 6.

A piezoelectric energy harvesting device operates in two primary modes: mode 33, where the applied stress aligns with the direction of polarization, and mode 31, where the applied stress is perpendicular to it, as illustrated in Fig. 4.10(A and B) respectively.

4.5.2 Piezoelectric materials

Biomicroelectromechanical systems (BioMEMS) integrate biological components with microelectromechanical systems to create devices that can perform various functions in biomedical applications. Piezoelectric materials are vital in this field due to their ability to convert mechanical energy into electrical signals and vice versa. This capability enables the development of sensors, actuators, and devices for monitoring and manipulating biological systems. The types of piezoelectric materials commonly used in BioMEMS can be broadly classified into crystalline materials, polymers, and composites.

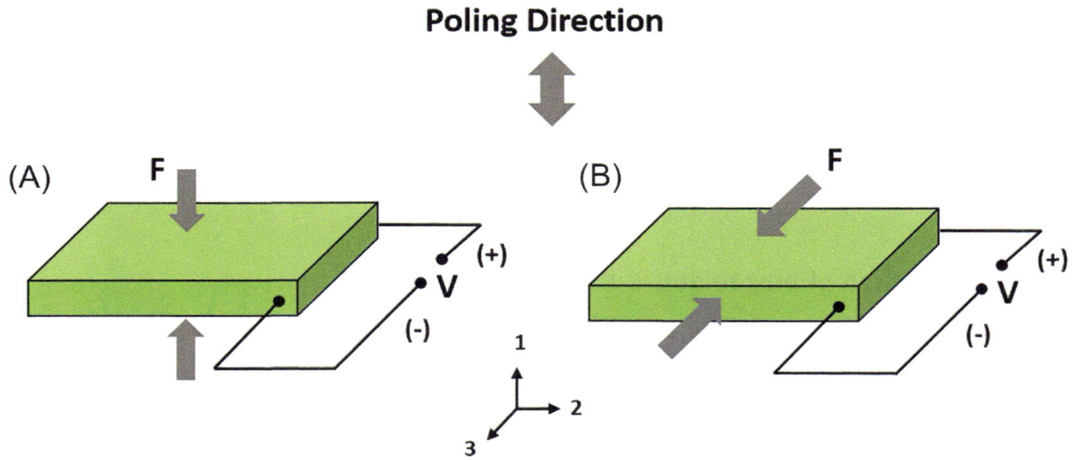

FIGURE 4.10

Operating modes of piezoelectric materials: (A) 33 mode and (B) 31 mode. This figure illustrates the two common operating modes of piezoelectric materials where these mode are critical in determining the behavior and performance of piezoelectric devices.

4.5.2(a) *Crystalline piezoelectric materials*

Crystalline piezoelectric materials are characterized by their well-defined structures that lack a center of symmetry, which is essential for their piezoelectric properties. Natural crystals, such as quartz (SiO_2), are often employed in biosensors and monitoring devices due to their stability, biocompatibility, and low cost. Quartz's ability to generate an electrical signal in response to mechanical stress makes it suitable for detecting physiological parameters such as pressure and strain. Synthetic crystalline materials, such as barium titanate ($BaTiO_3$) and lead zirconate titanate (PZT), offer enhanced piezoelectric coefficients and are used in applications requiring precise control (Jaffe et al., 1954; Shuai et al., 2020). For instance, PZT is commonly found in ultrasound transducers, making it invaluable for noninvasive imaging techniques in medical diagnostics.

4.5.2(b) *Piezoelectric polymers*

Polyvinylidene fluoride (PVDF) is one of the most widely used piezoelectric polymers in BioMEMS (Kawai, 1969). PVDF exhibits excellent flexibility and biocompatibility, making it ideal for wearable devices that monitor physiological signals such as heart rate, muscle activity, and other vital signs. Its piezoelectric properties can be enhanced through various processing techniques, enabling it to function effectively in thin-film applications. PVDF's lightweight nature allows for integration into flexible electronics, facilitating the development of devices that can conform to the body's surface, thus improving patient comfort and compliance.

4.5.2(c) *Composite materials*

Composite piezoelectric materials combine the advantageous properties of ceramics and polymers, offering a balance between sensitivity and flexibility. For instance, combining PZT with PVDF creates a composite that harnesses the high piezoelectric response of ceramics while maintaining the adaptability of polymers (Ahn et al., 2013; Sharma et al., 2016). These composites are particularly useful in applications such as soft robotics and implantable devices, where flexibility and biocompatibility are crucial. By adjusting the composition and fabrication methods, researchers can tailor the properties of these composites to meet specific application requirements, leading to improved performance in BioMEMS.

4.5.3 Application of piezoelectric-based sensor

4.5.3(a) *In medical*

The piezoelectric effect has significant implications in the biomedical field. This technology is increasingly utilized in various medical devices and applications due to its unique properties, such as energy harvesting, sensitivity, and biocompatibility. Among the application of piezoelectric effects in biomedical field are ultrasound imaging, blood pressure and heart rate monitoring, biosensor for virus detection, and obstructive sleep apnea (OSA) screening.

Piezoelectric-based biosensors play a crucial role in ultrasound imaging equipment by serving as transducers that convert electrical signals into ultrasonic waves and vice versa. When an alternating voltage is applied to piezoelectric materials, they vibrate, generating high-frequency sound waves that penetrate the body. These waves encounter different tissues and boundaries, leading to reflections that are captured by the same piezoelectric transducer. The reflected waves are then converted back into electrical signals, which are processed to create detailed images of internal structures, such as organs and tissues. This technology is particularly advantageous in medical diagnostics, as it is noninvasive and safe for patients, including pregnant women, making it a fundamental tool in prenatal care and various other medical imaging applications.

Park et al. (2023) presents an innovative method for generating dual-frequency ultrasound using a ferroelectric piezoelectric micromachined ultrasound transducer (PMUT). This technique utilizes polarization switching in ferroelectric thin films, allowing for tunable ultrasound frequencies from a single transducer by adjusting a direct current (DC) bias. This capability enables the PMUT to effectively produce both low- and high-frequency ultrasounds, which is beneficial for advanced imaging techniques such as harmonic imaging and image-guided therapy. The design simplifies the complexity of transducer arrays while enhancing imaging resolution and depth.

Physiological signals, such as body pulse waves, encompass two key indicators of human health, which are blood pressure and heart rate. These signals can be effectively measured by assessing the local pressure and expansion of the radial or carotid arteries utilizing the piezoelectric effect. Heart rate can be obtained from various cardiac pulsating waves that reflect the movements of the human heart, including pulse waves, electrocardiogram, photoplethysmogram, seismocardiogram , and ballistocardiogram. Among these, pulse waves are the most commonly utilized signals in piezoelectric sensing techniques. Pulse wave analysis is an emerging method for evaluating arterial elasticity and has become crucial for the early detection of arteriosclerosis. By applying piezoelectric biosensors to the skin over an artery, the local vibrations caused by cardiac movements can be detected, and this mechanical force is subsequently recorded and converted into electrical signals. This cardiovascular function index relies on the time-domain characteristics of the pulse waveform and related pathophysiological information (Samartkit et al., 2022).

Noninvasive continuous blood pressure measurement is increasingly important for monitoring cardiovascular health. Blood pressure (BP) is closely related to cardiac activity, as arterial blood vessels experience periodic pressure fluctuations during circulation. BP is divided into systolic blood pressure (SBP) and diastolic blood pressure (DBP), with arterial pulses creating detectable pressure changes. A typical pulse waveform consists of an ascending branch, caused by arterial expansion during ventricular contraction, and a descending branch, due to ventricular diastole. Wang and Lin (2020) reported on a wearable piezoelectric system for continuous beat-to-beat blood pressure measurement. Similar to heart rate, BP relies on arterial pulses, resulting in comparable waveforms. However, blood pressure measurements can be influenced by factors like the transmission of pressure from arteries to the skin and body posture. The piezoelectric biosensor's output can reflect two correlations with blood pressure: one based on voltage changes and integration during pressure application, and another based solely on the integration of the output voltage. Adjusting the thickness of the piezoelectric layer alters these correlations, enhancing measurement accuracy.

Piezoelectric-based biosensors are powerful tools for real-time, label-free detection of biological interactions, making them particularly valuable in point-of-care (POC) testing and other diagnostic applications. Jothimani et al. (2020) described a novel biosensing device designed for the rapid and accurate detection of antibodies against the SARS-CoV-2 virus. This POC biosensor employs a piezoelectric sensor made from a lithium niobate wafer, configured as a multithreaded comb with cantilever beams. This design enhances the sensitivity and selectivity for detecting specific antigen–antibody interactions. The operation of this biosensor relies on the piezoelectric effect, which is the ability of certain materials to generate an electric charge in response to mechanical stress. When antibodies bind to the sensor's surface, they induce changes in mass and pressure that affect the mechanical vibrations of the cantilever beams. These changes are then detected as variations in the electric signals generated by the piezoelectric material, facilitating the identification of SARS-CoV-2 antibodies. Overall, this biosensor represents a significant advancement in diagnostic technology, providing a quick, noninvasive method for detecting COVID-19 antibodies, which is essential for monitoring infection spread and vaccine efficacy.

Erdenebayar et al. (2017) explore the development and use of a piezoelectric sensor system designed to screen for OSA. This noninvasive device monitors respiratory patterns by detecting vibrations generated by airflow during sleep. The piezoelectric sensor captures these vibrations and converts them into electrical signals, which are then analysed to identify irregularities in breathing patterns that may indicate OSA. The operation of this biosensor is based on the piezoelectric effect; when the piezoelectric material experiences mechanical stress from breathing vibrations, it produces an electric charge proportional to the applied force. This characteristic enables the sensor to accurately detect changes in pressure and movement, making it effective for real-time monitoring of physiological signals. Overall, the incorporation of piezoelectric sensors in OSA screening marks a significant advancement in sleep medicine, offering a convenient and effective method for the early detection of this condition. The ability of piezoelectric materials to convert mechanical energy into electrical signals enhances the sensor's sensitivity and reliability, making it a valuable tool for continuous health monitoring.

4.5.4 Conclusion

The piezoelectric effect, where certain materials generate an electric charge in response to mechanical stress, has significant implications in the biomedical field, particularly through its application in piezoelectric-based biosensors. These sensors are crucial in various medical devices, including ultrasound imaging systems, where they convert electrical signals into ultrasonic waves and vice versa. By applying an alternating voltage, piezoelectric materials vibrate to produce high-frequency sound waves that penetrate the body, allowing for detailed imaging of internal structures. Additionally, piezoelectric sensors are employed in noninvasive monitoring of physiological signals such as blood pressure and heart rate by detecting vibrations from arterial movements. This technology enables real-time, label-free detection of biological interactions, making it invaluable for point-of-care testing, including rapid detection of SARS-CoV-2 antibodies and screening for conditions such as OSA. Overall, the versatility and sensitivity of piezoelectric materials enhance diagnostic capabilities and patient care across various biomedical applications.

4.6 Heat-transfer biosensors

4.6.1 Theory and mechanism of heat transfer biosensors

Heat transfer-based sensors are crucial in various applications, measuring the rate of heat energy transfer through surfaces. These sensors can be classified into several types, each with unique characteristics and applications. Below is an overview of their functionalities and uses, particularly in biomedical contexts.

Heat-transfer-based sensors operate on the principle that changes in temperature or thermal conductivity can be correlated with physical parameters. When a temperature differential exists between a sensing element and its environment, heat transfer occurs via conduction, convection, or radiation. The rate of heat transfer can be measured, and this information is used to infer the desired physical parameter.

Heat transfer sensors operate based on the principles of conduction, convection, and radiation to measure temperature differences and heat flow. Conduction involves the transfer of heat through solid materials as it moves from hotter to cooler areas via direct contact. Convection refers to the transfer of heat through fluids (liquids or gases) as they circulate, allowing sensors to assess heat loss or gain in dynamic environments. Additionally, radiation enables non-contact temperature sensors to detect thermal energy emitted by objects through electromagnetic waves, providing a versatile approach to measuring temperature and heat transfer in various applications.

Temperature sensors function based on two primary physical principles, which are voltage measurement and resistance change. In thermocouples and semiconductor sensors, a voltage is generated at the junction of two different metals or diodes, which varies with temperature changes; this voltage is then measured and converted into a readable temperature format. Conversely, resistance temperature detectors (RTDs) and thermistors operate by detecting changes in electrical resistance as temperature fluctuates; this resistance is measured and translated into temperature readings using calibration curves or lookup tables.

4.6.2 Types of heat biosensor

Temperature measurement is a fundamental aspect of various industrial, scientific, and everyday applications. Heat-based sensors play a crucial role in providing accurate and reliable temperature readings, enabling effective monitoring and control of processes. These sensors can be broadly categorized into

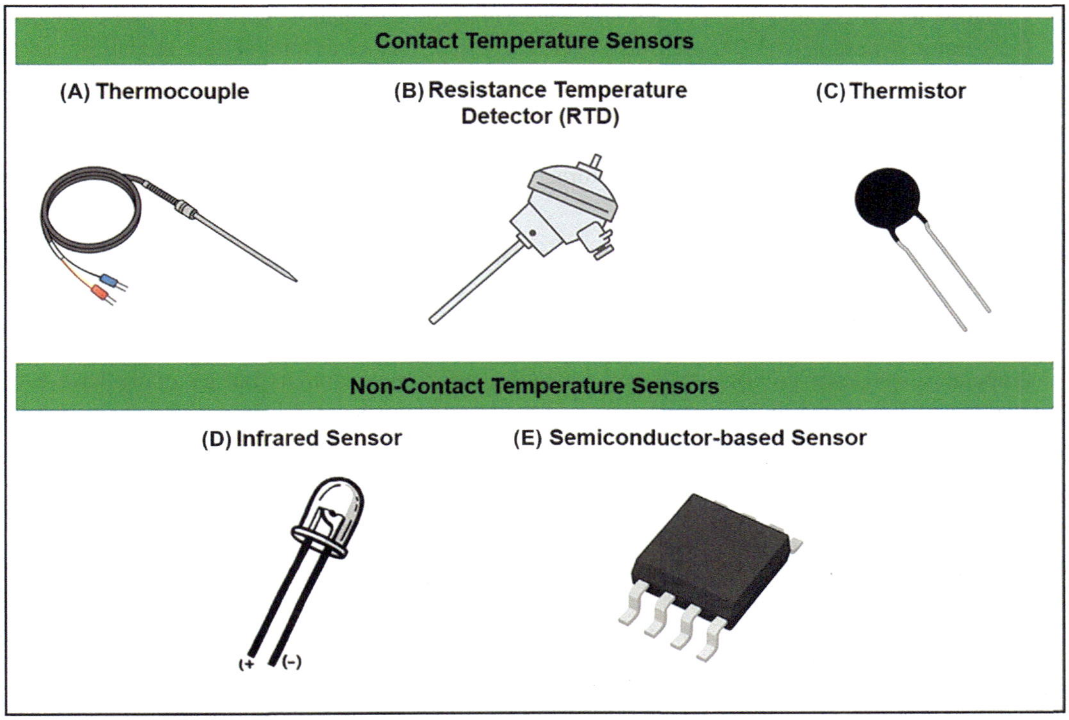

FIGURE 4.11

Images of contact temperature sensors: (A) Thermocouple, (B) Resistance Temperature Detector (RTD), (C) Thermistor, and non-contact temperature sensors: (D) Infrared sensor, and (E) Semiconductor-based sensor. This figure illustrates the contact and non-contact temperature sensors.

two main types: contact and non-contact temperature sensors. Fig. 4.11 shows the images of contact temperature sensors and non-contact temperature sensors.

4.6.2(a) *Contact temperature sensors*

These sensors measure temperature by direct contact with the object. They are widely used for liquids, solids, and gases.

- **Thermocouples**: Composed of two different metals joined at one end, thermocouples generate a voltage that correlates with temperature differences. They are favored for their wide temperature range (−200°C to 1750°C) and cost-effectiveness but may have lower accuracy (0.5– 5°C) compared to other types.
- **Resistance Temperature Detectors (RTDs)**: RTDs consist of materials (often platinum) that change resistance with temperature. They provide high accuracy and stability over a range of −200°C to 600°C but are more expensive than thermocouples.

- **Thermistors**: These are resistors whose resistance changes significantly with temperature. Negative Temperature Coefficient (NTC) thermistors decrease in resistance as temperature increases, making them sensitive and accurate for lower temperature ranges (−50°C to 250°C).

4.6.2(b) *Non-contact temperature sensors*

These sensors measure temperature without direct contact, making them ideal for hazardous environments.

- **Infrared Sensors:** These sensors detect thermal radiation emitted from objects, allowing for temperature measurement from a distance. They are commonly used in medical applications and industrial processes.
- **Semiconductor-Based Sensors:** These sensors utilize integrated circuits (ICs) with diodes that have temperature-sensitive voltage characteristics. They offer linear outputs but generally have lower accuracy and slower response times compared to thermocouples and RTDs.

4.6.3 Application of heat-transfer based sensor

4.6.3(a) *In medical*

Heat-transfer based micromechanical sensors have several promising applications in the biomedical field. These sensors leverage principles of thermal conductivity and heat transfer to monitor physiological parameters, offering noninvasive solutions for health monitoring. Among the application of heat-transfer based sensor in biomedical field includes body temperature monitoring, temperature-controlled hyperthermia, diseased tissue ablation, and whole-body cryotherapy (WBC).

Geng et al. (2022) present a novel wearable sensor utilizing conductive polymer composites (CPCs) designed for accurate body temperature measurement. The authors highlight the advantages of CPCs, including flexibility, lightweight nature, and high sensitivity, making them ideal for integration into wearable health devices. The sensor conforms closely to the skin, enabling real-time temperature readings essential for monitoring health conditions such as fever or hypothermia. The study emphasizes the importance of responsiveness and durability in wearable sensors, addressing challenges such as maintaining sensitivity during physical activity and ensuring stable performance under various environmental conditions. The sensor operates primarily through thermal conduction, allowing efficient heat transfer from the skin to the sensing element, which ensures rapid detection of temperature changes. Additionally, the design considers other heat transfer mechanisms such as convection and radiation, managing heat dissipation to enhance measurement accuracy. This technology has significant implications for continuous health monitoring, providing valuable data for healthcare providers and patients alike. Overall, the research contributes to advancing smart wearable technologies that improve personal health management through enhanced thermal sensing capabilities.

Wang et al. (2023) presents a novel approach for temperature-controlled hyperthermia, utilizing high-intensity focused ultrasound (HIFU) combined with non-invasive temperature monitoring via speed of sound (SoS) imaging. The authors propose that the speed of sound in biological tissues serves as a reliable indicator of temperature changes, enabling precise control during cancer therapies. This method is particularly relevant for health temperature monitoring, as it allows for real-time assessment of tissue temperature during treatments, ensuring that targeted areas reach therapeutic temperatures while protecting surrounding healthy tissues from overheating. The study highlights the significance of understanding heat transfer mechanisms, as HIFU generates localized heating through acoustic energy, which is monitored by assessing how sound propagates through heated tissue. By continuously evaluating the SoS, clinicians can optimize treatment parameters, enhancing the efficacy and safety of hyperthermia therapies. Overall, this innovative approach integrates advanced imaging techniques with thermal management to improve cancer treatment outcomes and patient care.

Research conducted by Hu et al. (2021) explores the use of flexible electrodes for radiofrequency (RF) tissue ablation, focusing on the electrothermal dynamics and bio-heat transfer mechanisms involved in the process. The authors analyze how these injectable electrodes can provide localized heating while considering factors such as blood perfusion and tissue properties that affect heat distribution. This research is particularly relevant for health temperature monitoring, as precise temperature control during RF ablation is crucial to ensure effective treatment while minimizing damage to surrounding healthy tissues. The study emphasizes the importance of bio-heat transfer models, particularly Pennes' bioheat equation, which accounts for thermal conduction, blood perfusion, and metabolic heat generation in biological tissues. By examining how RF energy generates heat through resistive heating and how that heat is transferred via conduction, the authors aim to optimize electrode design and RF application parameters for better control over temperature profiles during ablation procedures. Overall, this work enhances the safety and efficacy of RF ablation techniques in clinical settings by integrating advanced thermal analysis with innovative electrode technologies.

Elfahem et al. (2024) examine the heat transfer dynamics during whole-body cryotherapy (WBC) using a three-dimensional thermodynamic modeling framework. The study emphasizes how individual differences, such as body composition and metabolic rate, influence thermal responses during cryotherapy sessions, which are commonly employed for recovery and pain management. By developing a detailed 3D model, the researchers simulate temperature changes across various body regions, allowing for a better understanding of how different factors affect heat transfer. This modeling approach aims to optimize cryotherapy protocols tailored to individual physiological characteristics, enhancing treatment effectiveness. The findings indicate that personalized cryotherapy can improve recovery outcomes by considering these variations. Overall, the research provides valuable insights into the thermodynamic principles governing WBC and highlights the potential for advanced modeling techniques to inform clinical practices, ultimately leading to more effective and safer cryotherapy treatments tailored to individual patient needs.

4.6.4 Conclusion

Heat-transfer based micromechanical sensors are revolutionizing the biomedical field by providing innovative solutions for monitoring and managing health parameters. Their ability to leverage thermal conductivity principles enables noninvasive applications such as body temperature monitoring, temperature-controlled hyperthermia, diseased tissue ablation, and whole-body cryotherapy. Research highlights the development of wearable sensors utilizing CPCs for accurate body temperature measurement, showcasing their flexibility and sensitivity for real-time health monitoring. Additionally, novel approaches combining high-intensity focused ultrasound with speed of sound imaging enhance temperature control during cancer therapies, ensuring effective treatment while protecting healthy tissues. The use of flexible electrodes in radiofrequency tissue ablation emphasizes the importance of precise temperature management to optimize treatment efficacy and safety. Furthermore, advanced thermodynamic modeling in whole-body cryotherapy illustrates how individual variability can influence thermal responses, paving the way for personalized treatment protocols. Collectively, these advancements underscore the significant potential of heat-transfer based sensors to improve patient care, enhance therapeutic outcomes, and contribute to the development of smart healthcare technologies that prioritize individual needs and safety.

4.7 Sound based biosensors

4.7.1 Microphone biosensor

Microelectromechanical systems (MEMS) microphones are advanced audio sensors that leverage micromachining technology to convert sound waves into electrical signals. These devices have become increasingly popular in various applications, including consumer electronics, medical devices, and environmental monitoring, due to their compact size, low-power consumption, and high performance.

4.7.1(a) Principle mechanism of microphone

MEMS microphones are similar to the standard electret condenser microphones (ECMs) found in modern consumer electronics, except that the components are built onto a single chip using CMOS technology, rather than assembled from discrete parts. There are two main categories of MEMS microphones, which are analog and digital. Analog MEMS microphones convert a sound pressure input into an analog voltage output. Whereas digital MEMS microphones convert a sound pressure input into a digital output signal, typically in pulse density modulation format (PDM).

Microphone sensors operate as transducers, converting sound energy (acoustic pressure) into electrical energy (Spiegel et al., 2014). They are designed to detect sound waves within the audible range (20 Hz to 20 kHz). A typical MEMS microphone consists of two important components, which are diaphragm and backplates. Diaphragm is a thin flexible membrane that vibrates in response to sound waves. Whereas backplates always positioned closed to the diaphragm. There are three primary types of microphone-based sensors with different physical mechanisms.

4.7.1(a)(i) Capacitive

In capacitive microphones (condenser microphones), sound waves cause a diaphragm to vibrate, changing the distance between the diaphragm and a backplate as shown in Fig. 4.12.

The backplate is positioned close to the diaphragm and works like capacitor. As sound waves cause the diaphragm to move, the capacitance between the diaphragm and backplate varies. This change in capacitance is converted into an electrical signal that corresponds to the sound wave's amplitude and frequency (Martin et al., 2007). This change in capacitance causes a change in voltage according to:

$$C = \frac{Q}{V}$$

Whereas C is the capacitance (Farads), Q is the charge (Coulumbs), and V is voltage (Volts).

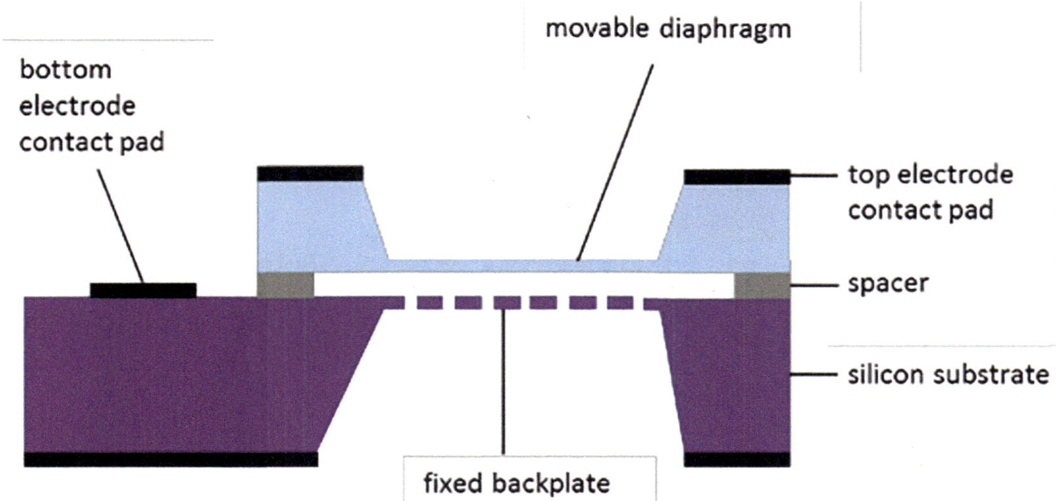

FIGURE 4.12

Basic structure of a micro-electro-mechanical systems (MEMS) capacitive microphone. This figure illustrates the basic structure of a MEMS capacitive microphone.

4.7.1(a)(ii) Piezoelectric

Some microphones use piezoelectric (PZT) materials that generate an electrical charge in response to mechanical stress from sound waves as shown in Fig. 4.13.

There are several materials have the potential as piezoelectric materials such as quartz, ceramic, and certain polymer. The PZT thin film is integrated with the diaphragm and the metal layers (electrodes) collect the electric charge generated by the piezoelectric element and transfer it to an external circuit for further processing. This mechanism need an external preamplifier, which does not require power source, to reduce the equivalent noise level (Royer et al., 1983).

4.7.1(a)(iii) Electromagnetic-induced

The use of electromagnetic in microphones able to detect and convert sound waves into electrical signals as shown in Fig. 4.14. This mechanism is well known in dynamic microphones. The induction coil captures the changes in electromagnetic field generated by sound waves or other environmental stimuli. When sound waves interact with the coil, they induce a current due to the changing magnetic field. The induced electrical signal is then amplified and processed to produce an audible sound or to analyze specific frequencies relevant to biomedical applications.

4.7.1(b) Applications of MEMS microphone as biosensors

MEMS microphones serve as versatile biosensors in the medical field, enabling innovative solutions for monitoring respiratory and cardiac functions noninvasively. Their integration into wearable devices and telemedicine applications enhances patient care by providing real-time health data while minimizing discomfort and risk. As technology continues to advance, the role of MEMS microphones in healthcare

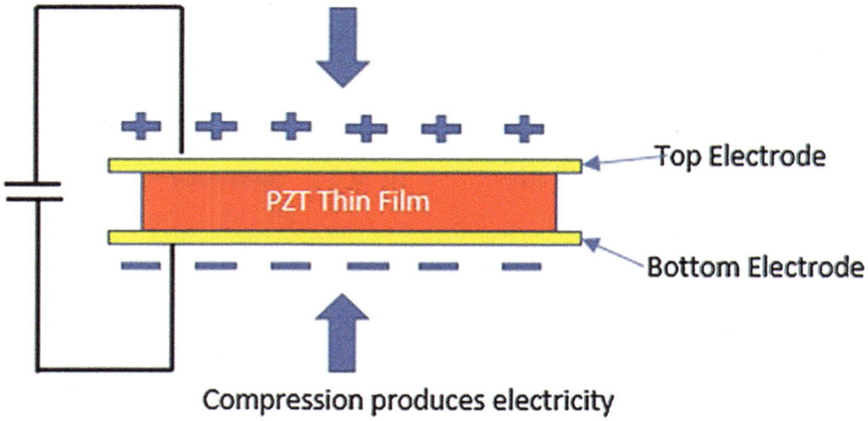

FIGURE 4.13

Basic structure of a micro-electro-mechanical systems (MEMS) PZT-based microphone. This figure illustrates the basic structure of a MEMS PZT-based microphone, which utilizes a piezoelectric material.

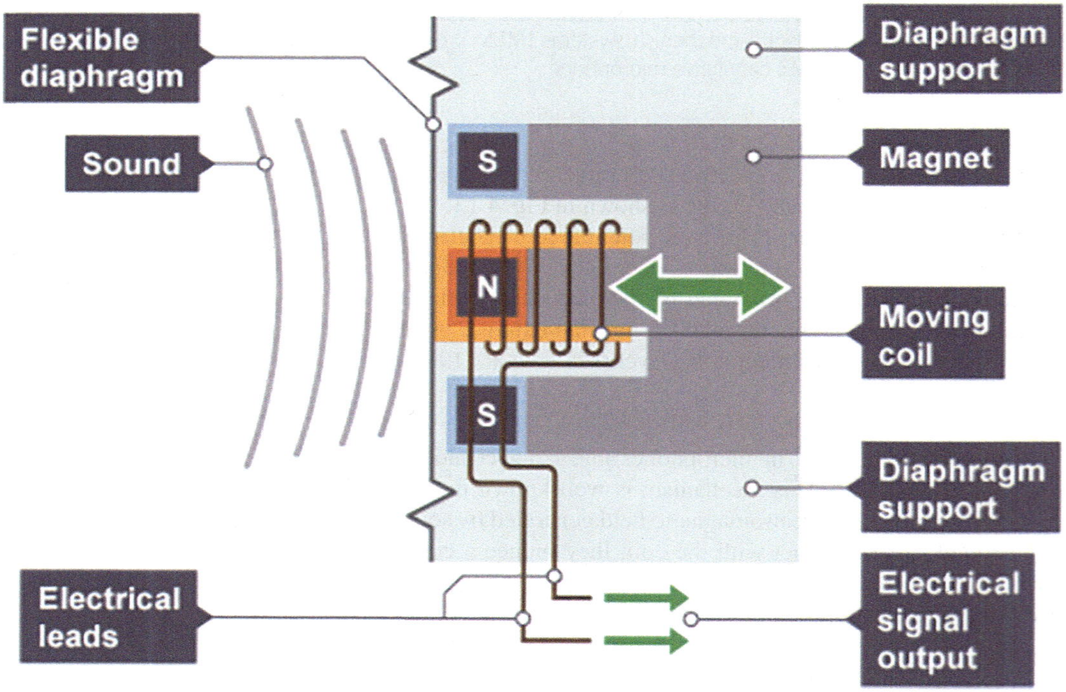

FIGURE 4.14

Basic structure of a micro-electro-mechanical systems (MEMS) electromagnetic-induced microphone. This figure shows the basic structure of a MEMS electromagnetic-induced microphone.

is expected to expand further, contributing significantly to personalized medicine and remote patient monitoring.

MEMS microphones can effectively capture respiratory sounds, making them useful for monitoring respiratory conditions such as sleep apnea and chronic obstructive pulmonary disease. They can be integrated into wearable devices that continuously track breathing patterns, helping healthcare providers assess respiratory health in real time. This application is particularly beneficial in neonatal care units, where the small size of MEMS microphones minimizes discomfort for infants while providing critical monitoring capabilities (Sezen et al., 2005). MEMS microphones can also be employed to monitor heart sounds, such as heartbeats and murmurs. By analyzing the acoustic signals generated by the heart, these sensors can assist in diagnosing various cardiac conditions. The ability to wirelessly transmit data allows for remote monitoring, enhancing patient care and enabling timely interventions (Qu et al., 2021).

The compact size and low-power consumption of MEMS microphones make them ideal for telemedicine applications. They can be used in portable diagnostic devices that enable healthcare professionals to remotely assess patients' health by analyzing body sounds, such as heart and lung activity. This capability is crucial for providing medical care in underserved areas or during emergencies (Muhammad et al., 2024). MEMS microphones facilitate noninvasive diagnostics by capturing sound waves produced by various bodily functions without the need for invasive procedures. This approach reduces patient discomfort and risk while providing valuable data for diagnosis and monitoring (Chircov & Grumezescu, 2022).

The miniaturization capabilities of MEMS technology enable the development of wearable health devices that monitor vital signs through sound analysis. These devices can provide continuous feedback on a patient's condition, promoting proactive healthcare management (Dias & Paulo Silva Cunha, 2018). MEMS microphones hold significant potential for a wide range of applications in the medical field. This highlights how advancements in MEMS technology are enhancing health management.

4.8 Ultrasound biosensors

Ultrasound or ultrasonic sensors are devices that utilize high-frequency sound waves, beyond human hearing sound, which typically above 20kHz, to measure distances, detect objects, and monitor various parameters.

4.8.1 Principle mechanism of ultrasound biosensors

The basic principle of ultrasound sensor is based on the principles of sound wave propagation and reflection. Ultrasound sensor contains transducers that could emits ultrasonic sound waves, which converts electrical energy into sound energy. The sensor responds to mass, viscosity, or electrical properties of the analytes, which reflect as variation in the acoustic wave amplitude, phase/velocity, or frequency in the sensor output (Fig. 4.15).

The transducers will emit sound waves that travel through the air or any medium until they encounter (reflection). On the same transducers (or a separate one), they will receive the reflected sound waves

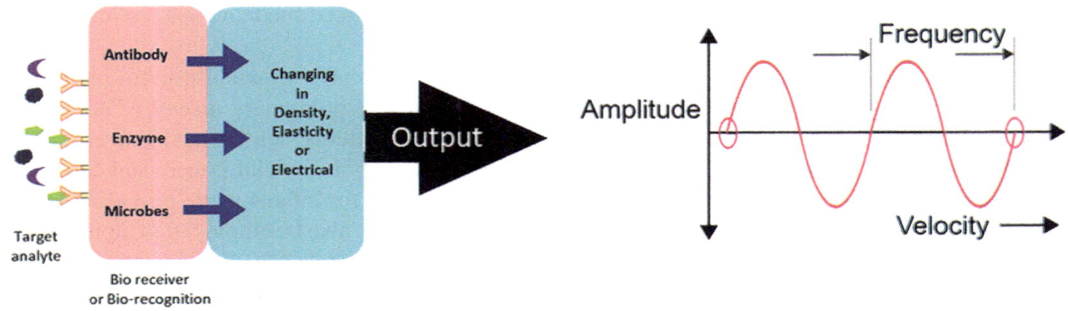

FIGURE 4.15

Basic principle of ultrasound biosensor. The sensor responds to mass, viscosity or electrical properties of the analytes which reflect as variation in the acoustic wave amplitude, phase/velocity, or frequency in the sensor output.

(reception). The distance calculation for the sound waves emits and return used the formula below:

$$D = \frac{TxC}{2}$$

Whereas D = distance to the object, T = time taken for the echo to return and C = speed of sound in medium (example air = 343 m/s at room temperature).

4.8.2 Application of ultrasound as biosensors

Ultrasonic sensors have emerged as valuable tools in the field of biosensing, leveraging their ability to utilize sound waves for various biological applications. As technology continues to evolve, the integration of ultrasonic sensing with other modalities promises to further expand their utility in the life sciences.

Ultrasonic sensors can function as acoustic biosensors to monitor enzyme activity within living organisms. A notable advancement involves genetically encodable acoustic biosensors that "light up" during ultrasound imaging in response to specific protease activities. These sensors utilize air-filled protein nanostructures known as gas vesicles, which produce nonlinear ultrasound signals when activated by enzymatic reactions (Lakshmanan et al., 2020). This capability allows researchers to visualize dynamic biochemical processes in real-time, enhancing our understanding of biological functions.

Ultrasonic sensors are widely used in medical imaging, particularly in ultrasound technology. They provide noninvasive methods to generate detailed images of internal body structures, making them essential for diagnostics such as prenatal monitoring and tumor detection. Their ability to penetrate deep tissues without the scattering issues associated with light-based imaging techniques makes them particularly advantageous (Delorme, 2006).

MEMS ultrasonic sensors could be applied to detect pulmonary embolism, blood clot that develops in a blood vessel elsewhere in the body, which can form blockage in the artery. The use of PMUT has been simulated to measure the presence of solid objects at different medium (Ponnuthurai et al., 2023).

Recent studies have highlighted the role of ultrasonic sensors in assessing mechanical and chemical properties of cells and tissues. This includes applications in elastography, where ultrasonic waves measure tissue stiffness, aiding in the diagnosis of conditions like liver fibrosis or cancer. By analyzing how sound waves interact with biological tissues, ultrasonic sensors can provide critical information regarding tissue health and disease states.

Example of elastography is liver elastography. Liver elastography is a non-invasive imaging technique used to assess liver stiffness, which serves as an important indicator of liver health and disease progression. The application of capacitive micromachined ultrasonic transducer (CMUT) technology for liver elastography, proposing a novel dual-mode CMUT array designed for both high-intensity focused ultrasound (HIFU) and imaging operations. The CMUT array facilitates the transmission of focused ultrasound waves to temporarily deform liver tissue, allowing for the measurement of elasticity through shear wave propagation. The research includes analytical calculations and finite element analysis to optimize the design of a 64 × 64 CMUT array operating at a center frequency of 7.5 MHz, demonstrating its effectiveness in detecting liver fibrosis and potentially improving diagnostic capabilities in clinical settings (Debnath, no date).

4.9 Pressure biosensors

Pressure sensors, also known as pressure transducers, are devices designed to measure the pressure exerted by gases or liquids. They convert the physical force exerted by a fluid onto a surface into an electrical signal, which can then be used for various applications, including monitoring and control systems.

4.9.1 Principle of mechanical of pressure biosensors

The fundamental principle behind pressure sensors is the measurement of force per unit area. When a fluid exerts pressure on the sensor, it causes a deformation in a sensing element (often a diaphragm). This deformation is then converted into an electrical signal that corresponds to the pressure level.

The principle of operation of pressure sensor is mostly similar to sound sensor. A thin, flexible membrane made from materials such as silicon or polymers is used as the diaphragm. Different from sound sensor, the membrane are deformed when pressure exerted from gases or liquids.

4.9.2 Type of pressure biosensors

Several types of pressure sensors exist, each utilizing different mechanisms for pressure measurement.

4.9.2(a) Strain gauge sensors

These sensors use strain gauges attached to a diaphragm. When pressure is applied, the diaphragm deforms due to stress such as elongation or compression, causing a change in resistance in the strain gauges. This change is measured and converted into an electrical signal as shown in Fig. 4.16. Semiconductor materials such as silicon or germanium are used to create MEMS strain gauges. They offer higher sensitivity compared to metallic gauges, but are more susceptible to temperature fluctuations and are generally more expensive.

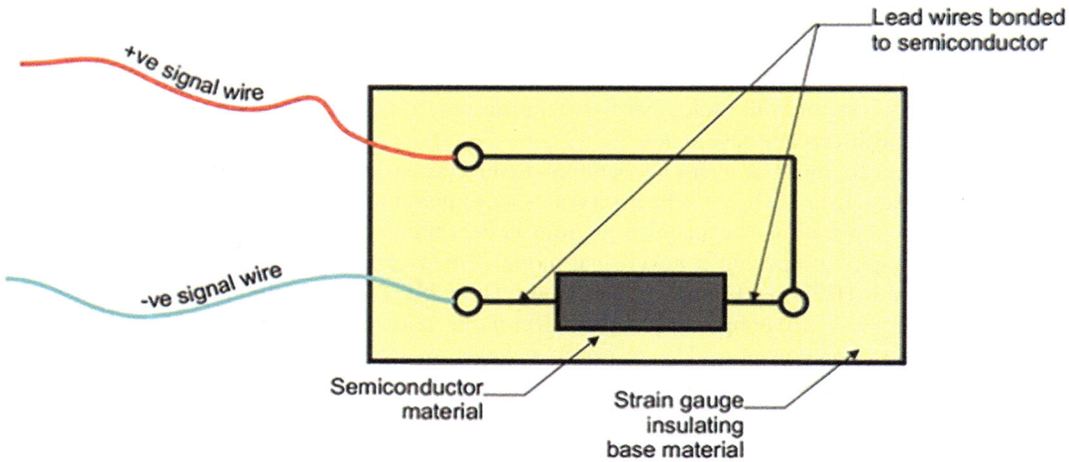

FIGURE 4.16

Basic components of strain gauge semiconductor. This figure illustrates the basic components of a strain gauge semiconductor.

4.9.2(b) *Capacitive sensors*

In capacitive pressure sensors, the diaphragm acts as one plate of a capacitor as shown in Fig. 4.17. As the diaphragm deflects under pressure, the distance between the plates changes, altering the capacitance. The deflection of the diaphragm alters the capacitance between the plates, which can be converted into an electrical signal corresponding to the applied pressure.

4.9.2(c) *Piezoresistive sensors*

Many MEMS pressure sensors utilize the piezoresistive effect, where the resistance of semiconductor materials changes in response to mechanical stress. This change in resistance is measured using a Wheatstone bridge (Fig. 4.18B) configuration, allowing for precise pressure readings.

4.9.3 Application of MEMS pressure biosensor

4.9.3(a) *In medical field*

Pressure (surface) sensors have significant applications as biosensors in the medical field, leveraging their ability to measure pressure changes within biological systems.

Pressure sensors are integral to both invasive and noninvasive blood pressure monitoring systems. They provide real-time measurements of arterial pressure, central venous pressure, and other critical parameters. These data are crucial for diagnosing hypertension and managing cardiovascular diseases effectively. Panwar et al. (2017) proposed a design and functionality of a MEMS piezoelectric blood pressure sensor, which utilizes micro-electromechanical systems technology to provide a noninvasive method for continuous blood pressure monitoring. The sensor operates based on the piezoelectric effect, where mechanical stress applied to piezoelectric materials generates an electrical voltage proportional to

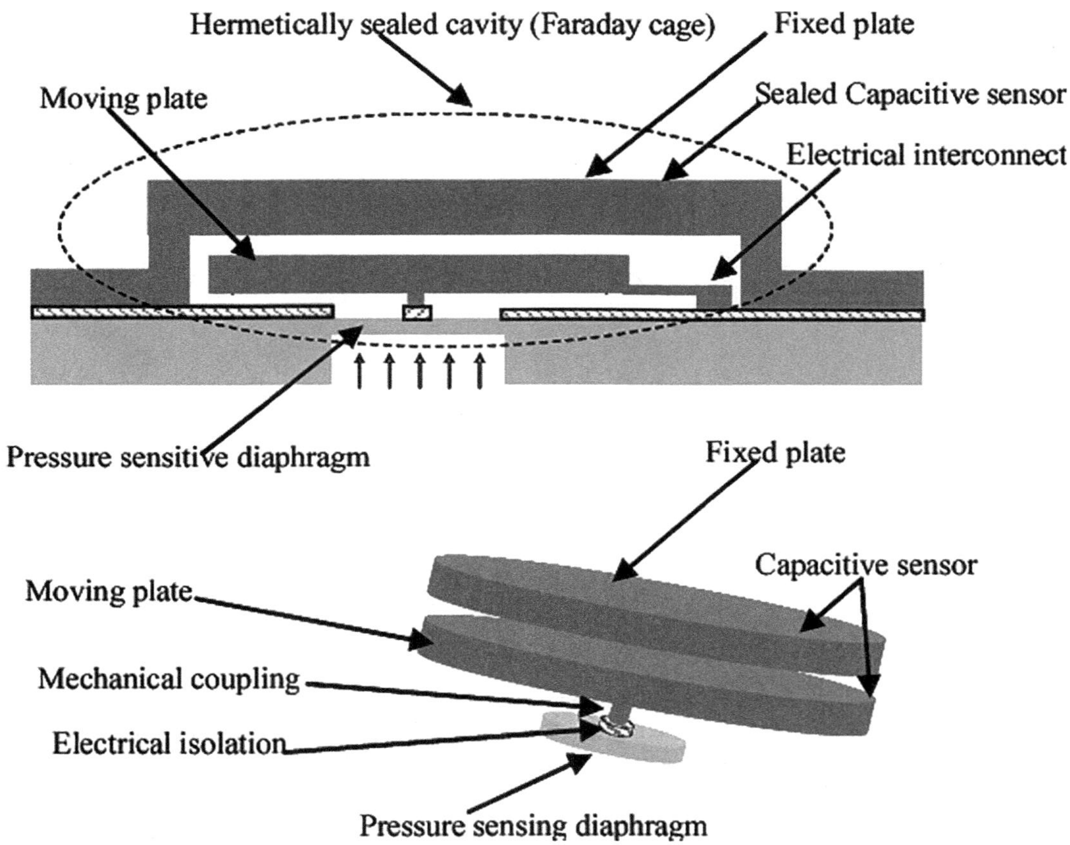

FIGURE 4.17

Basic components of MEMS capacitive sensor. This figure illustrates the basic components of MEMS capacitive sensors.

the pressure exerted by blood on the sensor. Specifically, a cantilever structure made from piezoelectric materials is employed; when blood pressure changes, it causes deflection in the cantilever, leading to variations in the generated voltage. This output voltage is then used to determine the blood pressure levels accurately.

In ventilators, pressure sensors ensure that patients receive the correct amount of air or oxygen. They continuously monitor airway pressure, triggering adjustments to airflow as needed. This functionality is vital for patients with respiratory challenges, allowing healthcare providers to respond promptly to any issues that arise during treatment. Han et al., 2023 discussed the minute differential pressure sensor, designed with a "beam-membrane-island" structure that achieves high sensitivity (up to 66 μV/V/kPa) and a natural frequency of 11.3 kHz, making it suitable for precise pressure monitoring in ventilators. Various sensor designs, including resonant pressure sensors and integrated systems, emphasizing the importance of these technologies in enhancing the performance of ventilators by enabling accurate

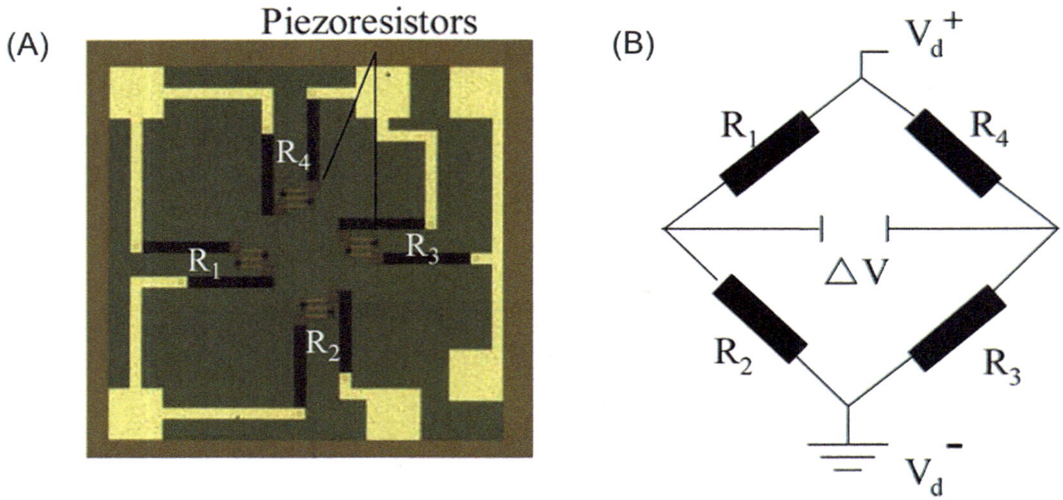

FIGURE 4.18

Basic components of MEMS piezoresistive sensor. This figure illustrates the fundamental structure of MEMS piezoresistive sensor.

monitoring of patient respiratory parameters and optimizing ventilation strategies. Furthermore, the fabrication techniques that allow for the production of ultrathin and compact sensor chips are essential for integration into modern medical devices while ensuring reliability under varying environmental conditions.

During dialysis, pressure sensors help regulate fluid delivery and monitor system integrity. They detect pressure fluctuations that may indicate blockages or malfunctions, allowing for immediate corrective actions. This ensures patient safety and the effectiveness of the treatment. The application of MEMS pressure sensors for monitoring dialysis access flow through a novel geometry-independent approach. (Weitzel et al., 2008) explores the use of differential pressure (ΔP) measurements between two dialysis needles to estimate blood flow rates (Q) within arteriovenous grafts. By employing a fluid circuit model, they demonstrated that ΔP increases with both flow rate and needle separation distance, achieving a high correlation ($R^2 > 0.98$) across various graft diameters and configurations. The study emphasizes the potential of MEMS sensors to provide real-time ΔP data, which can be crucial for detecting low or reversed flow conditions, thereby enhancing patient safety and reducing the risk of vascular access failure. The findings suggest that with further development of needle designs and MEMS integration, this method could facilitate more frequent and accurate monitoring of dialysis access, ultimately improving clinical outcomes for patients with end-stage renal disease.

Miniature pressure sensors are often integrated into catheters to measure internal pressures, such as blood pressure or intracranial pressure. This capability is particularly beneficial for long-term monitoring and reduces the need for multiple invasive procedures. (Brancato et al., 2016) discussed on the integration of MEMS pressure sensors in catheters for monitoring hemodynamic parameters, particularly focusing on their application in intravascular settings. The design and packaging of a

MEMS pressure sensor that is compact, biocompatible, and capable of providing real-time pressure measurements within catheters used for various medical applications. The sensor is constructed using a piezoresistive element mounted on a ceramic substrate and encapsulated with polydimethylsiloxane and parylene-C coatings to enhance its durability and hemocompatibility. This configuration allows the sensor to withstand the harsh conditions of blood flow while minimizing thrombus formation. The study emphasizes the importance of accurate pressure monitoring in improving patient outcomes by enabling better management of vascular access during procedures such as dialysis and cardiac support, where maintaining optimal blood flow is critical. The results indicate that these MEMS sensors can significantly enhance the functionality and safety of catheter systems by providing continuous feedback on pressure changes, thus facilitating timely medical interventions.

MEMS pressure (surface) sensors serve as vital biosensors across various medical applications, enhancing patient care through accurate monitoring of physiological parameters. Their integration into medical devices not only improves functionality but also contributes significantly to patient safety and treatment efficacy. As technology advances, the role of these sensors in healthcare is expected to expand further, offering new possibilities for monitoring and managing health conditions.

4.10 Chapter summary

This chapter presents an in-depth exploration of advanced micromechanical biosensors, emphasizing their pivotal role in modern diagnostics, monitoring, and industrial applications. At its core, the discussion highlights mass-sensitive biosensors, particularly QCM and SAW devices, which detect minute mass variations through shifts in resonance frequency. These sensors stand out for their label-free, real-time detection capabilities, making them indispensable in biomedical research, food safety assurance, and environmental monitoring.

The chapter begins by situating biosensors within a broader framework of biorecognition elements and transducing mechanisms, underscoring how mechanical signals are converted into quantifiable outputs. QCM technology is examined in detail, with attention to its piezoelectric foundation, structural design, and the governing Sauerbrey equation. Applications span diverse domains: in the food industry, QCM biosensors detect pathogens, allergens, spoilage indicators, and contaminants; in environmental monitoring, they track pollutants in air, water, and soil; and in medicine, they enable early disease diagnosis, drug screening, and therapeutic monitoring.

SAW biosensors are introduced as versatile devices operating across MHz–GHz frequencies, valued for their ruggedness, compactness, and wireless accessibility. Their principle mechanism relies on interdigitated transducers generating acoustic waves on piezoelectric substrates, enabling highly sensitive detection of surface changes. Originally developed for communication technologies, SAW devices have evolved into powerful biosensing tools, now widely applied in food safety, medical diagnostics, and industrial monitoring.

Beyond QCM and SAW, the chapter surveys other micromechanical sensor classes that extend the functional landscape: cantilever-based biosensors harness nanoscale deflections for protein detection and drug discovery; magnetic biosensors employ nanoparticles for ultra-sensitive, low-noise detection; piezoelectric biosensors underpin ultrasound imaging and wearable health monitoring; heat-transfer sensors leverage thermal conductivity for innovative therapeutic and wearable applications; and sound-based, ultrasonic, and pressure biosensors expand diagnostic versatility through acoustic and mechanical signal detection.

Together, these technologies illustrate the breadth and depth of micromechanical sensing, from fundamental principles to real-world applications. The chapter concludes by emphasizing the convergence of these sensors with nanotechnology, artificial intelligence, and Internet of Things platforms, pointing toward a future where biosensors will enable precision medicine, sustainable environmental practices, and enhanced global health monitoring.

References

Agostini, M., & Cecchini, M. (2021). Ultra-high-frequency (UHF) surface-acoustic-wave (SAW) microfluidics and biosensors. *Nanotechnology, 32*(31), 312001. https://doi.org/10.1088/1361-6528/abfaba.

Ahn, Y., Lim, J. Y., Hong, S. M., Lee, J., Ha, J., Choi, H. J., & Seo, Y. (2013). Enhanced piezoelectric properties of electrospun poly(vinylidene fluoride)/multiwalled carbon nanotube composites due to high β-phase formation in poly(vinylidene fluoride). *Journal of Physical Chemistry C, 117*(22), 11791–11799. https://doi.org/10.1021/jp4011026.

Akbari Nakhjavani, S., Khalilzadeh, B., Samadi Pakchin, P., Saber, R., Ghahremani, M. H., & Omidi, Y. (2018). A highly sensitive and reliable detection of CA15-3 in patient plasma with electrochemical biosensor labeled with magnetic beads. *Biosensors and Bioelectronics, 122*, 8–15. https://doi.org/10.1016/j.bios.2018.08.047.

Akgonullu, S., Ozgur, E., & Denizli, A. (2022). Recent advances in quartz crystal microbalance biosensors based on the molecular imprinting technique for disease-related biomarkers. *Chemosensors, 10*(3), 106. https://doi.org/10.3390/chemosensors10030106.

Alanazi, N., Almutairi, M., & Alodhayb, A. N. (2023). A review of quartz crystal microbalance for chemical and biological sensing applications. *Sensing and Imaging, 24*(1), 10. https://doi.org/10.1007/s11220-023-00413-w.

Ali, A. A., Altemimi, A. B., Alhelfi, N., & Ibrahim, S. A. (2020). Application of biosensors for detection of pathogenic food bacteria: A review. *Biosensors, 10*(6), 58. https://doi.org/10.3390/BIOS10060058.

Anbalagan, S., Lakshminarayanan, S., & Girija, K. G (2024). Design optimization of surface acoustic wave devices and investigation of fabricated devices under temperature, humidity and steam environments. *Physica Scripta, 99*(7), 075970. https://doi.org/10.1088/1402-4896/ad547b.

Arvand, M., & Hassannezhad, M. (2014). Magnetic core-shell Fe3O4@SiO2/MWCNT nanocomposite modified carbon paste electrode for amplified electrochemical sensing of uric acid. *Materials Science and Engineering C, 36*(1), 160–167. https://doi.org/10.1016/j.msec.2013.12.014.

Basu, A. K, & Basu, A. (2020). Shantanu Bhattacharya, Micro/Nano fabricated cantilever based biosensor platform: A review and recent progress. *Enzyme and Microbial Technology, 139*, 109558. https://doi.org/10.1016/j.enzmictec.2020.109558.

Boisen, A., & Thundat, T. (2009). Design & fabrication of cantilever array biosensors. *Materials Today, 12*(9), 32–38. https://doi.org/10.1016/s1369-7021(09)70249-4.

Bonyadi, F., Kavruk, M., Ucak, S., Cetin, B., Bayramoglu, G., Dursun, A. D., & Ozalp, V. C. (2023). Real-time biosensing bacteria and virus with quartz crystal microbalance: recent advances, opportunities, and challenges. *Critical reviews in analytical chemistry*, 1–12. https://doi.org/10.1080/10408347.2023.2211164.

Bonyadi, F., Kavruk, M., Ucak, S., Cetin, B., Bayramoglu, G., Dursun, A. D., … Ozalp, V. C. (2024). Real-Time Biosensing Bacteria and Virus with Quartz Crystal Microbalance: Recent Advances, Opportunities, and Challenges. *Critical Reviews in Analytical Chemistry,, 54*(8), 2888–2899. http://doi.org/10.1080/10408347.2023.2211164.

Bragazzi, N. L, Amicizia, D., Panatto, D., Tramalloni, D., Valle, I., & Gasparini, R. (2015). Quartz-crystal microbalance (QCM) for public health: An overview of its applications. *Advances in protein chemistry and structural biology, 101*, 149–211. https://doi.org/10.1016/bs.apcsb.2015.08.002.

Brancato, L., Keulemans, G., Verbelen, T., Meyns, B., & Puers, R. (2016). An implantable intravascular pressure sensor for a ventricular assist device. *Micromachines,, 7*(8), 135. https://doi.org/10.3390/mi7080135.

Bruckner, G., & Bardong, J. (2019). Wireless readout of multiple saw temperature sensors. *Sensors, 19*(14), 3077. https://doi.org/10.3390/s19143077.

Buckner, C. A., Lafrenie, R. M., Dénommée, J. A., Caswell, J. M., Want, D. A., Gan, G. G., … Mathijssen, R. H. J. (2016). We are IntechOpen, the world's leading publisher of Open Access books Built by scientists, for scientists TOP 1 %. *Intech, 11*, 179–198. doi:10.5772/17205.

Cai, Y. u, Zhang, S., Dong, C., Yang, J., Ma, T., Zhang, H., Cui, Y., & Hui, W. (2021). Lateral flow immunoassay based on gold magnetic nanoparticles for the protein quantitative detection: Prostate-specific antigen. *Analytical Biochemistry, 627*, 114265. https://doi.org/10.1016/j.ab.2021.114265.

Castillo-Henríquez L., Brenes-Acuña M., Castro-Rojas A., Cordero-Salmerón R., Lopretti-Correa M., & Vega-Baudrit J.R., Biosensors for the detection of bacterial and viral clinical pathogens. Sensors (Switzerland), *20*(23), 1–26. https://doi.org/10.3390/s20236926.

Chen C., & Wang J. (2020). Optical biosensors: An exhaustive and comprehensive review. Analyst, *145*(5), 1605–1628. https://doi.org/10.1039/c9an01998g.</bib>

Chircov, C., & Grumezescu, A. M (2022). Microelectromechanical Systems (MEMS) for biomedical applications. *Micromachines, 13*(2), 164. https://doi.org/10.3390/mi13020164.

Debnath, S. (2021). A 2D cmut array for liver elastography. University of Windsor, Canada.

Della Ventura, B., Mauro, M., Battaglia, R., & Velotta, R. (2021). Quartz crystal microbalance sensors: New tools for the assessment of organic threats to the quality of water. *ICT for Smart Water Systems: Measurements and Data Science*, 315–342. *3030619729* https://doi.org/10.1007/698_2019_390 .

Delorme, S. (2006). Contrast-enhanced ultrasound for examining tumor biology. *Cancer Imaging, 6*(1), 148–152. https://doi.org/10.1102/1470-7330.2006.0023.

Deng, W., Jin, L., & Yang, W. (2023). Piezoelectric materials design for high-performance sensing. *Crystals, 13*(7), 1063. https://doi.org/10.3390/cryst13071063.

Dias, D., & Paulo Silva Cunha, J. (2018). Wearable health devices—vital sign monitoring, systems and technologies. *Sensors (Basel, Switzerland), 18*(8), 2414. https://doi.org/10.3390/s18082414.

Elfahem, R., Bouchet, B., Abbes, B., Legrand, F., Polidori, G., & Beaumont, F. (2024). Investigating heat transfer in whole-body cryotherapy: A 3D thermodynamic modeling approach with participant variability. *Fluids, 9*(3), 61. https://doi.org/10.3390/fluids9030061.

Erdenebayar, U., Park, J. U., Jeong, P., & Lee, K. J. (2017). Obstructive sleep apnea screening using a piezo-electric sensor. *Journal of Korean Medical Science, 32*(6), 893–899. https://doi.org/10.3346/jkms.2017.32.6.893.

Ersoy, S., & Waqar, T. (2023). In *Surface acoustic wave sensing: 1-4* (pp. 129–141). Elsevier BV. https://doi.org/10.1016/b978-0-12-822548-6.00103-5.

Etayash, H., Jiang, K., Azmi, S., Thundat, T., & Kaur, K. (2015). Real-time detection of breast cancer cells using peptide-functionalized microcantilever arrays. *Scientific Reports, 5*(1), 13967. https://doi.org/10.1038/srep13967.

Fritz, J., Baller, M. K., Lang, H. P., Rothuizen, H., Vettiger, P., Meyer, E., Güntherodt, H.-J., Gerber, C. h., & Gimzewski, J. K. (2000). Translating biomolecular recognition into nanomechanics. *Science, 288*(5464), 316–318. https://doi.org/10.1126/science.288.5464.316.

Geng, Y., Cao, R., Tendo Innocent, M., Hu, Z., Zhu, L., Wang, L. e, Xiang, H., & Zhu, M. (2022). A high-sensitive wearable sensor based on conductive polymer composites for body temperature monitoring. *Composites Part A: Applied Science and Manufacturing, 163*, 107269. https://doi.org/10.1016/j.compositesa.2022.107269.

Ghosh, G. (2020). Early detection of cancer: Focus on antibody coated metal and magnetic nanoparticle-based biosensors. *Sensors International, 1*, 100050. https://doi.org/10.1016/j.sintl.2020.100050.

Hagan, M. F., Majumdar, A., & Chakraborty, A. K. (2002). Nanomechanical forces generated by surface grafted DNA. *Journal of Physical Chemistry B, 106*(39), 10163–10173. https://doi.org/10.1021/jp020972o.

Han, X., Huang, M., Wu, Z., Gao, Y. i, Xia, Y., Yang, P., … Jiang, Z. (2023). Advances in high-performance MEMS pressure sensors: design, fabrication, and packaging. *Microsystems & Nanoengineering, 9*(1), 156. https://doi.org/10.1038/s41378-023-00620-1.

Hashemi Goradel, N., Mirzaei, H., Sahebkar, A., Poursadeghiyan, M., Masoudifar, A., & Malekshahi, Z. V. (2018). Biosensors for the detection of environmental and urban pollutions. *Journal of cellular biochemistry, 119*(1), 207–212. https://doi.org/10.1002/jcb.26030.

Hathaway, H. J., Butler, K. S., Adolphi, N. L., Lovato, D. M., Belfon, R., Fegan, D., Monson, T. C., Trujillo, J. E., Tessier, T. E., Bryant, H. C., Huber, D. L., Larson, R. S., & Flynn, E. R. (2011). Detection of breast cancer cells using targeted magnetic nanoparticles and ultrasensitive magnetic field sensors. *Breast Cancer Research, 13*(5), 1–13. https://doi.org/10.1186/bcr3050.

Hu, M., Feng, Z., Chu, Y., & Li, Y. (2021). Electrothermal analysis of radiofrequency tissue ablation with injectable flexible electrodes considering bio-heat transfer. *Theoretical and Applied Mechanics Letters, 11*(4), 100258. https://doi.org/10.1016/j.taml.2021.100258.

Jaffe, B., Roth, R. S., & Marzullo, S. (1954). Piezoelectric properties of lead zirconate-lead titanate solid-solution ceramics. *Journal of Applied Physics, 25*(6), 809–810. https://doi.org/10.1063/1.1721741.

Jain, A., Prashanth, K. J., Sharma, A. K., Jain, A., & P. n, R. (2015). Dielectric and piezoelectric properties of PVDF/PZT composites: A review. *Polymer Engineering and Science, 55*(7), 1589–1616. https://doi.org/10.1002/pen.24088.

Jothimani, D., Kailasam, E., Danielraj, S., Nallathambi, B., Ramachandran, H., Sekar, P., Manoharan, S., Ramani, V., Narasimhan, G., Kaliamoorthy, I., & Rela, M. (2020). COVID-19: Poor outcomes in patients with zinc deficiency. *International Journal of Infectious Diseases, 100*, 343–349. https://doi.org/10.1016/j.ijid.2020.09.014.

Justino, C. I., & Duarte, A. C. (2017). Recent progress in biosensors for environmental monitoring: A review. *Sensors, 17*(12), 2918. https://doi.org/10.3390/s17122918.

Kambhampati, D., & Fritz, J. ü. (2004). Nanomechanical cantilever sensors for microarrays protein microarray technology. *Wiley, 1*, 195–213. https://doi.org/10.1002/3527601554.ch9.

Kawai, H. (1969). The Piezoelectricity of Poly (vinylidene Fluoride). *Japanese Journal of Applied Physics, 8*(7), 975. https://doi.org/10.1143/JJAP.8.975.

Kirstein, K. U., Li, Y., Zimmermann, M., Vancura, C., Volden, T., Song, W. H., Lichtenberg, J., & Hierlemannn, A. (2005, December 12). Switzerland Cantilever-based biosensors in CMOS technology II. In *Proceedings –Design, Automation and Test in Europe, DATE '05* (pp. 1340–1341). 15301591. doi:10.1109/DATE.2005.90.

Koo, K. M., Soda, N., & Shiddiky, M. J. A. (2021). Magnetic nanomaterial–based electrochemical biosensors for the detection of diverse circulating cancer biomarkers. *Current Opinion in Electrochemistry, 25*. https://doi.org/10.1016/j.coelec.2020.100645.

Kryuk, R., Mukhsin, M., Kurbanova, M., & Kryuk, V. (2023). Biosensors in food industry. *BIO Web of Conferences, EDP Sciences, 64*, 01006. doi:10.1051/bioconf/20236401006.

Kukkar, D., Zhang, D., Jeon, B. H., & Kim, K. i-H. (2022). Recent advances in wearable biosensors for non-invasive monitoring of specific metabolites and electrolytes associated with chronic kidney disease: Performance evaluation and future challenges. *TrAC Trends in Analytical Chemistry, 150*, 116570. https://doi.org/10.1016/j.trac.2022.116570.

Kumar, N., Hu, Y., Singh, S., & Mizaikoff, B. (2018). Emerging biosensor platforms for the assessment of water-borne pathogens. *Analyst, 143*(2), 359–373. doi:10.1039/C7AN00983F.

Lakshmanan, A., Jin, Z., Nety, S. P., Sawyer, D. P., Lee-Gosselin, A., Malounda, D., Swift, M. B., Maresca, D., & Shapiro, M. G. (2020). Acoustic biosensors for ultrasound imaging of enzyme activity. *Nature Chemical Biology, 16*(9), 988–996. https://doi.org/10.1038/s41589-020-0591-0.

Lam, C. S., Wang, C. Y., & Wang, S. M. (2010, March 3). A review of the recent development of temperature stable cuts of quartz for SAW applications. In *Proceedings of the Fourth International Symposium on Acoustic Wave Devices for Future Mobile Communication Systems 3 5*. Citeseer.

Li, D., Wang, J., Wang, R., Li, Y., Abi-Ghanem, D., Berghman, L., Hargis, B., & Lu, H. (2011). A nanobeads amplified QCM immunosensor for the detection of avian influenza virus H5N1. *Biosensors and Bioelectronics, 26*(10), 4146–4154. https://doi.org/10.1016/j.bios.2011.04.010.

Li, C., Zhang, M., Zhang, Z., Tang, J., & Zhang, B. (2019). Microcantilever aptasensor for detecting epithelial tumor marker Mucin 1 and diagnosing human breast carcinoma MCF-7 cells. *Sensors and Actuators B: Chemical, 297*, 126759. https://doi.org/10.1016/j.snb.2019.126759.

Liang, R. P., Yao, G. H., Fan, L. X., & Qiu, J. D. (2012). Magnetic Fe 3O 4@Au compositeenhanced surface plasmon resonance for ultrasensitive detection of magnetic nanoparticleenriched α-fetoprotein. *Analytica Chimica Acta, 737*, 22–28. https://doi.org/10.1016/j.aca.2012.05.043.

Lozano, M. G., García, Y. P., Gonzalez, J. A. S., Bañuelos, C. V. O., Escareño, M. P. L., & Balagurusamy, N. (2018). Biosensors for food quality and safety monitoring: Fundamentals and applications. In *Enzymes in Food Biotechnology: Production, Applications, and Future Prospects* (pp. 691–709). Mexico: Elsevier. Available from:. http://www.sciencedirect.com/science/book/9780128132807 . doi:10.1016/B978-0-12-813280-7.00040-2.

Ma, M., Yang, X., Ying, X., Shi, C. e, Jia, Z., & Jia, B. (2023). Applications of gas sensing in food quality detection: A review. *Foods, 12*(21), 3966. https://doi.org/10.3390/foods12213966.

Mandal, D., & Banerjee, S. (2022). Surface acoustic wave (SAW) sensors: Physics, materials, and applications. *Sensors, 22*(3), 820. https://doi.org/10.3390/s22030820.

Martin, D. T., Liu, J., Kadirvel, K., Fox, R. M., Sheplak, M., & Nishida, T. (2007). A micromachined dual-backplate capacitive microphone for aeroacoustic measurements. *Journal of Microelectromechanical Systems, 16*(6), 1289–1302. https://doi.org/10.1109/JMEMS.2007.909234.

Mason, W. P. (1981). Piezoelectricity, its history and applications. *Journal of the Acoustical Society of America, 70*(6), 1561–1566. https://doi.org/10.1121/1.387221.

Morales, M. A., & Halpern, J. M (2018). Guide to selecting a biorecognition element for biosensors. *Bioconjugate Chemistry, 29*(10), 3231–3239. https://doi.org/10.1021/acs.bioconjchem.8b00592.

Muhammad, M., Li, M., Lou, Y., & Lee, C.-S. (2024). Continual monitoring of respiratory disorders to enhance therapy via real-time lung sound imaging in telemedicine. *Electronics, 13*(9), 1669. https://doi.org/10.3390/electronics13091669.

Mujahid, A., Afzal, A., Dickert, F. L, et al. (2019). An overview of high frequency acoustic sensors—QCMs, SAWs and FBARs—chemical and biochemical applications. *Sensors, 19*(20), 4395–4424. doi:10.3390/s19204395.

Mujahid, A., & Dickert, F. L. (2017). Surface acoustic wave (SAW) for chemical sensing applications of recognition layers. *Sensors, 17*(12), 2716. https://doi.org/10.3390/s17122716.

Mustafa, F., & Andreescu, S. (2018). Chemical and biological sensors for food-quality monitoring and smart packaging. *Foods, 7*(10), 168. https://doi.org/10.3390/foods7100168.

Na Songkhla, S., & Nakamoto, T. (2021). Overview of quartz crystal microbalance behavior analysis and measurement. *Chemosensors, 9*(12), 350. https://doi.org/10.3390/chemosensors9120350.

Okahata, Y., Kawase, M., Niikura, K., Ohtake, F., Furusawa, H., & Ebara, Y. (1998). Kinetic measurements of DNA hybridization on an oligonucleotide-immobilized 27-MHz quartz crystal microbalance. *Analytical Chemistry, 70*(7), 1288–1291. https://doi.org/10.1021/ac970584w.

Ozalp, V. C., Bayramoglu, G., Erdem, Z., & Arica, M. Y. (2015). Pathogen detection in complex samples by quartz crystal microbalance sensor coupled to aptamer functionalized core-shell type magnetic separation. *Analytica Chimica Acta, 853*(1), 533–540. https://doi.org/10.1016/j.aca.2014.10.010.

P J, J., Prabakaran, K., Luo, J., & Derry Holaday, M. G (2021). Effective utilization of quartz crystal microbalance as a tool for biosensing applications. *Sensors and Actuators A: Physical, 331*, 113020. https://doi.org/10.1016/j.sna.2021.113020.

Pandya, A., & Singh, V. (2022). Progress molecular biology and translational science. *Micro/Nanofluidics and Lab-on-Chip Based Emerging Technologies for Biomedical and Translational Research Applications-Part B, 187*. doi:10.1016/S1877-1173(22)00005-9.

Panwar, L. S., Panwar, V., Kala, S., Panwar, S. S., & Sharma, S. (2017). Design of MEMS piezoelectric blood pressure sensor 2018-. In *Proceedings – 2017 3rd International Conference on Advances in Computing, Communication and Automation (Fall), ICACCA 2017* (pp. 1–7). India: Institute of Electrical and Electronics Engineers Inc. 9781509064038. doi:10.1109/ICACCAF.2017.8344698.

Park, J. S., Jung, S. Y., Kim, D. H., Park, J. H., Jang, H. W., Kim, T. G., Baek, S. H., & Lee, B. C. (2023). Dual-frequency piezoelectric micromachined ultrasound transducer based on polarization switching in ferroelectric thin films. *Microsystems and Nanoengineering, 9*(1). https://doi.org/10.1038/s41378-023-00595-z.

Pohanka, M. (2021). Quartz crystal microbalance (QCM) sensing materials in biosensors development. *International Journal of Electrochemical Science, 16*(12), 211220. https://doi.org/10.20964/2021.12.15.

Ponnuthurai, D. R., Amirah Alim, A., Mohd Razip Wee, M. F., & Latif, R. (2023 January 1). Equivalent Circuit Model and Simulation of 2D Asymmetrical PMUT for Non-Destructive Testing. In *Proceedings – 2023 IEEE Regional Symposium on Micro and Nanoelectronics, RSM 2023* (pp. 36–39). Malaysia: Institute of Electrical and Electronics Engineers Inc. 9798350323689. http://ieeexplore.ieee.org/xpl/mostRecentIssue.jsp?punumber=10326449 . doi:10.1109/RSM59033.2023.10327201.

Prakash, A., Chandra, S., & Bahadur, D. (2012). Structural, magnetic, and textural properties of iron oxide-reduced graphene oxide hybrids and their use for the electrochemical detection of chromium. *Carbon, 50*(11), 4209–4219. https://doi.org/10.1016/j.carbon.2012.05.002.

Qu, M., Yang, D., Chen, X., Li, D., Zhu, K., & Xie, J. (2021 January 25). Heart Sound Monitoring Based on a Piezoelectric Mems Acoustic Sensor 2021-. In *Proceedings of the IEEE International Conference on Micro Electro Mechanical Systems (MEMS)* (pp. 59–63). China: Institute of Electrical and Electronics Engineers Inc. 10.1109/MEMS51782.2021.9375350.

Ravalli, A., & Marrazza, G. (2015). Gold and magnetic nanoparticles-based electrochemical biosensors for cancer biomarker determination. *Journal of Nanoscience and Nanotechnology, 15*(5), 3307–3319. https://doi.org/10.1166/jnn.2015.10038.

Royer, M., Holmen, J. O., Wurm, M. A., Aadland, O. S., & Glenn, M. (1983). ZnO on Si integrated acoustic sensor. *Sensors and Actuators, 4*(I), 357–362. https://doi.org/10.1016/0250-6874(83)85044-6.

Samartkit, P., Pullteap, S., & Bernal, O. (2022). A non-invasive heart rate and blood pressure monitoring system using piezoelectric and photoplethysmographic sensors. *Measurement, 196*, 111211. https://doi.org/10.1016/j.measurement.2022.111211.

Sezen, A. S., Sivaramakrishnan, S., Hur, S., Rajamani, R., Robbins, W., & Nelson, B. J. (2005). Passive wireless MEMS microphones for biomedical applications. *Journal of Biomechanical Engineering, 127*(6), 1030–1034. https://doi.org/10.1115/1.2049330.

Sharma, A., Badea, M., Tiwari, S., & Marty, J. L. (2021). Wearable biosensors: An alternative and practical approach in healthcare and disease monitoring. *Molecules, 26*(3). https://doi.org/10.3390/molecules26030748.

Sharma, M., Srinivas, V., Madras, G., & Bose, S. (2016). Outstanding dielectric constant and piezoelectric coefficient in electrospun nanofiber mats of PVDF containing silver decorated multiwall carbon nanotubes: assessing through piezoresponse force microscopy. *RSC Advances, 6*(8), 6251–6258. https://doi.org/10.1039/c5ra25671b.

Shuai, C., Liu, G., Yang, Y., Qi, F., Peng, S., Yang, W., He, C., Wang, G., & Qian, G. (2020). A strawberry-like Ag-decorated barium titanate enhances piezoelectric and antibacterial activities of polymer scaffold. *Nano Energy, 74*, 104825. https://doi.org/10.1016/j.nanoen.2020.104825.

Smagin, N., Vanotti, M., Duquennoy, M., Rousseau, L., Alhousseini, H., Blondeau-Patissier, V., & Herth, E. (2022). Electrical and optical characterization of SAW sensors coated with parylene C and their analysis using the coupling-of-modes (COM) theory. *Sensors,, 22*(22), 8611. https://doi.org/10.3390/s22228611.

Spiegel, B. M. R., Kaneshiro, M., Russell, M. M., Lin, A., Patel, A., Tashjian, V. C., Zegarski, V., Singh, D., Cohen, S. E., Reid, M. W., Whitman, C. B., Talley, J., Martinez, B. M., & Kaiser, W. (2014). Validation of an acoustic gastrointestinal surveillance biosensor for postoperative ileus. *Journal of Gastrointestinal Surgery, 18*(10), 1795–1803. https://doi.org/10.1007/s11605-014-2597-y.

Srinivasan, B., Li, Y., Jing, Y., Xing, C., Slaton, J., & Wang, J. P. (2011). A three-layer competition based giant magnetoresistive assay for direct quantification of endoglin from human urine. *Analytical Chemistry, 83*(8), 2996–3002. https://doi.org/10.1021/ac2005229.

Tetyana, P., Shumbula, P. M, & Njengele-Tetyana, Z. (2021). *Biosensors: Design, development and applications.* IntechOpen https://doi.org/10.5772/intechopen.97576.

Wang, H., Sun, Y., Wang, Y., Chen, Y., Ge, Y., Yuan, J., & Carson, P. (2023). Temperature-controlled hyperthermia with non-invasive temperature monitoring through speed of sound imaging. *Applied Sciences (Switzerland), 13*(12). https://doi.org/10.3390/app13127317.

Wang, T. W., & Lin, S. F. (2020). Wearable piezoelectric-based system for continuous beat-to-beat blood pressure measurement. *Sensors (Switzerland), 20*(3). https://doi.org/10.3390/s20030851.

Weitzel, W. F., Cotant, C. L., Wen, Z., Biswas, R., Patel, P., Panduranga, H., Gianchandani, Y. B., & Rubin, J. M. (2008). Analysis of novel geometry-independent method for dialysis access pressure-flow monitoring. *Theoretical Biology and Medical Modelling, 5*. https://doi.org/10.1186/1742-4682-5-22.

Wu, Guanghua, Datar, Ram H, Hansen, Karolyn M, Thundat, Thomas, Cote, Richard J, & Majumdar, Arun (2001). Bioassay of Prostate-Specific Antigen (PSA) Using Microcantilevers. *Nature Biotechnology, 19*(9), 856–860. https://doi.org/10.1038/nbt0901-856.

Xue, Y., Liu, Y., Zhou, C., & Zhang, X. Y. (2024). Acoustic wave devices based on piezoelectric/ferroelectric thin films for high-frequency communication systems and sensing applications. *Ceramics International, 50*(23), 52051–52058. https://doi.org/10.1016/j.ceramint.2024.03.212.

Yan, T.-R., Lee, C.-F. a, & Chou, H.-C. (2012). *QCM as cell-based biosensor.* InTech https://doi.org/10.5772/34261.

Ye, Y., Guo, H., & Sun, X. (2019). Recent progress on cell-based biosensors for analysis of food safety and quality control. *Biosensors and Bioelectronics, 126*, 389–404. https://doi.org/10.1016/j.bios.2018.10.039.

Yılmaz, M, Bakhshpour, M, Göktürk,, I, Pişkin, A, K, Denizli, A, et al. (2021). Quartz crystal microbalance (QCM) based biosensor functionalized by HER2/Neu antibody for breast cancer cell detection. *Chemosensors, 9*(4), 80–95. doi:10.3390/chemosensors9040080.

Zafirah, T. N, Triqadafi, A. O, Kusuma, F. J, & Sakti, S. P (2021). Frequency response of Quartz Crystal Microbalance under uniaxial force. *Journal of Physics: Conference Series, 1951*(1), 012041. https://doi.org/10.1088/1742-6596/1951/1/012041.

Zhao, X., Gao, H., Hou, Y., Gbologah, L., Zeng, X., & Wang, Y. (2022). Analysis of crystallization and deposition process using electrochemical-quartz crystal microbalance: A review. *Journal of Electroanalytical Chemistry, 904*(115936). https://doi.org/10.1016/j.jelechem.2021.115936.

Zhu, M., Yi, Z., Yang, B., & Lee, C. (2021). Making use of nanoenergy from human – Nanogenerator and self-powered sensor enabled sustainable wireless IoT sensory systems. *Nano Today, 36*, 101016. https://doi.org/10.1016/j.nantod.2020.101016.

Zu, P., Chan, C. C., Koh, G. W., Lew, W. S., Jin, Y., Liew, H. F., Wong, W. C., & Dong, X. (2014). Enhancement of the sensitivity of magneto-optical fiber sensor by magnifying the birefringence of magnetic fluid film with Loyt-Sagnac interferometer. *Sensors and Actuators, B: Chemical, 191*, 19–23. https://doi.org/10.1016/j.snb.2013.09.085.

CHAPTER

5

BioMEMS and dielectrophoresis for bio-cells and particle separation, manipulation, and filtration

Muhamad Ramdzan Buyong[a], Celine Elie-Caille[b], Arash Zulkarnain[a], Muhammad Akmal Suhaimi[c] and Noraziah Mohamad Zin[d]

[a]*Institut Kejuruteraan Mikro dan Nanoelektronik, Universiti Kebangsaan Malaysia, Bangi, Selangor, Malaysia,* [b]*Institut FEMTO-ST, UMR 6174 CNRS, Université de Franche-Comté, Besançon, France,* [c]*Institute of Microengineering and Nanoelectronics (IMEN), Universiti Kebangsaan Malaysia, Bangi, Selangor, Malaysia,* [d]*Center for Diagnostic, Therapeutic and Investigative Studies, Faculty of Health Sciences, Universiti Kebangsaan Malaysia, Jalan Raja Muda Abdul Aziz, Kuala Lumpur, Malaysia*

5.1 Evolution of biological detection technologies

The ability to detect and characterize biological particles has shaped modern medicine, environmental protection, and biotechnology manufacturing. Historically, throughout the decades, Fig. 5.1 shows all microbiological detection times for detection and their sensitivity. For healthcare diagnostics, environmental quality monitoring, and food and beverages quality, the microbiological detection relied primarily on culture-based techniques known as the plate count method, which involves growing microorganisms under controlled laboratory conditions. These methods remain highly reliable and widely accepted as diagnostic gold standards toward body fluid samples, environmental samples, and food and beverages for microbiological quality assessment. However, these culture techniques method, such as plate count method, require incubation periods ranging from several hours 24 hours to multiple days, limiting their usefulness in time-critical diagnostic scenarios.

Biochemical identification methods were later introduced to improve detection speed through metabolic profiling and enzymatic activity measurement. Biochemical identification deemed rapid identification for bacterial species than culture methods. Although biochemical tests often suffer from limited specificity as it distinguishing closely related species such as coagulase test for *Staphylococcus* species identification to differentiate *Staphylococcus aureus* and *Staphylococcus epidermidis*. The emergence of molecular diagnostics, including polymerase chain reaction (PCR) and sequencing technologies, marked a major leap forward in detection sensitivity and specificity. These techniques allow the identification of genetic material directly from biological samples without requiring organism growth.

Despite these advancements, molecular diagnostics present several limitations. They typically require expensive reagents, specialized instrumentation, and trained personnel. In addition, molecular detection cannot always distinguish between viable and nonviable biological particles, which is particularly important in infection diagnostics and environmental safety monitoring.

BioMEMS Devices. DOI: https://doi.org/10.1016/B978-0-443-24015-7.00013-9

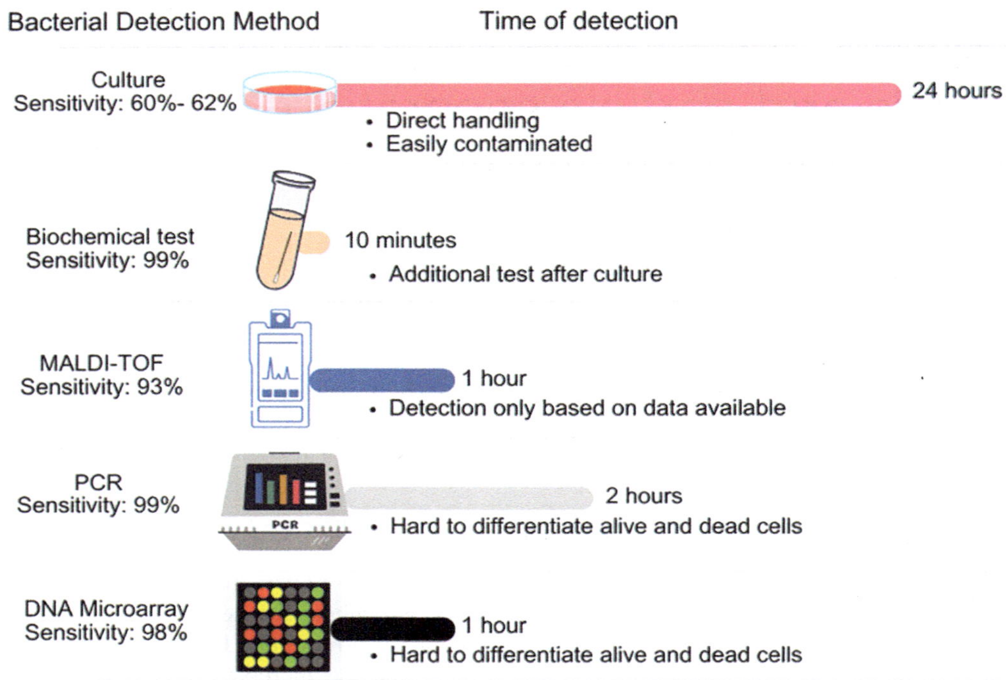

FIGURE 5.1

Microbial detection technique, time of detection, sensitivity and disadvantages. Figure show different microbial detection techniques.

These challenges have driven the development of physical and biophysical detection technologies. Electrical characterization techniques, including impedance spectroscopy and dielectrophoresis (DEP), offer label-free detection mechanisms based on intrinsic biological properties rather than biochemical markers. These methods are capable of providing rapid characterization while preserving biological sample integrity.

DEP has emerged as one of the most promising physical characterization techniques due to its ability to manipulate particles directly based on dielectric property differences. DEP allows separation, concentration, enrichment, and phenotypic characterization of biological particles using electrical field gradients (Castellanos et al., 2003; Jones, 1995; Morgan & Green 2003; Pohl, 1978). This capability has positioned DEP as a core technology in emerging BioMEMS-based diagnostic platforms.

The theoretical and experimental foundation of DEP originates from classical electromechanical particle theory and electrokinetic physics, where particle motion in nonuniform electric fields is governed by induced dipole polarization and the Clausius–Mossotti factor, which determines positive and negative DEP behavior across frequency domains (Jones, 1995; Morgan & Green 2003; Pohl, 1978). Early dielectric characterization studies established that biological cells behave as multi-shell dielectric particles, where membrane capacitance, cytoplasmic conductivity, and structural integrity determine frequency-dependent polarization behavior and interfacial charge accumulation (Asami, 2002; Markx & Davey, 1999; Schwan, 1957). These dielectric principles form the basis for modern DEP-based biological particle manipulation and separation technologies, which have been widely demonstrated for mammalian cells, circulating tumor cell enrichment, and dielectric phenotyping of heterogeneous biological populations (Gascoyne & Shim, 2014; Gascoyne & Vykoukal, 2002).

The integration of DEP into microsystems and BioMEMS platforms has been enabled through electrohydrodynamic system modeling and microelectrode field engineering, allowing precise particle manipulation within microfluidic environments (Castellanos et al., 2003; Cummings & Singh, 2003; Hughes, 2002). The evolution of microfluidic BioMEMS systems has further accelerated DEP translation into biomedical diagnostics, where microfluidic lab-on-chip platforms provide low sample volume, high throughput, and portable diagnostic capability (Chin et al., 2012; Drain et al., 2014; Sackmann et al., 2014; Whitesides, 2006). These developments are supported by advances in microfabrication technologies, particularly CMOS-compatible MEMS fabrication, enabling high-resolution electrode patterning, scalable manufacturing, and integration of sensing and actuation within single microelectrode platforms (Franssila, 2010; Madou, 2011). Electrode material selection, particularly the use of gold and biocompatible conductive materials, has further improved DEP device stability and long-term biological compatibility in complex biological media (Grieshaber et al., 2008).

Recent experimental studies have expanded DEP application toward clinically relevant pathogen detection and rapid microbial screening, including ESKAPE pathogen detection through crossover-frequency dielectric analysis and conductivity–permittivity correlation mapping. Advances in microelectrode engineering, particularly tapered microelectrode DEP systems, have demonstrated improved electric field gradient control, enabling enhanced particle trapping efficiency, DEP filtration, and simultaneous sensing-actuation mechanisms for lab-on-chip DEP applications. These device engineering approaches are supported by dielectric modeling and computational simulation tools such as MyDEP, enabling predictive polarization spectrum analysis based on experimentally derived dielectric shell parameters.

DEP-based biological analysis has been widely applied across multiple biological scales. At the microbial level, DEP has demonstrated strong capability for bacterial separation, physiological state characterization, antibiotic susceptibility testing, and strain-level discrimination based on dielectric property differences linked to cell wall composition and membrane permeability. These approaches are further supported by structural microbiology studies demonstrating correlation between antibiotic resistance and bacterial cell wall structural modification, which directly influences dielectric polarization behavior. In parallel, DEP has been successfully applied for nanoscale biological particle manipulation, including extracellular vesicle and exosome isolation, enabling emerging diagnostic approaches for precision medicine and biomarker discovery. DEP-based microbial detection methods are also positioned relative to conventional microbiology technologies including MALDI-TOF mass spectrometry and molecular PCR workflows, highlighting DEP's role as a rapid physical phenotyping complement to molecular diagnostics.

DEP applications have further expanded into mammalian cell engineering and biomedical therapeutic platforms, including wound healing and tissue regeneration systems using keratinocyte-directed DEP manipulation and particle synchronization for controlled cell migration and biological interaction studies. Integration of DEP with advanced microfluidic flow control, including flow focusing and channel geometry optimization, further improves DEP manipulation efficiency and system scalability. Low-frequency DEP behavior characterization and single-particle electrokinetic modeling further

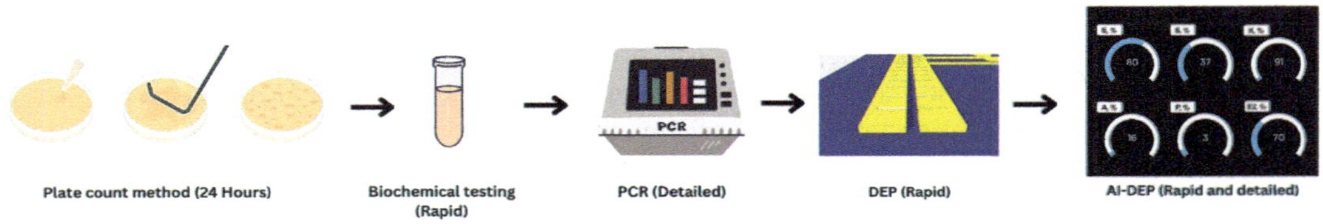

FIGURE 5.2

Evolution of microbial technique.

strengthen theoretical–experimental linkage between polarization response and real particle motion under DEP fields [48]. The broader field of bioelectric-based antimicrobial technologies further supports the growing importance of electric-field-driven biological manipulation in future biomedical treatment platforms.

More recently, the integration of artificial intelligence (AI) and machine learning into DEP signal analysis has enabled automated particle tracking, polarization spectrum interpretation, dielectric phenotype classification, and predictive diagnostic modeling as shown in Fig. 5.2 AI-assisted medical imaging and digital health integration frameworks provide the computational foundation for future DEP systems capable of autonomous biological diagnostics and real-time biological state monitoring (Litjens et al., 2017). These developments position DEP as a key enabling technology for next-generation digital healthcare ecosystems, global diagnostic deployment platforms, and sustainable low-resource diagnostic solutions. All these will be discuss throughout the whole book chapter.

5.2 Emergence of BioMEMS and microscale biological engineering

BioMEMS technology integrates microelectronics, microfluidics, and biological interfaces into compact microscale devices. These systems are capable of performing complex biological analysis with minimal sample volume and high automation potential. BioMEMS devices have enabled significant advances in point-of-care diagnostics, wearable biosensors, and portable environmental monitoring tools.

Microelectronics refers to the design and fabrication of extremely small electronic circuits and components, typically built using semiconductor manufacturing technologies such as photolithography and thin-film deposition. In BioMEMS devices, microelectronics provides the electrical infrastructure required for sensing, actuation, signal processing, and data transmission. In biological applications, microelectronic components commonly include microelectrodes for generating electric fields example microelectrode for DEP device and sensors for detecting electrical specifically dielectric in DEP. Microelectronics enables precise electrical control at micronscale spatial resolution, which is essential for manipulating cells, bacteria, or biomolecules using dielectric parameter.

Microfluidics refers to the science and engineering of controlling fluid flow in channels with dimensions typically ranging from tens to hundreds of micrometers. At this scale, fluid behavior is dominated by laminar flow rather than turbulence, allowing highly predictable fluid motion and precise sample handling. Microfluidic systems perform important functions in BioMEMS devices especially DEP to Isolate or concentrate biological particles. Due to microfluidic systems require extremely small sample volumes, this system significantly reduce reagent consumption and allow rapid reaction kinetics due to one small space region for region of interest.

The miniaturization of biological analysis systems provides several advantages. Reduced sample volume requirements lower reagent costs and enable testing in resource-limited environments. Small diffusion region within microfluidic channels accelerates reaction kinetics and improves detection speed. Integration of multiple analytical steps within a single device reduces contamination risk and improves reliability.

Electrical manipulation techniques such as DEP are particularly well-suited for BioMEMS integration because DEP can be implemented using microfabricated electrode structures. These electrodes can generate localized electric field gradients capable of manipulating biological particles at microscale dimensions. DEP is uniquely compatible with BioMEMS fabrication processes. Standard microfabrication methods, including photolithography and thin film deposition, can produce highly precise electrode geometries optimized for specific DEP applications.

5.3 Fundamental physics of dielectrophoresis

DEP describes the motion of polarizable particles in nonuniform electric fields. Unlike electrophoresis, which depends on net particle charge, DEP operates on induced dipole interactions and therefore can manipulate neutral particles. When a biological particle is exposed to an alternating electric field, charges within the particle redistribute, forming an induced dipole. The interaction between this dipole and the spatial gradient of the electric field produces a net force that drives particle movement. The DEP force magnitude is governed by: Eq. 5.1

The force increases with particle size, medium permittivity, and electric field gradient strength. The Clausius–Mossotti factor determines the direction of particle motion and is influenced by particle and medium dielectric properties. Biological particles exhibit frequency-dependent polarization behavior due to interfacial charge accumulation at dielectric boundaries. These mechanisms are described by Maxwell–Wagner polarization theory and are fundamental to DEP-based biological characterization. The theoretical framework described above is consistent with widely accepted DEP modeling approaches used for microbial and cellular dielectric analysis.

5.4 Frequency-dependent biological polarization mechanisms on multi-shell dielectric model

Biological particles model will be discussed in this section where all biological particles contain multiple dielectric interfaces, including cell membrane, cell wall, cytoplasm, and extracellular coatings. Each interface contributes to overall polarization behavior (Asami, 2002; Markx & Davey, 1999; Schwan, 1957). At low frequencies, electric fields interact primarily with outer structural layers. Cell membranes act as capacitive barriers that limit field penetration. At intermediate frequencies, electric fields begin penetrating membrane structures and interacting with intracellular components. At high frequencies, electric fields can fully penetrate cellular structures and interact with cytoplasmic conductivity. This multi-frequency response allows DEP to distinguish biological particles based on

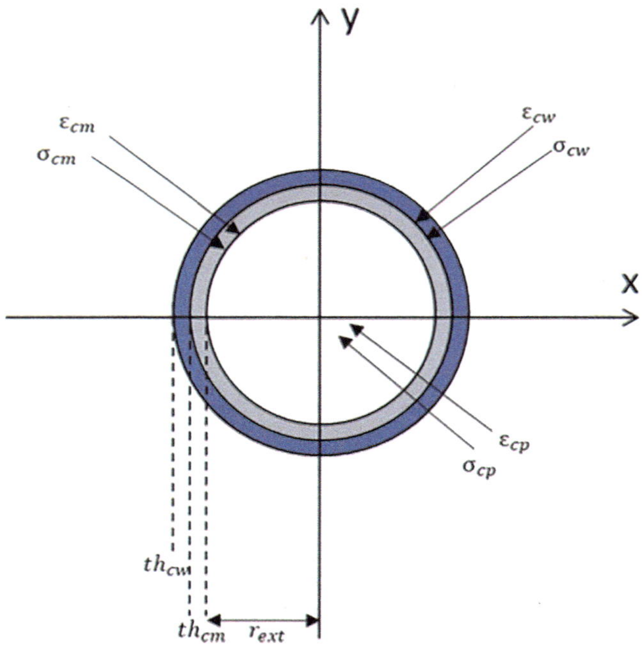

FIGURE 5.3

Multi-shell model consist of thickness, conductivity, and permittivity of cell wall, thickness, conductivity, and permittivity of cell membrane and conductivity and permittivity of cytoplasm.

structural integrity, physiological state, and composition. Changes in membrane permeability, intracellular conductivity, and cell wall structure can significantly shift polarization behavior. Bacteria typically require multi-shell dielectric modeling especially double shell model as shown in Fig. 5.3 due to their complex envelope structure. Most bacteria contain at least three major dielectric regions: cytoplasm, cell membrane, and cell wall.

Red blood cells have a simpler dielectric structure compared to bacteria. Mature red blood cells lack nuclei and most internal organelles. Their dielectric behavior is dominated by the plasma membrane and hemoglobin-rich cytoplasm. At low frequencies, polarization is dominated by membrane capacitance. At intermediate frequencies, electric fields penetrate the membrane and interact with cytoplasmic hemoglobin conductivity. At high frequencies, intracellular conductivity dominates polarization behavior. Fig. 5.4 shows single-shell dielectric model and red blood cells are often modeled using a single-shell dielectric model because the membrane is the dominant dielectric barrier.

White blood cells contain complex internal structures including nucleus, mitochondria, and cytoplasmic organelles. These internal structures introduce additional dielectric interfaces that influence polarization. At low frequencies, polarization is dominated by plasma membrane capacitance. At intermediate frequencies, nuclear membrane and cytoplasmic organelles contribute additional

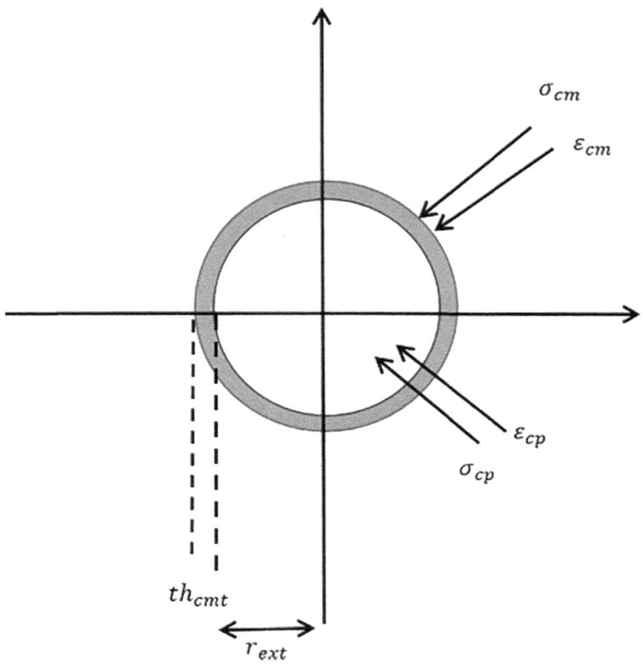

FIGURE 5.4

Single-shell model consist of thickness, conductivity and permittivity of cell membrane and conductivity and permittivity of cytoplasm.

polarization transitions. At high frequencies, intracellular conductivity dominates. Due to internal complexity, white blood cells often require multi-shell dielectric modeling for accurate DEP prediction.

Viruses are structurally simpler than cells but more complex than proteins. Many viruses contain a nucleic acid core surrounded by a protein capsid, and some contain an additional lipid envelope. Nonenveloped viruses can often be approximated as homogeneous particles or single-shell particles depending on capsid thickness and internal conductivity contrast. Enveloped viruses require multi-shell modeling because lipid envelopes act as dielectric barriers similar to cellular membranes.

Proteins are typically nanoscale biological molecules without internal compartmentalization. Their dielectric behavior is dominated by their overall molecular permittivity and hydration shell. Because proteins lack internal dielectric layering, they are typically modeled as homogeneous dielectric particles or effective spheres. At low frequencies, polarization is dominated by surface charge and hydration shell conductivity. At higher frequencies, polarization becomes dominated by bulk molecular permittivity.

Extracellular vesicles are lipid bilayer particles containing internal biomolecular cargo. Their structure is similar to simplified cells but without organelles. Polarization is dominated by lipid bilayer membrane at low frequency and internal cargo conductivity at higher frequency. Table 5.1 shows the biological particles and type of model shell for dielectric properties.

Table 5.1 Biological particles and type of model.

Biological particle	Structural Complexity	Type of Model
Bacteria	Very high	Double / Triple shell
Red blood cell	Moderate	Single shell
White blood cell	High	Multi shell
Virus (nonenveloped)	Low	Homogeneous / Single shell
Virus (enveloped)	Moderate	Double shell
Protein	Very low	Homogeneous sphere
Exosome	Moderate	Single shell

5.5 DEP device optimization and microfluidic integration in DEP system

This section discussed on DEP devices optimization into several major categories to address the diverse challenges associated with biological particle manipulation. Due to biological samples vary significantly in conductivity, composition, particle size, and required processing throughput, different DEP device designs have been developed to optimize performance under specific experimental or application conditions. While all DEP systems as shown in Table 5.2 operate based on the polarization of particles in nonuniform electric fields, the method used to generate electric field gradients and control particle motion differs between device types. Modern DEP systems are generally classified into electrode-based DEP devices, insulator-based DEP devices, traveling wave DEP systems, and microfluidic-integrated DEP platforms, each offering unique advantages depending on the target biological application (Castellanos et al., 2003; Cummings & Singh, 2003; Hughes, 2002).

Electrode-based DEP devices generate electric field gradients directly using patterned microelectrodes fabricated on substrates such as silicon, glass, or polymer materials. When an alternating current voltage is applied across these microelectrodes, strong localized electric field gradients form near electrode edges and narrow interelectrode gaps. Biological particles entering these regions experience DEP forces that cause them to move according to their dielectric properties relative to the surrounding medium. The magnitude and spatial distribution of these electric field gradients can be precisely controlled by modifying electrode geometry, electrode spacing, applied voltage amplitude, and signal frequency. Researchers frequently use electrode-based DEP because these devices offers high spatial precision, strong force generation capability, and excellent frequency selectivity. These systems are particularly valuable when precise positioning of particles is required, such as trapping individual cells, concentrating bacterial cells into sensing regions, or performing DEP-based cytometry. However, electrode-based DEP devices can experience electrode fouling when exposed to complex biological fluids, and high conductivity media may increase Joule heating and reduce long-term electrode stability.

Insulator-based DEP (iDEP) devices generate electric field gradients using insulating microstructures embedded within conductive microchannels rather than relying on patterned metal electrodes to shape the field. In these systems, the electric field becomes concentrated around geometric constrictions created by insulating pillars or channel features, producing strong local field gradients that induce DEP

Table 5.2 DEP devices.

DEP device type	Operating principle	Advantage	Disadvantage	References	Applications
Electrode-based DEP	Electric field gradients generated directly from patterned microelectrodes	High field precision; strong DEP force; low voltage operation; compatible with CMOS fabrication	Electrode fouling; electrolysis risk in high conductivity samples	(Castellanos et al., 2003; Jones, 1995; Morgan & Green, 2003; Pohl 1978)	Bacterial detection, cell sorting, DEP quantification, EV trapping
Insulator-based DEP (iDEP)	Electric field distorted using insulating channel geometries instead of exposed electrodes	Reduced electrode degradation; works in high conductivity samples; robust for environmental samples	Requires higher voltages; complex channel fabrication	(Cummings & Singh, 2003; Lapizco-Encinas, 2011)	Environmental microbial separation, pathogen concentration
Traveling wave DEP (twDEP)	Phase-shifted electrodes generate moving electric field waves producing directional particle transport	Continuous particle transport; dynamic separation based on dielectric phase lag	Complex electrode driving circuitry; higher system cost	(Castellanos et al., 2003; Hughes, 2002; Morgan & Green, 2003)	Continuous cell sorting, stem cell fractionation
Insulator-based DEP (iDEP)	Electric field distorted using insulating channel geometries instead of exposed electrodes	Reduced electrode degradation; works in high conductivity samples; robust for environmental samples	Requires higher voltages; complex channel fabrication	(Cummings & Singh, 2003; Lapizco-Encinas, 2011)	High-throughput cell capture, CTC isolation

(continued on next page)

Table 5.2 DEP devices—cont'd

DEP device type	Operating principle	Advantage	Disadvantage	References	Applications
3D DEP / Vertical DEP	Electric fields generated in vertical or 3D electrode geometries	Strong trapping volume; increased throughput; improved capture probability	More complex fabrication; alignment challenges	(Gascoyne & Vykoukal, 2002)	Point-of-care diagnostics, antimicrobial testing, DEP quantification systems
Microfluidic-integrated DEP	DEP combined with hydrodynamic flow control	Continuous processing; high throughput; PoC compatible; automation ready	Requires precise force balancing (DEP vs. drag vs. EO flow)	(Chin et al., 2012; Sackmann et al., 2014; Whitesides, 2006)	Rapid pathogen detection, DEP quantification, crossover frequency diagnostics
Capacitive DEP sensing-actuation hybrid	Combines DEP manipulation with capacitive sensing feedback	Simultaneous manipulation + detection; enables real-time monitoring	More complex electronics integration	(Ghafar-Zadeh & Sawan, 2007)	Smart diagnostics, automated microbial quantification, digital health monitoring

forces on suspended particles. Researchers often choose iDEP when working with high conductivity biological samples such as blood or complex environmental fluids because electrodes can be placed outside the primary sample region, reducing direct electrode exposure to biological material and minimizing fouling effects. This configuration improves device durability and enables longer continuous operation times. Insulator-based DEP systems are commonly applied in blood cell separation, environmental particle purification, and continuous biological sorting applications. Despite these advantages, iDEP devices often require higher applied voltages compared to electrode-based systems and may provide less localized field precision.

Traveling wave DEP systems represent another important DEP architecture in which phase-shifted alternating current signals are applied across multiple electrode arrays to generate moving electric field gradients. These traveling electric fields create directional particle transport along electrode surfaces, allowing particles to be moved across device regions without relying on fluid flow. Traveling wave DEP is particularly useful for applications requiring continuous particle transport while maintaining gentle manipulation conditions that preserve biological viability. These systems are frequently used in automated cell handling, continuous cell sorting, and micro total analysis systems. However, traveling wave DEP systems typically require more complex electronic control circuitry and precise phase synchronization, increasing device design complexity.

Microfluidic DEP systems represent the most advanced and application-ready form of DEP technology by integrating electrical particle manipulation with controlled fluid transport. In microfluidic DEP devices, biological samples are transported through microchannels into DEP active regions where electric field gradients perform selective trapping, repulsion, or separation (Chin et al., 2012; Sackmann et al., 2014; Whitesides, 2006). Particle motion within microfluidic DEP systems is governed by the balance between DEP force, hydrodynamic drag force, electroosmotic flow, and Brownian motion. Successful microfluidic DEP device operation requires careful optimization of channel geometry, fluid flow velocity, medium conductivity, and electric field strength to ensure that dielectrophoretic forces dominate particle behavior within the active separation region.

Researchers increasingly focus on microfluidic DEP systems because they enable continuous sample processing, high throughput biological analysis, and automated operation with minimal reagent consumption. These characteristics make microfluidic DEP particularly attractive for point-of-care diagnostic applications where rapid detection and minimal sample preparation are essential. Microfluidic DEP systems are currently being explored for rapid pathogen detection, liquid biopsy analysis, environmental monitoring, and portable diagnostic device development. In these applications, microfluidic integration enables translation of DEP from laboratory-scale experiments to real-world clinical and field-deployable diagnostic platforms.

The balance between fluid forces and dielectrophoretic forces is critical in microfluidic DEP system design. Due to that, in our case study, to solve fluid flow velocity causing particles may escape DEP trapping regions before separation can occur, we introduced small stationary droplet for uniform movement and manipulation for before DEP exposure and when after DEP exposure. Conversely, even electric field gradients can be controlled by wave station generator based on voltage peak to peak, to control the joule heating, and when DEP electric fields are too weak, DEP forces will be sufficient to overcome hydrodynamic drag forces at stationary droplet. As a result, researchers can apply and carefully optimize operating parameters including flow velocity, channel dimensions, electrode spacing, and buffer conductivity to achieve stable and efficient separation performance.

5.6 DEP in clinical and environmental applications

In this section, we discuss on how DEP can be implemented in clinical and environmental application. DEP technology has emerged as a powerful analytical and manipulation tool across multiple biomedical, environmental, and industrial sectors due to its ability to exploit intrinsic dielectric properties of biological particles. Unlike conventional detection methods that rely on biochemical labeling, molecular amplification, or metabolic activity measurements, DEP enables label-free physical characterization and separation based on electrical polarization behavior. This capability allows rapid, real-time analysis of biological particles while preserving their structural integrity, making DEP particularly valuable in applications where speed, sensitivity, and minimal sample preparation are critical.

In clinical microbiology, DEP technology provides a promising alternative to traditional culture-based and molecular diagnostic methods. Conventional microbiological detection typically requires bacterial culturing, which can take 24 to 72 h, or molecular amplification methods such as PCR, which require extensive sample preparation and reagent use. DEP enables direct manipulation and concentration of bacterial cells from clinical samples such as blood, urine, or wound swabs by exploiting dielectric differences between bacterial cells and host biological components. Because bacterial membrane integrity, cytoplasmic conductivity, and cell wall structure influence DEP response, DEP systems can potentially provide information about bacterial physiological state in addition to presence detection. This capability may allow rapid screening for bacterial infection and early assessment of antimicrobial treatment response. Furthermore, DEP-based enrichment can be integrated upstream of molecular diagnostics, increasing target concentration and improving detection sensitivity in low bacterial load samples.

In cancer cell detection, DEP technology is increasingly explored for isolation and characterization of circulating tumor cells and tumor-derived extracellular vesicles. Cancer cells often exhibit altered membrane composition, cytoplasmic conductivity, and intracellular ion distribution due to metabolic and structural abnormalities. These changes result in distinct dielectric signatures compared to normal healthy cells. DEP can selectively enrich cancer cells from complex biological fluids such as blood without requiring antibody labeling or chemical modification. This label-free capability preserves cell viability and allows downstream molecular or functional analysis. In addition, DEP-based cancer cell phenotyping may enable monitoring of disease progression and treatment response by detecting dielectric changes associated with apoptosis, necrosis, or drug resistance development. These capabilities align with emerging precision medicine strategies focused on real-time monitoring of disease state and treatment efficacy.

In water quality monitoring, DEP provides rapid detection and concentration of microbial contaminants directly from environmental samples. Traditional environmental microbiological analysis often requires sample transport to centralized laboratories and lengthy culture-based detection protocols. DEP-based environmental sensing systems can potentially enable near real-time detection of microbial contamination by concentrating bacteria, protozoa, or algae from water samples and directing them into detection regions. Because DEP relies on physical dielectric properties rather than metabolic activity, it can detect microorganisms regardless of growth phase, potentially improving early detection of contamination events. DEP systems can also be integrated into portable environmental monitoring devices for field deployment, enabling rapid response to contamination risks in drinking water supplies, wastewater treatment systems, and natural water bodies.

In biopharmaceutical manufacturing, DEP technology offers valuable capabilities for cell sorting, purification, and quality control in biologics production processes. Many biopharmaceutical products rely on living cell cultures, such as mammalian cells used for monoclonal antibody production or stem cells used for regenerative medicine therapies. DEP enables label-free cell sorting based on dielectric phenotype, reducing the need for chemical markers that may alter cell function or introduce contamination risk. DEP-based sorting systems can maintain high cell viability and reduce mechanical stress compared to mechanical filtration or centrifugation methods. Additionally, DEP can be used for real-time monitoring of cell physiological state during manufacturing, providing early detection of cell stress or contamination. These capabilities support the development of automated and closed-loop cell manufacturing systems, which are essential for large-scale and reproducible biopharmaceutical production.

Overall, DEP technology provides a unique platform that bridges physical particle manipulation with biological analysis. By enabling rapid, label-free detection and separation based on intrinsic electrical properties, DEP supports the development of next-generation diagnostic systems, environmental monitoring tools, and advanced biomanufacturing technologies. As microfluidic integration, microfabrication techniques, and signal processing technologies continue to advance, DEP is expected to play an increasingly important role in translational biomedical and industrial biotechnology applications.

5.7 Engineering limitations and solutions

This section discussed on how DEP systems face several fundamental engineering challenges that arise from the interaction between electric fields, electrode materials, and complex biological sample environments. Among the most critical limitations are Joule heating, electrode degradation, and variability in sample conductivity. These challenges directly influence DEP force stability, biological sample viability, and device lifespan. Joule heating occurs when electrical current passes through conductive media, resulting in resistive heat generation within the sample solution. In DEP systems, especially those operating in conductive biological buffers or physiological fluids, excessive Joule heating can lead to temperature increases that alter medium conductivity, disturb electric field distribution, and potentially damage biological particles through thermal stress.

However, Joule heating can be mitigated through careful electric field engineering. One effective strategy is reducing the microelectrode gap spacing as shown in Fig. 5.5, which increases electric field gradient strength without requiring high applied voltage. Because DEP force is proportional to the electric field gradient rather than simply the applied voltage, reducing electrode spacing allows researchers to maintain strong DEP manipulation forces while lowering peak-to-peak voltage. This approach reduces total current flow through the medium and therefore minimizes resistive heating effects. In bacterial DEP applications, optimized microelectrode spacing combined with lower applied voltage has been demonstrated to maintain strong particle manipulation while preserving sample thermal stability.

Electrode degradation represents another major limitation in DEP device longevity and reproducibility. Biological samples contain proteins, salts, and cellular debris that can adsorb onto electrode surfaces, altering electrical impedance and reducing field uniformity. Over time, electrochemical reactions at electrode interfaces may cause corrosion or surface roughening, degrading device performance. Material selection is therefore critical for maintaining electrode stability in biological environments. Gold is

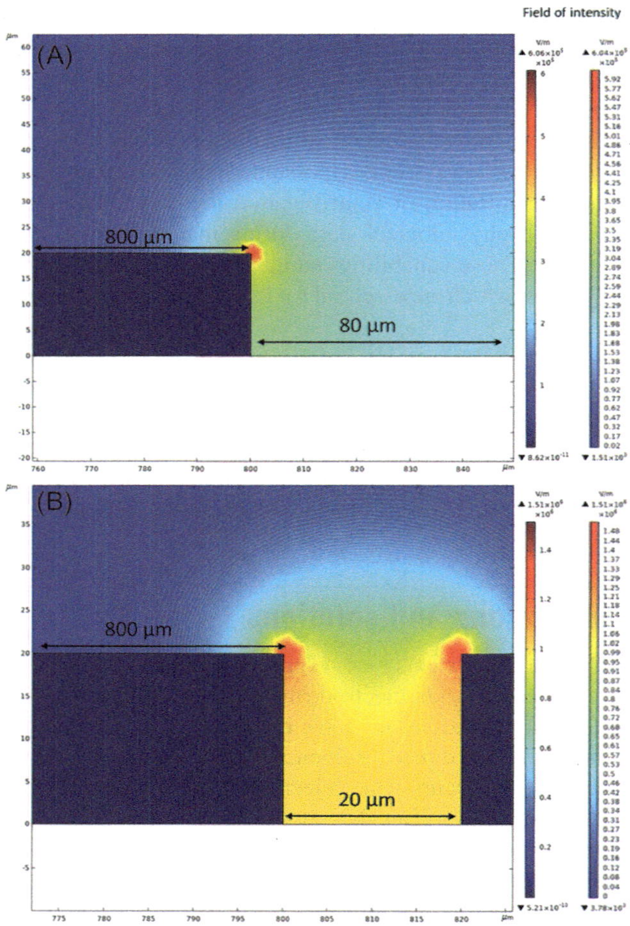

FIGURE 5.5

COMSOL simulation: Interelectrode gap spacing electric field simulation between 80 µm and 20 µm.

widely used as a microelectrode material due to its excellent electrical conductivity, high chemical stability, and strong biocompatibility. Gold surfaces resist oxidation and corrosion in aqueous biological media and exhibit minimal electrochemical reactivity compared to other conductive metals. This allows gold microelectrodes to maintain consistent electrical performance while minimizing adverse biological interactions. In addition, thin adhesion layers such as chromium or titanium can be used to improve electrode structural stability without compromising electrical performance. Surface passivation and protective coatings are also being explored to further enhance electrode durability in long-term DEP operation.

Sample conductivity variability is another major challenge, particularly when working with biological fluids such as blood, serum, or environmental water samples. Variations in ionic concentration directly influence medium permittivity and conductivity, which in turn alter the Clausius–Mossotti factor and DEP force magnitude. High conductivity samples can reduce polarization contrast between particles and medium, weakening DEP forces and reducing separation efficiency. This challenge can be addressed through sample preparation strategies such as controlled dilution, conductivity buffering, and concentration optimization. By introducing serial dilution protocols and carefully selecting DEP buffer composition, researchers can tune medium conductivity to maximize dielectric contrast between target particles and surrounding fluid. In bacterial DEP studies, optimization of medium conductivity through dilution and buffer selection has been shown to significantly improve DEP manipulation efficiency and reproducibility. Maintaining conductivity within an optimal operating window ensures stable polarization behavior while minimizing Joule heating risk.

Beyond these primary mitigation strategies, additional advances are improving DEP system robustness. In addition, microfluidic channel design can be optimized to improve heat dissipation through convective cooling and increased surface-to-volume heat transfer efficiency, further reducing Joule heating risk. Advanced electrode materials including graphene, platinum alloys, and conductive polymers are being explored to enhance electrode durability and reduce biofouling. Microfluidic cooling strategies, including passive thermal dissipation structures and active fluid heat exchange, are being developed to control temperature during high-field DEP operation. Adaptive signal control systems that dynamically adjust voltage amplitude and frequency based on real-time sample conductivity measurements are also emerging, enabling automated optimization of DEP operating conditions. Together, these advances are enabling DEP systems to operate reliably in increasingly complex biological environments while maintaining high manipulation precision and biological sample integrity.

5.8 Emerging trends and future DEP technologies

This section focus on the emerging trends that can be integrated with DEP. One of the most transformative future directions in DEP technology is the integration of AI with real-time DEP signal interpretation and particle motion analysis. Traditional DEP analysis relies on manual observation or simplified image processing methods to determine whether particles exhibit positive DEP, negative DEP, or crossover behavior. However, DEP particle behavior contains significantly richer information than simple positional classification. The trajectory, velocity, acceleration, and spatial distribution of particles under DEP fields encode detailed information about particle dielectric properties, polarization factor, and interaction with surrounding microfluidic forces. AI-assisted image tracking systems can now extract this dynamic motion information in real time and translate it into quantitative dielectric characterization metrics.

In AI-assisted DEP trajectory interpretation, high-speed imaging combined with deep learning particle tracking algorithms can reconstruct full particle motion paths across time. These trajectories provide velocity vector fields that describe particle response under specific electric field conditions. Because DEP force is directly related to the real component of the Clausius–Mossotti factor, particle velocity measured experimentally can be correlated to polarization factor magnitude. By combining AI-tracked velocity data with known fluid viscosity and electric field gradient models, real DEP force acting on each particle can be estimated. This allows direct translation of observed particle motion into dielectric property estimates without requiring extensive simulation fitting. In this framework, AI becomes a bridge between experimental observation and theoretical DEP simulation, allowing real biological particle polarization behavior to be mapped directly to Clausius–Mossotti factor predictions.

Furthermore, AI tracking systems enable real-time extraction of velocity distribution across particle populations. Instead of measuring average DEP response qualitatively, AI can quantify particle-by-particle velocity and acceleration behavior, enabling high-resolution DEP phenotyping. This is particularly valuable when working with heterogeneous biological populations where different physiological states produce subtle dielectric differences. AI-assisted trajectory analysis can therefore improve DEP-based classification accuracy and enable automated identification of particle subpopulations based on dielectric phenotype signatures.

Beyond trajectory analysis, AI integration enables a major shift in DEP quantification capability. Traditional DEP quantification methods typically rely on estimating concentration changes within regions of interest or measuring percentage response to DEP forces. However, AI-based particle detection and tracking enables direct absolute particle counting in real time. By continuously tracking particle entry, exit, trapping, and release events within DEP regions, AI systems can calculate the total number of cells interacting with DEP forces over time. This enables conversion of DEP response into absolute cell number quantification rather than relative concentration estimation. In practical applications, this allows DEP to move beyond qualitative concentration monitoring into quantitative enumeration of biological particles per unit sample volume. Such capability is particularly valuable for microbial load measurement, circulating tumor cell enumeration, and environmental contamination quantification, where absolute particle counts are more clinically and industrially meaningful than relative concentration metrics.

AI can also be integrated into DEP-based force modeling for microfluidic systems. In microfluidic DEP devices, particle motion is influenced not only by DEP force but also by hydrodynamic drag, electroosmotic flow, gravitational sedimentation, and Brownian motion. AI-driven multi-physics modeling can be trained to predict particle behavior by learning from real experimental datasets rather than relying solely on theoretical fluid dynamics equations. This enables adaptive prediction of particle trajectories under varying flow rates, buffer conditions, and electric field strengths. Such adaptive models can dynamically adjust DEP operating parameters to maintain optimal separation efficiency even under changing sample conditions.

The integration of AI with DEP device operation also enables automated device optimization. Real-time monitoring of particle motion patterns can be used to adjust voltage amplitude, frequency selection, and flow rate dynamically. This closed-loop optimization approach allows DEP systems to maintain optimal operating conditions without manual intervention. In clinical or environmental monitoring applications, this capability could allow DEP diagnostic systems to automatically adapt to sample variability.

In addition to signal interpretation and quantification, AI can support hybrid DEP-molecular diagnostic platforms. DEP can be used to enrich target biological particles, while AI can analyze downstream molecular or optical signals to improve diagnostic accuracy. AI can also assist in distinguishing between true biological DEP responses and artifacts caused by electroosmosis or debris movement, improving diagnostic reliability.

Looking further into future DEP applications, wearable DEP biosensors represent an emerging frontier. Miniaturized DEP devices integrated with AI signal processing and wireless communication may enable continuous monitoring of biological fluids such as sweat, saliva, or interstitial fluid. AI algorithms could analyze DEP signal variations over time to detect early biomarkers of disease or physiological stress.

Overall, the integration of AI into DEP systems represents a paradigm shift from static measurement platforms to adaptive, intelligent bioanalytical systems. AI-assisted DEP platforms are expected to enable real-time dielectric phenotyping, absolute biological particle quantification, automated device optimization, and multi-modal diagnostic integration. As imaging technology, computational power, and microfabrication techniques continue to advance, AI-enhanced DEP systems are likely to play a central role in next-generation biomedical diagnostics, environmental sensing, and precision biotechnology manufacturing.

5.9 Microfabrication material for DEP devices

DEP devices are commonly fabricated using MEMS manufacturing technologies that are highly compatible with conventional complementary metal–oxide–semiconductor (CMOS) fabrication processes (Franssila, 2010; Madou, 2011). Because DEP relies on microscale electrode patterning to generate precise electric field gradients, fabrication methods developed for integrated circuit manufacturing provide an ideal technological foundation for DEP device construction. Standard semiconductor fabrication techniques such as photolithography, thin-film deposition, thermal oxidation, and metal patterning allow precise fabrication of microelectrode arrays with micron-scale resolution. These fabrication capabilities are essential because DEP force strength is highly dependent on electrode geometry and spacing, requiring highly controlled manufacturing tolerances to ensure consistent device performance.

Silicon and glass are among the most commonly used substrates for DEP devices due to their mechanical rigidity, thermal stability, and compatibility with cleanroom photolithography processes. Silicon substrates provide excellent dimensional stability and are widely used when integration with microelectronic circuits is required. Glass substrates are frequently used for DEP devices designed for optical microscopy observation because of their optical transparency and chemical inertness. Both materials allow high-resolution electrode fabrication while maintaining structural stability during electrical operation and fluidic exposure.

Metal electrode layers in DEP devices are typically fabricated using thin-film deposition techniques such as sputtering or thermal evaporation. These techniques allow precise control of electrode thickness, surface roughness, and conductivity. Gold is one of the most widely used electrode materials in DEP applications due to its exceptional chemical stability, corrosion resistance, and biocompatibility (Grieshaber et al., 2008). Gold does not readily oxidize in aqueous biological environments, allowing long-term stable electrical performance during DEP operation. Additionally, gold exhibits minimal electrochemical reactivity compared to other conductive metals, reducing unwanted electrochemical reactions that could damage biological samples or alter electrode performance. From a biological compatibility perspective, gold surfaces exhibit low cytotoxicity and minimal protein denaturation, making them suitable for direct interaction with biological fluids, cells, and microorganisms. These properties make gold particularly attractive for DEP applications involving clinical samples, bacterial suspensions, or biological fluids.

However, gold does not adhere strongly to silicon or glass surfaces without an adhesion layer. Therefore, thin adhesion layers such as chromium or titanium are commonly deposited prior to gold deposition. These adhesion layers provide strong bonding between the substrate and gold electrode while maintaining electrical conductivity. This layered metal stack approach is widely used in MEMS devices including capacitive micromachined ultrasonic transducers (CMUTs), biosensors, and microelectrode arrays. In CMUT and MEMS sensor fabrication, similar thin-film material stacks are used to ensure mechanical stability, electrical conductivity, and long-term reliability. These fabrication parallels highlight the compatibility between DEP devices and broader MEMS device manufacturing ecosystems.

Photolithography enables precise patterning of electrode geometries required for DEP manipulation. Because DEP force is highly sensitive to electric field gradients, even small deviations in electrode spacing or edge geometry can significantly influence device performance. High-resolution photolithography allows fabrication of electrode gaps down to micron or sub-micron scale, enabling strong localized electric field gradients without requiring extremely high applied voltage. This precision is critical for minimizing Joule heating while maintaining effective DEP force generation.

In addition to metal electrode fabrication, polymer microfluidic channel integration is commonly achieved using soft lithography techniques. Polydimethylsiloxane (PDMS) is widely used due to its optical transparency, flexibility, and biocompatibility. PDMS allows rapid prototyping of microfluidic channel designs and easy bonding to glass or silicon substrates. However, PDMS can absorb small hydrophobic molecules and biomolecules, which may affect long-term device stability or quantitative analysis. As a result, alternative polymer materials such as cyclic olefin copolymer (COC), polymethyl methacrylate (PMMA), and SU-8 photoresist structures are being explored to reduce molecular absorption and improve chemical resistance in long-term DEP device operation.

Insulator-based DEP devices require different material considerations compared to electrode-based DEP systems. Instead of relying on metal electrode geometry to generate electric field gradients, iDEP devices rely on insulating microstructures such as polymer pillars, glass constrictions, or etched silicon features to shape electric field distribution. Materials used for iDEP structures must exhibit high electrical resistivity to prevent current leakage and maintain strong field gradient formation around geometric constrictions. Common materials used in iDEP include PDMS, SU-8 photoresist, glass microstructures, and silicon dioxide insulating layers. The advantage of iDEP is that electrodes can be placed outside the main sample region, reducing electrode fouling and improving long-term device reliability. This makes iDEP particularly useful for high conductivity biological samples such as blood or serum. However, iDEP devices often require higher applied voltage to achieve equivalent field gradients compared to electrode-based DEP systems, which can increase Joule heating risk and power consumption.

Each DEP fabrication material system presents specific advantages and limitations. Gold electrodes offer excellent biocompatibility and chemical stability but increase fabrication cost compared to aluminum or copper electrodes. Silicon substrates provide high integration potential but require more complex fabrication infrastructure. Polymer microfluidic systems allow rapid prototyping but may suffer from long-term chemical stability challenges. Insulator-based DEP structures provide improved fouling resistance but require higher operating voltages. Therefore, DEP device fabrication must be carefully tailored to application requirements, balancing performance, cost, and long-term device reliability.

Advances in microfabrication technologies continue to expand DEP device capabilities. Emerging approaches include graphene electrodes for enhanced conductivity and reduced electrochemical reactivity, nanostructured electrode surfaces for improved field localization, and hybrid silicon-polymer microfluidic integration for scalable manufacturing. These advances are expected to further improve DEP device performance while enabling large-scale manufacturing for clinical and industrial applications.

5.10 DEP signal analysis and polarization spectrum interpretation versus other electrokinetic manipulation techniques

DEP response data are often analyzed by measuring particle movement across frequency sweeps. These measurements generate polarization spectra that provide insights into particle dielectric properties. Biological cells behave electrically as multi-layer dielectric particles, where each structural component contributes independently to frequency-dependent polarization behavior. The cytoplasm forms the conductive core, typically exhibiting high ionic conductivity and moderate permittivity. Surrounding the cytoplasm, the lipid membrane acts as a thin capacitive insulating barrier characterized by low conductivity and high capacitance. In bacteria, an additional cell wall layer introduces another dielectric shell with its own thickness, permittivity, and conductivity contribution.

These layered dielectric interfaces determine how electric fields interact with cells across frequency domains. At low frequencies, electric field penetration is limited because the membrane capacitance dominates impedance behavior. Charge accumulates at membrane interfaces, producing strong interfacial polarization and typically resulting in nDEP behavior for many viable cells. As frequency increases, capacitive impedance decreases, allowing electric fields to penetrate membrane layers and interact with intracellular conductivity. At higher frequencies, cytoplasmic electrical properties dominate polarization behavior, often producing pDEP depending on cytoplasmic conductivity relative to medium conductivity.

Mathematically, this behavior is expressed through the Clausius–Mossotti factor, where the real component determines DEP direction. The transition between negative and positive DEP corresponds to the crossover frequency, which is highly sensitive to membrane integrity, cytoplasmic conductivity, and structural damage.

Polarization spectrum analysis is performed by sweeping frequency while monitoring particle motion. The resulting curve maps frequency against CMF real component or observed DEP response. At low frequencies, many viable cells exhibit nDEP because membrane insulation prevents strong polarization relative to the medium. As frequency increases, CMF increases toward zero and reaches crossover frequency, where DEP force approaches zero and particles show minimal net movement. At higher frequencies, CMF becomes positive, producing pDEP behavior where particles move toward high electric field gradient regions such as electrode edges.

This frequency-dependent behavior enables dielectric phenotyping as shown in Fig. 5.6 Membrane damage shifts crossover frequency downward because membrane insulation decreases. Cytoplasmic conductivity changes shift high-frequency polarization magnitude. Cell size also influences CMF scaling and crossover position.

Within DEP devices, polarization behavior directly translates into particle motion inside the region of interest. When CMF is negative, particles are repelled from electrode high-field regions and migrate toward field minima zones. When CMF becomes positive, particles migrate toward electrode edges or high field gradient regions, producing trapping or accumulation behavior. The number of particles entering or remaining within the ROI therefore represents a physical manifestation of dielectric polarization behavior at that frequency.

DEP response progression across frequency typically follows a structural hierarchy. Early frequency response is dominated by membrane and outer shell dielectric behavior. Mid-frequency response reflects combined membrane and internal conductivity interactions. High frequency response reflects

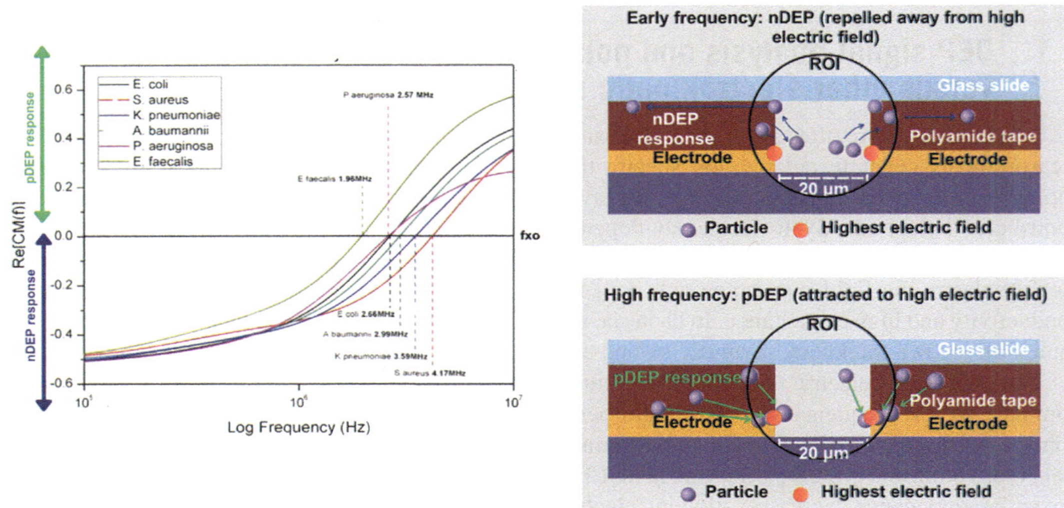

FIGURE 5.6

Re [CM(f)] of bacteria and trajectory of bacterial cells during nDEP and pDEP response.

cytoplasmic dielectric properties and intracellular ionic conductivity. This progression allows DEP to detect structural integrity changes, metabolic activity shifts, and physiological stress through polarization spectrum shifts. Particle velocity under DEP can be expressed as:

$$v \propto \text{CMF} \times \nabla E^2/\text{fluid drag}$$

Thus, velocity magnitude is directly proportional to polarization factor magnitude. By measuring particle trajectory and velocity across frequency sweeps, DEP response can be converted into quantitative dielectric characterization metrics.

Trajectory analysis allows classification of particle behavior as strong pDEP, weak pDEP, crossover, weak nDEP, or strong nDEP. Velocity magnitude further refines this classification by providing quantitative force strength estimation. DEP quantification can be performed by calculating particle percentage within ROI at each frequency. At low frequency, ROI particle count is low due to nDEP repulsion. At crossover, particle distribution becomes random. At high frequency, ROI particle count increases due to pDEP trapping. Mapping ROI occupancy versus frequency produces experimental polarization spectrum approximations, allowing biological state comparison between samples.

DEP differs fundamentally from other electrokinetic manipulation methods because it depends on particle polarization rather than net charge. Electrophoresis relies on net surface charge and is primarily useful for charged molecules such as nucleic acids. However, electrophoresis cannot effectively manipulate neutral biological particles. DEP can manipulate both charged and neutral particles because it depends on induced dipole polarization rather than static charge.

Electroosmosis generates bulk fluid movement due to electric field interaction with charged channel surfaces. While useful for fluid transport, electroosmosis lacks particle selectivity and therefore cannot replace DEP for separation based on dielectric properties. Electrothermal flow arises from temperature gradients produced by electric fields in conductive media. While typically considered a disturbance in DEP systems, electrothermal flow can be used for fluid mixing but does not provide selective particle manipulation. DEP remains unique because it directly probes intrinsic dielectric properties, enabling label-free biological phenotyping and separation independent of net particle charge.

5.11 Microfluidic channels separation

Microfluidic channel separation within precisely engineered microscale environments has emerged as an important component of BioMEMS technology due to its ability to manipulate biological samples such as whole blood, urine, saliva, and seminal fluid. Blood plasma is the most commonly used biofluid in disease diagnostic and biomedical analysis because it contains a broad range of biomarkers associated with physiological and pathological conditions. As a result, the development of microfluidic plasma separation platforms has become a trend research in lab-on-a-chip and point-of-care diagnostic systems. Conventional centrifugation remains the laboratory gold standard for plasma extraction. However, the requirement for bulky instrumentation and lengthy processing times limits its suitability for portable point-of-care systems. Microfluidic channel platforms address this limitation by performing plasma separation directly on-chip that selectively direct plasma while restricting cellular components. Unlike conventional macroscale filtration systems, microfluidic platforms operate under laminar flow conditions where fluid motion is highly predictable and particle trajectories can be controlled through channel geometry, hydrodynamic interactions, and size-dependent transport mechanisms. These characteristics have made microfluidic channels one of the most widely explored approaches for rapid blood plasma separation (Al-Aqbi et al., 2021).

Several passive microfluidic separation strategies have been reported, including microfiltration, sedimentation, bifurcation flow, wettability control, and curved channel designs (Wang et al., 2021). Among these approaches, microfiltration-based systems are particularly attractive because the separation mechanism can be achieved through geometric constraints incorporated directly into the channel structure (Al-Aqbi et al., 2021). The dimensions of the filtration region are designed to permit the passage of plasma and small molecules while preventing larger biological particles such as erythrocytes, leukocytes, and platelets from entering the collection channel. Since the process relies primarily on channel geometry, no external pumps, magnetic fields, or optical systems are required, simplifying device operation and reducing system complexity.

One example of this approach is a nano-junction microfluidic filtration device consisting of a V-shaped microchannel connected to a narrow filtration channel through a nano-junction formed by dielectric breakdown as illustrated in Fig. 5.7. The nano-junction acts as a highly selective transport barrier that permits plasma and small molecular species to cross while restricting blood cells and larger plasma proteins. Using this configuration, blood plasma extraction yields of approximately 62% were achieved while maintaining plasma quality suitable for downstream analytical applications. The device highlights how microscale channel design can simultaneously achieve filtration, sample purification, and biomolecule preparation within a single integrated platform.

The importance of microfluidic channel separation extends beyond blood plasma processing. Similar channel architectures have been utilized for cell sorting, particle enrichment, bacterial isolation, and environmental sample preparation. By tailoring channel dimensions and flow pathways, microfluidic

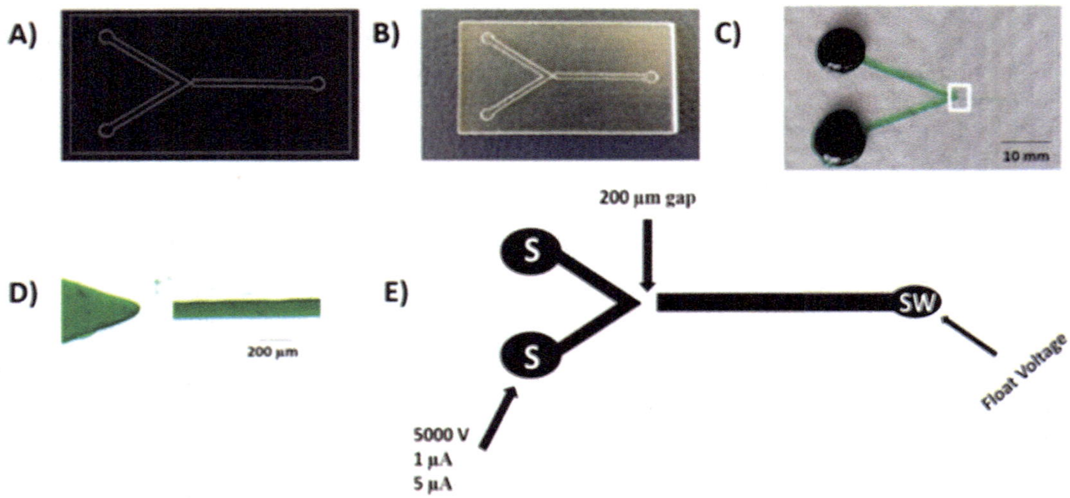

FIGURE 5.7

(A) An AutoCAD design of the filtration device (B) photograph image of the negative 3D printed portable device using an Eden 3D printer (C) Photograph image of the hot embossed filtration poly (methyl methacrylate) (PMMA)/adhesive tape device filled with green food dye. Scale bar = 10 mm. (D) Zoomed-in image of the V channel and the filtration. Adapted from Al-Aqbi et al. (2021).

devices can selectively manipulate biological particles over a wide size range, from cells and microorganisms to extracellular vesicles and nanoscale biomolecules. These capabilities are particularly relevant to BioMEMS systems where sample preparation often represents a major bottleneck before electrical, optical, or biochemical detection can be performed.

5.12 MEMS microvalves

Microvalves form the foundation of many different microfluidic systems. Microvalves control routing, timing, and separation of fluids within a microfluidic device and are crucial for designs with complex functionality. This control is achieved through microvalves, which function as miniature flow regulators capable of opening, closing, or redirecting fluid pathways within a microfluidic device. Similar to valves used in conventional fluid systems, microvalves determine when and where fluids are allowed to move and operate at dimensions ranging from a few micrometers to several hundred micrometers and are designed to handle extremely small sample volumes. Microvalves are commonly integrated with microchannels, micropumps, sensors, and reaction chambers to create automated lab-on-a-chip systems. In biomedical applications, multiple processing steps such as sample introduction, reagent mixing, incubation, washing, and detection must occur in a controlled environment and sequence. By incorporating microvalves directly onto the chip, fluid routing can be automated while maintaining compact device dimensions without extensive external tubing and manual intervention. Microvalves

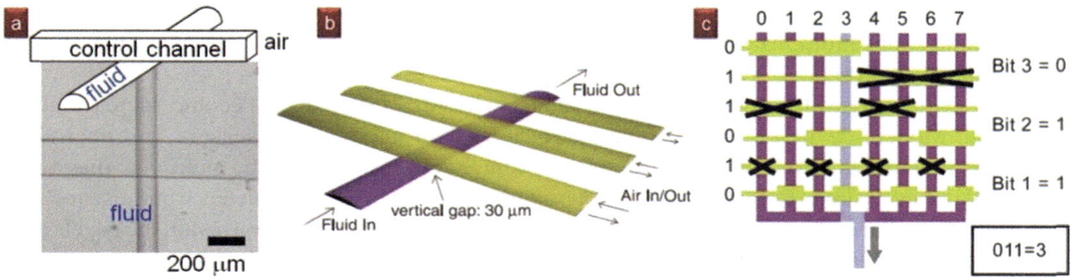

FIGURE 5.8

Schematic illustration of a pneumatic membrane microvalve. Application of pressure through the control channel deforms the flexible membrane and interrupts fluid flow in the underlying microchannel, while pressure release restores fluid transport. Adapted from Au et al. (2011).

are therefore frequently regarded as the fluidic equivalent of electrical switches in integrated circuits (Au et al., 2011; Zheng et al., 2025).

Microvalve actuation mechanisms has been researched and developed based on its crucial to complete and operate on microfluidic devices, including pneumatic, electrokinetic, phase-change, thermal and mechanically actuated designs. Among these approaches, pneumatic membrane microvalves showed one of the most applied for microvalve based on relatively simple fabrication, reliable operation and compatibility with soft lithography processes. Pneumatic microvalve as shown in Fig. 5.8 consist of pressure applied through a control channel causes a flexible membrane to deform and obstruct fluid flow within an adjacent microchannel. Releasing the pressure allows the membrane to return to its original position, restoring fluid transport. This simple operating principle has enabled the development of highly integrated microfluidic systems capable of performing complex biological assays with minimal user intervention (Au et al., 2011).

Beyond conventional flow regulation, microvalves have also been investigated as responsive components in biomedical drug delivery systems. Microvalve can be utilized with stimulus-sensitive hydrogels as actuating materials. Unlike externally powered valves, hydrogel-based microvalves respond directly to changes in their surrounding chemical environment. Variations in pH, temperature, or glucose concentration induce swelling or shrinking of the hydrogel structure, producing mechanical deformation that can open or close a microfluidic pathway. This behavior allows fluid flow to be regulated without external power sources, electronic controllers, or complex mechanical actuators. Such autonomous operation is particularly attractive for implantable and wearable biomedical devices (Dzulkefli et al., 2008; Tehranirokh et al., 2006).

One of its application has explored for the use of glucose-sensitive hydrogels as actuators for microvalve systems intended for controlled insulin delivery. In these devices, a hydrogel layer is positioned between a porous membrane and a flexible diaphragm. As glucose concentration increases, the hydrogel absorbs water and expands, generating sufficient force to deform the membrane and activate the valve mechanism. Numerical simulations demonstrated that hydrogel swelling could produce membrane deflections large enough to permit fluid transport through the microchannel, allowing insulin release to be regulated according to glucose concentration. This approach illustrates how smart materials can transform chemical signals directly into mechanical motion, creating self-regulating BioMEMS devices with potential applications in disease management and personalized medicine (Tehranirokh et al., 2006; Tehranirokh et al., 2009).

The role of microvalves continues to expand as BioMEMS systems become increasingly sophisticated. Modern devices often combine microvalves with micropumps, sensors, and microfluidic processing units to achieve automated sample handling, cell culture control, molecular diagnostics, and point-of-care testing. In many cases, a micropump is essentially constructed from a sequence of microvalves operating in a coordinated manner to generate directional flow. As a result, the development of reliable microvalve technologies has become a critical step toward fully integrated micro total analysis systems (μTAS) and next-generation lab-on-a-chip platforms. The continuing advancement of responsive materials, soft lithography fabrication methods, and microactuation mechanisms is expected to further improve the functionality and versatility of microvalves in future biomedical applications.

5.13 MEMS microfilters

Microfilters represent one of the most important functional components in BioMEMS and lab-on-a-chip systems. Microfilters provide the ability to selectively separate particles, microorganisms and bio-cells according to their physical dimensions. While microfluidic channels guide fluid movement as discussed in section 5.11 and microvalves regulate fluid pathways in section 5.12, this section will discussed on the actual fractionation process by allowing selected components to pass while retaining unwanted particles using microfilters. This capability is crucial, where biological samples often contain a complex mixture of cells, proteins and microorganisms that must be isolated prior to detection or downstream processing. The integration of commercial membrane filters into BioMEMS platforms remains challenging due to differences in materials, fabrication processes, and device architectures. Many polymer-based filters exhibit irregular pore geometries and broad pore size distributions, resulting in inconsistent filtration performance and limited compatibility with microfabrication technologies (Nam et al., 2019; Fissell et al., 2009). These limitations have motivated the development of MEMS-based microfilters capable of producing highly uniform pore structures with precisely controlled dimensions. Highly uniform pore structure filters can be directly integrated with microchannels, micropumps, sensors, and microvalves to form compact micro-total-analysis systems (μTAS).

MEMS microfilters are generally fabricated using silicon micromachining, electrochemical etching, thin-film deposition, electroplating, or polymer microfabrication techniques (Hamzah et al., 2013). MEMS fabrication provides precise control over pore geometry, pore spacing, and membrane thickness (Hamzah et al., 2013; Fissell et al., 2009). This level of execution for fabrication allows the filtration characteristics to be tailored according to the target application, whether the objective is the removal of blood cells, bacterial isolation, protein fractionation, nanoparticle separation, or sample purification. As a result, filtration performance can be optimized while maintaining low pressure drop and high throughput within a miniaturized device. Based on this research, uniformly distributed micropores with dimensions in the submicron range were produced through controlled electroplating and membrane release processes using the fabrication of microporous membranes using electroplated copper and electrochemically etched silicon structures for biological fluid microfiltration (Hamzah et al., 2013). The resulting membranes enabled selective transport of nanoscale particles while maintaining structural integrity and compatibility with MEMS fabrication processes. Figure 5.9 illustrates the evolution

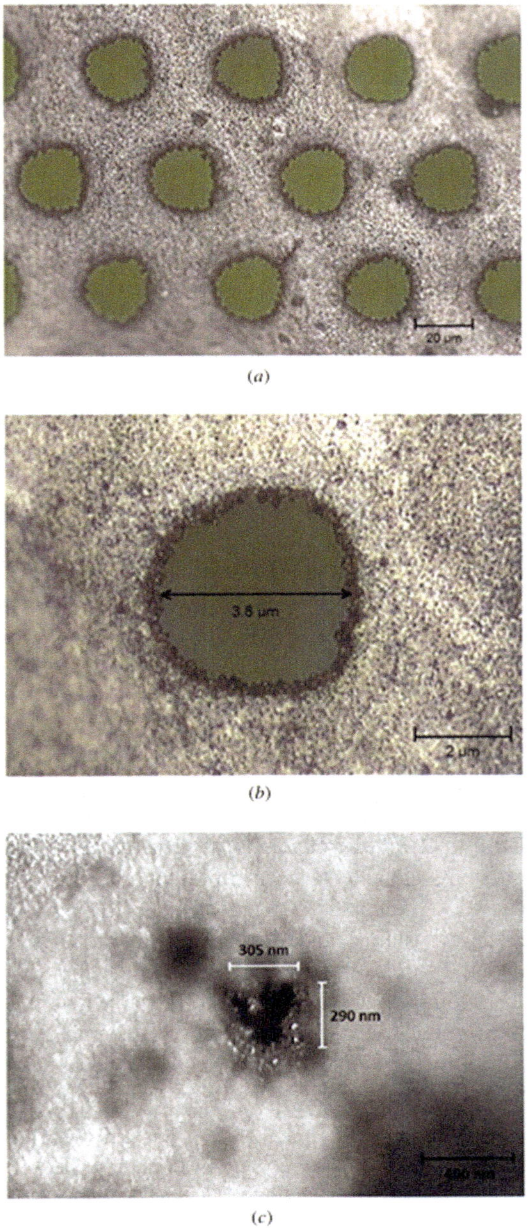

FIGURE 5.9

Optical and microscopic images of electroplated copper membranes with controlled micropore formation after different electroplating durations. Uniformly distributed micropores were produced to enable selective filtration of biological particles and contaminants. Adapted from Hamzah et al. (2013).

of micropore formation within the electroplated membrane, highlighting the ability to control pore dimensions through the fabrication process.

Beyond microscale filtration, advances in silicon micromachining have enabled the development of nanoporous membranes with substantially improved transport characteristics (Fissell et al., 2009). The high-performance silicon nanopores hemofiltration membrane developed for renal replacement applications (Fissell et al., 2009). The ability to accurately define pore geometry at the nanoscale provides enhanced hydraulic permeability while maintaining strict molecular selectivity. Consequently, high fluid transport rates can be achieved without compromising the rejection of larger biomolecules. Unlike conventional polymer membranes that contain irregular and polydisperse pores, silicon nanopores membranes employ highly uniform slit-shaped nanopores fabricated using MEMS processing techniques. The significance of this approach becomes particularly evident in hemofiltration and artificial kidney technologies. Traditional dialysis membranes often require relatively large membrane areas because pore irregularities increase hydraulic resistance and reduce transport efficiency. Silicon nanopores membranes address this limitation by utilizing ultrathin membrane structures combined with monodisperse slit-shaped pores. The uniformity of the pores reduces flow resistance and promotes predictable molecular transport behavior. Furthermore, surface modification using poly(ethylene glycol) (PEG) significantly reduces protein adsorption, membrane fouling, and thrombus formation, thereby improving long-term blood compatibility. Experimental studies demonstrated stable filtration performance and maintained molecular selectivity during prolonged hemofiltration of anticoagulated whole blood, highlighting the potential of MEMS-based filtration technologies for implantable and wearable medical devices.

Microfilter technology continues to evolve alongside advances in microfabrication and nanofabrication. Recent developments include nanoporous silicon membranes, PDMS membrane filters, nanoslit filtration devices and hybrid microfluidic filtration platforms capable of separating particles ranging from blood cells to bacteria and extracellular vesicles (Nam et al., 2019; Xie et al., 2023). In bacterial analysis, membrane-based microfilters have been utilized to concentrate microorganisms prior to downstream detection, improving analytical sensitivity while reducing processing time (Nam et al., 2019). The integration of microfilters with microfluidic channels, microvalves, and sensing elements has enabled highly compact analytical systems capable of performing sample preparation and detection within a single device (Hamzah et al., 2013; Nam et al., 2019). As BioMEMS technology moves toward portable diagnostics, wearable healthcare systems, and implantable therapeutic devices, the role of MEMS microfilters is expected to become increasingly important. Their ability to provide precise size-selective separation, low sample consumption, and seamless integration with other microsystem components makes them indispensable in modern lab-on-chip platforms. Future developments will likely focus on improving membrane durability, antifouling performance, and nanoscale filtration precision while maintaining high throughput and biocompatibility for long-term biomedical applications.

5.14 Integrated applications of dielectrophoresis in biomedical and environmental

DEP has emerged as one of the most versatile electrokinetic manipulation technologies across modern biomedical engineering, microbiology, environmental science, and biotechnology manufacturing.

Unlike conventional biological analysis techniques that rely on biochemical labeling, molecular amplification, or culture-based growth, DEP operates by exploiting intrinsic dielectric differences between biological particles and their surrounding medium. Because dielectric properties are directly linked to cell membrane structure, intracellular conductivity, macromolecular composition, and physiological state, DEP provides a powerful label-free platform capable of simultaneously performing particle separation, enrichment, characterization, and phenotyping. This capability allows DEP to function not only as a separation tool but also as a physical diagnostic technology capable of detecting subtle biological state changes without chemical modification of the sample.

In mammalian cell engineering, DEP has become an increasingly important tool for label-free cell sorting, phenotyping, and physiological monitoring (Gascoyne & Shim, 2014). Mammalian cells exhibit complex dielectric behavior due to multilayer structures consisting of plasma membrane, cytoplasm, and internal organelles such as the nucleus and mitochondria. Changes in membrane lipid composition, intracellular ion concentration, and cytoplasmic conductivity directly influence polarization behavior across frequency domains. Cancer cells, for example, often exhibit altered membrane capacitance and cytoplasmic conductivity resulting from metabolic reprogramming and structural remodeling. DEP can exploit these dielectric differences to selectively isolate circulating tumor cells from complex biological fluids such as whole blood without requiring antibody labeling. This label-free capability preserves cell viability and allows downstream molecular and functional analysis. In stem cell engineering, DEP has been used to separate stem cell populations based on differentiation state. As stem cells differentiate, membrane composition, cytoplasmic conductivity, and organelle density change, resulting in measurable shifts in DEP polarization spectra. This enables noninvasive monitoring of cell state transitions, which is particularly important in regenerative medicine and cell therapy manufacturing. DEP is also being explored for immune cell phenotyping, where dielectric property shifts associated with immune activation states can be detected through frequency-dependent polarization analysis.

Beyond mammalian cells, DEP is increasingly being applied in bacterial detection and characterization. Bacterial dielectric behavior is strongly influenced by cell wall structure, membrane permeability, cytoplasmic conductivity, and physiological growth state. Because these properties change dynamically during growth, stress response, and antibiotic exposure, DEP can be used for rapid bacterial screening and antimicrobial susceptibility testing (Lapizco-Encinas, 2011; Markx & Davey, 1999; Yafouz, Kadri, & Ibrahim, 2014). By measuring frequency-dependent polarization spectra, DEP can distinguish bacterial species and physiological states without requiring culture amplification or molecular labeling. Engineering optimization of DEP electrode geometry and medium conductivity has significantly improved bacterial manipulation efficiency. Simulation and experimental studies have demonstrated that reducing microelectrode spacing increases electric field gradient strength, enhancing DEP force and particle velocity. These improvements are critical for translating DEP bacterial detection systems into continuous microfluidic diagnostic platforms capable of real-time bacterial monitoring.

DEP applications are also rapidly expanding into viral and extracellular vesicle research. Viruses and exosomes are significantly smaller than bacterial cells and require high electric field gradient DEP systems for effective manipulation. DEP has been successfully applied to isolate exosomes from biological fluids for diagnostic biomarker analysis (Chen et al., 2015). Because exosomes contain nucleic acids and proteins derived from parent cells, DEP-based exosome isolation provides strong potential for cancer diagnostics, neurological disease monitoring, and precision medicine applications. DEP offers particular advantages in exosome isolation because it avoids chemical labeling that could alter vesicle surface structure or biological function.

In environmental monitoring and water quality testing, DEP provides a rapid and sensitive platform for detecting microbial contamination (Yafouz, Kadri, & Ibrahim, 2014). Traditional environmental microbiology relies heavily on laboratory-based culture analysis, which can delay contamination detection by days. DEP-based environmental sensing systems enable rapid concentration and detection of microbial contaminants directly from water samples. Because DEP relies on physical dielectric properties rather than metabolic activity, it can detect microorganisms regardless of growth state. DEP systems can be integrated into portable environmental monitoring devices for field deployment, enabling real-time water quality surveillance and early detection of contamination events.

DEP is also emerging as a powerful tool in biopharmaceutical manufacturing and cell therapy production. Modern biologics manufacturing relies heavily on living cell cultures for therapeutic protein production and regenerative medicine applications. DEP provides gentle, label-free cell separation methods that reduce chemical contamination risk and preserve cell viability (Chin et al., 2012; Drain et al., 2014). This is particularly important in stem cell manufacturing, where maintaining cell phenotype integrity is essential. DEP-based sorting systems are being developed for integration into automated, closed-loop cell manufacturing platforms capable of performing continuous quality control and cell state monitoring during production.

Future DEP systems are expected to integrate AI, automated microfluidic control, and wireless data transmission to enable fully autonomous biological analysis platforms. AI-assisted DEP systems can analyze particle trajectory and velocity data to extract dielectric phenotype information in real time (Litjens et al., 2017). Hybrid DEP-molecular diagnostic platforms are also being developed, combining physical enrichment with genetic detection to improve diagnostic accuracy and sensitivity.

The transition of DEP from laboratory research to clinical deployment requires improvements in device robustness, automation, and clinical workflow integration. Clinical samples present additional complexity compared to laboratory samples due to heterogeneous biological matrices. Disposable cartridge-based DEP systems are increasingly being developed to reduce contamination risk and simplify clinical workflow. Clinical validation requires demonstration of analytical sensitivity, specificity, reproducibility, and performance across diverse sample populations.

DEP also shows strong promise in rapid infectious disease diagnostics. Traditional pathogen detection methods often require culture amplification or nucleic acid extraction. DEP-based systems can potentially identify pathogens directly from biological samples by exploiting dielectric differences between microbial cells and host biological components. DEP also enables physical phenotyping of pathogens, allowing detection of membrane damage or physiological stress. Integration of DEP with molecular detection methods such as PCR and sequencing may enable hybrid diagnostic workflows combining physical enrichment and genetic confirmation.

In oncology diagnostics and precision medicine, DEP is being explored for circulating tumor cell enrichment and cancer cell phenotyping. Because cancer cells exhibit altered dielectric signatures due to metabolic and structural changes, DEP can selectively enrich tumor cells from blood samples. DEP-based phenotyping may also enable real-time monitoring of cancer treatment response by detecting dielectric changes associated with apoptosis or drug resistance.

Industrial scaling of DEP technology requires development of multi-channel parallel microfluidic architectures capable of processing large sample volumes while maintaining uniform electric field distribution. Advances in roll-to-roll microfabrication and printed electronics may enable low-cost mass production of DEP devices for widespread clinical and environmental deployment.

Commercial adoption of DEP technology depends on demonstrating clear advantages over existing diagnostic and separation technologies. Early commercialization efforts are focusing on niche applications such as environmental monitoring and specialized research instrumentation. As device robustness and automation improve, DEP is expected to expand into broader clinical diagnostic markets.

Regulatory approval of DEP diagnostic systems requires demonstration of analytical validity, clinical performance, manufacturing reproducibility, and biocompatibility. Because DEP devices interact directly with biological samples, regulatory evaluation includes electrical safety, material compatibility, and long-term device stability testing.

Overall, DEP represents a convergence of electrical engineering, microfabrication technology, and biological science. By enabling label-free biological manipulation and characterization based on intrinsic dielectric properties, DEP provides a powerful platform for next-generation diagnostic systems, environmental monitoring technologies, and advanced biotechnology manufacturing processes. As microfabrication, AI, and microfluidic integration technologies continue to advance, DEP is expected to become a core enabling technology in future biomedical and environmental monitoring systems.

5.15 Long-term technology roadmap

Over the next two decades, DEP technologies are expected to transition from primarily laboratory-based analytical tools into fully integrated, intelligent, and globally deployable bioanalytical platforms. This evolution will be driven by simultaneous advances in microfabrication, microfluidic automation, AI integration, and digital healthcare connectivity. Because DEP operates based on intrinsic dielectric properties of biological particles rather than biochemical labeling or reagent-based detection, it is uniquely positioned to support future diagnostic systems that are faster, more portable, and less resource-intensive than conventional laboratory methods.

In the near-term period from approximately 2025 to 2030, DEP development is expected to focus on integrated lab-on-chip diagnostic platforms. These systems will combine DEP-based particle separation and enrichment with downstream detection modules such as optical imaging, electrical impedance sensing, or fluorescence detection. In these integrated platforms, DEP will function as the front-end physical enrichment technology, rapidly concentrating target biological particles such as bacteria, circulating tumor cells, or extracellular vesicles into detection regions. This physical concentration step can dramatically improve sensitivity of downstream detection systems while reducing sample preparation complexity. The integration of DEP into lab-on-chip devices will also reduce analysis time from hours or days to minutes, enabling rapid clinical decision-making. In addition, these integrated systems will support multiplexed analysis, allowing simultaneous detection of multiple biological targets from a single sample, which is particularly important in infectious disease diagnostics and cancer biomarker screening.

In the mid-term period between approximately 2030 and 2035, DEP technologies are expected to expand into wearable biosensing platforms capable of continuous biological monitoring. Advances in flexible electronics, low-power microelectrode fabrication, and miniaturized microfluidic sampling systems will enable DEP sensors to be integrated into wearable health monitoring devices. In these systems, DEP could be used to continuously analyze biological particles present in sweat, saliva, or interstitial fluid. Continuous DEP monitoring could allow early detection of infection biomarkers, inflammatory markers, or metabolic disease indicators before symptoms become clinically visible.

These wearable DEP systems would represent a major shift from reactive medicine toward predictive and preventative healthcare models. Because DEP does not require chemical reagents or labeling, wearable DEP systems could operate continuously for extended periods without requiring consumable replacement, significantly reducing long-term monitoring cost.

In the long-term period extending toward 2040, DEP systems are expected to evolve into fully autonomous diagnostic platforms integrated into digital healthcare infrastructure. These systems would combine automated sample handling, AI-assisted signal interpretation, wireless data transmission, and cloud-based diagnostic decision support. In such systems, DEP could be used as a universal front-end biological analysis module capable of performing rapid physical characterization of biological particles. AI-driven analysis would automatically interpret polarization spectra, particle trajectories, and dielectric phenotypes to generate diagnostic outputs without requiring specialist operator interpretation. Integration into digital healthcare networks would enable remote monitoring, telemedicine diagnostics, and real-time epidemiological surveillance.

Beyond technological advancement, DEP systems offer strong potential for sustainable and globally accessible diagnostic deployment. Traditional diagnostic methods often require cold-chain reagent storage, complex laboratory infrastructure, and highly trained technical personnel. DEP-based diagnostic platforms, by contrast, require minimal chemical reagents and can be designed as portable, battery-powered devices. This significantly reduces operational cost and infrastructure requirements, making DEP particularly suitable for deployment in low-resource healthcare environments. In developing regions where laboratory infrastructure may be limited, portable DEP diagnostic systems could enable point-of-care testing for infectious diseases, reducing diagnostic delay and improving treatment outcomes.

DEP environmental monitoring systems also offer strong potential for global health applications, particularly in water quality monitoring and outbreak prevention. Waterborne pathogens remain a major global health risk, particularly in regions with limited water treatment infrastructure. DEP-based environmental sensors could enable real-time detection of microbial contamination in drinking water supplies, allowing early intervention before outbreaks occur. Because DEP can detect microorganisms based on physical dielectric properties rather than growth-based detection, DEP environmental sensors can provide faster contamination alerts compared to conventional culture-based methods.

In addition to healthcare and environmental monitoring, DEP technologies support sustainable biotechnology manufacturing. Because DEP enables label-free cell separation and monitoring, it reduces reliance on chemical reagents used in conventional separation technologies. This reduces chemical waste and environmental impact in biopharmaceutical production processes. DEP-based monitoring systems can also improve manufacturing efficiency by enabling real-time monitoring of cell physiological state, reducing batch failure and improving resource utilization.

DEP technologies also support sustainability through reduced energy consumption compared to large laboratory instrumentation. Miniaturized DEP systems can perform biological analysis using low electrical power, particularly when optimized microelectrode geometry and microfluidic flow systems are used. The ability to perform rapid diagnostics at point-of-care locations also reduces transportation-related carbon footprint associated with centralized laboratory testing.

Looking toward the long-term future, DEP systems are expected to play a major role in global digital health ecosystems. Integration with AI will enable adaptive diagnostic systems capable of continuously learning from global datasets. This could allow early detection of emerging infectious diseases, monitoring of antimicrobial resistance patterns, and real-time tracking of environmental pathogen distribution. Such capabilities will be critical for global pandemic preparedness and public health monitoring.

Overall, the long-term DEP technology roadmap reflects a transition from specialized laboratory instrumentation toward globally distributed, intelligent, and sustainable bioanalytical platforms. By enabling rapid, label-free biological analysis using minimal resources, DEP technology is uniquely positioned to support next-generation healthcare delivery, environmental protection, and sustainable biotechnology manufacturing. As microfabrication technologies continue to improve and AI becomes increasingly integrated into bioanalytical systems, DEP is expected to become a foundational technology in global health infrastructure and future digital diagnostic networks.

5.16 Future outlook

The rapid evolution of biological detection technologies has been driven by the increasing need for faster, more sensitive, and more accessible diagnostic platforms capable of addressing modern global healthcare and environmental challenges. While traditional microbiological and molecular detection techniques remain essential in laboratory-based diagnostics, their reliance on biological amplification, chemical reagents, and centralized laboratory infrastructure limits their suitability for rapid and decentralized testing. In this context, DEP has emerged as a powerful complementary technology that enables direct physical characterization and manipulation of biological particles based on intrinsic dielectric properties.

At the fundamental level, DEP provides a unique capability to probe biological structure through frequency-dependent polarization behavior governed by Maxwell–Wagner interfacial polarization mechanisms. The multi-shell dielectric nature of biological particles, including bacteria, mammalian cells, viruses, and extracellular vesicles, allows DEP to detect subtle structural and physiological differences that are not easily measurable using conventional biochemical methods. By linking dielectric behavior to biological structure, DEP enables label-free phenotyping that preserves biological integrity while providing rapid analytical output.

The integration of DEP within BioMEMS platforms has further accelerated its translation toward real-world applications. The combination of microelectronics, microfluidics, and biological interfaces enables precise electrical manipulation of biological particles at microscale resolution while maintaining minimal sample volume requirements. Microfluidic integration allows DEP to transition from static laboratory experiments into continuous-flow diagnostic platforms capable of automated sample processing and high-throughput analysis. This integration is essential for the development of portable point-of-care diagnostic systems and environmental monitoring devices.

Engineering optimization has played a critical role in improving DEP device reliability and performance. Addressing challenges such as Joule heating, electrode degradation, and sample conductivity variability has enabled DEP operation in increasingly complex biological environments. Advances in electrode materials, microfluidic cooling strategies, and adaptive signal control have significantly improved long-term device stability and reproducibility. Microfabrication technologies compatible with CMOS manufacturing have enabled high-precision electrode fabrication while supporting potential large-scale manufacturing of DEP diagnostic devices.

DEP signal analysis has evolved from qualitative observation of particle movement to quantitative polarization spectrum interpretation. By analyzing particle velocity, trajectory, and region-of-interest occupancy across frequency sweeps, DEP enables detailed dielectric phenotyping of biological particles. The integration of AI further enhances DEP signal interpretation by enabling automated particle tracking, real-time dielectric property estimation, and adaptive system optimization.

Application-driven development has demonstrated DEP's versatility across multiple sectors. In mammalian cell engineering, DEP supports label-free cell sorting and phenotyping for cancer diagnostics, stem cell manufacturing, and immune cell monitoring. In microbiology, DEP enables rapid bacterial detection and physiological characterization without requiring culture amplification. In viral and extracellular vesicle research, DEP enables isolation of nanoscale biological particles critical for precision diagnostics. Environmental applications highlight DEP's ability to provide rapid water quality monitoring and pathogen detection in field-deployable formats. In biopharmaceutical manufacturing, DEP supports label-free cell separation and real-time cell state monitoring, improving manufacturing efficiency and product quality.

Looking toward the future, DEP is expected to transition from specialized research instrumentation into widely deployed diagnostic and monitoring infrastructure. The integration of AI, automated microfluidics, and wireless communication will enable fully autonomous DEP diagnostic systems capable of real-time biological analysis. These systems will support digital healthcare ecosystems by enabling remote diagnostics, real-time epidemiological monitoring, and predictive disease surveillance.

From a global health perspective, DEP technologies offer unique advantages for deployment in low-resource environments. Because DEP requires minimal chemical reagents and can be implemented in portable, low-power devices, it provides a sustainable diagnostic platform for developing regions. DEP-based environmental monitoring systems may play critical roles in water safety monitoring and outbreak prevention, contributing to global public health infrastructure.

Ultimately, DEP represents a convergence of electrical engineering, microfabrication technology, microfluidics, and biological science. Its ability to provide rapid, label-free, and information-rich biological characterization positions DEP as a foundational technology for next-generation diagnostics, environmental monitoring, and biotechnology manufacturing. As technological advances continue to reduce device cost, improve automation, and enhance integration with digital healthcare networks, DEP is expected to play an increasingly central role in future global health and biotechnology ecosystems.

5.17 Chapter summary

This chapter presented DEP as a powerful BioMEMS-based technology for label-free manipulation, separation, and characterization of biological particles using intrinsic dielectric properties. The fundamental principles of frequency-dependent polarization and Clausius–Mossotti factor behavior were discussed to explain how biological structures such as membranes, cell walls, and cytoplasm determine positive and negative DEP responses. The chapter reviewed major DEP device architectures, including electrode-based, insulator-based, traveling wave, and microfluidic-integrated systems, highlighting their operational principles and application suitability.

Key engineering considerations including Joule heating, electrode material stability, and sample conductivity optimization were discussed alongside CMOS-compatible microfabrication strategies and microfluidic integration approaches. Advances in DEP signal interpretation, particularly polarization spectrum analysis and emerging AI-assisted particle tracking, were highlighted as enabling technologies for quantitative dielectric phenotyping and automated diagnostics. Applications across mammalian cell engineering, bacterial detection, viral and extracellular vesicle isolation, environmental monitoring, and biopharmaceutical manufacturing demonstrate DEP's versatility as a cross-disciplinary bioanalytical platform. Looking forward, integration with AI, automated microfluidics, and digital healthcare networks is expected to transform DEP into intelligent, portable diagnostic systems. Due to its low reagent requirement, portability, and rapid analysis capability, DEP shows strong potential for sustainable deployment in global health monitoring and low-resource diagnostic environments.

Acknowledgment

The authors acknowledge the Technology Development Fund (TDF), grant number TDF09241800, funded by Ministry of Science, Technology and Innovation (MOSTI) and Geran Universiti Penyelidikan (GUP), grant number GUP-2024-070, funded by Universiti Kebangsaan Malaysia (UKM).This work was performed in part at the Micro Nano Research Facility of RMIT University in the Victorian Node of the Australian National Fabrication Facility (ANFF). We acknowledge the project funding from the Australian Research Council through IH210100040. The author would like to thank Professor Sharath Sriram, Professor Madhu Bhaskaran, and Professor Sumeet Walia, who is leading the Functional Materials and Microsystems Research Group at RMIT University, for allocating resources and technical expertise in micro-and nanofabrication activities for this project.

$$F_{DEP} = 2\pi r_{ext}^3 \varepsilon_0 \varepsilon_m Re[f_{CM}] \nabla E^2 \tag{5.1}$$

References

Al-Aqbi, Z. T., Al-Mamoori, M. J., Breadmore, M. C., Guijt, R. M., & Shallan, A. I. (2021). A PMMA nano-junction microfluidic device for blood plasma filtration and collection. *Micromachines, 12*(6), 675. https://doi.org/10.3390/mi12060675.

Asami, K. , (2002). Characterization of biological cells by dielectric spectroscopy. *Journal of Non-Crystalline Solids, 305*(1-3), 268–277. https://doi.org/10.1016/s0022-3093(02)01110-9.

Au, A. K., Lai, H., Utela, B. R., & Folch, A. (2011). Microvalves and micropumps for BioMEMS. *Micromachines,, 2*(2), 179–220. https://doi.org/10.3390/mi2020179.

Castellanos, A., Ramos, A., Gonzalez, A., Green, N. G., & Morgan, H. (2003). Electrohydrodynamics and dielectrophoresis in microsystems. *Journal of Physics D, 36*(20), 2584–2597. https://doi.org/10.1088/0022-3727/36/20/023.

Chen, J., Zhang, H., Liu, Y., Wang, Z., Huang, C., & Zhang, Z. (2015). Exosome separation by dielectrophoresis. *Lab on a Chip, 15*(17), 3554–3562. doi:10.1039/C5LC00617F.

Chin, C. D., Laksanasopin, T., Cheung, Y. K., Steinmiller, D., Linder, V., Parsa, H., … Sia, S. K. (2012). Microfluidics-based diagnostics. *Nature Medicine, 18*(8), 1133–1143. doi:10.1038/nm.2841.

Cummings, E. B., & Singh, A. K. (2003). Dielectrophoresis in microchips containing arrays of insulating posts. *Analytical Chemistry, 75*(18), 4724–4731. https://doi.org/10.1021/ac0340612.

Drain, P. K., Hyle, E. P., Noubary, F., Freedberg, K. A., Wilson, D., Bishai, W. R., … Bassett, I. V. (2014). Point-of-care diagnostics for infectious diseases. *Lancet Infectious Diseases, 14*(3), 239–249. doi:10.1016/S1473-3099(13)70250-0.

Dzulkefli, N. A., Tehranirokh, M., & Majlis, B. Y. (2008). Preparation of glucose hydrogel as actuator in microvalve. In *Proceedings of the IEEE International Conference on Semiconductor Electronics (ICSE 2008)* (pp. 496–499).

Fissell, W. H., Dubnisheva, A., Eldridge, A. N., Fleischman, A. J., Zydney, A. L., & Roy, S. (2009). High-performance silicon nanopore hemofiltration membranes. *Journal of Membrane Science, 326*(1), 58–63. https://doi.org/10.1016/j.memsci.2008.09.039.

Franssila, S. (2010). *Introduction to microfabrication*. Wiley https://10.1002/9781119990413.

Gascoyne, P. R. C., & Shim, S. (2014). Isolation of circulating tumor cells by dielectrophoresis. *Cancers, 6*(1), 545–579. https://doi.org/10.3390/cancers6010545.

Gascoyne, P. R. C., & Vykoukal, J. (2002). Particle separation by dielectrophoresis. *Electrophoresis, 23*(13), 1973–1983. https://doi.org/10.1002/1522-2683(200207)23:13⟨1973::AID-ELPS1973⟩3.0.CO;2-1.

Ghafar-Zadeh, E., & Sawan, M. (2007). A Hybrid Microfluidic/CMOS Capacitive Sensor Dedicated to Lab-on-Chip Applications. *IEEE Transactions on Biomedical Circuits and Systems, 1*(4), 270–277. https://doi.org/10.1109/TBCAS.2007.910966.

Grieshaber, D., MacKenzie, R., Vörös, J., & Reimhult, E. (2008). Electrochemical biosensors - sensor principles and architectures. *Sensors,, 8*(3), 1400–1458. https://doi.org/10.3390/s8031400.

Hamzah, A. A., Abidin, H. E. Z., Majlis, B. Y., Mohd Nor, M., Ismardi, A., Sugandi, G., Tiong, T. Y., Dee, C. F., & Yunas, J. (2013). Electrochemically deposited and etched membranes with precisely sized micropores for biological fluids microfiltration. *Journal of Micromechanics and Microengineering, 23*(7), 074007. https://doi.org/10.1088/0960-1317/23/7/074007.

Hughes, M. P. (2002). Strategies for dielectrophoretic separation in laboratory-on-a-chip systems. *Electrophoresis, 23*(16), 2569–2582. https://doi.org/10.1002/1522-2683(200208)23:16⟨2569::AID-ELPS2569⟩3.0.CO;2-M.

Jones, T. B. (1995). *Electromechanics of particles*. Cambridge University Press.

Lapizco-Encinas, B. H. (2011). DEP for microorganism separation. *Electrophoresis, 32*(3–4), 247–257. https://doi.org/10.1002/elps.201000461.

Litjens,, G., Kooi, T., Bejnordi, B. E, Setio, A. A. A., Ciompi, F., Ghafoorian, M., van der Laak, J. A. W. M., Ginneken, B., & Sánchez, C. I. (2017). A survey on deep learning in medical image analysis. *Medical Image Analysis, 42*(2017), 60–88. https://doi.org/10.1016/j.media.2017.07.005.

Madou, M. (2011). *Fundamentals of microfabrication and nanotechnology*. CRC Press.

Markx, G. H., & Davey, C. L. (1999). The dielectric properties of biological cells at radiofrequencies. *Enzyme and Microbial Technology, 25*(3-5), 161–171. https://doi.org/10.1016/S0141-0229(99)00008-3.

Morgan, H., & Green, N. G. (2003). *AC electrokinetics: Colloids and nanoparticles*. Research Studies Press.

Nam, Y. H., Lee, S. K., Kim, J. H., & Park, J. H. (2019). PDMS membrane filter with nano-slit array fabricated using three-dimensional silicon mold for the concentration of particles with bacterial size range. *Microelectronic Engineering, 215*, 111008. https://doi.org/10.1016/j.mee.2019.111008.

Pohl, H. A. (1978). *Dielectrophoresis: The behavior of neutral matter in nonuniform electric fields*. Cambridge University Press.

Sackmann, E. K., Fulton, A. L., & Beebe, D. J. (2014). The present and future role of microfluidics in biomedical research. *Nature, 507*(7491), 181–189. https://doi.org/10.1038/nature13118.

Schwan, H. P. (1957). Electrical properties of tissue and cell suspensions. *Advances in Biological and Medical Physics, 5*, 147–209. https://doi.org/10.1016/b978-1-4832-3111-2.50008-0.

Tehranirokh, M., Majlis, B. Y., & Bais, B. (2009). Design and simulation of a normally closed glucose sensitive hydrogel based microvalve. *Microsystem Technologies, 15*(5), 753–762. https://doi.org/10.1007/s00542-008-0783-2.

Tehranirokh, M., Bais, B., & Majlis, B. Y. (2006). A hydrogel-based microvalve for insulin delivery application. In *Proceedings of the IEEE International Conference on Semiconductor Electronics (ICSE 2006)* (pp. 418–421).

Wang, Z., Zhang, X., Wang, C., & Li, J. (2021). Passive microfluidic blood plasma separation for point-of-care testing: A review. *Micromachines, 12*(10), 1207. https://doi.org/10.3390/mi12101207.

Whitesides, G. M. (2006). The origins and the future of microfluidics. *Nature, 442*(7101), 368–373. https://doi.org/10.1038/nature05058.

Xie, P., Yao, X., Chu, Z., Yang, Y., Li, H., Tan, S., Tang, H., Zhou, J., & Jin, W. (2023). Homoporous polydimethylsiloxane membrane microfilter for ultrafast label-free isolation and recognition of circulating tumor cells in peripheral blood. *Iscience, 26*(11), 108246. https://doi.org/10.1016/j.isci.2023.108246.

Yafouz, B., Kadri, N. A., & Ibrahim, F. (2014). Dielectrophoretic separation of bacteria from environmental samples. *Biosensors and Bioelectronics, 61*(1), 1–7. doi:10.1016/j.bios.2014.04.036.

Zheng, Y., Shi, H., Tan, Z., Xu, W., Zhao, R., Dan, Z., Dai, Z., Liao, J., & Wu, H. (2025). Micropumps and microvalves for biomedical applications. *Trends in Analytical Chemistry, 188*, 118236. https://doi.org/10.1016/j.trac.2025.118236.

CHAPTER

BioMEMS for drug delivery (micropumps, microneedles and micromixers)

6

Roer Eka Pawinanto[a], Abdul Manaf Hashim[b], Budi Mulyanti[c], Norazreen Abd Aziz[d], Jumril Yunas[e] and Arifah Syahirah Abdul Rahman[e]

[a]*Study Program of Industrial Automation Engineering Education and Robotics, Universitas Pendidikan Indonesia, Bandung, Indonesia,* [b]*Malaysia-Japan International Institute of Technology, Universiti Teknologi Malaysia, Kuala Lumpur, Malaysia,* [c]*Study Program of Electrical Engineering, Universitas Pendidikan Indonesia, Bandung, Indonesia,* [d]*Faculty of Engineering and Built Environment, Universiti Kebangsaan Malaysia, Bangi, Selangor, Malaysia,* [e]*Institute of Microengineering and Nanoelectronics (IMEN), Universiti Kebagsaan Malaysia, Bangi, Selangor, Malaysia*

6.1 Introduction

In micro-electro-mechanical systems (MEMS), it is possible to integrate mechanical components, micron-scale sensors and actuators, microelectronic parts, and three-dimensional structures onto a single substrate (such as silicon, glass, or polymer) to create a microfluidic system. MEMS devices incorporating microfluidic systems can be fabricated using the same methods and equipment as those used in semiconductor technology, similar to the production of integrated circuits (ICs). This compatibility makes the development of biomedical MEMS (bioMEMS) technology straightforward, requiring only adaptation within the existing semiconductor industry. Consequently, MEMS technology has garnered significant attention in the manufacturing sector and has rapidly advanced.

One promising application of microfluidic systems in the biomedical sector is drug delivery. These systems are crucial in medical treatments and healthcare because they influence the patient treatment process and are directly linked to the patient's recovery potential. BioMEMS devices are instrumental in the advancement of drug delivery systems, which are designed to provide and administer drugs in a controlled manner. This integrated microsystem facilitates controlled, precise, cost-effective, and compact drug delivery systems, making them ideal for portable biomedical systems.

Drug delivery can be achieved by injecting a liquid that administers drugs locally into blood vessels, tissues, and organs. Injections are necessary when oral drug delivery is not feasible. BioMEMS technology-based drug delivery devices are typically very small, sensitive, and responsive, allowing them to accurately manage the delivery of critical drug doses without the direct involvement of healthcare professionals. Micron-sized BioMEMS drug delivery devices, such as micropumps, micromixers, and microneedles (MNs), enable the device to concentrate solely on the body area and target site requiring treatment, facilitating local therapy or situations where the specific dose of drug needs to be introduced into the body gradually and continuously over a specified period.

BioMEMS Devices. DOI: https://doi.org/10.1016/B978-0-443-24015-7.00004-8

Microfluidic systems for drug delivery generally comprise three types of devices: micropumps, micromixing devices, and microneedle devices. These three devices can be integrated into a single microfluidic system that includes an inlet/outlet, microchannel, and microchamber. In this chapter, these three devices are discussed and introduced, with a particular emphasis on the electromagnetically driven micromixing system owing to its benefits and the growing industrial demand for intelligent and automated drug delivery systems in the future.

6.2 Micropumps

There are three main parts that play an important role in drug delivery systems: drug storage reservoirs, fluidic channels, and injectors that function to enable drug delivery to the target site through the catheter tube. Fluidic pumping is the most important mechanism in the operation of drug delivery devices because the pump provides the energy required to move the drug fluid from the drug storage reservoir through the microfluidic system to release the drug fluid into the human body.

Microscale pumps fabricated using MEMS technology were first created for use in controlling the dose of insulin fluid to maintain blood sugar levels in diabetic patients. These pumps use piezoelectric materials as their actuators. Currently, technology is developing, and the use of micro pumps is becoming more widespread and important when they become part of lab-on-a-chip (LoC) systems to obtain precise chemical and biological fluid movement at the micro and nano scale. In LoC systems, MEMS micropumps can be combined with biosensors and integrated on a chip (Amirouche et al., 2009). This integration makes the drug delivery system a controlled, precise, and complete method of drug delivery, especially for portable systems, without the need to take the patient to the hospital.

Generally, micropumps are operated mechanically by applying pressure to the fluid through the movement of the actuator, allowing the fluid to flow continuously in the microchannel. Most mechanical micropumps use membrane actuators. Examples of actuators for these mechanical micropumps include electrostatic, piezoelectric, thermal pneumatic, electromagnetic (EM), and shape-memory alloy (SMA) actuators (Amirouche et al., 2009; Hamid et al., 2016).

Micropump devices driven by piezoelectric actuators were the first actuator mechanism used in micropumps and are the most widely used in micropump technology, dominating drug delivery systems, and biomedical applications. Piezoelectric micropumps convert the piezoelectric tension caused by an electric field on a piezoelectric crystal to obtain the bending displacement of the actuator.

The structure of a membrane-driven micropump has four important components: a flexible membrane with a movement space for the membrane, an actuator material (magnetic, piezo, or capacitive material), and a power supply. This part is integrated with the fluid channel, chamber, and inlet and outlet valves, which are commonly referred to as microfluidic parts.

Electrostatic actuators are widely used in MEMS devices. They rely on the columbic attraction of two electrostatic plates with materials that have opposite charges. The attraction can be calculated from the electrostatic energy stored in the devices. The electrostatic force on the plates is caused by the attraction and repulsion of electric field charge. Similar charges repel each other, whereas dissimilar charges attract each other. Electrostatic micropumps usually have a two-plate parallel actuator structure, where one fixed plate and one movable plate are the actuators for pumping liquid

Thermopneumatic actuators have a structure consisting of a heater, membrane, and sealed chamber. The deflection of the thermo-pneumatic actuator is caused by a change in the amount of induced heat

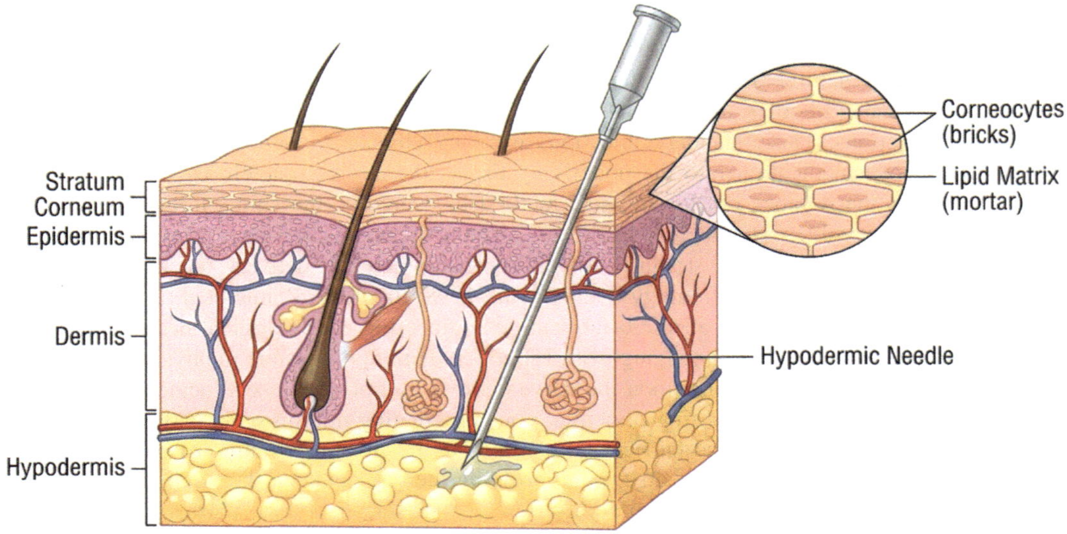

FIGURE 6.1

Cross-section of human skin showing the layers of the epidermis, dermis, and hypodermis. A conventional hypodermic needle is shown penetrating these layers to deliver medication.

or a change in the phase of the liquid in the chamber, which causes a change in pressure to deflect the membrane

EM micropumps use the concept of repulsion and attraction of magnetic fields caused by the interaction between the EM microcoil and a permanent magnet, thereby generating magnetic power (Zhou & Amirouche, 2011).

The flow of fluid by the EM micropump is produced by the movement of an actuator membrane in an insulated chamber that obtains a pressure force from the magnetic field (Said et al., 2016). This pressure force is produced by the microcoil and the magnetic field of the permanent magnet. The change in the magnetic field in the microcoil causes attraction and repulsion of the flexible actuator membrane, where this membrane is attached to the permanent magnet. The force obtained by the membrane depends on the input current and number of turns of the coil.

6.3 Microneedles in MEMS for drug delivery

6.3.1 Fundamentals of transdermal drug delivery

Human skin is designed primarily as a biological shield, a function that simultaneously presents the greatest challenge to transdermal drug delivery. This protective capability resides in the epidermis's outermost layer, the stratum corneum. Although only 10–20 μm thick, this layer forms a formidable barrier through its brick-and-mortar architecture, composed of dead, keratinized corneocytes embedded within a dense lipid matrix, as illustrated in Fig. 6.1. This highly organized structure severely restricts passive transport, effectively blocking most therapeutic agents except for small, lipophilic molecules with a molecular weight under 500 Daltons.

Because of this natural defense system, the delivery of vital, larger macromolecules such as insulin, monoclonal antibodies, and various vaccines has historically required the use of conventional hypodermic injections. These standard injections work by physically piercing through the stratum corneum to deposit medication into the underlying tissue, thereby completely bypassing the skin's barrier. However, the reliance on this invasive method to overcome the skin's limited permeability creates a significant problem statement in healthcare, driving the necessity for a less invasive alternative that can still effectively breach the stratum corneum for macromolecule delivery, a need that has led to the emergence of microneedle technology (Henry et al., 1998; Ita, 2017; Zheng et al., 2023)

MNs represent a convergence of micro-fabrication engineering and biological necessity. By engineering structures with dimensions ranging from 150 to 1500 μm, it is possible to physically breach the stratum corneum without impinging upon the nerve endings and blood vessels located deeper within the dermis. The fundamental operating principle relies on creating micron-scale pathways that allow macromolecules to bypass the lipid barrier entirely. Unlike the stochastic nature of passive diffusion in patches, MNs offer a mechanical entry route, bridging the gap between the noninvasive nature of a transdermal patch and the high bioavailability of a hypodermic injection. This approach not only facilitates the delivery of large-molecule drugs but also addresses patient compliance by rendering the administration process virtually pain-free (Alimardani et al., 2021; Guillot et al., 2020; Maderuelo et al., 2019)

6.3.2 Classification of MEMS microneedles

6.3.2(a) Solid microneedles

The evolution of microneedle technology has yielded diverse architectures, each tailored to specific pharmacokinetic requirements and release mechanisms, as illustrated in Fig. 6.2. The simplest iteration, solid MNs, operates on a "Poke and Patch" strategy (Fig. 6.2A). These structures, typically fabricated from silicon or metal, are applied to the skin to puncture the stratum corneum (Li et al., 2017; Pradeep Narayanan & Raghavan, 2017; Senel et al., 2019). Following insertion, they are removed, and a conventional drug formulation is applied over the transiently created micropores. While effective for skin pretreatment, this two-step process introduces variability in dosing, as the microchannels may close before sufficient drug absorption occurs (Kim et al., 2020).

To integrate the drug directly with the device, engineers developed coated MNs utilizing a "Coat and Poke" strategy (Fig. 6.2B). This approach uses a solid structural core, often metal for mechanical strength layered with a water-soluble drug formulation. Upon insertion, the coating dissolves into the interstitial fluid (Balmert et al., 2020; Cormier et al., 2004; Gill & Prausnitz, 2007). However, the payload capacity is geometrically limited by the surface area of the needle tip, making this approach suitable only for highly potent compounds. Conversely, dissolvable MNs represent a "Poke and Release" strategy (Fig. 6.2C). These are fabricated entirely from biodegradable polymers or sugars that encapsulate the therapeutic agent (Balashanmugam et al., 2008; Park et al., 2005; Sullivan et al., 2010). The device acts as both the storage medium and the delivery vehicle, dissolving completely within the skin to release the payload, thereby eliminating the generation of sharp biohazardous waste.

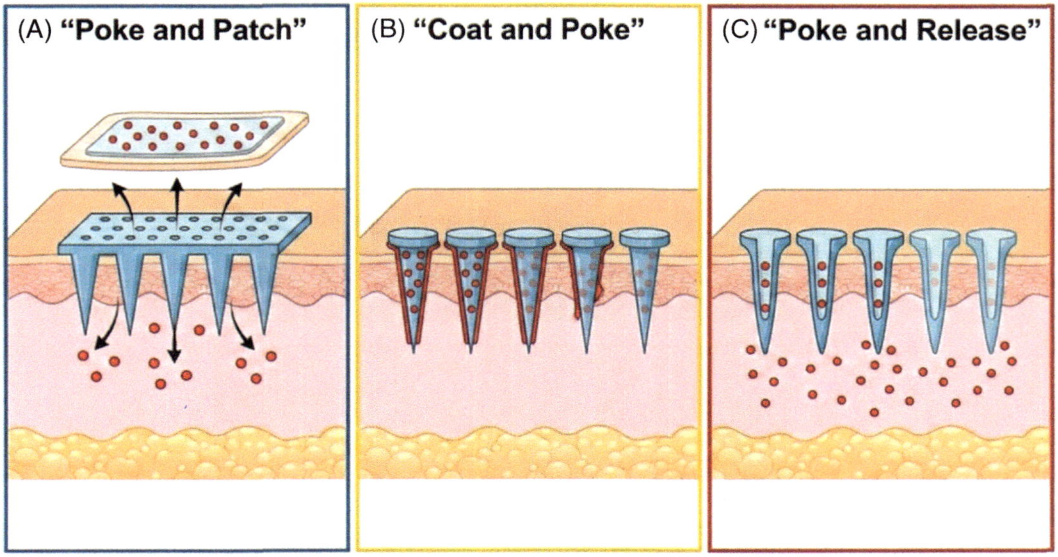

FIGURE 6.2

Strategies for drug delivery using microneedles: (A) The "Poke and Patch" strategy using solid microneedles for pretreatment; (B) the "Coat and Poke" strategy where drug formulations are coated onto solid cores; and (C) the "Poke and Release" strategy using dissolving microneedles.

6.3.2(a)(i) Geometry and insertion performance

Beyond material composition and drug loading strategy, geometric design remains a key determinant of mechanical strength and insertion efficiency in solid MNs. Hamzah et al. reported concave solid silicon MNs fabricated by HNA etching where changes in HNA composition, mask window size, temperature, and bath agitation produced clear differences in morphology and aspect ratio. As shown in Fig. 6.3, the structures all show sharp tips and smooth sidewalls, which are widely accepted to be critical for reducing insertion force and limiting tissue disruption during penetration. Under the tested conditions, it is reported that aspect ratios ranged from 1.33 to 1.44, suggesting that careful optimization of the etching parameters can produce relatively high aspect ratio MNs with increased penetration capability, which further enhances the reliability of solid MNs in skin pretreatment applications and supports their in vitro and in vivo applications (Hamzah et al., 2012).

6.3.2(a)(ii) Mechanical integrity and insertion performance

Solid MNs may fracture during insertion and thus their mechanical performance determines the feasibility of insertion without fracture. Fig. 6.4 discusses this with complementary experimental insertion tests and numerical modelling. Prepared rat skin was loaded, using a controlled external force, directly on the microneedle array and post-insertion images revealed sufficient penetration through the dyed tissue, indicating adequate tip sharpness and structural integrity during loading. Erny et al. reported that finite element analysis provides a quantitative understanding of the same process by mapping

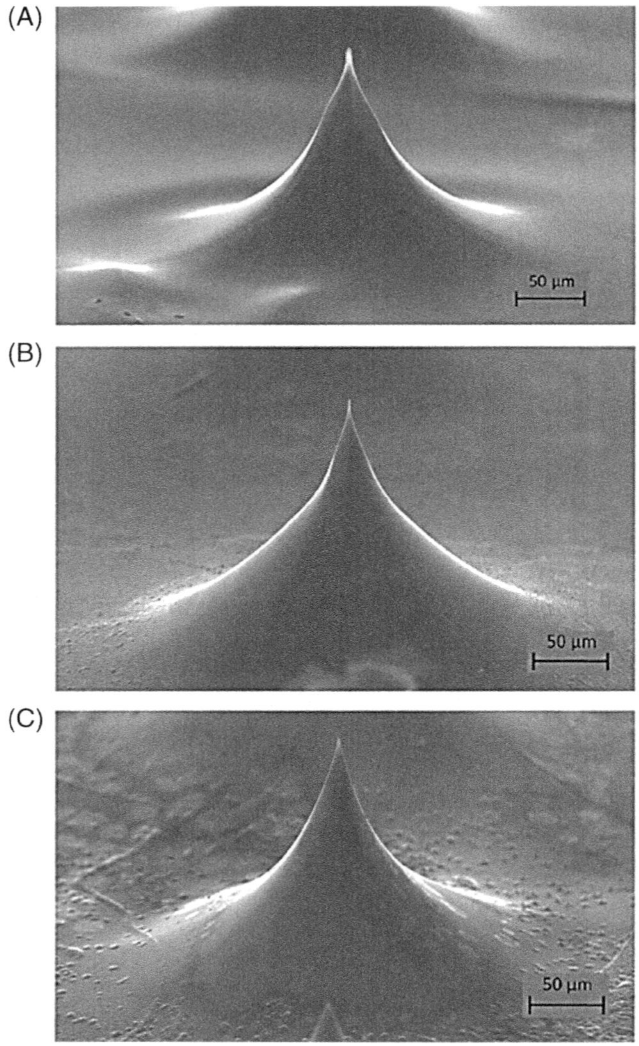

FIGURE 6.3

SEM images of concave, cone-shaped solid microneedles made under three different HNA etching conditions: (A) 400 rpm agitation, 500 µm mask window, HNA 2:7:1 (aspect ratio 1.35); (B) 450 rpm agitation, 400 µm mask window, HNA 3:6:1 (aspect ratio 1.33); and (C) 450 rpm agitation, 500 µm mask window, HNA 2:7:1 (aspect ratio 1.44).

Reproduced with permission from Hamzah, A. A., Aziz, N. A., Majlis, B. Y., Yunas, J., & Dee, C. F. (2012). Optimization of HNA etching parameters to produce high aspect ratio solid silicon microneedles. Journal of Micromechanics and Microengineering, *22(9), 95017. http://doi.org/10.1088/0960-1317/22/9/095017. Copyright (2012), reproduced with permission from IOP Publishing.*

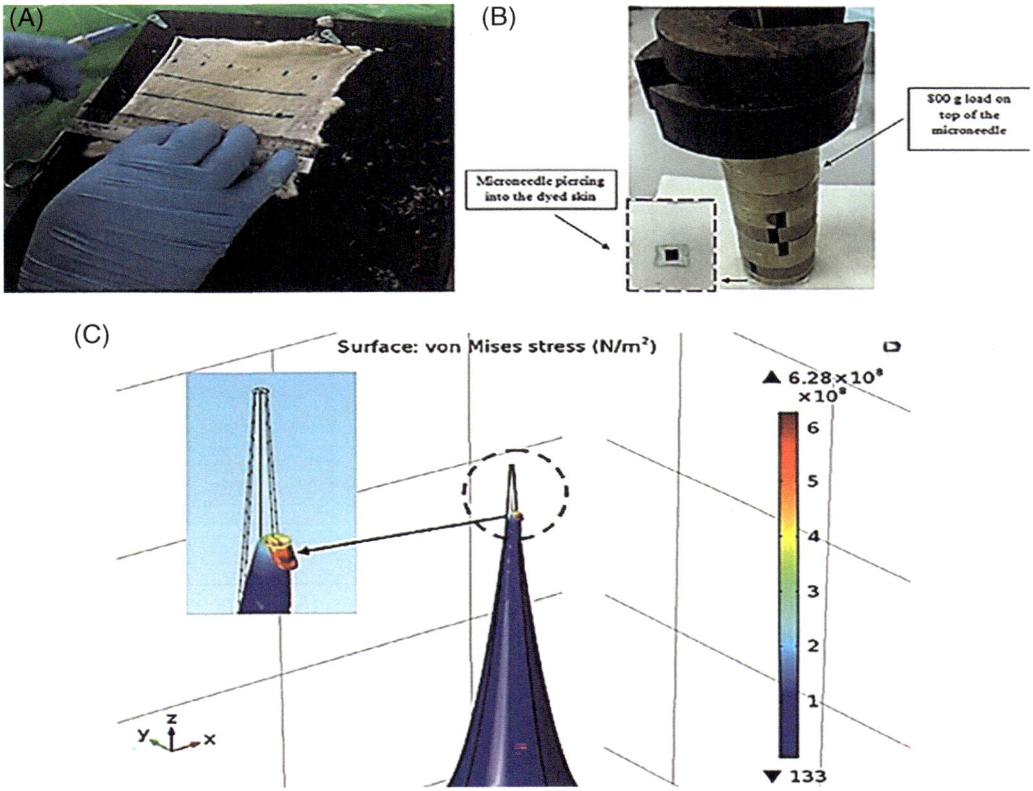

FIGURE 6.4

(A) Rat skin prepared by cutting into 2 × 2 cm sections. (B) Insertion setup showing an 800 g load applied to the microneedle array on dyed skin, the inset highlights microneedle penetration into the stained tissue. (C) von Mises stress distribution under a 50 g applied load, with the inset showing a magnified view of tip deformation.

Reproduced with permission from Zainal Abidin, H. E., Ooi, P. C., Tiong, T. Y., Marsi, N., Ismardi, A., Noor, M. M., Fathi, N. A. F. N. Z., Abd Aziz, N., Sahari, S. K., Sugandi, G., Yunas, J., Dee, C. F., Majlis, B. Y., & Hamzah, A. A. (2020). Stress and deformation of optimally shaped silicon microneedles for transdermal drug delivery. Journal of Pharmaceutical Sciences, 109, *2485–2492.* *Copyright (2020), reproduced with permission from Elsevier.*

stress under the applied load, where peak stresses are localized at the microneedle tip (consistent with the initial point of contact during insertion) and deformation remains localized near the tip, with the shaft remaining mechanically stable. Overall, these results indicate that the silicon MNs are capable of withstanding insertion forces without mechanical failure, a crucial result because consistent penetration supports predictable delivery performance (Erny et al., 2020).

6.3.2(b) Hollow microneedles

Hollow MNs represent a significant step forward in MEMS drug delivery, functioning essentially as microscopic hypodermic needles. They operate on a "Poke and Flow" principle, once the needle

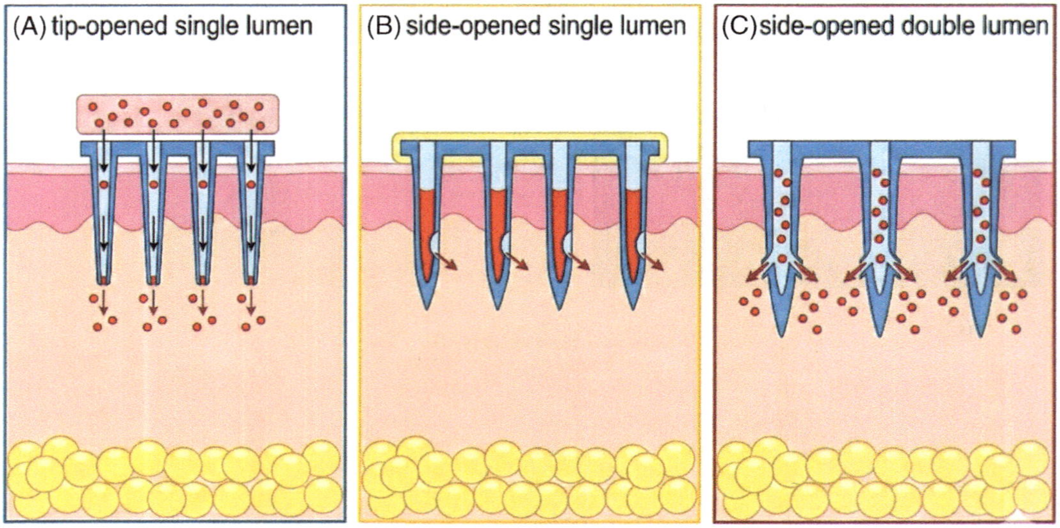

FIGURE 6.5

Illustration of hollow microneedle designs classified by their bore openings: (A) Tip-opened single lumen, (B) side-opened single lumen designed to prevent blockage during insertion, and (C) side-opened double lumen for improved flow consistency.

penetrates the dermis, the drug solution stored within the internal reservoir is actively driven into the skin (Anbazhagan et al., 2023; Bolton et al., 2020; Datta et al., 2023; Economidou et al., 2021). This delivery can be controlled through various mechanisms, such as applied pressure, diffusion, or even electrical stimulation, allowing the payload to reach the deeper, vascularized layers of the epidermis. The design of the internal bore is critical for performance. As illustrated in Fig. 6.5, these MNs are generally classified by their lumen geometry. The most basic configuration is the tip-opened single lumen (Fig. 6.5A), where the bore exits at the very apex of the needle. Engineers developed these alternative designs to prevent the needle tip from getting clogged with skin tissue, a common issue known as coring. These include the side-opened single lumen (Fig. 6.5B) and the side-opened double lumen (Fig. 6.5C).

The side-opened double-lumen design is particularly effective; by splitting the flow path, it reduces the likelihood of occlusion and allows for a more consistent flow rate. There is, however, a trade-off in fabrication. Expanding the hollow lumen to reduce flow resistance naturally compromises the mechanical strength of the needle walls. To address this, researchers often reinforce the tips with metal coatings to ensure they are durable enough to penetrate the stratum corneum without fracturing. Despite these structural challenges, the ability of hollow MNs to accommodate larger volumes makes them ideal candidates for delivering high-molecular-weight biotherapeutics, such as proteins, oligonucleotides, and vaccines.

6.3.2(c) *Dissolvable microneedles*

Dissolvable MNs provide an alternative to conventional patches and injections because the drug can be incorporated into the microneedle matrix during fabrication. This supports a single-step process and avoids additional coating or post-fabrication loading steps (Jalal et al., 2026). Lignocaine is a widely used local anesthetic with rapid onset and an established safety profile. Incorporating lignocaine into dissolvable MNs may support localized delivery for dermatological indications and minor surgical procedures (Gordh et al., 2010). However, drug loading can vary between needles and between batches, and many studies lack standardized, nondestructive methods to verify drug distribution within the needle matrix. This may contribute to variability in delivered dose. Rahman et al. reported maltose-based dissolvable MNs containing lignocaine prepared using a simplified single-step method. They evaluated microneedle geometry and mechanical performance and assessed drug distribution using confocal microscopy to visualize dispersion within the microneedle matrix (Rahman et al., 2026).

6.3.2(c)(i) Fabrication by micro-molding

Micromolding is a simple and scalable process of producing dissolvable MNs (Rahman et al., 2026). This article describes maltose-based formulation methods, which begin with dissolving the components in deionized water and then casting the mixture into silicone micromolds. The method was then followed by centrifugation to achieve complete cavity filling and minimize trapped air, and subsequent drying under atmospheric pressure to solidify the MNs. The patches were finally demolded and stored in sealed containers to limit moisture exposure. In this study, maltose is used as both a microneedle matrix and the backing in this system; an effective fill efficiency of 90 or more was used to ensure homogeneous formation of each array of needles.

6.3.2(c)(ii) Geometrical and surface analysis

Field emission scanning electron microscopy (FE-SEM), a qualitative method, was used to observe the surface morphology and structural integrity of drug embedded microneedle. Micrographic images in Fig. 6.6 show a uniform distribution of conical MNs with smooth surfaces and clearly defined ends, which confirms the correct mold replication and justifies a regulated drying procedure. The mean heights of the measurements obtained by dimensional analysis were close to 418.5, 423.1, and 413.9 μm. Base width was 178.5, 171, and 179.2 μm. The low total dispersion, as well as the uniform cavity filling, indicates the presence of a minor contraction of the matrix in the drying phase.

The insignificant difference in height and base width indicates that the fabrication process is reproducible, which also implies good shape retention after drying. The geometry of the conical forms is effective with the sharp tips help reduce the force required for insertion. Additionally, broader bases increase resistance to buckling under compressive load (Hamzah et al., 2004). The medicine is delivered more reliably and consistently when the MNs in an array are dimensionally uniform, they tend to penetrate the skin to a similar depth, which helps deliver the drug more consistently and predictably. Overall, the integrated geometric and mechanical results prove that the MNs meet the basic structural requirements of transdermal uses.

FIGURE 6.6

SEM micrographs of maltose-based lignocaine microneedle arrays. The images show regularly spaced microneedles with sharp tips and consistent geometry. (A) Tip (vertical) height measurements for representative needles (H1 = 418.5 µm, H2 = 423.1 µm, H3 = 413.9 µm; mean = 418.5 µm, n = 3). (B) Base width measurements for representative needles (V1 = 178.5, 171.0 and 179.2 µm; mean = 176.2 µm, n = 3), indicating well-defined structures supportive of mechanical robustness during skin insertion.
Scale bar: 200 µm.

Reproduced with permission from Rahman, A. S., Cheah, F. C., Nashriby, M. E. A., Bastion, M. L. C., Dee, C. F., Buyong, M. R., Mohamed, M. A., Chua, X. Y., Ooi, P. C., Abdul Jalal, M. I., Lam, C., Mun, Y. Y., Goh, C. S., Ismail, A. G., & Hamzah, A. A. (2026). Optical and mechanical characterization of lignocaine-impregnated maltose-based dissolvable microneedles. Engineering Proceedings, 110*(1), 1–9. https://doi.org/10.3390/engproc2025110007, published by MDPI under the Creative Commons Attribution (CC BY 4.0) license.*

6.3.2(c)(iii) Mechanical characterization and insertion performance

Mechanical strength is one of the critical parameters of microneedle dissolution because microneedle must be able to withstand compression when inserted without fracturing or permanently deforming. Nanoindentation is commonly used to determine fracture force and elastic modulus. In actually, around 0.1 N per needle is often quoted as the minimum insertion force required to penetrate the stratum corneum consistently (Irfan et al., 2025).

Rahman et al reported that the fracture force of maltose-made MNs around 0.6 N/needle. Elastic recovery is also possible for formulations containing 25% and 50% lignocaine through load-displacement testing, which suggests that inclusion of lignocaine does not always affect mechanical integrity, but instead, makes the sugar matrix stronger, and this finding is consistent with the reports where dissolvable MNs can still be used to maintain mechanical functionality following drug loading of the needle (Rahman et al., 2026). Furthermore, the conical geometry and the optimized formulation allowing alignment of the microneedle tip with the conical geometry of the stratum corneum and thus facilitating reliable penetration into the skin (Erny et al., 2020).

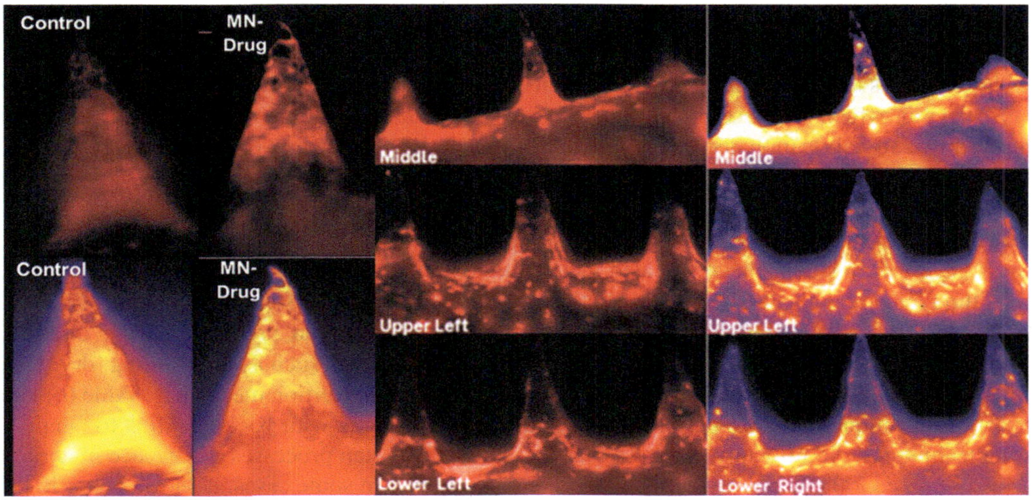

FIGURE 6.7

Fluorescence microscopy of maltose-based microneedle showing lignocaine distribution. (A) Control microneedles without drug show no detectable fluorescence, whereas drug-loaded microneedles exhibit intense fluorescence in pseudo-colored images, indicating successful drug loading. (B) The obtained images of different parts of the patch (center, upper-left, and lateral sector) show no discriminative intensity of fluorescence in the array, which implies a uniform distribution of lignocaine.

Reproduced with permission from Rahman, A. S., Cheah, F. C., Nashriby, M. E. A., Bastion, M. L. C., Dee, C. F., Buyong, M. R., Mohamed, M. A., Chua, X. Y., Ooi, P. C., Abdul Jalal, M. I., Lam, C., Mun, Y. Y., Goh, C. S., Ismail, A. G., & Hamzah, A. A. (2026). Optical and mechanical characterization of lignocaine-impregnated maltose-based dissolvable microneedles. Engineering Proceedings, 110*(1), 1–9. https://doi.org/10.3390/engproc2025110007, published by MDPI under the Creative Commons Attribution (CC BY 4.0) license.*

6.3.2(c)(iv) Drug distribution and release behavior

The fabrication approach used to allow the drug distribution to be modulated, allowing it to be uniformly loaded the entirety of the microneedle or it can be biased to concentrate at the ends of the needles. Confocal laser scanning microscopy of the system showed that there was a uniform distribution of lignocaine across the microneedle (Fig. 6.7). The drug-loaded MNs showed strong and stable fluorescence response, and control patches showed no fluorescence response, thus confirming that the observed fluorescence was due to lignocaine. Besides, the uniformity of the intensity of the fluorescence in various areas of the patch suggests even distribution across the array, and loading uniformity is reported to be greater than 85%. To explain further, the microneedle matrix begins to dissolve after being exposed to hydration by interstitial fluid. Drug release is dependent on solubility in matrices and hydration kinetics rather than diffusion in a polymer matrix because sugar-based formulations typically take minutes to hydrate and dissolve (Monou et al., 2025).

6.3.2(c)(v) Performance of dissolvable microneedle fabricated using various technique

Various platforms for Dissolvable MNs have been documented in the scientific literature, with emphasis on the ways in which the performance of the patch is influenced by the differences in the material composition and fabrication techniques. Various production methods were used to produce MNs based on polymers such as polyvinyl polyethylene or polyvinyl alcohol mixtures, systems based on carboxymethyl cellulose and sugar compounds in particular iso-maltulose. These techniques include mold casting, 3D printing, vacuum-assisted casting, and double-casting. Computational modeling and process engineering studies have revealed that drying conditions, needle geometry configuration, and mold-filling dynamics all significantly impact tip sharpness and insertion strength (Rahman et al., 2026).

Table 6.1 presents a comparison between the current fabrication process and the dissolvable microneedle systems that have previously been documented. Factors such as mold filling, drying conditions, and needle geometry strongly influence tip sharpness and insertion mechanics. Table 6.1 shows fabrication techniques ranging between multistep, layered designs with backing films to single-step casting techniques. Additionally, it shows that low-cost, single-step processes can reduce material use and lower procedural complexity and still achieve comparable. In some cases, it shows better in terms of mechanical performance and drug-loading uniformity as more complex processes. These findings suggest that streamlined fabrication strategies are well suited for the scalable production of dissolvable MNs.

Dissolvable MNs are an economical, user-friendly drug delivery method that has been shown to be clinically feasible in recent translational studies (Jalal et al., 2026). The fact that they dissolve after the insertion eliminates the necessity of sharps waste, as well as the use of complicated infusion apparatus, but still practical limitations exist, such as dose restrictions associated with minute needle volume, the need to store hygroscopic or sugar-based preparations in moisture-controlled containers, and the possibility of stability damage to heat-sensitive or desiccating drugs during the manufacturing process. Despite these challenges, the dissolution of MNs can offer a favorable trade-off between performance and manufacturability especially when manufactured using scalable micromolding processes that can ensure geometric precision and consistent distribution of drugs, resulting to more predictable release profiles which can be advantageous for microneedle-based delivery platforms.

6.3.3 MEMS fabrication materials and techniques

The fabrication of MNs demands high-precision micro-manufacturing techniques adapted from the semiconductor industry, with material selection dictating the available processing methods. Silicon remains the substrate of choice for high-fidelity prototyping due to its well-characterized mechanical properties and the maturity of silicon micromachining protocols. However, its brittle nature poses a safety risk of fracture within the skin, necessitating rigorous mechanical characterization or the exploration of alternative materials such as medical-grade metals or biocompatible polymers such as SU-8 and cyclic olefin copolymer (COC) (Mishra et al., 2018).

Fabricating the sharp, high-aspect-ratio geometries required for skin penetration typically involves subtractive manufacturing. In silicon processing, anisotropic wet etching is utilized to shape the needle exterior. By exposing single-crystal silicon to etchants such as potassium hydroxide (KOH) or tetramethylammonium hydroxide (TMAH), engineers exploit the varying etch rates of different

Table 6.1 Lignocaine/Lidocaine microneedles compared with conventional topical anesthetics.

Study	Material	Drug	Fabrication Method	Mechanical Strength (N/Needle)	Drug Uniformity	Notable Features
Rahman et al. (2026)	Maltose	Lignocaine	Micromolding	~0.6 N	≥85% (CLSM verified)	Simple, low-cost process; no backing layer
Jiang et al. (2024)	Isomaltulose (ISO)	Lidocaine hydrochloride (LIDH)	3D-printing-supported mold microneedle casting	Up to ~80%	Not reported	High drug loading
Mao (2024)	PVP/PVA/HA (1:2:1)	Lidocaine hydrochloride	Micromolding	>0.1 N/needle	Visual but not quantified	Only confirms loading amount, not distribution uniformity
Loizidou et al. (2015)	CMC/SUC, CMC/MAL, CMC/TRD	Propranolol	Vacuum mold casting	Up to 7.42 GPa (modulus	Not assessed	CLSM used only for skin insertion
Jin et al. (2025)	PVP/PVA	Lidocaine HCl	Double casting	0.32 N/needle	Quantified by HPLC only	OCT showed insertion, no CLSM

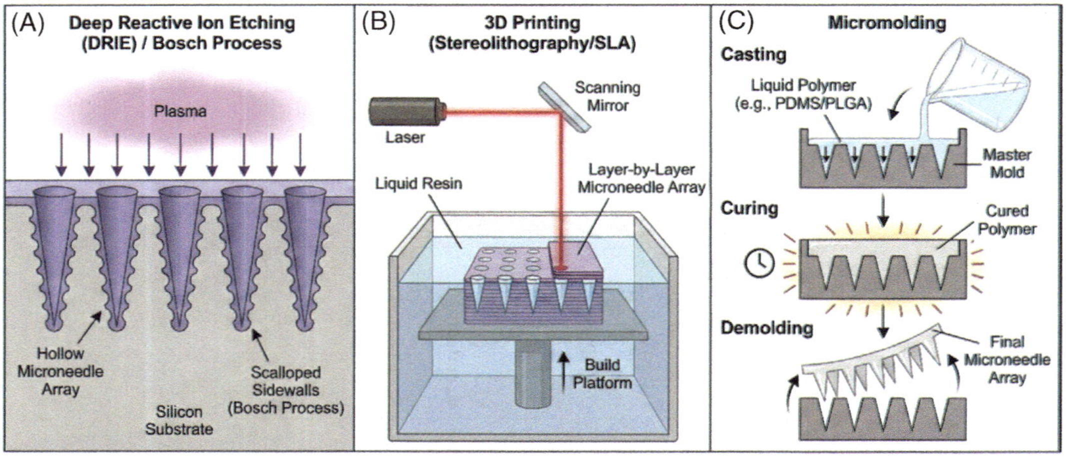

FIGURE 6.8

Schematic illustration of common microneedle fabrication techniques. (A) Deep reactive ion etching (DRIE) utilizing the Bosch process to create high-aspect-ratio hollow silicon microneedles, characterized by scalloped sidewalls. (B) Additive manufacturing via stereolithography (SLA) 3D printing, where the array is constructed layer-by-layer from liquid resin. (C) Micromolding process flow involving the casting of liquid polymer (e.g., PDMS or PLGA) into a master mold, followed by curing and demolding to release the final array.

crystal planes. This crystallographic dependency naturally results in the formation of atomically sharp, pyramidal tips ideal for reducing insertion force (Wilke et al., 2005).

For hollow MNs, the challenge lies in creating the internal lumen. The selection of materials for hollow MNs is the primary determinant of the fabrication methodology, with silicon serving as the foundational material in the field due to the maturity of MEMS processing. Specifically, processes like the Bosch technique (Fig. 6.8A) enable the etching of high-aspect-ratio lumens through the wafer, although the intrinsic brittleness of silicon and the high cost of cleanroom fabrication have driven the search for alternatives. Borosilicate glass has also been utilized, primarily in research settings, to visualize fluid flow within the needle bore; however, like silicon, its fragility limits widespread commercial application. Key studies have highlighted that while silicon remains relevant for high-precision prototyping, its limitations necessitate rigorous mechanical reinforcement or alternative material choices for clinical safety.

To address the mechanical failure modes associated with brittle materials, medical-grade metals and high-strength ceramics have been adopted for their superior durability and fracture toughness. Metals such as stainless steel and titanium possess high Young's moduli (approximately 180 GPa for stainless steel), making them resistant to deformation and breakage during skin insertion. The fabrication of metallic MNs typically bypasses complex masking and etching steps in favor of laser machining and electrical discharge machining (EDM), which streamlines the production of robust, side-opened or bevel-tipped designs. Concurrently, ceramic materials such as alumina (Al_2O_3) and zirconia (ZrO_2) are employed for their excellent chemical stability and compression strength, often processed via cost-effective micromolding of ceramic slurries, though their brittleness remains a consideration.

Recent advancements have shifted focus toward biocompatible polymers and additive manufacturing techniques to enable scalable, cost-efficient production. Polymers such as SU-8, poly(methyl methacrylate) (PMMA), and biodegradable options like polylactic-co-glycolic acid (PLGA) offer tunable mechanical properties and safety for biological applications (Park et al., 2005), (Larrañeta et al., 2016). These are frequently fabricated using solvent casting or micromolding (Fig. 6.8C), which allows for rapid replication from master molds. Furthermore, disruptive technologies like stereolithography (SLA) (Fig. 6.8B) and two-photon polymerization (2PP) 3D printing have emerged, enabling the direct fabrication of complex, bio-inspired geometries such as mosquito-proboscis mimics that reduce insertion force and are impossible to achieve with traditional lithography.

6.4 Micromixers

6.4.1 Introduction to micromixers in BioMEMS

Micromixers represent a critical component in the evolution of BioMEMS technology, particularly in the context of drug delivery systems. Laminar flow, rather than turbulent mixing, governs fluid behavior at the microscale, making the challenge of achieving efficient mixing essential. This subchapter explores the development of micromixers through detailed case studies, examining both theoretical foundations and practical implementations that have advanced the field of microfluidic drug delivery.

The importance of micromixers in biomedical applications cannot be overstated. In drug delivery systems, rapid and homogeneous mixing of therapeutic agents with carrier fluids, reagents, or biological samples is essential for ensuring dosage accuracy, reaction efficiency, and treatment efficacy. Unlike macroscale mixing, where turbulence naturally enhances it, microscale systems operate at low Reynolds numbers (typically $Re < 100$), where viscous forces dominate and mixing occurs primarily through molecular diffusion.

The development of micromixers has been driven by the need to overcome these physical constraints while maintaining the advantages of miniaturization, including reduced sample consumption, faster reaction times, and the potential for portable diagnostic and therapeutic devices. Therefore, this analysis presents a comprehensive overview of micromixer development through three distinct case studies, each highlighting different approaches to solving the microscale mixing challenge.

6.4.2 Fundamental principles of microscale mixing

6.4.2(a) The physics of microscale fluid dynamics

At the microscale, fluid behavior is fundamentally different from that observed at macroscopic scales. The Reynolds number (Re), a dimensionless quantity that characterizes fluid flow, is defined as Eq. 6.1:

$$Re = \frac{\rho v L}{\mu} \tag{6.1}$$

where ρ is the fluid density, v is the characteristic velocity, L is the characteristic length, and μ is the dynamic viscosity. In microfluidic systems, Re typically ranges from 0.01 to 100, indicating laminar flow conditions where mixing is limited to molecular diffusion.

The Peclet number (Pe) reveals the relative importance of convective transport versus diffusive transport, as described in Eq. 6.2:

$$Pe = \frac{vL}{D} \tag{6.2}$$

where D is the diffusion coefficient. The slow rate of molecular diffusion across fluid interfaces still limits mixing, but high Pe numbers (Pe >> 1) indicate that convective transport dominates.

6.4.2(b) Mixing mechanisms at the microscale

Mixing at the microscale is fundamentally different from mixing in conventional, macroscale fluid systems. In microfluidic and BioMEMS devices, fluid flows are typically characterized by very low Reynolds numbers, resulting in strictly laminar behavior with negligible turbulence. Under these conditions, the absence of turbulent eddies eliminates one of the primary drivers of mixing in larger systems. Consequently, microscale mixing relies on alternative mechanisms—primarily molecular diffusion and chaotic advection—to achieve homogenization of solutes, reagents, or biological species within confined microchannels. Understanding these mechanisms is essential for designing efficient micromixers for biochemical reactions, drug delivery systems, point-of-care diagnostics, and LoC technologies.

Molecular diffusion represents the most fundamental mode of mixing at the microscale. Driven by the random thermal motion of molecules, diffusion allows species to move from regions of high concentration to regions of low concentration, gradually evening out concentration gradients across the microchannel. In laminar microflows, where fluid streams frequently flow parallel to one another with negligible transverse convection, diffusion emerges as the primary mechanism for mass transport across flow interfaces. Although diffusion is highly effective over very short distances—an advantage in microfluidics—it can be slow when mixing of larger biomolecules or high-viscosity fluids is required. Therefore, while diffusion ensures predictable and controllable mass transport, its reliance on molecular-scale motion can limit mixing efficiency in microchannels lacking supplementary enhancement strategies.

To overcome the limitations of pure diffusion, microfluidic systems often employ chaotic advection as an additional or alternative mixing mechanism. Chaotic advection does not need high Reynolds numbers like turbulence does. Instead, it happens when fluid elements are stretched, folded, and reoriented over and over again as they move through structured channel features such as herringbone grooves, serpentine paths, staggered obstacles, or flow fields that change over time (Hasanah et al., 2022). This continual deformation enhances the interfacial area between fluid streams and reduces characteristic diffusion lengths, enabling molecular diffusion to operate more effectively. The synergy between chaotic advection-induced stretching and molecular diffusion-driven homogenization makes this approach particularly powerful for rapid mixing in microscale environments.

Molecular diffusion and chaotic advection are the main ways that mixing happens in microscale systems. Effective micromixer design often requires balancing the predictable, physics-driven diffusion process with strategically engineered flow perturbations that induce chaotic advection. Microfluidics can achieve rapid and reliable mixing even under strictly laminar conditions by tailoring channel geometry, flow rates, and material properties. These mechanisms enable efficient microreactors and diagnostic devices and support advances in drug delivery, biosensing, and complex lab-on-chip operations where precise control over fluid interactions is essential.

6.4.2(c) Performance metrics for micromixers

Evaluating the performance of a micromixer requires a comprehensive understanding of several key metrics that quantify how effectively and efficiently mixing occurs within a microfluidic environment. Due to the laminar nature of microflows and the dominant role of diffusion-driven transport, the design and optimization of micromixers rely on defined criteria that can be consistently measured and compared across different configurations. Among the most widely used performance metrics are mixing efficiency, mixing time or mixing length, pressure drop, device footprint, and fabrication complexity. Together, these parameters inform the suitability of a micromixer for specific applications in chemical synthesis, biomedical diagnostics, drug delivery, and LoC systems.

Mixing efficiency is the primary metric used to assess how completely two or more fluid streams become homogenized within the micromixer. It is typically quantified through concentration distribution analysis at the channel outlet or at designated cross-sections along the channel. Commonly, mixing efficiency is expressed as a percentage derived from statistical measures such as the coefficient of variation (CoV), standard deviation of concentration profiles, or normalized intensity fluctuations in optical measurements. Eq. (6.3) illustrates the degree of mixing by utilizing the standard deviation, which is the concentration at a particular cross-section of the microfluidic mixer (Pawinanto et al., 2020). In this context, the variable represents the total number of samples mixed in the cross-section, another variable denotes the mass fraction at the sampling volume, and a third variable indicates the average mass fractions. The mixing efficiency can be calculated from Eq. (6.4), where is the initial standard deviation at the inlet. A higher mixing efficiency indicates better homogenization, although the target value depends on the application. For example, biochemical assays may require near-complete mixing ($\geq 90\%$), whereas preliminary pre-mixing stages in chemical processing may tolerate moderate levels of mixing. Because microscale flows are laminar, mixing efficiency is strongly influenced by diffusion coefficients, channel geometry, and flow rates, as well as any implemented mixing enhancement strategies such as chaotic advection or electrokinetic perturbations.

$$\sigma = \sqrt{\frac{1}{N}\sum_{i=1}^{N}{}_{(c_i - c_m)}2} \tag{6.3}$$

$$MI = 1 - \sqrt{\frac{\sigma^2}{\sigma_{max}^2}} \tag{6.4}$$

Closely related to mixing efficiency are the metric mixing time and mixing length, which reflect the spatial and temporal requirements for achieving a desired level of mixing. Mixing time refers to the duration needed for species to reach a predefined level of homogeneity, while mixing length denotes the channel length required to achieve that same outcome. These parameters are essential for ensuring compatibility with high-throughput systems, continuous-flow reactors, or microfluidic assays where reaction times or flow residence times must be tightly controlled. Shorter mixing times or lengths indicate more efficient micromixer designs, particularly when space is limited or when rapid biochemical reactions are required. Laminar flow constraints often make diffusion the rate-limiting step, but strategic design elements—such as serpentine channels, herringbone grooves, or induced vortices—can substantially reduce the effective mixing length by increasing interfacial contact between fluid streams.

Another critical performance metric is pressure drop, which measures the resistance to flow as fluid passes through the micromixer. Pressure drop is influenced by channel geometry, flow rate, surface roughness, and the presence of obstacles or mixing structures. While aggressive geometric modifications such as sharp bends or embedded ridges can significantly improve mixing, they also tend to increase pressure drop, leading to greater energy consumption and higher demands on pumping systems. Excessive pressure drop can compromise device portability, increase operational costs, or limit compatibility with delicate biological samples that cannot withstand high shear stresses. Therefore, effective micromixer design requires a balance between enhanced mixing features and acceptable hydrodynamic resistance, ensuring that mixing performance does not come at the cost of impractical operating pressures.

Footprint, or the physical size of the micromixer, is another important factor, particularly in portable or highly integrated microfluidic systems. A smaller footprint enables efficient use of chip real estate and supports the integration of multiple functional components—such as sensors, valves, and reaction chambers—on a single microfluidic platform. Designs that achieve high mixing efficiency within a compact form factor are especially valuable in LoC and point-of-care applications, where device miniaturization is essential. However, reducing the footprint often increases design complexity, as shorter channels must compensate for limited space by employing more sophisticated mixing mechanisms, such as chaotic advection or multi-layer structures, to achieve adequate performance.

Finally, fabrication complexity serves as a practical performance metric that influences the overall feasibility, cost, and scalability of a micromixer design. Simple designs such as T-junctions, Y-junctions, or straight channels with minimal geometric perturbations are easy to fabricate using standard lithographic or soft-lithography techniques and are suitable for rapid prototyping. In contrast, advanced micromixers incorporating 3D features, multilayer architectures, or precise surface patterning may require sophisticated fabrication methods such as deep reactive ion etching (DRIE), two-photon polymerization, or multi-material 3D printing (Yunas et al., 2020). While these complex designs often offer superior mixing capabilities, they may also increase production cost, extend fabrication time, and limit compatibility with mass-manufacturing techniques. Therefore, the choice of micromixer architecture must consider not only functional performance but also fabrication practicality, especially for commercial or large-scale deployment.

These performance metrics—mixing efficiency, mixing time or length, pressure drop, footprint, and fabrication complexity—together make up a complete set of tools for judging and comparing micromixer designs. By understanding the trade-offs among these parameters, researchers and engineers can select or tailor micromixer configurations that meet the specific requirements of their intended applications while balancing performance, practicality, and manufacturability.

6.4.3 Case study in development of electromagnetic micromixer

Typically, a magnetic membrane-driven microfluidic mixing system comprises two primary parts: the EM actuation part and the microfluidic part. The schematic of the EM microfluidic mixer is depicted in Fig. 6.9. In summary, the working principle developed for the studied system can be described as follows. Two distinct fluid samples are introduced into a single microchannel, namely microchannel 2, originating from two separate inlets, specifically microchannel 1a and microchannel 1b. Here, the optimization of the angle of the input channel needs to be performed to adjust the mixture flow rate

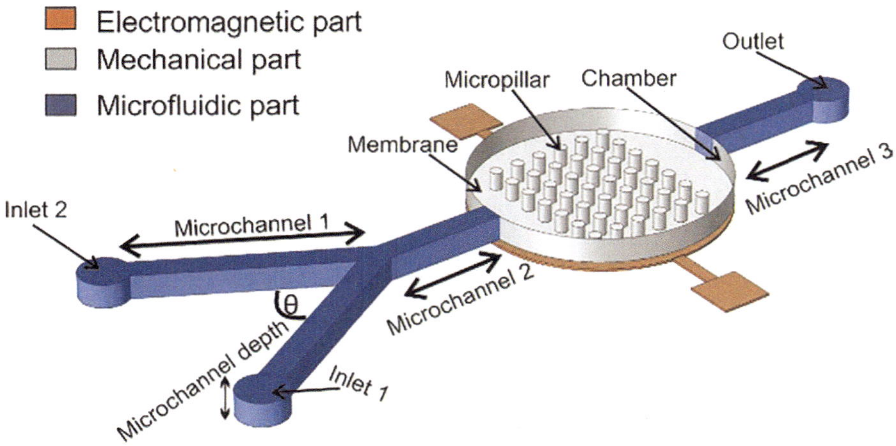

FIGURE 6.9

A schematic of electromagnetic active microfluidic mixer.

entering microchannel 2. Upon the liquid's departure from microchannel 2 and subsequent entry into the microchamber, the EM part concurrently generates force, serving as an input parameter for the displacement of the mechanical membrane. The mechanical membrane part consists of a thin layer of polydimethyl siloxane (PDMSe) membrane, an array of micropillars on top of the PDMS membrane, and a permanent magnet pasted beneath the membrane. In addition, the micropillars on the flexible membrane can be bent through the shape of the membrane indentation. Finally, the sample derived from the process of mixing will flow through microchannel 3.

6.4.3(a) ***Design of electromechanical actuator***

The EM actuator part consists of a polymer membrane with a micro-pillar structure, a planar micro-coil, and a permanent magnet connected to the membrane. The primary purpose of an EM microcoil is to generate a magnetic flux by applying an electrical input current to the coil system. The EM coil interacts with the magnetic field of a permanent magnet to create the Lorentz force (Fz), which is also called the magnetic force. The Biot–Savart law, as shown in Eq. (6.5), can be used to figure out the magnetic force that is created.

$$F_z = B_r A_m \int_{z}^{z+h_m} \frac{\partial H_z}{\partial z} dz \tag{6.5}$$

In this context, the variables represent the following: the magnetic force (Lorentz force) at the z position of the coil, the magnetic remanence, the surface area of the permanent magnet, the magnitude of the magnetic field produced by the input current on the coil, and the thickness of the permanent magnet.

The pressure from the Lorentz force on the membrane causes it to bend. The height of the curve depends on the strength of the magnetic field induction, the material properties, and the shape of the

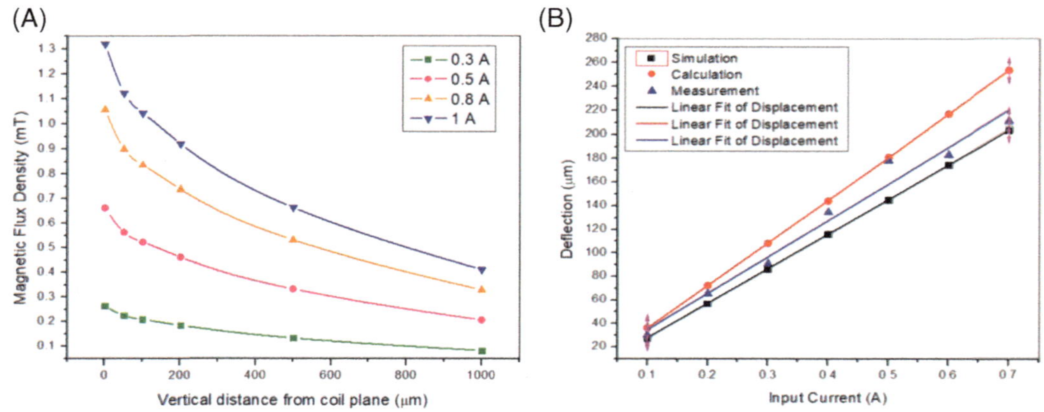

FIGURE 6.10

Effect of input current on actuator performance: (A) Magnetic flux density of micro coil, (B) membrane curvature height at various input currents.

membrane, as stated in Eq. (6.6) (Pawinanto et al., 2019);

$$d_z = C\frac{F_z\, l_m^{\,2}}{D} \tag{6.6}$$

where l_m is the membrane length, D is the bending stiffness of the membrane material, and C is a constant based on the membrane's shape. Furthermore, D is defined as the mechanical properties of the membrane; that is, E represents the Young's modulus, hm signifies the membrane's thickness, and n signifies the membrane's Poisson's ratio.

The results of a theoretical study of the characteristics of the magnetic field and its impact on the membrane in response to an input current provided to the actuator are illustrated in Fig. 6.10. The study revealed that the amplitude of the input current significantly influences the deformation characteristics of the membrane, the magnetic field strength, and the distance separating the coil and the membrane (see Fig. 6.10A). Fig. 6.10B depicts the membrane curvature resulting from the input current. The relationship between the curvature of the membrane and the input current can be demonstrated to be positively correlated. The resulting outcome fulfills the correlation between electric current, magnetic field generation, and membrane curvature, as indicated by Eqs. 6.5 and 6.6.

Fig. 6.11A-B shows a schematic diagram of membrane bending and micropillars caused by the Lorentz force acting on the membrane. Pressure on the membrane causes the micropillars to rotate from their original axis. The micropillars move up (when the membrane is in a push state) (Fig. 6.11A) and down (in a pull state) (Fig. 6.11B), which depends on the orientation of the membrane deflection.

The swivel angle, δ the micropillars can be evaluated by the pitch of the pillars. If the turning angle changes linearly with the micropillar distance, the micropillars in the chamber's center only move up and down. The further the distance of the pole from the center the greater the change in the turning

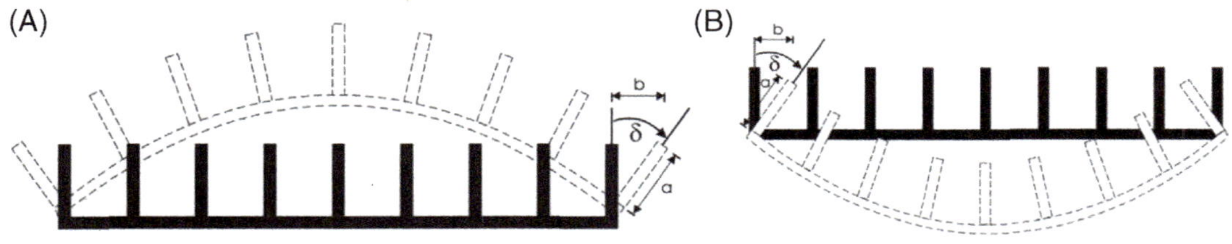

FIGURE 6.11

Micropillar swivel in (A) an upwards and (B) downwards direction.

FIGURE 6.12

Deformation of membrane and micropillars in various shapes: (A) circular, (B) square, and (C) rectangular.

angle. Therefore, the maximum swivel angle can be calculated by Eq. 6.7:

$$\sin(\delta) = \frac{b}{a} \tag{6.7}$$

where a is the height of the micropillar and b is the tip displacement of the micropillar.

The study of membrane deformation is conducted using three commonly used geometric configurations, namely circular, square, and rectangular geometries, as shown in Fig. 6.12A-C. The number of micropillars present on the membrane is 37, covering an area of 4 mm^2. As the dimensions of the pole geometry, namely the radius and distance between poles, increase, there is a corresponding decrease in the quantity of micropillars. Consequently, the positioning of the micropillars also changes.

Fig. 6.13A illustrates the impact of micropillar diameter and spacing on the dynamic characteristics. The increase in distance between micropillars is shown to increase micropillar rotation. This change can be attributed to the scattering of pillar positions towards the membrane's border. On the other hand, when the diameter of the pole increases, there is a corresponding decrease in the turning angle, as illustrated in Fig. 6.13B. The observed outcomes of this research can be related to the reduced projected value (as indicated in Eq. 6.1), resulting in a decrease in the rotation of the micropole. An increase in the micropillar's diameter causes its size to expand, which in turn reduces its movement.

The two figures demonstrate that the circular-shaped membrane yields a higher rotation of the micropillars. Circular membranes exhibit a higher curvature compared to other types of membrane shapes.

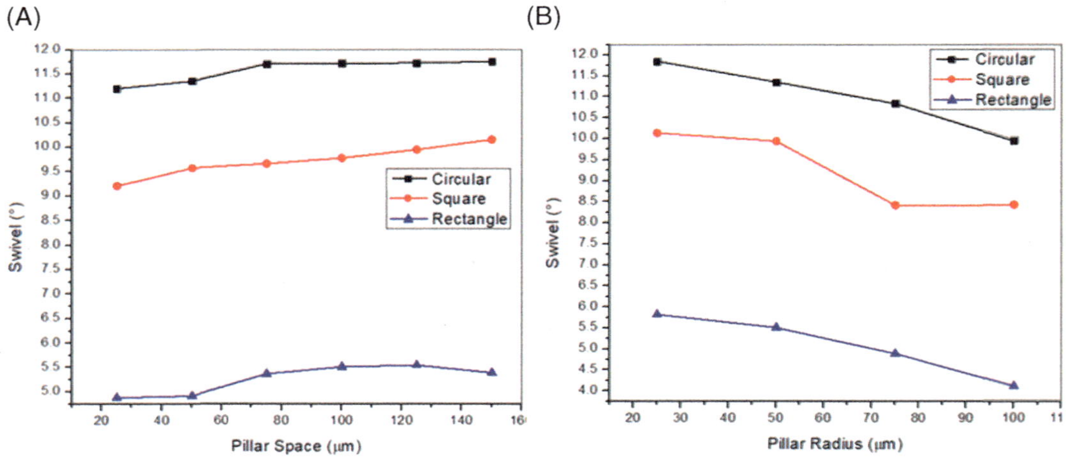

FIGURE 6.13

The tendency of micropillars on various membrane shapes: (A) effect of various pillar spacing at constant diameter and (B) effect of various pillar diameters at constant pillar spacing.

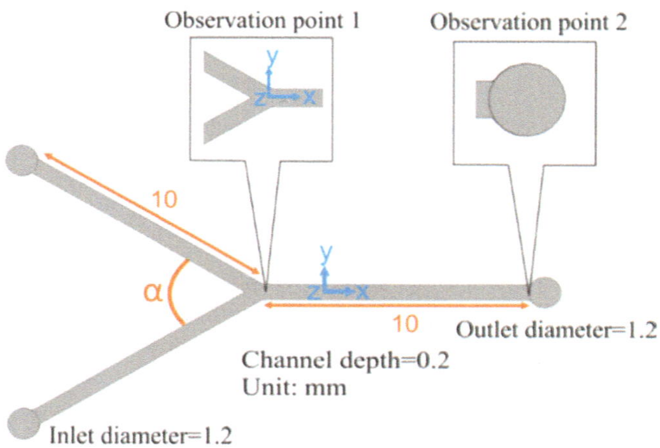

FIGURE 6.14

Microchannel configuration of microfluidic mixer (without reaction chamber).

6.4.3(b) *Design of microfluidic channel*

Fig. 6.14 illustrates the structure of microchannels in the microfluidic system. Generally, this system comprises an input system (inlet), an output (outlet), a microchannel, and a reaction chamber. The speed of the liquid needs to be controlled using a pump and valve system to obtain optimal mixing when it is used as a mixer. In addition, the angle of the inlet channel is also one of the important parameters that determine the homogeneity of the mixture (Pawinanto et al., 2019).

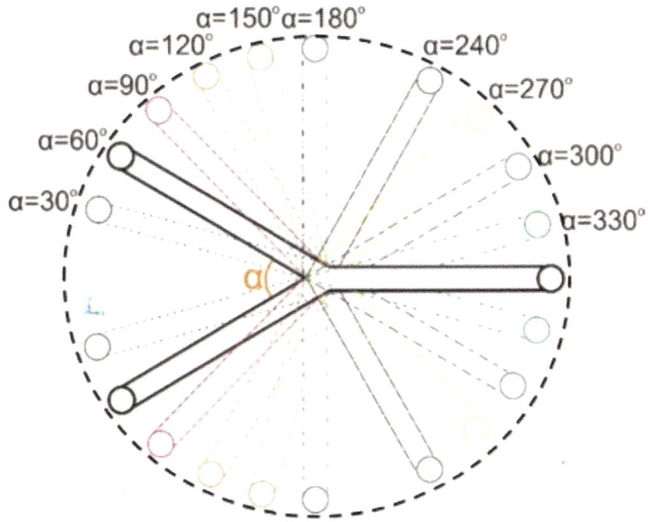

FIGURE 6.15

Mixing angles of microfluidic mixers.

Fig. 6.15 illustrates the configuration of the mixing channel angle α of the microfluidic mixer. The mixing channel angle α was varied starting at an angle of 30° up to 330°. The area of the device is determined depending on the size of the mixing angle α at the input channel. When the mixing channel angle α changes, then the x or y length will also change. This results in a change in the device area. The channel has a size of 200 μm with a depth of $d = 200$ μm. The inlet channel has a length of 10 cm, while the outlet channel is an observation area that has a length of 10 cm. All reported mixing channel lengths are from positions $x = 0$ cm (observation point 1), 5 cm (middle observation point), and $x = 10$ cm (observation point 2).

Based on the design, the mixing efficiency can be calculated as a mixing index using Eqs. (6.3) and (6.4). This study has been initially conducted on channel design in a passive state (no turbulence). The purpose is to find out the effect of the mixing channel angle (α) on mixing efficiency.

Fig. 6.16A shows the effect of inlet channel angle α on the mixing index at different observation point distances at position X along the channel, as shown schematically in Fig. 6.14. The mixing index rises as the channel length increases. This illustrates the impact of fluid particle diffusion in a non-turbulent state along the channel. The mixing index from the beginning microchannel to the middle microchannel is more affected by fluid collisions and vortices from the junction between the two inlets. Furthermore, the diffusion effect significantly influences the increase in the mixing index from the middle to the end of the microchannel. Gidde et al. (2018) reported almost identical results to ours. A longer interaction of two or more different fluids occurs due to the long channel length, which allows for an increase in the mixing index. The average mixing index for the angles studied at the starting point, middle point, and end point are 0.20, 0.44, and 0.67, respectively. The mixing index increased by 0.24 from the start to the middle and by 0.23 from the middle to the end. It shows a consistent increase.

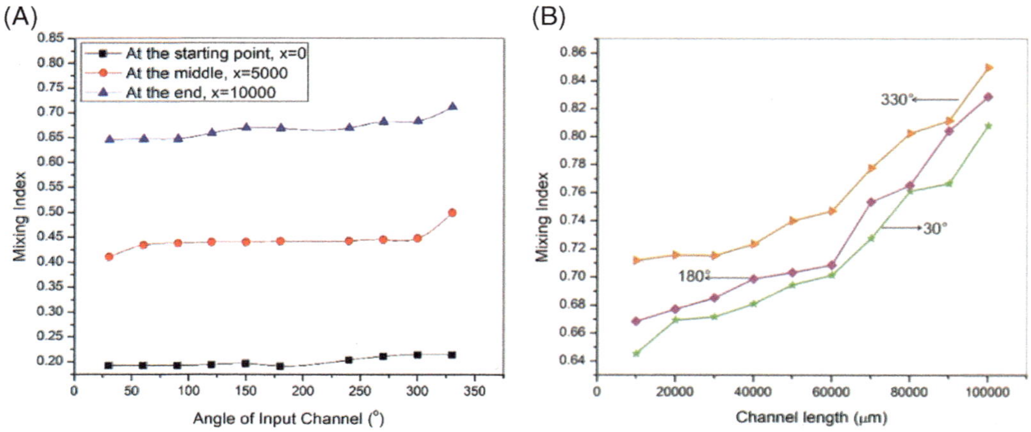

FIGURE 6.16

The mixing index performance with (A) different input channel angles and (B) different channel lengths.

Although the input channel angle of 330° produces the highest mixing index toward the end of the channel (specifically, at observation point 2), it is important to note that the achieved mixing index value does not approach 0.8. This shows that the fluid in the final channel is not homogeneous at an observation distance of up to 10 cm. Therefore, further studies are needed to determine how long the channel is required to enable homogeneous liquid mixing.

Fig. 6.16B presents a study of the mixing index on a 10 cm-long channel. Three input channel angle parameters are used, namely, 30°, 180°, and 330°, where these angles represent acute, straight, and obtuse angles. For an input channel angle of 330° (obtuse angle), a mixing index above 0.8 can be achieved at a channel length and input channel angle of 80 cm. Meanwhile, for inlet channel angles of 180° and 30°, a mixing index exceeding 0.8 can be achieved at a channel length of more than 9 cm. However, it should be noted that channels that are too long are not suitable for microscale mixers for lab-on-chip applications. Therefore, an active mixer that uses turbulence in the reaction chamber will significantly increase the mixing index and shorten the channel length.

Fig. 6.17A-D shows the microfluidic mixer design when using a reaction chamber in the middle of the outlet channel. The improvement in mixing performance is expected to increase significantly when the EM actuator works in the reaction chamber operating in active mode. This mode generates a simultaneous occurrence of diffusion and turbulence at a specific location. Hence, in order to examine the attributes of fluid mixing in an active state, a time-dependent analysis is employed. This implies that the mixing characteristics vary based on the duration of mixing at a fixed location within the reaction chamber.

Simulations show that a mixing index above 0.8 can be produced in less than 2 minutes. The findings of this study indicate that liquids can achieve correct and uniform mixing within a limited channel length and time frame with the application of turbulence in the reaction chamber.

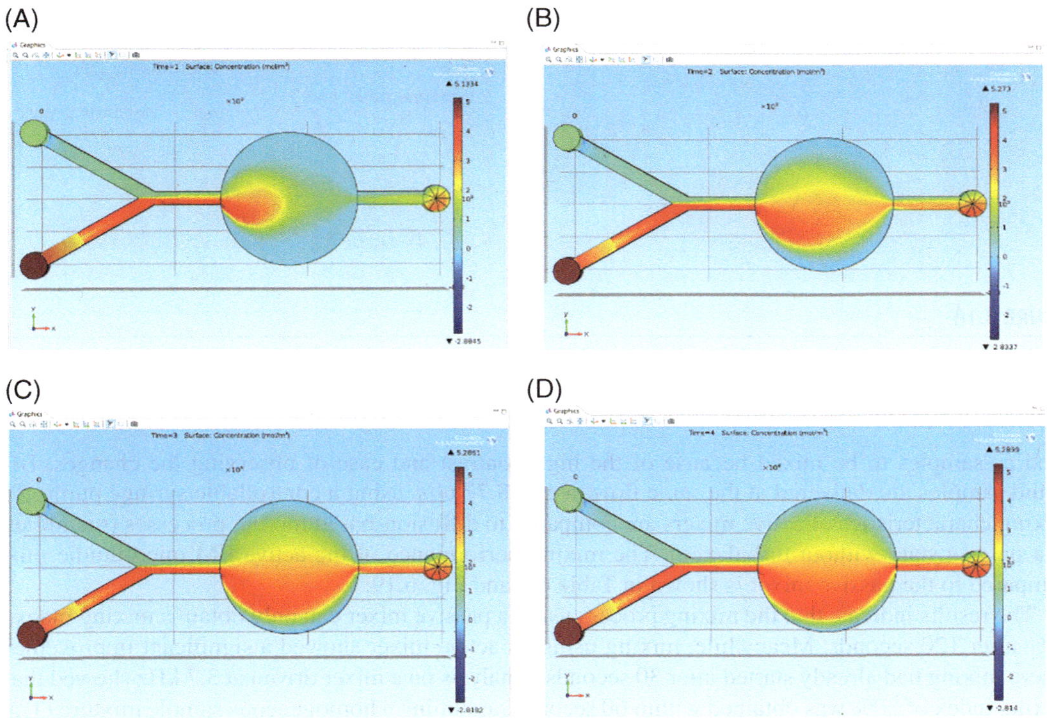

FIGURE 6.17

Concentration distribution on a channel length of 20 cm at various observation times: (A) 30 seconds, (B) 60 seconds, (C) 90 seconds, (D) 120 seconds.

6.4.3(c) *Integration of microfluidic parts and electromagnetic actuators*

The final part of the microfluidic mixer fabrication process is the integration of the actuator part with the microfluidic part by bonding these two parts with the help of corona discharge. Corona discharge with high voltage is used for the integration of ions between the two adhesion surfaces. It changes the surface properties of various materials to make them easier to bond with. Moreover, it is to ensure that the layer has a satisfactory seal and avoid liquid leakage. The final image of the microfluidic mixing device is shown in Fig. 6.18.

6.4.3(d) *Characterization of the mixing index*

Characterization of the mixing performance of the microfluidic mixer was performed by mixing two types of liquid samples into the reaction chamber. An AC electrical input source at three frequencies of 1.1 kHz, 2.3 kHz, and 3.7 kHz is applied to the EM coil to drive the vibration of the membrane and pole (mechanical part). This frequency is the frequency that produces the harmonic deformation of the membrane that allows the membrane to vibrate. Food coloring materials (red and green) are

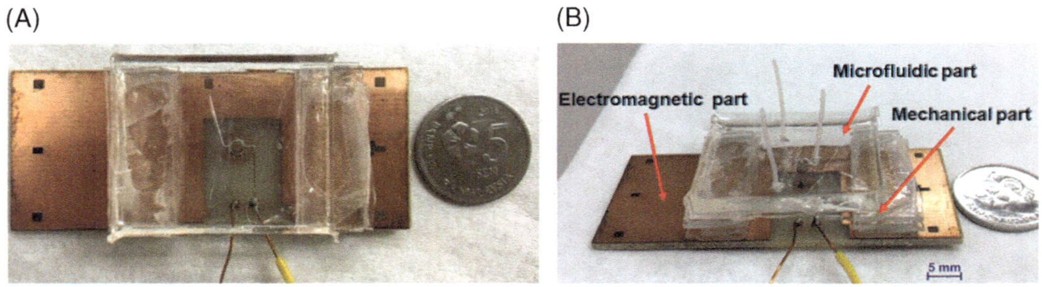

FIGURE 6.18

Photographic image of the microfluidic mixer device after integration between the actuator and the microfluidic part: (A) top view and (B) side view.

used as samples to be mixed because of the high contrast and ease of observing the changes. Both liquid samples are delivered at the same flow rate of 8.77 μl/s, using a controllable syringe pump. The mixing characteristics of active mixers are compared to diffusion-based mixing processes (membranes in a passive state without turbulence). The mixing performance of the active EM microfluidic mixer compared to the passive mixer is shown in Table 6.2 and Fig. 6.19.

The results indicate that the mixing process using a passive mixer can only obtain a mixing index of 35% after 120 seconds. Meanwhile, mixing using an active mixer showed a significant improvement, where mixing had already started after 30 seconds. Analysis on a mixer driven at 3.7 kHz showed that a mixing index of 80% was obtained within 60 seconds, assuming a homogeneous sample mixture (Tran-Minh et al., 2014). This mixing index increases with activation frequency. Therefore, it can be concluded that the vibration of the column and membrane drives a more chaotic fluid flow in the reaction chamber, which has significantly improved the mixing performance.

6.4.3(e) *Testing in the biomedical field*

Finally, a glucose detection procedure was performed to test the functionality of the developed mixer. This glucose concentration is determined by examining the absorption spectrum of the DNS reagent/glucose sample that has been mixed following the colorimetric reaction process. The DNS reagent consisted of 0.315 g of 3,5-dinitrosalicylic acid ($C_7H_4N_2O_7$), 2 M sodium hydroxide (NaOH) solution, 9.25 g of potassium sodium tartrate ($KNaC_4H_4O_6 \bullet 4H_2O$), and 50 ml of DI water. Meanwhile, a glucose solution with a concentration between 100 and 900 ppm is prepared by mixing glucose powder ($C_6H_{12}O_6$) and deionized water (DI Water) in appropriate quantities.

The DNS reagent and glucose solution are then injected into the microfluidic chip until the reaction chamber is full. After the mixing process is completed, the microfluidic chip is then inserted into a UV-Vis spectrophotometer to analyze its absorption spectrum. The entire procedure involves only two steps and can be completed in just 10 minutes (the total time for the microfluidic glucose detection method). Therefore, microfluidic mixing chip-based detection methods are faster and simpler than conventional methods and have lower reagent and sample consumption.

Fig. 6.20 shows the experimental results obtained to examine the correlation between absorption and glucose concentration and compare them when using conventional detection methods. It should be

Table 6.2 Comparison of the mixing process in grayscale images compared to natural diffusion.

Diffusion	Time (s)			
	30 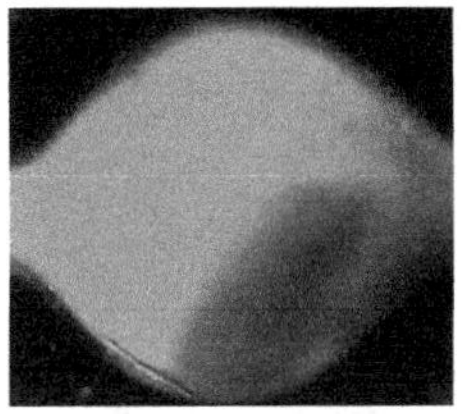	60 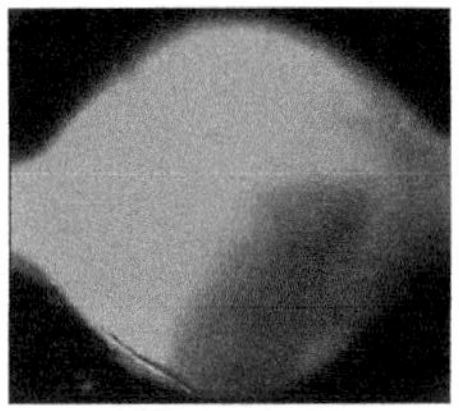	90 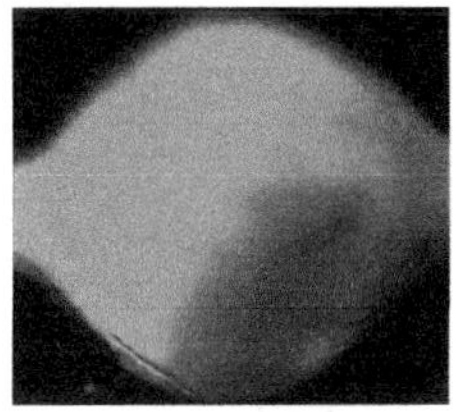	120 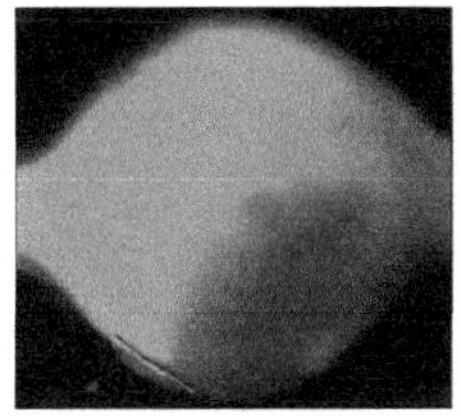
Frequency (kHz)	Time (s)			
1.1	1 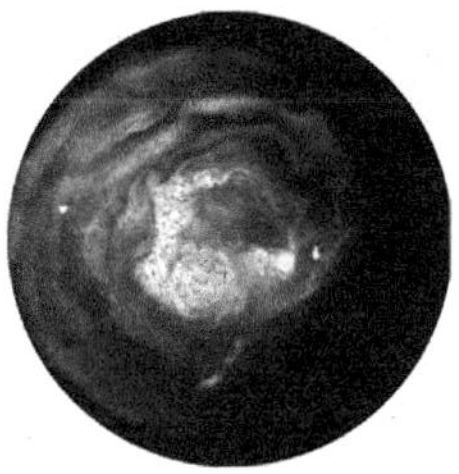	2 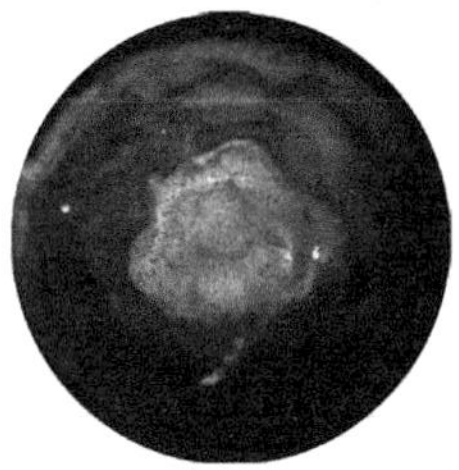	3 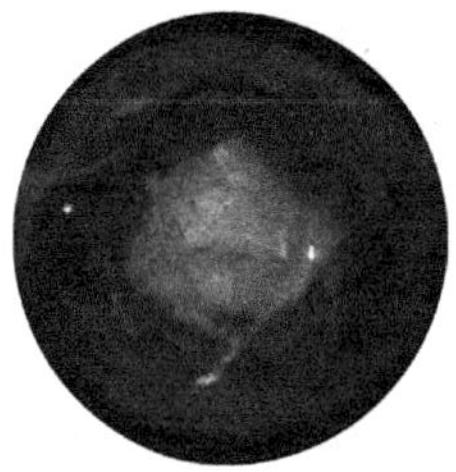	4 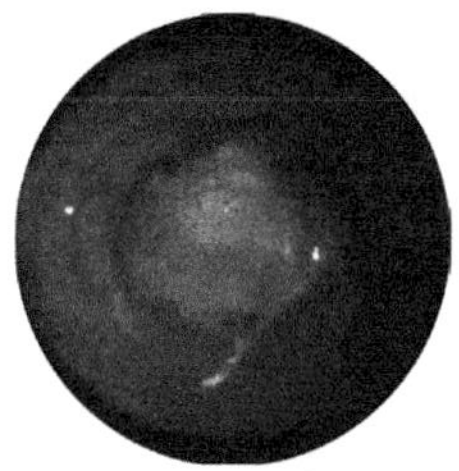

(continued on next page)

Table 6.2 Comparison of the mixing process in grayscale images compared to natural diffusion—cont'd

Diffusion	Time (s)			
2.3	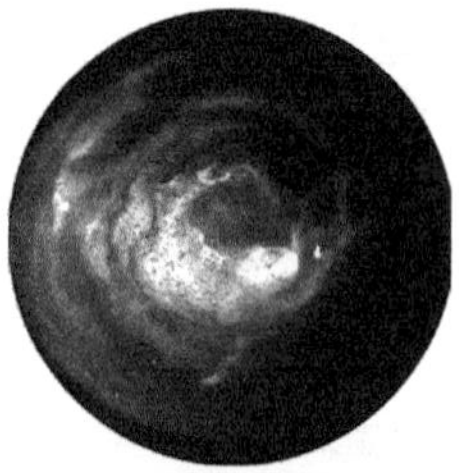	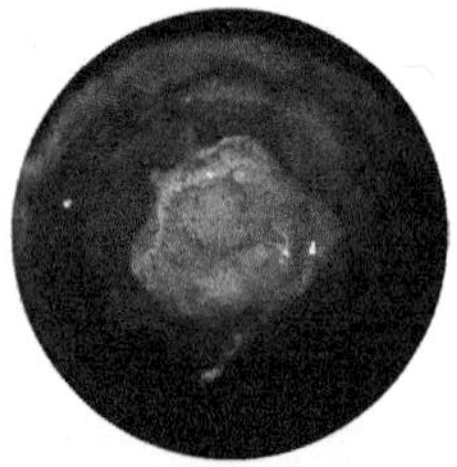	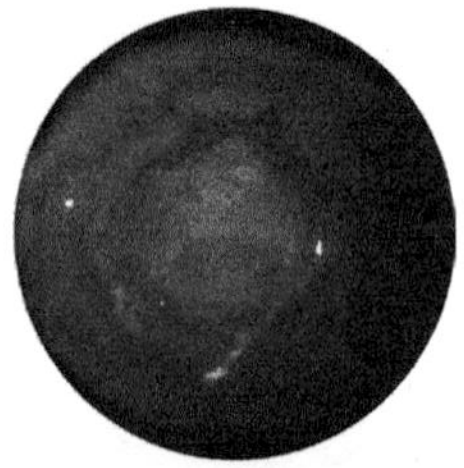	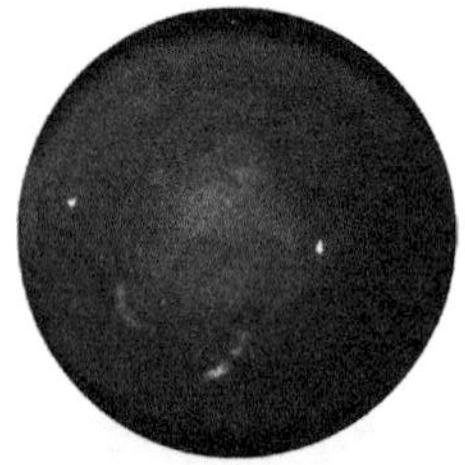
3.7	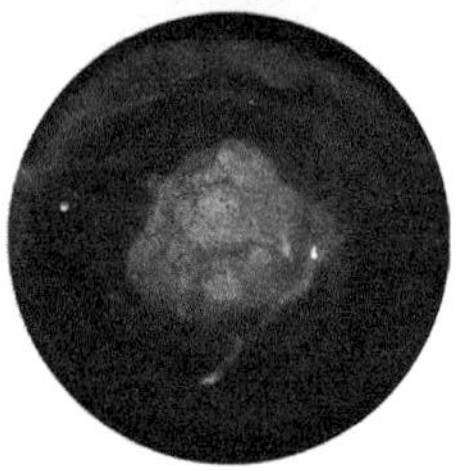	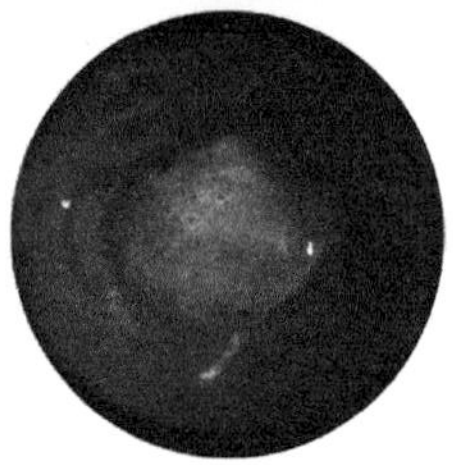		

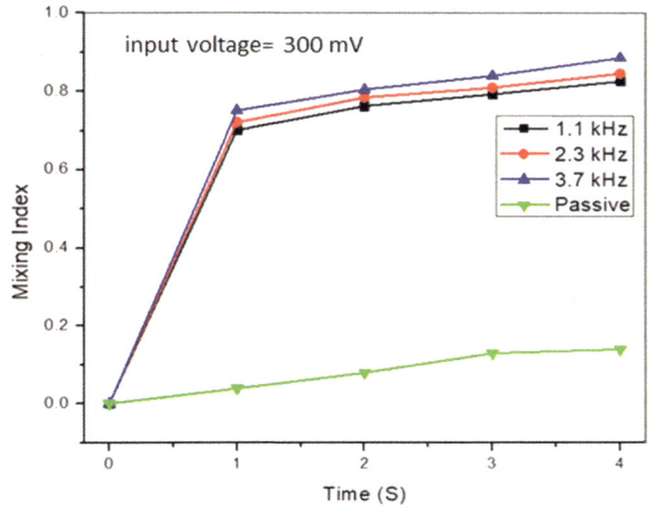

FIGURE 6.19

Mixing index along time with various frequencies.

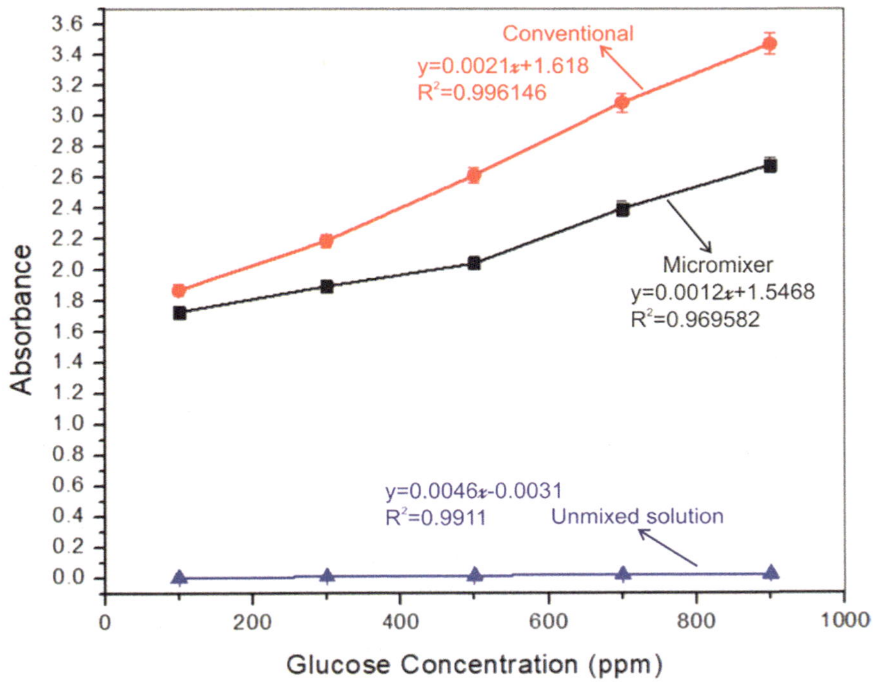

FIGURE 6.20

Linear regression analysis of the test results obtained from the conventional method and the microfluidic mixer chip method for the detection of glucose concentration in the range of 100–900 ppm.

noted that DNS reagent and glucose solution are mixed with a volume ratio of 1:1. If mixing occurs, an absorption signal should appear in the UV-Vis instrumentation. Here, the solution appears to be very stable and has an absorption maximum at a wavelength of 340 nm. The correlation coefficient was found to be $R^2 = 0.9961$ (conventional method) and $R^2 = 0.9696$ (active microfluidic mixer chip-based method). In other words, the accuracy of concentration measurements obtained using active microfluidic mixer chip-based methods is comparable to measurements obtained using conventional methods.

6.4.4 Future directions and emerging technologies

6.4.4(a) *Smart micromixers with embedded intelligence*

Advances in microfabrication, sensing technologies, and artificial intelligence have led to the emergence of smart micromixers—microfluidic mixing systems enhanced with embedded intelligence for autonomous operation, adaptive control, and real-time decision-making. Unlike conventional passive or active micromixers, which rely on fixed geometric structures or externally controlled actuators, smart micromixers integrate micro-sensors, data-processing units, and wireless communication interfaces to create systems capable of self-optimization and predictive operation(Loima et al., 2025). These intelligent devices are increasingly important in precision medicine, point-of-care diagnostics, pharmaceutical manufacturing, and automated laboratory systems, where accurate, reproducible, and dynamic mixing performance is essential. By embedding feedback mechanisms and computational intelligence directly into the micromixer, fluid handling can be adjusted to meet variable sample compositions, environmental conditions, or diagnostic requirements.

6.4.4(a)(i) Self-optimizing systems: Adaptive mixing based on real-time feedback

A defining feature of smart micromixers is their ability to perform self-optimization through closed-loop control. Embedded sensors—such as microelectrodes, optical detectors, temperature sensors, or pressure sensors—continuously monitor key parameters including flow rate, concentration gradients, viscosity changes, and temperature profiles (Kouhkord et al., 2024; Yin et al., 2026). These measurements are transmitted to an integrated microcontroller or edge-computing module that processes the data in real time. Machine learning algorithms or adaptive control strategies then dynamically adjust actuation parameters, such as the amplitude of electric fields, frequency of acoustic waves, or intensity of thermal gradients, to maintain optimal mixing efficiency. This capability is particularly beneficial for handling biological samples that exhibit variable properties, such as whole blood, saliva, or cell suspensions. Self-optimizing micromixers thus overcome the limitations of static designs by ensuring consistent performance under diverse operating conditions.

6.4.4(a)(ii) Predictive maintenance: Condition monitoring and failure prediction

Smart micromixers also incorporate predictive maintenance capabilities, enabling long-term reliability in environments where downtime or equipment failure is unacceptable. Condition monitoring is carried out by recording metrics such as electrical impedance changes at electrodes, acoustic transducer performance, mechanical resonance frequency shifts, or temperature fluctuations in micro-heaters. Over time, data-driven models—often based on machine learning techniques such as anomaly detection or degradation modeling—identify early signatures of component wear, fouling, delamination, or actuator fatigue. This allows users or automated systems to schedule maintenance or component replacement before catastrophic failure occurs. Predictive maintenance is especially advantageous in clinical settings, continuous bioprocessing, and high-throughput testing platforms, where equipment stability directly affects diagnostic accuracy and product quality. By transitioning from reactive to predictive management, smart micromixers significantly extend system lifespan and reduce operational costs.

6.4.4(c)(iii) Remote monitoring: IoT connectivity for telemedicine applications

Another transformative feature of smart micromixers is their ability to support remote monitoring through Internet of Things (IoT) connectivity (Kulkarni et al., 2022). By integrating low-power wireless modules such as Bluetooth Low Energy (BLE), Wi-Fi, or NB-IoT, micromixers can transmit performance data, diagnostic outputs, and operational status to external devices or cloud platforms. In telemedicine applications, this connectivity enables real-time oversight of point-of-care diagnostic tools operating in remote or resource-limited regions. Healthcare providers can track sample processing progress, verify proper mixing for biochemical assays, or remotely troubleshoot device malfunctions. Furthermore, IoT connectivity facilitates large-scale deployment of microfluidic devices for epidemiological monitoring, environmental sensing, and decentralized laboratory automation. Cloud-based analytics can aggregate data from thousands of micromixers, enabling population-level insights and system-wide optimization.

6.4.4(b) Advanced materials and fabrication techniques

Advances in materials science and microfabrication technologies have significantly expanded the design space and performance capabilities of micromixers. While early microfluidic systems were predominantly fabricated using silicon, glass, or polydimethylsiloxane (PDMS), the field has rapidly evolved to incorporate a wider range of functional, biocompatible, and stimuli-responsive materials. These innovations enable micromixers with enhanced durability, chemical stability, and tailored surface properties, supporting increasingly demanding applications in diagnostics, drug development, and chemical processing. In parallel, additive manufacturing and advanced microfabrication techniques now allow the production of intricate three-dimensional structures that were previously unachievable with traditional planar lithography. Together, these developments have presented fresh possibilities for creating highly efficient, application-specific micromixers with improved integration and multifunctionality.

6.4.4(b)(i) Novel materials

The introduction of novel materials has transformed micromixer engineering by enabling improved mechanical, chemical, and biological performance. Beyond conventional PDMS, emerging polymers such as COC, cyclic olefin polymer (COP), and thermoplastic polyurethane (TPU) offer superior chemical resistance, lower water absorption, and enhanced optical clarity—properties crucial for chemical analysis and optical sensing (Ameri et al., 2022; Rodrigues et al., 2023). Biocompatible photopolymers and hydrogels have also gained prominence in biomedical micromixers due to their tunable stiffness, permeability, and ability to mimic biological environments. Additionally, stimuli-responsive materials, including magneto-responsive elastomers, thermo-responsive polymers, and piezoelectric films, allow dynamic actuation and controlled deformation within micromixer structures, enabling hybrid passive–active mixing modes. Composite materials incorporating nanoparticles or nanofillers further enhance mechanical strength, electrical conductivity, and surface functionality, providing new opportunities for integrated sensing, heating, or magnetic actuation. These diverse material platforms support the development of micromixers tailored to specialized environments—from harsh chemical solvents to delicate biological samples—while improving long-term device stability.

6.4.4(b)(ii) 3D printing and additive manufacturing

3D printing and additive manufacturing have emerged as transformative fabrication methods for creating micromixers with unprecedented geometric freedom. Unlike traditional lithographic approaches that are inherently two-dimensional, additive manufacturing enables the construction of fully three-dimensional, multilayered architectures with complex internal geometries such as intertwined channels, embedded obstacles, and multi-material gradients(Garcia-Cardosa et al., 2022; Matining et al., 2025). Techniques such as SLA, digital light processing (DLP), two-photon polymerization (2PP), and fused deposition modeling (FDM) allow rapid prototyping of intricate micromixer designs with high spatial resolution. For example, SLA and DLP provide smooth channel walls and fine feature sizes suitable for microfluidic networks, while 2PP can produce submicron structures ideal for advanced mixing elements such as staggered ridges or helically twisted channels. Multimaterial printing further enables integration of flexible membranes, conductive traces, or magnetic components directly into the microfluidic device, reducing assembly steps and improving functional density.

6.4.4(b)(iii) Impact on micromixer design and applications

The adoption of advanced materials and 3D printing capabilities has significantly expanded the functional landscape of micromixers. Designers can now incorporate complex flow paths—such as fractal geometries (Hasanah et al., 2026), three-dimensional split-and-recombine networks (Molina et al., 2026), or biomimetic structures (Yang et al., 2022)—that dramatically enhance mixing efficiency without increasing pressure drop. Additive manufacturing also supports rapid iteration, making it possible to prototype and test multiple design variants in a short period of time. This accelerates the development cycle for application-specific micromixers used in personalized medicine, organ-on-chip platforms, and portable diagnostic instruments. Moreover, the ability to print hollow structures, flexible chambers, and integrated sensors enables seamless integration with pumps, valves, electrodes, and actuators, leading to fully functional microfluidic modules produced in a single manufacturing step. As material formulations and printing methods continue to advance, the future of micromixer fabrication is expected to shift increasingly toward customizable, on-demand manufacturing with enhanced performance and reliability.

Acknowledgment

The author would like to express their deepest gratitude to Universiti Kebangsaan Malaysia for the financial support provided under the Research University Grant (GUP-2024-048).

References

Alimardani, V., Abolmaali, S. S., Yousefi, G., Rahiminezhad, Z., Abedi, M., Tamaddon, A., & Ahadian, S. (2021). Microneedle arrays combined with nanomedicine approaches for transdermal delivery of therapeutics. *Journal of Clinical Medicine, 10*(2), 1–33. http://doi.org/10.3390/jcm10020181.

Ameri, A. R., Imanparast, A., Passandideh-Fard, M., & Shaegh, S. A. M. (2022). A whole-thermoplastic microfluidic chip with integrated on-chip micropump, bioreactor and oxygenator for cell culture applications. *Analytica Chimica Acta, 1221*, 340093. doi:10.1016/j.aca.2022.340093.

Amirouche, F., Zhou, Y. u, & Johnson, T. (2009). Current micropump technologies and their biomedical applications. *Microsystem Technologies, 15*(5), 647–666. http://doi.org/10.1007/s00542-009-0804-7.

Anbazhagan, G., Suseela, S. B., & Sankararajan, R. (2023). Effect of hollow microneedle geometry structure on mechanical stability and microfluidic flow for transdermal drug delivery applications. *Microfluid. Nanofluidics, 27*(4), 1–14. http://doi.org/10.1007/s10404-023-02636-5.

Balashanmugam, N., Krishna, P., & Kumar, M. (2008). Fabrication of polymeric micro needle array by micromachining & micro molding. *Certificate International Journal of Engineering and Innovative Technology, 9001*(6), 2277–3754.

Balmert, S. C., Carey, C. D., Falo, G. D., Sethi, S. K., Erdos, G., Korkmaz, E., & Falo, L. D. (2020). Dissolving undercut microneedle arrays for multicomponent cutaneous vaccination. *Journal of Controlled Release, 317*, 336–346. http://doi.org/10.1016/j.jconrel.2019.11.023.

Bolton, C. J. W., Howells, O., Blayney, G. J., Eng, P. F., Birchall, J. C., Gualeni, B., Roberts, K., Ashraf, H., & Guy, O. J. (2020). Hollow silicon microneedle fabrication using advanced plasma etch technologies for applications in transdermal drug delivery. *Lab on a Chip, 20*(15), 2788–2795. http://doi.org/10.1039/d0lc00567c.

Cormier, M., Johnson, B., Ameri, M., Nyam, K., Libiran, L., Zhang, D. D., & Daddona, P. (2004). Transdermal delivery of desmopressin using a coated microneedle array patch system. *Journal of Controlled Release, 97*(3), 503–511. http://doi.org/10.1016/j.jconrel.2004.04.003.

Datta, A., Biswas, S., Dhar, R., & Bhattacharyya, T. K. (2023). Design and development of a piezoelectric driven micropump integrated with hollow microneedles for precise insulin delivery. *Journal of Micromechanics and Microengineering, 33*(7), 75003. http://doi.org/10.1088/1361-6439/acd25f.

Economidou, S. N., Uddin, M. J., Marques, M. J., Douroumis, D., Sow, W. T., Li, H., Reid, A., Windmill, J. F. C., & Podoleanu, A. (2021). A novel 3D printed hollow microneedle microelectromechanical system for controlled, personalized transdermal drug delivery. *Additive Manufacturing, 38*, 101815. http://doi.org/10.1016/j.addma.2020.101815.

Erny H., Ooi P.C., Tiong T.Y., Marsi N., Ismardi A., Noor M.M., Fathi N.A.F.N.Z., Abd Aziz N., Sahari S.K., Sugandi G., Yunas J., Dee C.F., Majlis B.Y., & Hamzah A.A., Stress and deformation of optimally shaped silicon microneedles for transdermal drug delivery. Journal of Pharmaceutical Sciences, *109*, 2485–2492, (2020).

Garcia-Cardosa, M., Granados-Ortiz, F.-J., & Ortega-Casanova, J. (2022). A Review on Additive Manufacturing of Micromixing Devices. *Micromachines, 13*(1), 73. doi:10.3390/mi13010073.

Gidde, R. R., Pawar, P. M., Ronge, B. P., Misal, N. D., Kapurkar, R. B., & Parkhe, A. K. (2018). Evaluation of the mixing performance in a planar passive micromixer with circular and square mixing chambers. *Microsystem Technologies, 24*(6), 2599–2610. http://doi.org/10.1007/s00542-017-3686-0.

Gill, H. S., & Prausnitz, M. R. (2007). Coated microneedles for transdermal delivery. *Journal of Controlled Release, 117*(2), 227–237. http://doi.org/10.1016/j.jconrel.2006.10.017.

Gordh T., Gordh T.E., Lindqvist K., & Warner D.S., Lidocaine : The origin of a modern local anesthetic. Anesthesiology, *113*(6), 1433–1437, (2010).

Guillot, A. J., Cordeiro, A. S, Donnelly, R. F., Montesinos, M. C., Garrigues, T. M., & Melero, A. (2020). Microneedle-based delivery: An overview of current applications and trends. *Pharmaceutics, 12*(6), 569. http://doi.org/10.3390/pharmaceutics12060569.

Hamid, N. A., Ibrahim, M., Radzi, S. A., Chiew, W. Y., Yunas, J., & Majlis, B. Y. (2016). A stack bonded thermo-pneumatic micro-pump utilizing polyimide based actuator membrane for biomedical applications. *Microsystem Technologies, 23*(5), 75–81. http://doi.org/10.1007/s00542-016-2951-y.

Hamzah, A. A., Aziz, N. A., Majlis, B. Y., Yunas, J., & Dee, C. F. (2012). Optimization of HNA etching parameters to produce high aspect ratio solid silicon microneedles. *Journal of Micromechanics and Microengineering, 22*(9), 95017. http://doi.org/10.1088/0960-1317/22/9/095017.

Hasanah, L., Nurvadila, D., Pawinanto, R., Mulyanti, B., Wulandari, C., Aminudin, A., & Yunas, J. (2022). A passive micromixer with koch snowflakes fractal obstacle in microchannel. *Journal of Applied Fluid Mechanics, 15*(5), 1333–1344. https://www.jafmonline.net/article_2070.html.

Hasanah, L., Afifah, F. N., Pawinanto, R. E., Iqbal, M., Gumilar, G., Yusuf, M., ... Mulyanti, B. (2026). A stable high mixing performance of Koch fractal array obstacle-based micromixer. *International Journal of Thermofluids, 32*, 101550. doi:10.1016/j.ijft.2026.101550.

Henry, S., McAllister, D. V., Allen, M. G., & Prausnitz, M. R. (1998). Microfabricated microneedles: A novel approach to transdermal drug delivery. *Journal of Pharmaceutical Sciences, 87*(8), 922–925. http://doi.org/10.1021/js980042+.

Hamzah, A. A., Majlis, B. Y., & Ahmad, I. (2004). Deflection analysis of epitaxially deposited polysilicon encapsulation for MEMS devices. In 2004 IEEE International Conference on Semiconductor Electronics (ICSE 2004) (pp. 611–614). Kuala Lumpur, Malaysia. https://doi.org/10.1109/SMELEC.2004.1620960

Irfan M., Jalal A., Yen M.Y., & Chenshen L., The efficacy and safety of lignocaine-embedded dissolvable microneedle versus EMLA for topical analgesia in adults undergoing venepuncture : randomised clinical trial protocol in a tertiary care setting, PLoS One, *20*(11), 1–21, (2025). http://doi.org/10.1371/journal.pone.0335932

Ita, K. (2017). Dissolving microneedles for transdermal drug delivery: Advances and challenges. *Biomedicine & Pharmacotherapy, 93*, 1116–1127. http://doi.org/10.1016/j.biopha.2017.07.019.

Jalal, M. I. A., Ooi, K. S., Foo, K. C., et al. (2026). Randomized double-blind crossover trial of a transdermal maltose dissolvable microneedle patch for topical anesthesia before intravenous cannulation for thalassemia children. *Sci Rep, 16*, 671. doi:10.1038/s41598-025-30279-9.

Jiang, X., Wang, S., Zhang, L., Jiang, X., & Gou, M. (2024). Lidocaine hydrochloride loaded isomaltulose microneedles for efficient local anesthesia of the skin. *Chinese Chemical Letters, 35*(4), 108686. http://doi.org/10.1016/j.cclet.2023.108686.

Jin, S. Y., Park, E. J., Kwon, S. M., Jung, H., & Kim, D. W. (2025). Rapidly dissolving microneedles incorporating lidocaine hydrochloride : A PVP /PVA-based approach for local anesthesia. *Pharmaceutics, 17*(9), 1–19.

Kim, E., Erdos, G., Huang, S., Kenniston, T. W., Balmert, S. C., Carey, C. D., Raj, V. S., Epperly, M. W., Klimstra, W. B., Haagmans, B. L., Korkmaz, E., Falo, L. D., & Gambotto, A. (2020). Microneedle array delivered recombinant coronavirus vaccines: Immunogenicity and rapid translational development. *EBioMedicine, 55*. http://doi.org/10.1016/j.ebiom.2020.102743.

Kouhkord, A., Hassani, F., Amirmahani, M., Golshani, A., Naserifar, N., Moghanlou, F. S., & Ali Tarlani Beris, A. T. B. (2024). Controllable Microfluidic System through Intelligent Framework: Data-Driven Modeling, Machine Learning Energy Analysis, Comparative Multiobjective Optimization, and Experimental Study. *Industrial & Engineering Chemistry Research, 63*(30), 13326–13344. doi:10.1021/acs.iecr.4c00456v.

Kulkarni, M. B, Goyal, S., Dhar, A., Sriram, D., & Goel, S. (2022). Miniaturized and IoT Enabled Continuous-Flow-Based Microfluidic PCR Device for DNA Amplification. *IEEE Trans Nanobioscience, 21*(1), 97–104. doi:10.1109/tnb.2021.3092292.

Larrañeta, E., Lutton, R. E. M., Woolfson, A. D., & Donnelly, R. F. (2016). Microneedle arrays as transdermal and intradermal drug delivery systems: Materials science, manufacture and commercial development. *Materials Science and Engineering: R: Reports, 104*, 1–32. http://doi.org/10.1016/j.mser.2016.03.001.

Li, Q. Y., Zhang, J. N., Chen, B. Z., Wang, Q. L., & Guo, X. D. (2017). A solid polymer microneedle patch pretreatment enhances the permeation of drug molecules into the skin. *RSC Advances, 7*(25), 15408–15415. http://doi.org/10.1039/c6ra26759a.

Loima, T., Yoon, J.-Y., & Kaarj, K. (2025). Microfluidic Sensors Integrated with Smartphones for Applications in Forensics, Agriculture, and Environmental Monitoring. *Micromachines, 16*(7), 835. doi:10.3390/mi16070835.

Loizidou, E. Z., Williams, N. A., Barrow, D. A., Eaton, M. J., McCrory, J., Evans, S. L., & Allender, C. J. (2015). European journal of pharmaceutics and biopharmaceutics structural characterisation and transdermal delivery studies on sugar microneedles : Experimental and finite element modelling analyses. *European Journal of Pharmaceutics and Biopharmaceutics, 89*, 224–231. http://doi.org/10.1016/j.ejpb.2014.11.023.

Maderuelo, C., Lanao, J. M., & Zarzuelo, A. rá (2019). Enteric coating of oral solid dosage forms as a tool to improve drug bioavailability. *European Journal of Pharmaceutical Sciences, 138*, 105019. http://doi.org/10.1016/j.ejps.2019.105019.

Mao, Y. (2024). Fabrication of lidocaine-loaded polymer dissolving microneedles for rapid and prolonged local anesthesia. *Biomedical Microdevices, 26*, 1–19.

Matining, V., Colli, C., Mauri, E., & Moscatelli, D. (2025). Advances in microfluidics: From state-of-the-art to two-photon polymerization 3D printing perspectives. *Applied Materials Today, 47*, 102953. doi:10.1016/j.apmt.2025.102953.

Mishra, R., Maiti, T. K., & Bhattacharyya, T. K. (2018). Development of SU-8 hollow microneedles on a silicon substrate with microfluidic interconnects for transdermal drug delivery. *Journal of Micromechanics and Microengineering, 28*(10), 105017. http://doi.org/10.1088/1361-6439/aad301.

Molina, F. N., Paliwal, J., & Salimi, E. (2026). Monolithic 3D-printed split-and-recombine micromixer integrated into a microfluidic concentration gradient generator. *Lab on Chip, 26*, 2531–2542. doi:10.1039/D5LC01095K.

Monou, P. K., Saropoulou, E., Junqueira, L. A., Kolipaka, S. S., Andriotis, E. G., Tzimtzimis, E., Tzetzis, D., Bekiari, C., Bouropoulos, N., Harding, B., Katsamenis, O. L., Brambock, A., Treffer, D., Douroumis, D., & Fatouros, D. G. (2025). Fabrication and characterization of dissolving microneedles combining digital light processing and vacuum compression molding technique for the transdermal delivery of rivastigmine. *European Journal of Pharmaceutics and Biopharmaceutics, 210*(2024). http://doi.org/10.1016/j.ejpb.2025.114687.

Park, J. H., Allen, M. G., & Prausnitz, M. R. (2005). Biodegradable polymer microneedles: Fabrication, mechanics and transdermal drug delivery. *Journal of Controlled Release, 104*(1), 51–66. http://doi.org/10.1016/j.jconrel.2005.02.002.

Pawinanto, R. E., Yunas, J., & Hashim, A. M. (2019). Design optimization of active microfluidic mixer incorporating micropillar on flexible membrane. *Microsystem Technologies, 25*(4), 1203–1209. http://doi.org/10.1007/s00542-018-4134-5.

Pawinanto, R. E., Yunas, J., & Hashim, A. M. (2020). Micropillar based active microfluidic mixer for the detection of glucose concentration. *Microelectronic Engineering, 234*. http://doi.org/10.1016/j.mee.2020.111452.

Pradeep Narayanan, S., & Raghavan, S. (2017). Solid silicon microneedles for drug delivery applications. *The International Journal of Advanced Manufacturing Technology, 93*(1–4), 407–422. http://doi.org/10.1007/s00170-016-9698-6.

Rahman A.S., Cheah F.C., Nashriby M.E.A., Bastion M.L.C., Dee C.F., Buyong M.R., Mohamed M.A., Chua X.Y., Ooi P.C., Abdul Jalal M.I., Lam C., Mun Y.Y., Goh C.S., Ismail A.G., & Hamzah A.A., Optical and mechanical characterization of lignocaine- impregnated maltose-based dissolvable microneedles. Engineering Proceedings, 110(1), 1–9, (2026).

Rodrigues, R. G., Condelipes, P. G. M., Rosa, R. R., Chu, V., & Conde, J. P. (2023). Scalable Processing of Cyclic Olefin Copolymer (COC) Microfluidic Biochips. *Micromachines, 14*(10), 1837. doi:10.3390/mi14101837.

Said, M. M., Yunas, J., Pawinanto, R. E., Majlis, B. Y., & Bais, B. (2016). PDMS based electromagnetic actuator membrane with embedded magnetic particles in polymer composite. *Sensors and Actuators, A: Physical, 245*, 85–96. http://doi.org/10.1016/j.sna.2016.05.007.

Senel, M., Dervisevic, M., & Voelcker, N. H. (2019). Gold microneedles fabricated by casting of gold ink used for urea sensing. *Materials Letters, 243*, 50–53. http://doi.org/10.1016/j.matlet.2019.02.014.

Sullivan, S. P., Koutsonanos, D. G., del Pilar Martin, M., Lee, J. W., Zarnitsyn, V., Choi, S. O., Murthy, N., Compans, R. W., Skountzou, I., & Prausnitz, M. R. (2010). Dissolving polymer microneedle patches for influenza vaccination. *Nature Medicine, 16*(8), 915–920. http://doi.org/10.1038/nm.2182.

Tran-Minh, N., Dong, T., & Karlsen, F. (2014). An efficient passive planar micromixer with ellipse-like micropillars for continuous mixing of human blood. *Computer Methods and Programs in Biomedicine, 117*(1), 20–29. http://doi.org/10.1016/j.cmpb.2014.05.007.

Wilke, N., Mulcahy, A., Ye, S. R., & Morrissey, A. (2005). Process optimization and characterization of silicon microneedles fabricated by wet etch technology. *Microelectronics Journal, 36*(7), 650–656. http://doi.org/10.1016/j.mejo.2005.04.044.

Yang, L.-J., Joseph, V.-J., Unnam, N.-K., & Esakki, B. (2022). Design and Numerical Simulation of Biomimetic Structures to Capture Particles in a Microchannel. *Fluids, 7*(1), 32. doi:10.3390/fluids7010032.

Yin, B., Jiang, Z., Tan, G., Muhammad, R., Yang, Q., Zeng, S., … Lin, X. (2026). Automated design and performance prediction of micromixer chips via Convolutional Neural Network models for lab-on-a-chip detection. *Expert Systems with Applications, 319*, 132118. doi:10.1016/j.eswa.2026.132118.

Yunas, J., Mulyanti, B., Hamidah, I., Said, M. M, Roer Eka Pawinanto, Fikri Wan Ali, W. A, Subandi, A., Hamzah, A. A, Latif, R., & Majlis, B. Y (2020). Polymer-based MEMS electromagnetic actuator for biomedical application: A review. *Polymers, 12*(5), 1184. http://doi.org/10.3390/polym12051184.

Zheng, Y., Ye, R., Gong, X., Yang, J., Liu, B., Xu, Y., Nie, G., Xie, X. i, & Jiang, L. (2023). Iontophoresis-driven microneedle patch for the active transdermal delivery of vaccine macromolecules. *Microsystems & Nanoengineering, 9*(1), 1–14. http://doi.org/10.1038/s41378-023-00515-1.

Zhou, Y., & Amirouche, F. (2011). An electromagnetically-actuated all-PDMS valveless. *Micromachines, 2*(11), 345–355. http://doi.org/10.3390/mi2030345.

CHAPTER

7 Trends of point-of-care biosensors for diagnosing diseases

Santheraleka Ramanathan[a,b], Subash C.B. Gopinath[b,c,d], Azrul Azlan Hamzah[e], Sh. Nadzirah[b,e], Aziah Ismail[f], Mehmet Ozsoz[g] and Fahreddin Palaz[h]

[a]Department of Chemical Engineering, Faculty of Engineering, University of Malaya, Kuala Lumpur, Malaysia, [b]Institute of Nano Electronic Engineering, Universiti Malaysia Perlis (UniMAP), Kangar, Perlis, Malaysia, [c]Faculty of Chemical Engineering & Technology, Universiti Malaysia Perlis (UniMAP), Arau, Perlis, Malaysia, [d]Department of Neonatology, Saveetha Medical College and Hospital, Saveetha Institute of Medical and Technical Sciences, Chennai, Tamil Nadu, India, [e]Institute of Microengineering and Nanoelectronics (IMEN), Universiti Kebangsaan Malaysia, Bangi, Selangor, Malaysia, [f]Institute of Research in Molecular Medicine (INFORMM), Universiti Sains Malaysia, Health Campus, Kubang Kerian, Kelantan, Malaysia, [g]Department of Biomedical Engineering, Near East University, Mersin, TRNC, Turkey, [h]Department of Surgery, College of Medicine, University of Florida, Gainesville, FL, United States

7.1 Introduction

The field of biomedical sciences is integral to various aspects of daily life, continually evolving in response to the imperative for advancement. This dynamic progression has profoundly influenced the well-being of many individuals. Currently, human lifespan and longevity significantly exceed historical expectations, going beyond the most optimistic forecasts of past years, a clear indicator of the strides made in medical and scientific development. The global scale of mankind is confronted with considerable risks, with concerns about the health of individuals and the public being of utmost importance. Statistical assessments conducted by both public and private entities have demonstrated that incorrect diagnosis results in increased costs for patients and heightens the likelihood of the illness deteriorating (Grossi & Riccò, 2017; Ramanathan et al., 2023). The advancement of a disease following an infection exhibits exponential growth. In addition, investigations has been conducted, elucidating that the symptoms manifest significantly later following the infection. In conventional environments, unwell individuals are typically diagnosed using state-of-the-art devices commonly located in centralized laboratories. Typically, molecular and microbiological protocols are predominantly utilized. Consequently, the patient's state has already deteriorated by the time the clinical diagnosis is established (Cady et al., 2021; Iravani, 2020). While these solutions are robust and accurate, their implementation often requires a significant amount of time, rendering ineffective point-of-care (POC) scenarios. To address these issues, substantial efforts have been made to establish effective methods for diagnosing the disorders. Identifying the metabolic change during the asymptomatic transition phase is a crucial factor in combating any illness or condition. At this stage, it is crucial to address the disease in its early phases (Ramanathan et al., 2022a).

BioMEMS Devices. DOI: https://doi.org/10.1016/B978-0-443-24015-7.00009-7

Immediate implementation of equitable and dependable biosensors is crucial in clinical diagnostics to effectively address various clinical emergencies. Upon analyzing global health indicators, statisticians have forecasted a disconcerting scenario. Multiple research and communications emphasize various strategies to mitigate the crucial challenges of biosensors to achieve precise diagnostic outcomes. These techniques can detect biomarkers in the absence of any symptoms of a disease. Due to rising healthcare expenses and a global population that is getting older, there is a growing demand to remotely monitor the health status of patients. This encompasses scenarios when patients are not restricted to medical facilities. To tackle this problem, a range of POC biosensors have been developed (Deroco et al., 2021; Liu et al., 2020). These systems offer immediate input on the health status of patients, which can be accessed by either the patients themselves or healthcare providers through a medical database. POC systems offer a novel solution for managing and monitoring the health of patients with chronic diseases, elderly individuals, and postoperative rehabilitation patients. These devices fulfill duties that are typically carried out by laboratory testing (Tram et al., 2016; Xu et al., 2023).

POC systems used for health monitoring can consist of many types of biosensors that can be worn or implanted. The gathered data is subsequently transmitted via a wired or wireless connection to a central data acquisition node, such as a personal digital assistant, which can then relay the collected signals to a medical facility. These devices would not only decrease the amount of time it takes to respond, but would also provide the opportunity for testing in circumstances where laboratory testing is not possible. These systems have diverse applications in medical diagnostics and the detection of biological warfare agents. Additionally, the sensor configuration can be adjusted to accommodate various uses, such as ensuring food safety, monitoring the environment, and controlling industrial processes. Additional potential uses are now arising, including the detection of animal and plant infections, genetic studies, diagnostic testing for malarial strains, and water quality assessment. This chapter presents information on novel technologies, specifically POC biosensors, and their integration with suitable technology to offer medically significant data (Thompson et al., 2023). This integration aims to aid physicians and clinicians in the identification of diseases. The primary technological platform under consideration in this chapter is POC biosensors and their incorporation into POC systems for the examination of medically important biomarkers.

7.2 Components of point-of-care sensors

The compact dimensions and simple design of biological microelectromechanical systems (BioMEMS) make them highly suitable for the identification of biological and chemical agents. These systems are particularly well-suited for POC lab-on-a-chip platforms, as they can simultaneously detect various biomarkers in both environmental and human samples. Additionally, they can efficiently collect and analyze samples to obtain crucial data for evaluation purposes. Due to this rationale, BioMEMS has garnered significant interest in the biosensor research community in recent times. BioMEMS engineers primarily focus on applications related to diagnosis, sensing, and detection. The design, fabrication, and production procedures for generating POC sensors vary based on the specific diagnostic and the materials employed. The components of POC are categorized into three sections as discussed below, which are mainly considered to develop high-performance disease-diagnosing POC sensors.

7.2.1 Biomolecule recognition element

The effectiveness of POC biosensing and diagnostic platforms relies mostly on the precise interaction between the target molecule and the probe elements, which results in high sensitivity, specificity, stability, and rapid response time. However, any arrangement of biomolecules on a sensing platform that can identify a specific target is categorized as bio-recognition. Neurons, cells, antibodies, aptamers, enzymes, and nucleic acids are frequently employed as bio-recognition components for biomarker or biomolecular recognition. The recognition element in biosensors binds the analyte to provide precise detection. The efficacy of POC biosensors relies on the bio-recognition elements, which encompass entire cells, antibodies, enzymes, DNA, RNA, and proteins (Ramanathan et al., 2019a; Wei et al., 2023). Various functional biosensor technologies utilizing diverse bio-recognition elements have been effectively established. The recognition of DNA on a POC biosensor is critical for applications like genetic testing, biomarker and pathogen detection, and personalized medicine. POC biosensors rely on the specific binding between a DNA probe and its complementary DNA target. This binding process is based on the principles of complementary base pairing in DNA. In DNA, adenine (A) binds specifically to thymine (T), and cytosine (C) binds specifically to guanine (G). This inherent specificity allows for the design of DNA probes that will only bind to their complementary DNA sequences. To recognize a specific DNA sequence, a short, single-stranded DNA probe is designed. This probe is synthesized in the laboratory and labeled with a reporter molecule or a signal-generating tag, such as a fluorescent dye and enzymes. The isolated DNA target from the patient's sample is then exposed to the biosensor. The DNA probe on the surface of biosensor is complementary to a specific region of the target DNA. This binding event is detected and translated into a measurable signal, enabling the rapid and accurate detection of DNA sequences for various POC diagnostic purposes. In addition to DNA, proteins are essential biomolecules used in POC biosensors for their ability to recognize and interact with specific targets, such as other proteins, small molecules, or biomolecules like DNA. These biosensors utilize the unique binding properties of proteins to detect and quantify the presence of specific analytes. This recognition is often based on the principles of molecular recognition, including shape complementarity and chemical interactions. Another crucial category of bio-recognition elements used in POC biosensors is aptamers, which are functional nucleic acid molecules composed of either DNA or RNA, synthesized through an in vitro process of evolutionary selection known as SELEX (Systematic Evolution of Ligands by EXponential enrichment) (Ramanathan et al., 2020b, 2019b). Aptamers offer distinct benefits, including cost-effectiveness, low immunogenicity, and ease of preparation and modification, making them effective alternatives for proteins, particularly antibodies, for various applications. Aptamers can adopt a three-dimensional (3D) structure and can therefore interact with ligands through complementary shape interactions. This interaction is characterized by a strong affinity and selectivity for specific target molecules. Aptamers interact with targets by forming distinctive spatial conformations. Anti-aptamers are designed to inhibit the binding of aptamers to their respective targets. They are typically single-stranded DNA or RNA sequences that are engineered to have a shape and chemical properties that interfere with the aptamer's binding site. Anti-aptamers are used to boost the specificity of POC sensors (Cheen et al., 2017; Ramanathan et al., 2019b; Zhang et al., 2020). By using an anti-aptamer to prevent nonspecific binding or interference from interfering molecules, biosensors can provide more accurate and reliable results. In addition to these biomolecules, enzymes and small molecules are utilized as bio-elements to detect biomarkers in POC biosensors (Mahajan et al., 2023).

7.2.2 Transducer

Transducers transform biochemical interactions into measurable electronic signals. A transducer converts one type of energy to another. The choice of transducers for biosensors depends on the selection of material, specifications of the sensor device, and the specific signal conversion mechanism used. Device design has a substantial impact on the ultimate parameters. Transducers often amplify the relatively weak biological binding event signal into a more robust, measurable signal. This is particularly important when dealing with low concentrations of target analytes, as it improves the sensitivity of the biosensor (Back & Carroll, 2023; Polat et al., 2022). Transducers can provide real-time monitoring of the binding event, allowing clinicians to observe the kinetics of the interaction and capture dynamic changes in the sample over time. Regarding this, transducers are integral components of POC biosensors that facilitate the conversion of molecular interactions into measurable signals, enabling the detection, quantification, and analysis of specific biomolecules in clinical and diagnostic applications. They are key to the success of biosensor technology, particularly in providing rapid and accurate results in POC settings (Millner et al., 2009).

7.2.3 Electronics and signal processing unit

To accomplish accurate and sensitive detection, it is crucial to amplify the physicochemical changes during the identification process. This is necessary due to the low concentration of the target in real biological samples, as well as the weak conductivity and significant steric hindrance of normal biomolecules. Due to the poor conductivity of the commonly used glassy carbon electrode (GCE) or screen-printed carbon electrode, the limits of detection (LOD) for many POC biosensors are initially limited to the micromolar (μM) range. This does not meet the criteria for clinical diagnosis (Shafiee et al., 2018). Consequently, significant endeavors have been dedicated to the development of diverse techniques for enhancing signal strength in recent times. The electronics and signal processing unit in POC biosensors is a critical component that plays a central role in signal acquisition, processing, and data interpretation. This unit is responsible for converting the biological response of biomolecules binding into a measurable and interpretable signal, and it often includes various electronic components and software algorithms. Signal processing begins with signal acquisition, which converts the biological recognition event into an electrical, optical, or chemical signal. The acquired signal may require conditioning to filter out noise and make it suitable for processing. Components like filters, amplifiers, and analog-to-digital converters (ADCs) are used to condition and digitize the signal (Han et al., 2023). A microcontroller or microprocessor is the central processing unit that executes control functions, data processing, and decision-making tasks. It can also provide an interface for user interactions and data display. The microcontroller reads data from the transducer, which measures the biological or chemical response to the target analyte. The analog signal generated by the sensor is received by the microcontroller's ADC for digitization. POC biosensors are often equipped with a user interface, which can be a simple display screen or a smartphone app. This interface presents results to healthcare professionals in a user-friendly manner at POC diagnostics (Wang et al., 2022).

7.3 Types of point-of-care biosensors

One key aspect of POC biosensors in disease diagnosis is their diversity, with various types designed to suit different applications and detection needs. In this section, different types of POC biosensors, each tailored to address specific diagnostic challenges and objectives, are discussed. These biosensors encompass a range of technologies and detection methods, making them indispensable tools for healthcare providers in the field.

7.3.1 Electrochemical point-of-care biosensors

Biosensors equipped with electrochemical transducers are widely used and adaptable for POC applications. The transducer facilitates the generation of electrochemical species through the bio-interaction process, resulting in the production of an electrochemical signal. Electrochemical sensors can be categorized into potentiometric, impedimetric, voltammetic, conductometric, and amperometric sensors, based on the specific electrochemical feature they utilize (Okpara et al., 2023; Sumitha & Xavier, 2023). Potentiometric sensors quantify the electrical potential difference between the reference and working electrodes when they reach a state of balance. This measurement allows for the monitoring of charge variation occurring at the surface of the electrode, as defined by the Nernst equation. Amperometry biosensors quantify the current produced by the electrochemical redox reaction of electroactive molecules at an electrode coated with a bio-receptor while maintaining a constant voltage. Amperometry changes exhibit a direct and proportional relationship with the concentration of the analyte in the solution. The presence of redox mediators is necessary during analysis when the analytes are not capable of undergoing redox reactions with each other. Impedimetric biosensors are highly sensitive devices used to analyze the qualities of interfaces and measure both resistance and reactance in electrical impedance. The presence of a bioreceptor and analyte on the electrode surface hinders the movement of electrons in an alternating current circuit, leading to the development of capacitive resistance (Kiew et al., 2021; Ramanathan et al., 2022b; Yoon et al., 2017). Biorecognition can be performed without the need for any labeling, making impedimetric biosensors ideal for POC diagnostics. Electrochemical biosensors can be classified into two groups based on their detection strategy, which is label dependent and label-free. Label-dependent electrochemical biosensors rely on the presence of biological molecules such as enzymes, tagged antibodies, or physiochemical molecules like metallic nanoparticles and conductive polymers to generate electrical signals for measuring purposes. The label-free sensors rely on immobilizing the target analyte on the bio-receptor electrode and measuring changes in capacitance and resistance on the electrode surface (Gattani et al., 2019). Fig. 7.1 depicts a portable POC biosensor that is designed as an automated apparatus for the detection of human chorionic gonadotropin (hCG) in human urine samples, specifically to identify pregnancy. The utilization of gold and silver electrodes in sandwich assays greatly enhances the reaction rate on the surface of the POC sensor. Consequently, this enables the quick detection of extremely low concentrations of analyte. The immobilization of the antibodies was characterized using electrochemical impedance spectroscopy (EIS), while the sensor activities were determined using chronoamperometry (CA) measurements. A completely automated portable system, fitted with an electrochemical POC sensor, demonstrated exceptional precision, a low limit of detection, excellent reproducibility, and long-term stability in detecting hCG in urine samples (Yuksel et al., 2023).

7.3.2 Lateral flow assays point-of-care biosensors

Lateral flow assay (LFA) based biosensors are advanced devices used for detecting and measuring analytes from various specimens in POC diagnosis. They are considered primary and up to date in this field. In LFA, antibodies serve as the main capturing agents. However, other recognition biomolecules,

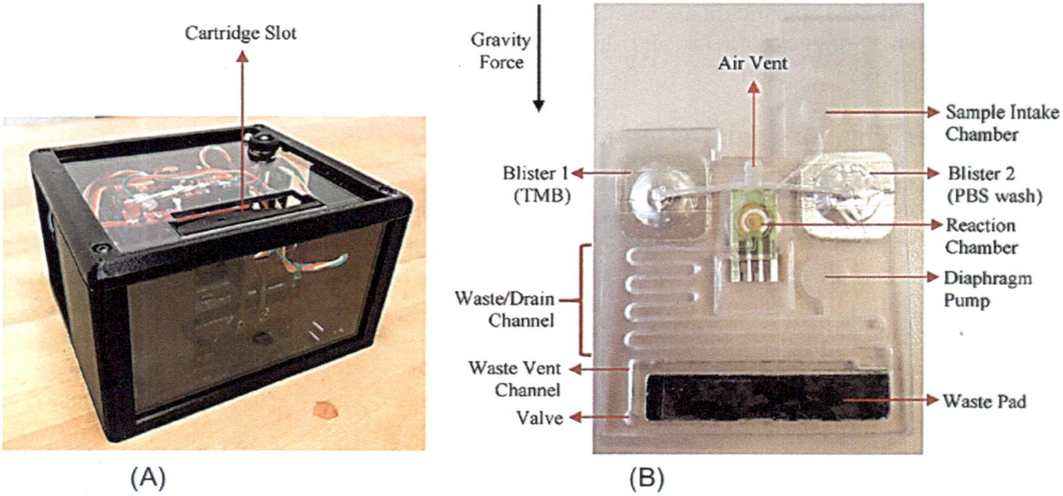

FIGURE 7.1

Digital image of (A) the developed automated portable POC device. (B) Electrochemical POC sensor equipped in the disposable cartridge of the automated device for detecting human chorionic gonadotropin (hCG).

Reproduced with permission from Yuksel et al. (2023) copyright 2023 Elsevier B.V.

such as DNA oligonucleotides, have also been utilized (Pei et al., 2023; Soh et al., 2020). The LFA comprises various components, such as a sample pad mostly made of glass fiber and a nitrocellulose membrane. Subsequently, the fluid travel through the porous membrane where the analyte, along with the reporter reagent, will bind to the test line, resulting in a detectable signal (Gumus et al., 2023; Zhou et al., 2021). The residual solution passes through the membrane, where a portion of the antibodies will bind to the control line, indicating the successful execution of the test. The absorbent pad regulates the pace of fluid flow and guarantees the continuous progression of the fluid. LFAs have high portability, ease of use, and significantly faster performance compared to conventional laboratory-based techniques. They also have reduced reliance on equipment and well-trained personnel, resulting in lower costs. These attributes render them highly valuable as POC sensors, particularly in developing or resource-limited environments, where primary health-care facilities lack access to advanced technology and skilled personnel (Hu et al., 2016). LFAs are highly adaptable and can be used with a wide range of biological samples, including saliva, urine, serum, whole blood, plasma, and swabs from different sources such as nasal, throat, or other relevant mucosal surfaces depending on the target analyte or pathogen being tested. The deployment of LFAs aids in the rapid detection of diseases, enabling timely medical intervention. This enhances disease surveillance and management, potentially leading to reduced use of expensive medications, lower morbidity and mortality rates, decreased local infection rates, and, ultimately, a reduction in overall healthcare costs and burdens. Furthermore, LFAs consist of separate paper-based elements that are merged in a single lamination process. LFA platforms possess high flexibility and versatility in terms of engineering and production, as the individual components can be assembled in numerous configurations to meet the requirements of different assays. The methods of

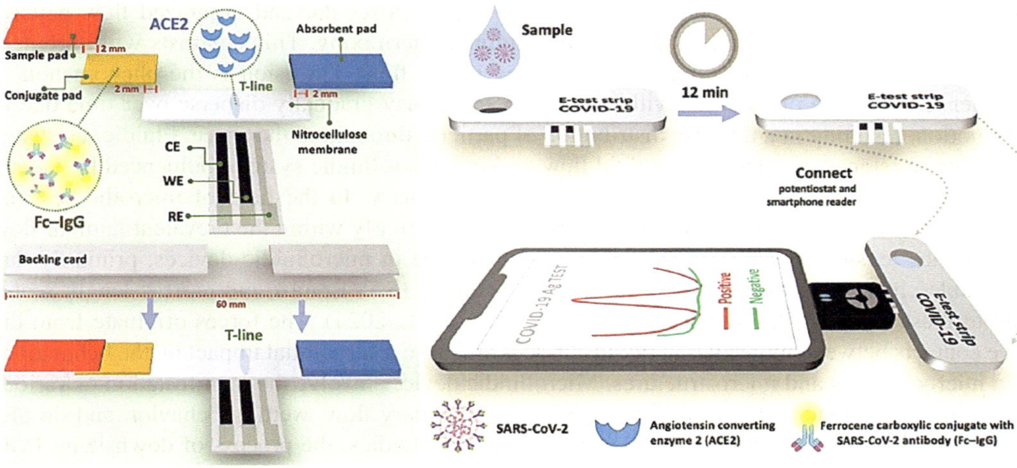

FIGURE 7.2

Schematic illustration of the E-test strip developed for detecting SARS-CoV-2 spike antigen using angiotensin-converting enzyme 2 (ACE2). The density of ACE2 immobilization on the POC sensor is visualized on smartphone as in differential pulse voltammetry signal transmitted through bluetooth networking.

Reproduced with permission from Deenin et al. (2023) copyright 2023 Elsevier B.V.

development and commercialization for LFA are highly structured and divided into modules (Hu et al., 2016; Niu et al., 2018). Fig. 7.2 illustrates the design of an E-test strip developed for detecting the SARS-CoV-2 spike protein. This detection is facilitated by employing angiotensin-converting enzyme 2 (ACE2) immobilized on a nitrocellulose membrane. Quantification of the detected spike protein is then accomplished using a smartphone application, integrating modern technology for rapid and accessible diagnostic capabilities. The ferrocene carboxylic acid, when bound to the SARS-CoV-2 antibody, functions as an electroactive substance. This binding occurs when the antibody attaches to the SARS-CoV-2 antigen in the sample. Subsequently, the sample flows continuously towards the ACE2-immobilization zone on the electrode. The E-test strip designed for COVID-19 diagnosis exhibits a direct proportionality to the signal observed on the smartphone, indicating the density of ACE2 immobilization (Deenin et al., 2023).

7.3.3 Microfluidic point-of-care biosensors

Microfluidics has become a revolutionary technique in the realm of POC diagnostics. This domain comprises a range of ideas and fundamental elements that facilitate the creation of very efficient microfluidic devices. Microfluidics presents several advantages in diagnostics, including rapid diagnosis, minimal sample volume demands, and effortless integration with diverse detection methods. To effectively navigate the realm of microfluidics, one must possess a comprehensive understanding of the fundamental physical phenomena that dictate processes at the microscale (Tavakoli et al., 2022). At the microscale, different pressures have a more significant impact than those experienced in ordinary situations. Laminar flow in microfluidics is characterized by a regular and organized flow pattern, where fluid particles move in parallel layers with minimal intermixing. This contrasts with turbulent flow, which involves random fluctuations and mixing of the fluid. Diffusion is the phenomenon in which a dense collection of particles within a certain space may gradually disperse over time due to random motion, resulting in a uniform distribution of particles throughout the space. Fluidic resistance quantifies the resistance encountered by fluid flow within a microfluidic system, influenced by factors such as channel diameters, fluid viscosity, and device geometry. In the field of microfluidics, the primary determinant of fluid flow resistance is viscosity, particularly within the prevalent laminar flow regime. Surface tension forces assume significant importance in microfluidic devices, primarily due to the notable characteristics of high surface area to volume (SAV) ratios and narrow length scales inherent to these devices (Bradley & Bhalla, 2023; Chen et al., 2024). The forces originate from the cohesive contacts between molecules at liquid surfaces and have a substantial impact on the behavior of fluids in microchannels and microstructures. Microfluidic devices involve the manifestation of various significant features of surface tension forces, including capillary flow, wetting behavior, and droplet manipulations. Through harnessing the capabilities of microfluidics, the process of downsizing POC diagnostic equipment offers notable benefits in a self-contained system that is fully automated. In contrast, many traditional technologies may require larger sample volumes to perform their analyses effectively. Microfluidic devices can analyze extremely small sample volumes, hence minimizing the requirement for lengthy sample preparation and the preservation of valuable or scarce materials. The rapid analysis time provided by microfluidic devices is crucial for making informed decisions regarding patient care in POC diagnostics. Microfluidic devices possess a high SAV and exhibit fine control over fluid movement, enabling them to accurately identify analytes even at extremely low concentrations. As a result, these devices are well-suited for a wide range of applications. Microfluidic devices offer the capability to integrate multiple functions, including sample preparation, mixing, separation, and detection, within a single device. This capability allows for the efficient analysis of numerous samples in a short timeframe, concurrently reducing both the time and costs associated with diagnostic testing. The portability of microfluidic devices is a notable benefit, allowing for their utilization in many situations where conventional laboratory equipment may be inconvenient or inaccessible. This capability enables healthcare personnel to do diagnostic procedures directly at the patient's bedside, in clinics, or even in remote locations. The prompt availability of this expedited diagnostic capability has the potential to significantly enhance patient outcomes by facilitating prompt treatment decisions (Cui et al., 2022; Mousaabadi et al., 2023). However, it is important to note that these devices also come with specific challenges related to their intricate manufacturing processes, strict uniformity requirements, regulatory considerations, and associated costs. Continuing research and technical progress in this area seek to overcome these constraints and further improve the functionalities of microfluidic devices for virus detection. The production of microfluidic devices can be intricate and necessitates specialized equipment and knowledge, posing a potential obstacle for certain researchers and diagnostic developers. In addition, the process of standardizing and obtaining regulatory approval for microfluidic devices can be complex due to the diverse range of designs and fabrication techniques employed, hence hindering the establishment of consistent quality control protocols (Khan et al., 2023). Fig. 7.3 depicts the microfluidic sensing chip, which isolates the process of capturing from the process of detecting, all the while preserving the sensor in its original and unaltered condition before measurement. The method enables the collection, cleansing, chemical reaction, and cultivation of the specimen in a specialized reaction chamber within the microfluidics system, ensuring that the sensor electrodes remain uncontaminated

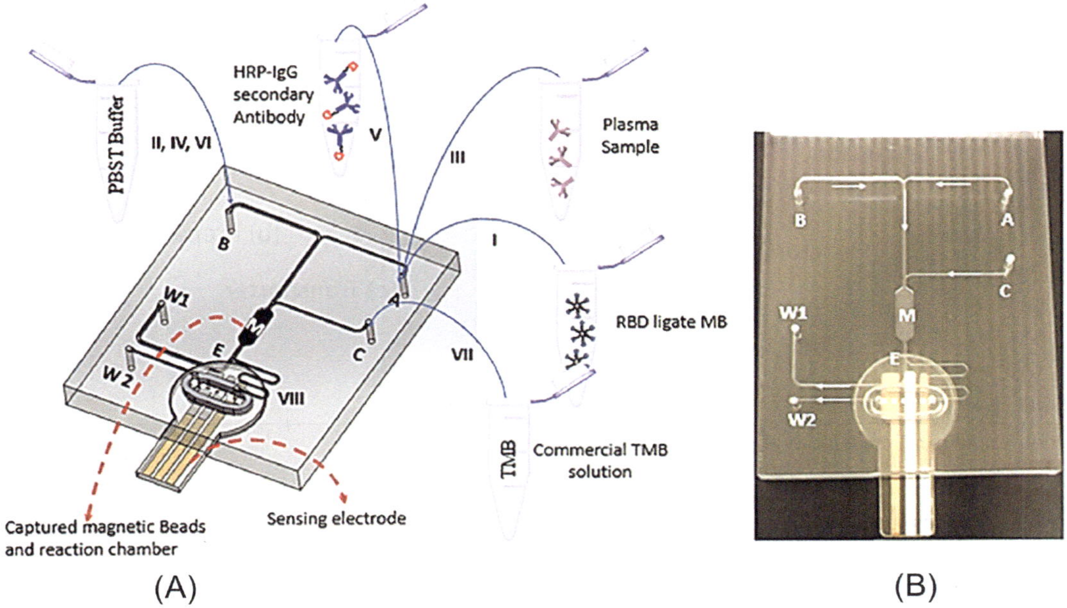

FIGURE 7.3

(A) Schematic of the microfluidic sensor chip developed to detect a capture from the sensing and the signal processing indicated by the Roman numerals (B) digital image of the developed microfluidic POC sensor equipped chip.

until the time of measurement. The process of isolating the capture step from the detection step was utilized in the electrochemical identification of anti-SARS-CoV-2 antibodies. The most effective arrangement demonstrated a strong association between the identification of COVID-19-positive blood samples from patients and control samples (Ko et al., 2022).

The integration of lab-on-printed circuit board (LoPCB) technology with electrochemical biosensors is a notable advancement in the realm of POC diagnostics. This integration consolidates numerous laboratory functionalities into a singular, compact chip (Zambry et al., 2023). These functionalities encompass key processes such as injection, preparation, transport, mixing, reaction, separation, detection, control, and power supply. This innovative approach, detailed in Fig. 7.4, significantly enhances the efficiency and accessibility of diagnostic technologies, offering a multifunctional platform for swift and accurate diagnostics across diverse settings.

7.3.4 Impedimetric point-of-care biosensors

Impedimetric POC biosensors are a novel electrochemical technology that is gaining recognition. Impedimetric sensing relies on the frequency-dependent reaction of an electrochemical system to a low-intensity sinusoidal voltage signal. It combines data regarding the capacitive and resistive characteristics

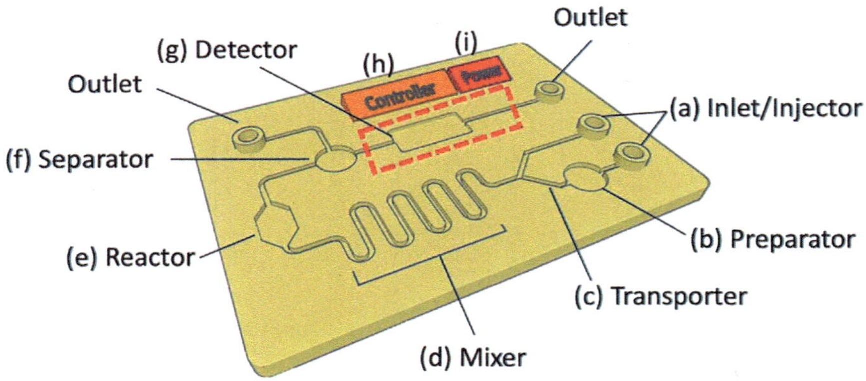

FIGURE 7.4

Schematic illustration of lab-on-chip components.

of materials. This technique has been effectively utilized to observe immunological binding events at the interface or electrode of an electrolyte. It has also been used in the synthesis of materials and in toxicological studies to monitor changes in cell morphology (Bahadir & Sezgintürk, 2016; Kirchhain et al., 2020). The impedimetric sensing principle involves the use of a sinusoidal potential in an electrochemical system under investigation. The resulting potential is then monitored with respect to frequency. The impedimetric measurements can be conducted using different electrodes arranged in various configurations. The most used configurations are often known as two-, three-, and four-electrode systems (Shoute et al., 2023). Practically, the measurement becomes increasingly intricate and accurate as the number of electrodes increases. Impedimetric biosensors commonly employ a three-electrode design. This electrochemical cell arrangement comprises a reference electrode, a working electrode, and a counter electrode. A biosensor utilizing impedance measurement for POC detection has antibodies that are immobilized and designed to specifically target a particular biomarker, complementary probe, or specific antibodies. This allows for the detection of genomes, proteins, and antigens at the working electrode. The selection of the most suitable approach is mostly contingent upon the anticipated use case. Within the faradaic classification, impedance arises from redox processes, whereas the non-faradaic classification pertains to impedance resulting from direct current, where the electrical characteristics are determined by the double-layer capacitance. Non-faradaic approaches have the benefit of being applicable in POC devices since the electrodes may be miniaturized, and there is no need for a redox pair. However, faradaic sensors are generally more sensitive (Cordeiro et al., 2020; Selimoğlu et al., 2023). An impedimetric biosensor is a highly sensitive method for detecting biomarkers without the need for labeling, allowing for simultaneous detection. Probes in impedimetric immunosensors are fixed onto electrodes, optical fibers, or semiconductor chips. These sensors have gained significant attention in recent years due to their potential uses in diverse fields. The probes and biomarkers are chemically linked, resulting in the formation of an immunocomplex. The electrode is then covered with a layer that prevents any non-specific binding. Consequently, electron transport undergoes alteration. Once biomarkers bind to the probes, the ability of the redox probes to access is significantly impeded compared to when there are no biomarkers present. As the Faradic reaction of a redox couple becomes

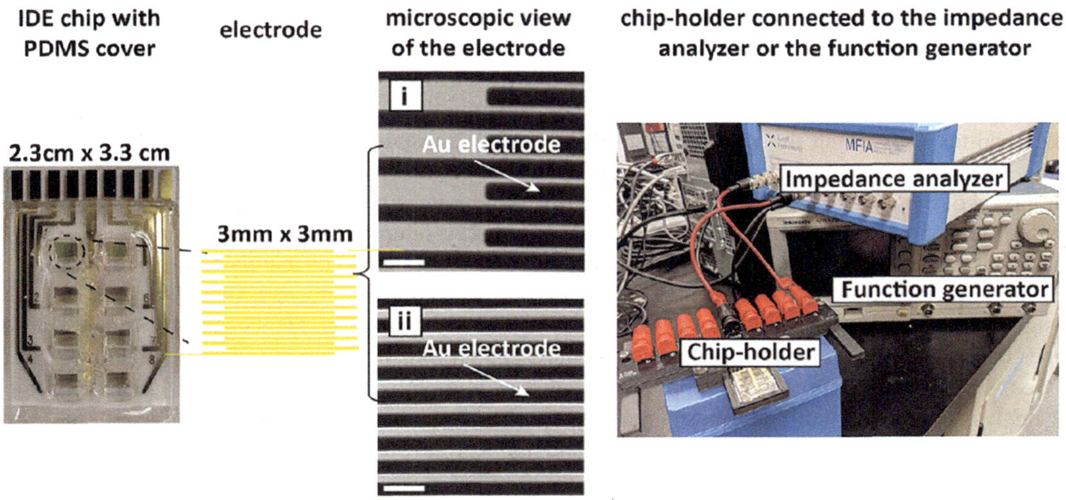

FIGURE 7.5

Components of the impedimetric COVID-19 serology sensor: IDE chip with PDMS mask, electrode layout, and chip holder. Digital images show the electrode edge (i) and center (ii).

Reproduced with permission from Zeng et al. (2022) copyright 2022 Elsevier B.V.

more obstructed, the resistance to electron transfer will rise, resulting in a decrease in capacitance. The crucial stage in the development of impedimetric biosensors is the immobilization of biomolecules on the electrode surface. This step is vital for developing biosensors that are stable, repeatable, and selective (Han et al., 2023; Shimizu et al., 2023). As an illustration, antibodies adhere to the gold surface through physical adsorption. However, directly attaching antibodies to the entire metal surface causes denaturation and reduces their ability to bind to biological substances. The activity of antibodies can be preserved when they are absorbed into the surface of noble metal nanoparticles, such as gold nanoparticles, due to the biocompatibility of these nanoparticles. Fig. 7.5 shows the impedimetric COVID-19 serology system that includes the IDE sensor with PDMS masking, its electrode layout, and the chip holder, along with DIC images illustrating the POC sensing setup for real-time diagnosis. The accompanying digital images highlight the structural features of the electrode fingers, capturing both the edge and the central region of the interdigitated pattern, which enhances the sensitivity of POC sensors (Zeng et al., 2022).

7.3.5 Paper-based point-of-care biosensors

Paper-based sensors have garnered growing interest over the past decade and have been widely adopted in diverse domains, including clinical diagnostics, food safety, and environmental monitoring. The capacity for on-site detection, which eliminates the need for traditional sampling and sample preparation procedures typical in clinical laboratories, renders the technology highly suitable for use as a POC sensor. This feature is surely driving numerous advancements in the development of lab-on-paper technology. Paper-based sensors are manufactured by strategically designing paper to establish channels, barriers, and specific areas for sensing through physical and chemical modification (Wang et al., 2023). Within the overarching category of paper-based sensors, two specific types of sensing systems might be categorized as paper-based assay devices. The initial system is based on traditional qualitative analytical tests, in which the identification of the desired substance can be carried out on filter paper using appropriate colorimetric and fluorescent agents. Paper is typically a pliable and unadorned sheet composed of compressed cellulose fibers. From this standpoint, a paper-based sensor is defined broadly as a porous membrane that utilizes capillary force to absorb sample solutions (Kuswandi et al., 2022). Paper was widely utilized as a platform for analyzing substances in the distant past. As an illustration, it was employed as litmus paper to identify acid-base reactions. In general, paper-based sensors comprise designated regions for sample intake, sample flow, and sensing areas. Hence, the manufacturing of paper-based sensors involves two primary procedures: patterning the paper and functionalizing the paper device for its specific purpose and application, such as immobilizing chemicals onto the paper device for analysis. Conventional patterning approaches involve employing computer-aided design to create device designs that are tailored to certain uses and applications, including the technologies utilized for patterning. Any drawing software can be utilized to create a device design, with preference given to popular drawing software that offers appropriate capabilities for device design (Xu et al., 2023). The objective of patterning paper is to create hydrophobic patterns on a paper surface, which act as barriers for the formation of microchannels and specific locations. The classification of paper patterning can primarily be divided into two categories: physical patterning, which involves cutting or plotting, and chemical patterning, which involves printing using hydrophobic polymers or inks to create channels as hydrophobic barriers (Hilda et al., 2024; Ma et al., 2023). Flexible paper with exceptional performance has been identified as a suitable substrate for supporting a POC sensing platform. It can rapidly and effortlessly operate, allowing for direct reading of changes. The high porosity and hydrophilicity of paper-based sensors are attributed to the interweaving of cellulose fibers and the presence of hydroxyl groups on the fibers. The inherent property of capillary action enables the transportation of liquids, eliminating the power supply (Murray & Mace, 2020). The elongation of the fiber in the paper-based sensor enhances its tensile strength, tear resistance, and break resistance. Additionally, it increases the flexibility of the paper, enabling it to withstand repeated bending, twisting, folding, and stretching on the sensing platform. The paper-based sensors possess characteristics that facilitate their biodegradable disposal in natural environments and incineration processes. This feature enhances their recyclability and contributes to the preservation of environmental health (Aguiar et al., 2023).

7.3.6 Acoustic point-of-care biosensors

Acoustic biosensors are mechanical waves that propagate on the surface of a material, with a shallow penetration depth of about one wavelength. Initially, acoustic biosensors were employed for electronic applications. Their distinct attributes and qualities render them well-suited for sensing applications in diverse domains, such as structural health monitoring, as well as the biological detection of biomarkers, poisons, and viruses. An acoustic biosensor device typically consists of interdigital transducers placed atop a piezoelectric substrate, allowing the surface acoustic wave to propagate. When the surface is subjected to various stimuli, it causes changes in the velocity of the acoustic wave, enabling the functioning of acoustic biosensors (Damasceno et al., 2023; Li et al., 2023). Acoustic biosensors provide a dual function as sensors for physical quantities and as communication devices, resulting in reduced

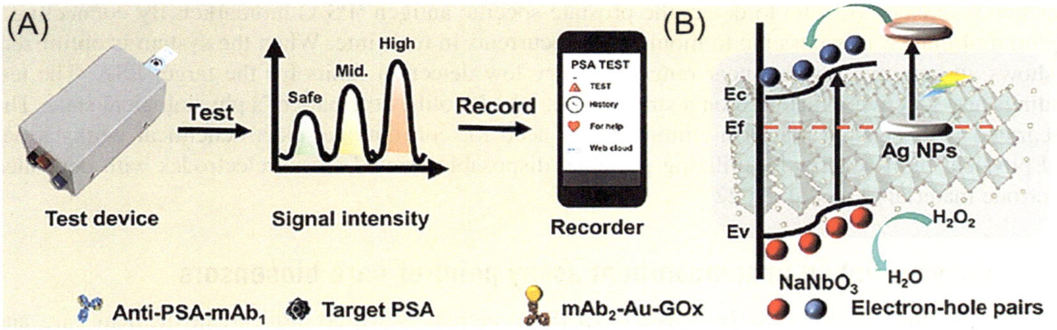

FIGURE 7.6

(A) Schematic diagram of the LSPR signal obtained from acoustic POC biosensor enhanced with silver nanoparticles, where the results are displayed on a smartphone. (B) Schematic diagram of the energy level structure and photoelectron transfer of the silver nanoparticles and the piezoelectric material sodium niobate.

Reproduced with permission from Xu et al. (2022) copyright 2022 Elsevier B.V.

prices, size, and weight. A piezoelectric structure can be combined with an antenna to exploit the relationship between the antenna's resonance frequency and the level of strain exerted, enabling stress measurement. Certain acoustic biosensors are wireless and battery-free, resulting in cost reduction and enabling the deployment of passive sensors in challenging locations where wires and batteries would not be feasible. Acoustic sensors employ several wave modes to fulfill diverse purposes (Mandal et al., 2022). The often-employed modes consist of Rayleigh waves, which exhibit a displacement component perpendicular to the acoustic wave and a shear component aligned with the surface of the substrate. This characteristic renders them well-suited for gas and microfluid sensing applications. Lamb waves possess both normal and shear components and are produced in thin substrates, rendering them well-suited for gas and liquid detection. Lamb waves are surface acoustic waves that propagate horizontally and have a shear component. They exhibit a high level of sensitivity, which makes them well-suited for detecting gases, liquids, and UV light. This has been demonstrated in the cited studies (Aslam et al., 2023; Pohanka, 2017). The operational mechanism of acoustic biosensor devices is primarily rooted in the piezoelectric phenomena, wherein the fixed dipoles of the piezoelectric materials undergo reorientation in response to the introduction of an external electrical field. The electrical signal is transformed into acoustic waves by the alternating polarity of the electrodes, causing the piezoelectric material to undergo periodic compression and expansion. The piezoelectric surface of the acoustic biosensors enables electromechanical interaction, wherein mechanical stress induces an electrical response (Pohanka, 2021). Piezoelectric sensors provide multiplexing, which refers to the simultaneous detection of numerous analytes. This skill is extremely beneficial for diagnosing complex disorders that involve many biomarkers (Howard-Quijano et al., 2023). Piezoelectric biosensors offer healthcare professionals a detailed overview of a patient's health condition, enabling them to make better-informed decisions about treatment and patient care. This leads to the implementation of more efficient and personalized healthcare strategies (Dekkers et al., 2022; Li et al., 2022; Wang et al., 2021). Fig. 7.6 depicts a newly created acoustic POC biosensor that has been improved with silver nanoparticles and the piezoelectric substance sodium niobate. This biosensor has been examined utilizing localized surface plasmon resonance (LSPR) to detect the prostate-specific antigen (PSA) biomarker. By connecting a digital multimeter, it is possible to monitor photocurrents in real-time. When the system is optimized, it shows an extremely wide linear range and very low detection limits for the target PSA. The test findings are effectively recorded on a smartphone, which notifies the patient's physiological state. The integrated POC sensing platform eliminates the need for conventional electrochemical workstations and physical light sources by utilizing portable, disposable printed screen electrodes with deposited electrode material (Xu et al., 2022).

7.3.7 Enzyme-linked immunosorbent assay point-of-care biosensors

The enzyme-linked immunosorbent assay (ELISA) method is widely utilized in medical care and clinical scientific research. Its increasing applicability in the health field is attributed to its intrinsic properties, including real-time detection, simplicity, low cost, automation, and high throughput. The ELISA technique is unique among other assays that rely on antibodies due to its distinctive process of separating specific and non-specific interactions through sequential binding to a solid surface. Additionally, it allows for the attainment of quantitative data (Carneiro et al., 2022; Singh & Satija, 2022). This method relies on the detection of an enhanced immunological reaction between a specific antigen and antibody, achieved by utilizing an enzyme-labeled antibody. Horseradish peroxidase (HRP) and alkaline phosphatase (ALP) are frequently employed enzymes in the ELISA technique. HRP is detected using 3,3′,5,5′ -tetramethylbenzidine (TMB) and 2,2′ -azino-di-[3-ethylbenzthiazoline- 6-sulfonic acid (ABTS)]. The fundamental concept underlying the ELISA reaction involves fixing the ligand, which can be either an antigen or a particular antibody, onto solid surfaces of multi-well plates, while an enzyme can be attached to another reagent. The ligand typically adheres to the plastic surface via non-covalent interactions between the hydrophobic areas of the ligand and the plastic surface. To prevent non-specific binding of the subsequent chemicals, the remaining uncoated surface of the well is saturated with an inert protein or polymer (Mello, 2021). During the tests, the solution to be examined is introduced into the wells, facilitating the attachment to the ligand, and subsequently followed by the introduction of an antibody-enzyme conjugate (Ramanathan et al., 2023, 2020a). The non-interacting molecules are eliminated by the process of washing. The identification of the interaction signal between chemicals is carried out using either labeling or an antibody-enzyme conjugation. During the experiment, it is necessary to introduce an enzyme substrate to induce the production of colorimetric or fluorescent signals from initially colorless products. The detection signal is achieved by quantifying the activity of the reporter enzyme by its reaction with the suitable substrate, which undergoes a color change upon modification by the enzyme. The strength of the color or light emitted is directly proportional to the concentration of the substance being analyzed in the sample. This is determined using microtiter plate readers that are calibrated to a specific absorbance value (Correa et al., 2021; Kumar et al., 2021; Peng et al., 2022). Despite its time-consuming nature and lack of user-friendliness for immediate diagnosis, ELISA is regarded as a valuable tool in POC diagnostics due to its ability to yield quantitative findings. This enables healthcare workers to ascertain the concentration of a particular biomarker in the sample. This information holds significant value in terms of illness staging and assessing the efficacy of treatment. Furthermore, ELISA can be tailored to identify a diverse array of target molecules, rendering it adaptable for the diagnosis of numerous ailments, such as infectious diseases, autoimmune conditions, and cancer. ELISA can be modified to perform multiplex tests, allowing for the concurrent identification of numerous biomarkers linked to a certain disease. This feature offers a more extensive perspective on the patient's health condition. ELISAs are widely employed in various

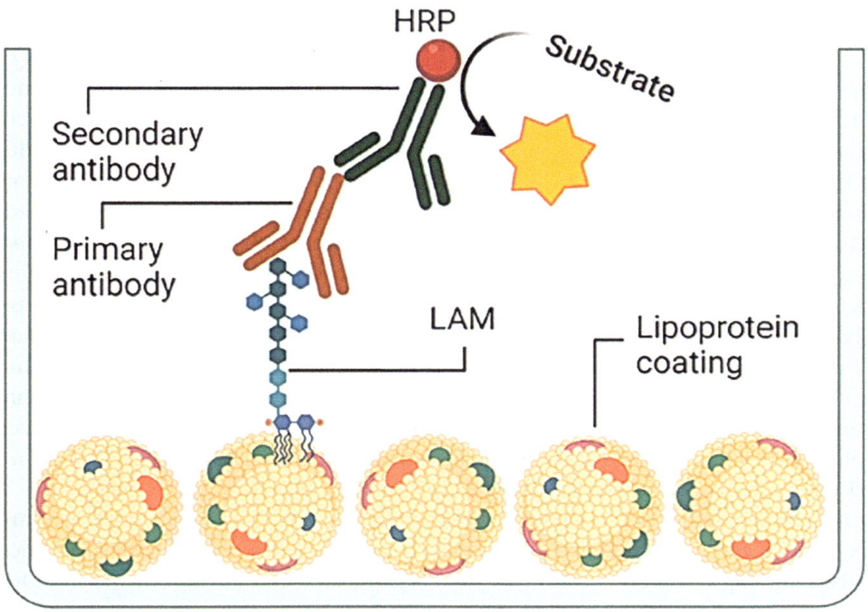

FIGURE 7.7

Schematic of ELISA designed for detecting LAM using lipoproteins as the target antigens.

Reproduced with permission from Lenz et al. (2022) copyright 2022 Elsevier B.V.

contexts (Bystryak & Acharya, 2016; Tiwari et al., 2020). HIV testing involves the collection of a blood or saliva sample for analysis, usually using indirect ELISA-based techniques. The ELISA is a method used to screen for the presence of HIV; however, it is not used to provide a definitive diagnosis. Additional testing using Western blot is necessary for diagnosis to decrease the risk of false-positive results. ELISA testing can be used to detect Molluscum contagiosum virus (MCV), a virus that often infects the skin of children and young people. The current evaluation in this scenario aims to assess the global seroprevalence of MCV by ELISA testing. The ELISA has significantly contributed to the advancement of allergy research and diagnosis in the field of food allergy. New iterations of the ELISA technique have been created to detect minuscule amounts of allergens, measuring in picograms. This is significant due to the potentially life-threatening impact that food allergies can have on public health (Sharma et al., 2022; Tiwari et al., 2013). Fig. 7.7 depicts the implementation of ELISA in a recent study, where it was utilized to capture lipoproteins and amphiphilic molecules, hence enhancing the detection of the biomarker for Mycobacterial infectious illness. The biomarker targeted in this study is the glycolipid lipoarabinomannan (LAM). The primary antibody utilized is anti-LAM, whereas the secondary antibody employed is HRP-streptavidin. The lipoprotein capture ELISA developed in this study presents a potential novel approach for identifying amphiphilic molecules in their native conformation, encompassing virulence factors that are highly conserved and released by bacteria during infection, along with other biomarkers of significance (Lenz et al., 2022).

7.4 Nanomaterial in point-of-care biosensors

POC devices have multiple limitations that can impact their sensitivity, selectivity, and accuracy. An inherent limitation is that POC devices may exhibit lower sensitivity or selectivity compared to diagnostic tests conducted in a laboratory setting, hence increasing the likelihood of producing false-negative or false-positive outcomes. Multiple variables may be influenced, such as the constraints of the technology employed in the POC device, the absence of suitable controls or benchmarks, and the existence of disruptive drugs in the examined sample (Aftab et al., 2023). POC devices may have reduced accuracy compared to laboratory-based tests due to variables exhibited in real-time settings. Furthermore, POC devices may lack validation or standardization procedures like those used for laboratory-based testing, thereby compromising their accuracy (Gopinath et al., 2022). Nanomaterials and nanotechnology could be valuable in addressing the limitations of POC devices, particularly by improving their precision, sensitivity, and specificity, thereby yielding a substantial effect. Nanomaterials possess distinct physical and molecular characteristics that set them apart from their larger counterparts. Nanotechnology encompasses the process of designing and manipulating materials to generate novel or enhanced products or technologies (Eivazzadeh-Keihan et al., 2022; Liu et al., 2023; Yang et al., 2024). Nanomaterial-based POC devices offer numerous benefits due to their increased surface area. Recently, there has been significant interest among researchers in the advancement of POC devices using nanomaterials. Ongoing research is examining the significance and advancements in POC platforms for COVID-19 detection that are based on nanomaterials (Gowri et al., 2021; Hashem et al., 2021). Analysis of the significance of POC diagnostics, particularly in the context of viral diseases such as COVID-19, HIV, Dengue, and Influenza. The present advancements, methodologies, and instruments for cancer biomarker identification have demonstrated the capability of POC platforms for disease surveillance. Nanotechnology has the potential to significantly transform the diagnosis and monitoring of several diseases in the realm of medicine using nanomaterials and nanotechnology. Nanomaterials improve the precision and accuracy of these tests by increasing their sensitivity and specificity. This enables the identification of disease biomarkers with a high level of accuracy (Gopinath et al., 2023, 2022). Nanomaterial-based POC assays can be developed for diverse purposes by the utilization of different methodologies, as mentioned above. A fundamental challenge in creating POC tests utilizing nanoparticles is the need to ensure the safety and compatibility of the nanomaterials used. Furthermore, current investigations are being conducted to comprehend the potential health and environmental consequences of nanomaterials, as well as to create reliable and secure diagnostic methods based on nanomaterials (Moulahoum et al., 2023). The future of the area hinges on ongoing progress in the design of nanomaterials, the development of innovative devices, and the investigation of novel applications. With the ongoing advancement of nanomaterials and nanotechnologies, there is a significant opportunity for enhancing POC platforms. This can lead to faster and more precise disease diagnosis, personalized treatment options, and better results for real-time medications.

7.5 Clustered regularly interspaced short palindromic repeats-based point-of-care diagnosis

Clustered regularly interspaced short palindromic repeats (CRISPR) and CRISPR-associated (Cas) systems, naturally occurring in bacteria as a defense mechanism against viruses, have revolutionized

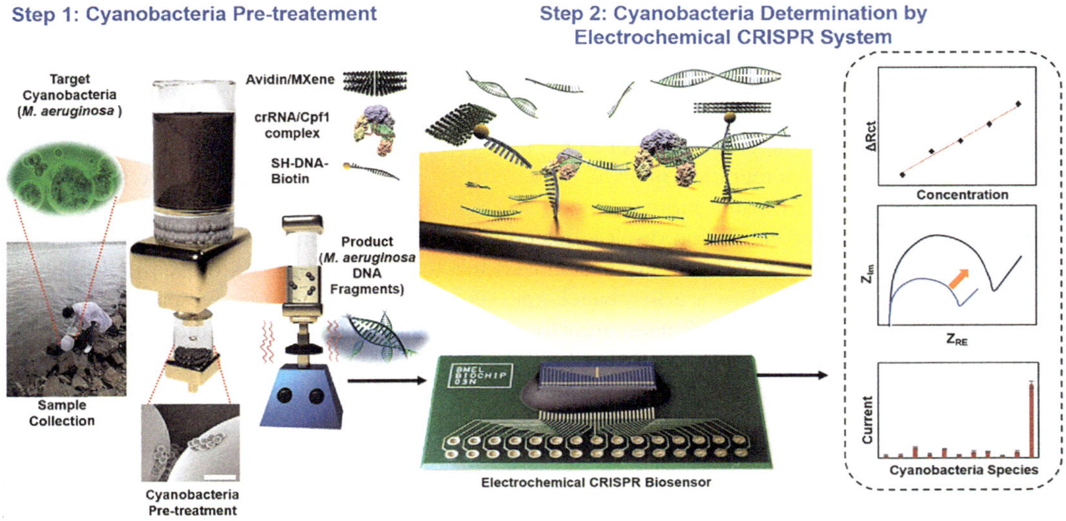

FIGURE 7.8

Detection of cyanobacteria in water samples using electrochemical CRISPR platform.

Reproduced with permission from Kim et al. (2023) copyright 2025 Elsevier B.V.

genetic engineering. Their ability to precisely target DNA sequences has broadened their applicability beyond mere genomic editing. These systems, notably the Cas9, Cas12, and Cas13 enzymes, have been adapted for diagnostic applications, displaying remarkable sensitivity and specificity in nucleic acid detection (Brogan & Akbari, 2022; Kaminski et al., 2021). Cas9 excels in DNA targeting, while Cas12 and Cas13 are distinguished by their ability to induce collateral cleavage in fluorescence reporters upon detecting target DNA and RNA, respectively, facilitating highly sensitive and specific nucleic acid detection. These properties enable their use in various diagnostic contexts, including rapid and accurate identification of infectious diseases. In POC settings, CRISPR diagnostics enable rapid, sensitive, and specific detection of pathogens. A prime example is the development of a Cas13a-based assay for the amplification-free detection of SARS-CoV-2 from nasopharyngeal swabs, integrated with a mobile-phone detection system. This method, detecting positive samples within 30 min, exemplifies cost-effectiveness and high diagnostic accuracy (Fozouni et al., 2021). CRISPR diagnostics can also be paired with electronic readouts. One innovative method utilizes deactivated Cas9 (dCas9) complexed with guide RNA, affixed to a graphene-based field-effect transistor, for detecting genomic DNA alterations. This approach, successfully applied to Duchenne muscular dystrophy cases, bypasses the need for labeled reporter molecules and delivers results in 15 min (Hajian et al., 2019). These examples illustrate the innovative ways in which CRISPR technology is being utilized for POC diagnostics, offering rapid, sensitive, and versatile solutions for a range of applications. Moreover, several CRISPR-based SARS-CoV-2 diagnostic kits have received Emergency Use Authorization (EUA) from the Food and Drug Administration (FDA), marking a significant milestone in the adoption of CRISPR technology in clinical settings (Palaz et al., 2021). Fig. 7.8 illustrates the operational principles of an electrochemical CRISPR platform for detecting cyanobacteria in water samples at 0.089 pg/μL of detection limit, which paves the way for multiple bacterial and biomarkers detection using the CRISPR system (Kim et al., 2023).

7.6 Chapter summary

In conclusion, the chapter provides a comprehensive overview of the rapidly evolving landscape of POC biosensors in disease diagnosis. It delves into the latest technological developments and their applications, highlighting the crucial role of POC biosensors in transforming healthcare delivery. The trends discussed from the integration of POC biosensing components in the innovative strides being made in the development of faster, more accurate, and increasingly accessible diagnostic tools. These trends hold the promise of enhancing disease detection and management in diverse clinical settings, from resource-constrained areas to highly specialized healthcare facilities. As biosensor technology continues to evolve, the path forward involves harnessing these trends to improve patient outcomes, reduce healthcare disparities, and ultimately transform the diagnosis and treatment methods at the POC. It is evident that the future of POC biosensors holds immense potential, and BioMEMs shed light on the path to realizing that potential for the betterment of global healthcare.

References

Aftab, S., Abbas, A., Iqbal, M. Z., Hussain, S., Kabir, F., Akman, E., Xu, F., & Hegazy, H. H. (2023). Recent advances in nanomaterials based biosensors. *TrAC - Trends in Analytical Chemistry, 167*. https://doi.org/10.1016/j.trac.2023.117223.

Aguiar, J. I. S., Rangel, A. O. S. S., & Mesquita, R. B. R. (2023). Salivary calcium determination with a specially developed microfluidic paper-based device for point-of-care analysis. *Talanta Open, 8*. https://doi.org/10.1016/j.talo.2023.100254.

Aslam, M. Z., Zhang, H., Sreejith, V. S., Naghdi, M., & Ju, S. (2023). Advances in the surface acoustic wave sensors for industrial applications: Potentials, challenges, and future directions: A review. *Measurement, 222*, 113657. https://doi.org/10.1016/j.measurement.2023.113657.

Back, M., & Carroll, C. (2023). Principles of pressure transducers, resonance, damping and frequency response. *Anaesthesia & Intensive Care Medicine, 24*(11), 720–724. https://doi.org/10.1016/j.mpaic.2023.08.003.

Bahadir, E. B., & Sezgintürk, M. K. (2016). A review on impedimetric biosensors. *Artificial Cells Nanomedicine and Biotechnology, 44*(1), 248–262. https://doi.org/10.3109/21691401.2014.942456.

Bradley, Z., & Bhalla, N. (2023). Point-of-care diagnostics for sepsis using clinical biomarkers and microfluidic technology. *Biosensors and Bioelectronics, 227*, 115181. https://doi.org/10.1016/j.bios.2023.115181.

Brogan, D. J., & Akbari, O. S. (2022). CRISPR diagnostics: Advances toward the point of care. *Biochemistry*. https://doi.org/10.1021/acs.biochem.2c00051.

Bystryak, S., & Acharya, C. (2016). Detection of HIV-1 p24 antigen in patients with varying degrees of viremia using an ELISA with a photochemical signal amplification system. *Clinica Chimica Acta, 456*, 128–136. https://doi.org/10.1016/j.cca.2016.02.022.

Cady, N. C., Tokranova, N., Minor, A., Nikvand, N., Strle, K., Lee, W. T., Page, W., Guignon, E., Pilar, A., & Gibson, G. N. (2021). Multiplexed detection and quantification of human antibody response to COVID-19 infection using a plasmon enhanced biosensor platform. *Biosensors and Bioelectronics, 171*. https://doi.org/10.1016/j.bios.2020.112679.

Carneiro, M. C. C. G., Rodrigues, L. R., Moreira, F. T. C., & Goreti, M. (2022). Paper-based ELISA for fast CA 15–3 detection in point-of-care. *Microchemical Journal, 181*. https://doi.org/10.1016/j.microc.2022.107756.

Cheen, O. C., Gopinath, S. C. B., Perumal, V., Arshad, M. K. M., Lakshmipriya, T., Chen, Y., Haarindraprasad, R., Rao, B. S., Hashim, U., & Pandian, K. (2017). Aptamer-based impedimetric determination of the human blood clotting factor IX in serum using an interdigitated electrode modified with a ZnO nanolayer. *Microchimica Acta, 184*(1), 117–125. https://doi.org/10.1007/s00604-016-2001-6.

Chen, Y., Zhu, Y., Wang, X., Peng, C., Wang, R., Wang, T., Che, Y., Wu, J., & Xu, J. (2024). Gravity-driven and rotation-controlled microfluidic chip for point-of-care nucleic acid detection in the fully closed environment. *Talanta, 267*, 125258. https://doi.org/10.1016/j.talanta.2023.125258.

Cordeiro, T. A. R., Martins, H. R., Franco, D. L., Santos, F. L. N., Celedon, P. A. F., Cantuária, V. L., de Lana, M., Reis, A. B., & Ferreira,, L. F. (2020). Impedimetric immunosensor for rapid and simultaneous detection of chagas and visceral leishmaniasis for point of care diagnosis. *Biosensors and Bioelectronics, 169*. https://doi.org/10.1016/j.bios.2020.112573.

Correa, V. A., Rodrigues, T. S., Portilho, A. I., Trzewikoswki de Lima, G., & De Gaspari, E. (2021). Modified ELISA for antibody avidity evaluation: The need for standardization. *Biomedical Journal, 44*(4), 433–438. https://doi.org/10.1016/j.bj.2020.10.009.

Cui, W., Ren, Z., Song, Y., & Ren, C. L. (2022). Development and potential for point-of-care heavy metal sensing using microfluidic systems: A brief review. *Sensors and Actuators A: Physical, 344*. https://doi.org/10.1016/j.sna.2022.113733.

Damasceno, B. S., Horta, I. M., de Oliveira, R. S., Pereira, R. M., Schatkoski, V. M., Bacher, G., Massi, M., Thim, G. P., André, A. L., da Silva Sobrinho, A. S., & Leite, D. M. G. (2023). Recent improvements on surface acoustic wave sensors based on graphenic nanomaterials. *Materials Science in Semiconductor Processing, 167*. https://doi.org/10.1016/j.mssp.2023.107811.

Deenin, W., Yakoh, A., Pimpitak, U., Pasomsub, E., Rengpipat, S., Crespo, G. A., & Chaiyo, S. (2023). Electrochemical lateral-flow device for rapid COVID-19 antigen-diagnostic testing. *Bioelectrochemistry, 152*. https://doi.org/10.1016/j.bioelechem.2023.108438.

Dekkers, G., Rosas, F., van Waterschoot, T., Vanrumste, B., & Karsmakers, P. (2022). Dynamic sensor activation and decision-level fusion in wireless acoustic sensor networks for classification of domestic activities. *Information Fusion, 77*, 196–210. https://doi.org/10.1016/j.inffus.2021.07.022.

Deroco, P. B., Wachholz Junior, D., & Kubota, L. T. (2021). Recent advances in point-of-care biosensors for the diagnosis of neglected tropical diseases. *Sensors and Actuators B: Chemistry, 349*. https://doi.org/10.1016/j.snb.2021.130821.

Eivazzadeh-Keihan, R., Bahojb Noruzi, E., Chidar, E., Jafari, M., Davoodi, F., Kashtiaray, A., Ghafori Gorab, M., Hashemi, S. M, Javanshir, S., Ahangari Cohan, R., Maleki, A., & Mahdavi, M. (2022). Applications of carbon-based conductive nanomaterials in biosensors. *Chemical Engineering Journal, 442*, 136183. https://doi.org/10.1016/j.cej.2022.136183.

Fozouni, P., Son, S., Díaz de León Derby, M., Knott, G. J., Gray, C. N., D'Ambrosio, M. V., Zhao, C., Switz, N. A., Kumar, G. R., Stephens, S. I., Boehm, D., Tsou, C. L., Shu, J., Bhuiya, A., Armstrong, M., Harris, A. R., Chen, P. Y., Osterloh, J. M., Meyer-Franke, A., … Ott, M. (2021). Amplification-free detection of SARS-CoV-2 with CRISPR-Cas13a and mobile phone microscopy. *Cell, 184*(2), 323. https://doi.org/10.1016/j.cell.2020.12.001.

Gattani, A., Singh, S. V., Agrawal, A., Khan, M. H., & Singh, P. (2019). Recent progress in electrochemical biosensors as point of care diagnostics in livestock health. *Analytical Biochemistry, 579*, 25–34. https://doi.org/10.1016/j.ab.2019.05.014.

Gopinath, S. C. B., Ramanathan, S., Mohd Yasin, M. N., Shapiai Razak, M. I., Ismail, Z. H., Salleh, S., Sauli, Z., Malarvili, M. B., & Subramaniam,, S. (2022). Failure analysis on silicon semiconductor device materials: optical and high-resolution microscopic assessments. *Journal of Materials Research and Technology, 21*, 3451–3461. https://doi.org/10.1016/j.jmrt.2022.10.116.

Gopinath, S. C. B., Ramanathan, S., More, M., Patil, K., Patil, S. J., Patil, N., Mahajan, M., & Madhavi, V. (2023). A review on graphene analytical sensors for biomarker-based detection of cancer. *Current Medicinal Chemistry,, 31*. https://doi.org/10.2174/0929867331666230912101634.

Gowri, A., Ashwin Kumar, N., & Suresh Anand, B. S. (2021). Recent advances in nanomaterials based biosensors for point of care (PoC) diagnosis of Covid-19 – A minireview. *TrAC - Trends in Analytical Chemistry*. https://doi.org/10.1016/j.trac.2021.116205.

Grossi, M., & Riccò, B. (2017). Electrical impedance spectroscopy (EIS) for biological analysis and food characterization: a review. *Journal of Sensors and Sensor Systems, 6*(2), 303–325. https://doi.org/10.5194/jsss-6-303-2017.

Gumus, E., Bingol, H., & Zor, E. (2023). Lateral flow assays for detection of disease biomarkers. *Journal of Pharmaceutical and Biomedical Analysis*. https://doi.org/10.1016/j.jpba.2022.115206.

Hajian, R., Balderston, S., Tran, T., deBoer, T., Etienne, J., Sandhu, M., Wauford, N. A., Chung, J. Y., Nokes, J., Athaiya, M., Paredes, J., Peytavi, R., Goldsmith, B., Murthy, N., Conboy, I. M., & Aran. , K. (2019). Detection of unamplified target genes via CRISPR–Cas9 immobilized on a graphene field-effect transistor. *Nature Biomedical Engineering, 3*(6), 427–437. https://doi.org/10.1038/s41551-019-0371-x.

Han, C., Xing, W., Li, W., Fang, X., Zhao, J., Ge, F., Ding, W., Qu, P., Luo, Z., & Zhang, L. (2023). Aptamers dimerization inspired biomimetic clamp assay towards impedimetric SARS-CoV-2 antigen detection. *Sensors and Actuators B: Chemical,, 380*. https://doi.org/10.1016/j.snb.2023.133387.

Han, D., Yang, K., Sun, S., & Wen, J. (2023). Signal amplification strategies in electrochemiluminescence biosensors. *Chemical Engineering Journal, 476*, 146688. https://doi.org/10.1016/j.cej.2023.146688.

Hashem, A., Hossain, M. A. M., Marlinda, A. R., Mamun, M. A., Simarani, K., & Johan, M. R. (2021). Nanomaterials based electrochemical nucleic acid biosensors for environmental monitoring: A review. *Applied Surface Science Advances, 4*. https://doi.org/10.1016/j.apsadv.2021.100064.

Hilda, L., Mutlaq, M. S., Waleed, I., Althomali, R. H., Mahdi, M. H., Abdullaev, S. S., Singh, R., Nasser, H. A., Mustafa, Y. F., & Alawadi, A. H. R. (2024). Genosensor on-chip paper for point of care detection: A review of biomedical analysis and food safety application. *Talanta, 268*. https://doi.org/10.1016/j.talanta.2023.125274.

Howard-Quijano, K., Saraf, K., Borgstrom, P., Baek, C., Wasko, M., Zhang, X., Zheng, Y., Saba, S., Mukkamala, R., Kaiser, W., & Mahajan, A. (2023). Evaluation of wearable acoustic sensors and machine learning algorithms for automated measurement of left ventricular ejection fraction. *American Journal of Cardiology, 200*, 87–94. https://doi.org/10.1016/j.amjcard.2023.04.047.

Hu, J., Zhang, Z., Wen, C., Tang, M., Wu, L., Liu, C., Zhu, L., Pang, D., Hu, J., Zhang, Z., Wen, C., Tang, M., Wu, L., Liu, C., & Zhu, L. (2016). Sensitive and quantitative detection of c-reaction protein based on immunofluorescent nanospheres coupled with lateral flow test strip. *Analytical Chemistry, 88*, 6577–6584. https://doi.org/10.1021/acs.analchem.6b01427.

Iravani, S. (2020). Nano- and biosensors for the detection of SARS-CoV-2: challenges and opportunities. *Materials Advances, 1*(9), 3092–3103. https://doi.org/10.1039/d0ma00702a.

Kaminski, M. M., Abudayyeh, O. O., Gootenberg, J. S., Zhang, F., & Collins, J. J. (2021). CRISPR-based diagnostics. *Nature Biomedical Engineering, 5*(7), 643–656. https://doi.org/10.1038/s41551-021-00760-7.

Khan, M., Zhao, B., Wu, W., Zhao, M., Bi, Y., & Hu, Q. (2023). Distance-based microfluidic assays for instrument-free visual point-of-care testing. *TrAC Trends in Analytical Chemistry, 162*, 117029. https://doi.org/10.1016/j.trac.2023.117029.

Kiew, L. V., Chang, C. Y., Huang, S. Y., Wang, P. W., Heh, C. H., Liu, C. T., Cheng, C. H., Lu, Y. X., Chen, Y. C., Huang, Y. X., Chang, S. Y., Tsai, H. Y., Kung, Y. A., Huang, P. N., Hsu, M. H., Leo, B. F., Foo, Y. Y., Su, C. H., Hsu, K. C., … Chang, C. C. (2021). Development of flexible electrochemical impedance spectroscopy-based biosensing platform for rapid screening of SARS-CoV-2 inhibitors. *Biosensors and Bioelectronics, 183*. https://doi.org/10.1016/j.bios.2021.113213.

Kim, G., Li, Y. G., Seo, Y., Baek, C., Choi, J.-H., Park, H., … Lee, T. (2023). Fabrication of graphene oxide-based pretreatment filter and Electrochemical-CRISPR biosensor for the field-ready cyanobacteria monitoring system. *Biosensors and Bioelectronics, 237*, 115474. https://doi.org/10.1016/j.bios.2023.115474.

Kirchhain, A., Bonini, A., Vivaldi, F., Poma, N., & Francesco, F. Di (2020). Latest developments in non-faradic impedimetric biosensors: Towards clinical applications. *TrAC Trends in Analytical Chemistry, 133*, 116073. https://doi.org/10.1016/j.trac.2020.116073.

Ko, D. H., Hosseini, A., Karaosmanoglu, H., Taredun, K., Jones, L., & Partridge, A. (2022). Microfluidic separation of capture from detection and its application for determination of COVID-19 antibodies. *Sensors and Actuators B: Chemical, 351*. https://doi.org/10.1016/j.snb.2021.130918.

Kumar, V., Ghosh, K., Chandran, A., Panwar, S., Bhat, A., Konaje, S., Das, S., Srikanta, S., Jaganathan, L., Prasad, S., Venkatesh, D. B., Shivaram, C., Krishnaswamy, P. R., & Bhat, N. (2021). Multicentric evaluation of a novel point of care electrochemical ELISA platform for SARS-CoV-2 specific IgG and IgM antibody assay. *Journal of Virological Methods, 298*, 114275. https://doi.org/10.1016/j.jviromet.2021.114275.

Kuswandi, B., Hidayat, M. A., & Noviana, E. (2022). Paper-based sensors for rapid important biomarkers detection. *Biosensors and Bioelectronics: X*,. https://doi.org/10.1016/j.biosx.2022.100246.

Lenz, K. D., Klosterman, K. E., Mukundan, H., & Kubicek-Sutherland, J. Z. (2022). Lipoprotein capture ELISA method for the sensitive detection of amphiphilic biomarkers. *Analytical Biochemistry, 652*. https://doi.org/10.1016/j.ab.2022.114747.

Li, H., Long, M., Su, H., Tan, L., Shi, X., Du, Y., Luo, Y., & Deng, H. (2022). Carboxymethyl chitosan assembled piezoelectric biosensor for rapid and label-free quantification of immunoglobulin Y. *Carbohydrate Polymers, 290*, 119482. https://doi.org/10.1016/j.carbpol.2022.119482.

Li, Y., Li, Y., Zhang, R., Li, S., Liu, Z., Zhang, J., & Fu, Y. (2023). Progress in wearable acoustical sensors for diagnostic applications. *Biosensors and Bioelectronics, 237*, 115509. https://doi.org/10.1016/j.bios.2023.115509.

Liu, D., Wang, J., Wu, L., Huang, Y., Zhang, Y., Zhu, M., Wang, Y., Zhu, Z., & Yang, C. (2020). Trends in miniaturized biosensors for point-of-care testing. *TrAC Trends in Analytical Chemistry, 122*, 115701. https://doi.org/10.1016/j.trac.2019.115701.

Liu, J. U, Xie, G., Lv, S., Xiong, Q., & Xu, H. (2023). Recent applications of rolling circle amplification in biosensors and DNA nanotechnology. *TrAC Trends in Analytical Chemistry, 160*, 116953. https://doi.org/10.1016/j.trac.2023.116953.

Ma, T., Ren, S., Wang, Y., Yu, H., Li, L., Li, X. u, Zhang, L., Yu, J., & Zhang, Y. (2023). Paper-based bipolar electrode electrochemiluminescence sensors for point-of-care testing. *Biosensors and Bioelectronics, 235*, 115384. https://doi.org/10.1016/j.bios.2023.115384.

Mahajan, A., Yadav, S., Singh, K., & Sharma, N. N (2023). Biomaterials and biomarkers for urinalysis using point of care testing. *Materials Today: Proceedings*. https://doi.org/10.1016/j.matpr.2023.07.122.

Mandal, D., Indaleeb, M. M., Younan, A., & Banerjee, S. (2022). Piezoelectric point-of-care biosensor for the detection of SARS-COV-2 (COVID-19) antibodies. *Sensing and Bio-Sensing Research, 37*. https://doi.org/10.1016/j.sbsr.2022.100510.

Mello, L. D. (2021). Potential contribution of ELISA and LFI assays to assessment of the oxidative stress condition based on 8-oxodG biomarker. *Analytical Biochemistry, 628*. https://doi.org/10.1016/j.ab.2021.114215.

Millner, P. A., Hays, H. C. W., Vakurov, A., Pchelintsev, N. A., Billah, M. M., & Rodgers, M. A. (2009). Nanostructured transducer surfaces for electrochemical biosensor construction—Interfacing the sensing component with the electrode. *Seminars in Cell & Developmental Biology, 20*(1), 34–40. https://doi.org/10.1016/j.semcdb.2009.01.011.

Moulahoum, H., Ghorbanizamani, F., Beduk, T., Beduk, D., Ozufuklar, O., Celik, E. G, & Timur, S. (2023). Emerging trends in nanomaterial design for the development of point-of-care platforms and practical applications. *Journal of Pharmaceutical and Biomedical Analysis, 235*, 115623. https://doi.org/10.1016/j.jpba.2023.115623.

Mousaabadi, K. Z, Vandishi, Z. T, Kermani, M., Arab, N., & Ensafi, A. A. (2023). Recent developments toward microfluidic point-of-care diagnostic sensors for viral infections. *TrAC Trends in Analytical Chemistry, 169*, 117361. https://doi.org/10.1016/j.trac.2023.117361.

Murray, L. P., & Mace, C. R. (2020). Usability as a guiding principle for the design of paper-based, point-of-care devices – A review. *Analytica Chimica Acta, 1140*, 236–249. https://doi.org/10.1016/j.aca.2020.09.063.

Niu, L., Zhao, F., Chen, J., Nong, J., Wang, C., Wang, J., Gao, N., Zhu, X., Wu, L., & Hu, S. (2018). Isothermal amplification and rapid detection of Klebsiella pneumoniae based on the multiple cross displacement amplification (MCDA) and gold nanoparticle lateral flow biosensor (LFB). *PLoS One, 13*(10). https://doi.org/10.1371/journal.pone.0204332.

Okpara, E. C., Ajiboye, T. O., Onwudiwe, D. C., & Wojuola, O. B. (2023). Optical and electrochemical techniques for Point-of-Care water quality monitoring: A review. *Results in Chemistry, 5*. https://doi.org/10.1016/j.rechem.2022.100710.

Palaz, F., Kalkan, A. K., Tozluyurt, A., & Ozsoz, M. (2021). CRISPR-based tools: Alternative methods for the diagnosis of COVID-19. *Clinical Biochemistry, 89*, 1–13. https://doi.org/10.1016/j.clinbiochem.2020.12.011.

Pei, F., Feng, S., Hu, W., Liu, B., Mu, X., Hao, Q., Cao, Y., Lei, W. U., & Tong, Z. (2023). Sandwich mode lateral flow assay for point-of-care detecting SARS-CoV-2. *Talanta, 253*, 124051. https://doi.org/10.1016/j.talanta.2022.124051.

Peng, P., Liu, C., Li, Z., Xue, Z., Mao, P., Hu, J., Xu, F., Yao, C., & You, M. (2022). Emerging ELISA derived technologies for in vitro diagnostics. *TrAC Trends in Analytical Chemistry, 152*, 116605. https://doi.org/10.1016/j.trac.2022.116605.

Pohanka, M. (2017). The Piezoelectric biosensors: Principles and applications, a review. *International Journal of Electrochemical Science, 12*(1), 496–506. https://doi.org/10.20964/2017.01.44.

Pohanka, M. (2021). Quartz crystal microbalance (QCM) sensing materials in biosensors development. *International Journal of Electrochemical Science, 16*(12), 211220. https://doi.org/10.20964/2021.12.15.

Polat, E. O., Cetin, M. M., Tabak, A. F., Güven, E. B., Uysal, B. Ö., Arsan, T., Kabbani, A., Hamed, H., & Gül, S. B. (2022). Transducer technologies for biosensors and their wearable applications. *Biosensors,, 12*(6). https://doi.org/10.3390/bios12060385.

Ramanathan, S., Gopinath, S. C. B., Arshad, M. d., & M. K. , P. (2019b). Multidimensional (0D-3D) nanostructures for lung cancer biomarker analysis: Comprehensive assessment on current diagnostics. *Biosensors Bioelectronics,, 141*, 111434. https://doi.org/10.1016/j.bios.2019.111434.

Ramanathan, S., Gopinath, S. C. B., & Arshad, M. K. (2020a). Aluminosilicate nanocomposites from incinerated Chinese holy joss fly ash: A potential nanocarrier for drug cargos. *Scientific Reports,, 10*. https://doi.org/10.1038/s41598-020-60208-x.

Ramanathan, S., Gopinath, S. C. B., Arshad, M. K. M., Poopalan, P., & Anbu, P. (2019a). A DNA based visual and colorimetric aggregation assay for the early growth factor receptor (EGFR) mutation by using unmodified gold nanoparticles. *Microchimica Acta,, 186*. https://doi.org/10.1007/s00604-019-3696-y.

Ramanathan, S., Gopinath, S. C. B., Hilmi Ismail, Z., & Subramaniam, S. (2022a). Nanodiamond conjugated SARS-CoV-2 spike protein: electrochemical impedance immunosensing on a gold microelectrode. *Microchimica Acta, 189*. https://doi.org/10.1007/s00604-022-05320-7.

Ramanathan, S., Gopinath, S. C. B., Ismail, Z. H., Md Arshad, M. K., & Poopalan, P. (2022b). Aptasensing nucleocapsid protein on nanodiamond assembled gold interdigitated electrodes for impedimetric SARS-CoV-2 infectious disease assessment. *Biosensors Bioelectronics, 197*. https://doi.org/10.1016/j.bios.2021.113735.

Ramanathan, S., Gopinath, S. C. B., Poopalan, P., & Perumal, V. (2020b). Nanoparticle synthetic methods: Strength and limitations. In S. C. B. Gopinath, & F. Gang (Eds.), *Nanoparticles in analytical and medical devices* (pp. 1–13). Elsevier.

Ramanathan, S., Malarvili, M. B., & Gopinath, S. C. B. (2023). Assessing respiratory complications by carbon dioxide sensing platforms: Advancements in infrared radiation technology and IoT integration. *Arabian Journal of Chemistry, 16*, 104478. https://doi.org/10.1016/j.arabjc.2022.104478.

Selimoğlu, F., Gür, B., Ayhan, M. E., Gür, F., Kalita, G., Tanemura, M., & Alma, M. H. (2023). Silver nanoparticle doped graphene-based impedimetric biosensor towards sensitive detection of procalcitonin. *Materials Chemistry and Physics, 297*. https://doi.org/10.1016/j.matchemphys.2023.127339.

Shafiee, A., Ghadiri, E., Kassis, J., Pourhabibi Zarandi, N., & Atala, A. (2018). Biosensing technologies for medical applications, manufacturing, and regenerative medicine. *Current Stem Cell Reports, 4*(2), 105–115. https://doi.org/10.1007/s40778-018-0123-y.

Sharma, P., Batheja, G., Verma, H. N., & Seth, P. (2022). Development of a novel molecular assay based on conventional PCR liquid hybridization and ELISA (NAT-ELISA) for the detection of HIV, HCV, and HBV in blood donors. *Indian Journal of Medical Microbiology, 40*(1), 122–126. https://doi.org/10.1016/j.ijmmb.2021.12.014.

Shimizu, F. M., de Barros, A., Braunger, M. L., Gaal, G., & Riul, A. (2023). Information visualization and machine learning driven methods for impedimetric biosensing. *TrAC - Trends in Analytical Chemistry, 165*. https://doi.org/10.1016/j.trac.2023.117115.

Shoute, L. C. T., Abdelrasoul, G. N., Ma, Y., Duarte, P. A., Edwards, C., Zhuo, R., Zeng, J., Feng, Y., Charlton, C. L., Kanji, J. N., Babiuk, S., & Chen, J. (2023). Label-free impedimetric immunosensor for point-of-care detection of COVID-19 antibodies. *Microsystems and Nanoengineering, 9*(1). https://doi.org/10.1038/s41378-022-00460-5.

Singh, M. M., & Satija, J. (2022). Enzyme-assisted metal nanoparticles etching based plasmonic ELISA: Progress and insights. *Analytical Biochemistry, 654*. https://doi.org/10.1016/j.ab.2022.114820.

Soh, J. H., Chan, H. M., & Ying, J. Y. (2020). Strategies for developing sensitive and specific nanoparticle-based lateral flow assays as point-of-care diagnostic device. *Nano Today*. https://doi.org/10.1016/j.nantod.2019.100831.

Sumitha, M. S., & Xavier, T. S. (2023). Recent advances in electrochemical biosensors – A brief review. *Hybrid Advances, 2*, 100023. https://doi.org/10.1016/j.hybadv.2023.100023.

Tavakoli, H., Mohammadi, S., Li, X., Fu, G., & Li, X. J. (2022). Microfluidic platforms integrated with nanosensors for point-of-care bioanalysis. *TrAC - Trends in Analytical Chemistry*. https://doi.org/10.1016/j.trac.2022.116806.

Thompson, E. J., Foote, H. P., Hill, K. D., & Hornik, C. P. (2023). A point-of-care pharmacokinetic/pharmacodynamic trial in critically ill children: Study design and feasibility. *Contemporary Clinical Trials Communications, 35*. https://doi.org/10.1016/j.conctc.2023.101182.

Tiwari, A. K., Upadhyay, A. P., Arora, D., Wadhwa, T., Aggarwal, G., Pabbi, S., Luthra, A. S., & Rawat, S. S. (2020). Head-to-head comparison of Enzyme Linked Immunosorbent Assay (ELISA) and Enhanced Chemiluminescence Immunoassay (ECLIA) for the detection of Transfusion Transmitted Disease (TTD) Markers; HIV, HCV and HBV in blood donors, in India. *Journal of Virological Methods, 285*. https://doi.org/10.1016/j.jviromet.2020.113962.

Tiwari, R. P., Jain, A., Khan, Z., Kumar, P., Bhrigu, V., & Bisen, P. S. (2013). Designing of novel antigenic peptide cocktail for the detection of antibodies to HIV-1/2 by ELISA. *Journal of Immunological Methods, 387*(1-2), 157–166. https://doi.org/10.1016/j.jim.2012.10.009.

Tram, D. T. N., Wang, H., Sugiarto, S., Li, T., Ang, W. H., Lee, C., & Pastorin, G. (2016). Advances in nanomaterials and their applications in point of care (POC) devices for the diagnosis of infectious diseases. *Biotechnology Advances, 34*(8), 1275–1288. https://doi.org/10.1016/j.biotechadv.2016.09.003.

Wang, Y., Li, B., Tian, T., Liu, Y., Zhang, J., & Qian, K. (2022). Advanced on-site and in vitro signal amplification biosensors for biomolecule analysis. *TrAC Trends in Analytical Chemistry, 149*, 116565. https://doi.org/10.1016/j.trac.2022.116565.

Wang, Y., Rong, Y., Ma, T., Li, L., Li, X. u, Zhu, P., Zhou, S., Yu, J., & Zhang, Y. (2023). Photoelectrochemical sensors based on paper and their emerging applications in point-of-care testing. *Biosensors and Bioelectronics, 236*, 115400. https://doi.org/10.1016/j.bios.2023.115400.

Wang, Y., Shi, Y., & Narita, F. (2021). Design and finite element simulation of metal-core piezoelectric fiber/epoxy matrix composites for virus detection. *Sensors and Actuators A: Physical,, 327*. https://doi.org/10.1016/j.sna.2021.112742.

Wei, L. N., Luo, L., Wang, B. Z., Lei, H. T., Guan, T., Shen, Y. D., Wang, H., & Xu, Z. L. (2023). Biosensors for detection of paralytic shellfish toxins: Recognition elements and transduction technologies. *Trends in Food Science and Technology, 133*, 205–218. https://doi.org/10.1016/j.tifs.2023.02.012.

Xu, J., Wang, M., Li, M., Yang, J., & Yang, L. (2023). Paper-based chiral biosensors using enzyme encapsulation in hydrogel network for point-of-care detection of lactate enantiomers. *Analytica Chimica Acta,, 341834*. https://doi.org/10.1016/j.aca.2023.341834.

Xu, M., Lin, L., Jin, G., Lin, Y., & Zhang, K. (2022). Two-in-one: Portable piezoelectric and plasmonic exciton effect-based co-enhanced photoelectrochemical biosensor for point-of-care testing of low-abundance cancer markers. *Biosensors and Bioelectronics, 211*, 114413. https://doi.org/10.1016/j.bios.2022.114413.

Xu, W., Althumayri, M., Mohammad, A., & Ceylan Koydemir, H. (2023). Foldable low-cost point-of-care device for testing blood coagulation using smartphones. *Biosensors Bioelectronics, 242*. https://doi.org/10.1016/j.bios.2023.115755.

Yang, L., Gu, X., Liu, J., Wu, L. i, & Qin, Y. (2024). Functionalized nanomaterials-based electrochemiluminescent biosensors and their application in cancer biomarkers detection. *Talanta, 267*, 125237. https://doi.org/10.1016/j.talanta.2023.125237.

Yoon, Y., Jo, J., Kim, S., Lee, I. G., Cho, B. J., Shin, M., & Hwang, W. S. (2017). Impedance spectroscopy analysis and equivalent circuit modeling of graphene oxide solutions. *Nanomaterials, 7*(12). https://doi.org/10.3390/nano7120446.

Yuksel, M., Luo, W., McCloy, B., Mills, J., Kayaharman, M., & Yeow, J. T. W. (2023). A precise and rapid early pregnancy test: Development of a novel and fully automated electrochemical point-of-care biosensor for human urine samples. *Talanta, 254*. https://doi.org/10.1016/j.talanta.2022.124156.

Zambry, N. S., Awang, M. S., Beh, K. K., Hamzah, H. H., Bustami, Y., Obande, G. A., Khalid, M. F., Ozsoz, M., Manaf, A. A., & Aziah, I. (2023). A label-free electrochemical DNA biosensor used a printed circuit board gold electrode (PCBGE) to detect SARS-CoV-2 without amplification. *Lab on a Chip,, 23*(6), 1622–1636. https://doi.org/10.1039/d2lc01159j.

Zeng, J., Duarte, P. A., Ma, Y., Savchenko, O., Shoute, L., Khaniani, Y., Babiuk, S., Zhuo, R., Abdelrasoul, G. N., Charlton, C., Kanji, J. N., Babiuk, L., Edward, C., & Chen, J. (2022). An impedimetric biosensor for COVID-19 serology test and modification of sensor performance via dielectrophoresis force. *Biosensors and Bioelectronics, 213*. https://doi.org/10.1016/j.bios.2022.114476.

Zhang, J., Lakshmipriya, T., & Gopinath, S. C. B. (2020). Electroanalysis on an interdigitated electrode for high-affinity cardiac troponin i biomarker detection by aptamer-gold conjugates. *ACS Omega, 5*, 25899–25905. https://doi.org/10.1021/acsomega.0c03260.

Zhou, Y., Wu, Y., Ding, L. u, Huang, X., & Xiong, Y. (2021). Point-of-care COVID-19 diagnostics powered by lateral flow assay. *TrAC Trends in Analytical Chemistry, 145*, 116452. https://doi.org/10.1016/j.trac.2021.116452.

CHAPTER 8

BioMEMS in medical implants

Mohd H.S. Alrashdan
Department of Electrical Engineering, College of Engineering, Al-Hussein Bin Talal University, Ma'an, Jordan

8.1 Introduction

In the rapidly evolving landscape of healthcare, a unique fusion of biological elements and cutting-edge technology has given rise to a novel field known as biological micro-electro-mechanical systems (BioMEMS). BioMEMS comprise minuscule device that skillfully combine biological components found within living organisms with miniature electronic and mechanical elements. This emerging technology is helping in a new era for medical implant. This introductory section provides an overview of the historical context, significant relevance, diverse applications, and advanced technologies that support the functionality of BioMEMS within the domain of medical implants. BioMEMS originated in the 1980s when researchers began blending miniature technology with biological components. The concept revolved around creating exceptionally small systems capable of interacting with living organism. Over time, this concept has grown and is currently applied in the development of medical implant designed to help in the diagnosis, monitoring, and treatment of various health conditions.

The history of BioMEMS for medical implants is marked by remarkable invention. During the 1980s and 1990s, particularly at institutions such as the University of Michigan, scientists pioneered the development of minuscule silicon-based instrument known as microelectrode arrays (MEAs). These tiny devices enabled researcher to study and even communicate with the nerve cells in the brain. They have found applications in both experimental studies and medical intervention. During the late 1990s and early 2000s, researchers introduced compact machine capable of regulating and monitoring the administration of medications within the body. These microfluidic devices enable precise drug delivery, ensuring that the right dosage is dispensed when required. In the case of human with diabetes, the development of minuscule insulin pump has been ongoing since the 1990s. These devices are designed to enhance the efficiency and comfort of insulin administration for those in need. As technology advances, BioMEMS for medical implant continues to evolve, becoming increasingly sophisticated and practical. Scientist continually devise innovative ways to apply these tiny tool in the realm of healthcare. BioMEMS is a collaborative effort that draws on knowledge from various fields, including microengineering, materials science, biology, and medicine. While the initial BioMEMS tool for medical purpose began emerging in the late 20th century, the field has since made significant progress, providing an array of implantable device to address diverse medical issues (Grayson et al., 2004).

BioMEMS Devices. DOI: https://doi.org/10.1016/B978-0-443-24015-7.00008-5

BioMEMS for medical implants hold huge significance in healthcare. They possess the potential to transforms healthcare by improving patient healthcare, enhancing treatment comfort, and tailoring medical care to individual need. These little device are capable of continuous body monitoring, precise medication administrations, and seamless integration with the body's natural systems. Therefore, they empower medical specialists to develop highly personalized treatment that are optimally effective for each patient.

BioMEMS finds application in various medical scenario. They help us connect electronics to our nervous system, which can be a helpful in treating issues like paralysis and epilepsy. These tiny devices also help us give medicine with super accuracy, so we get better without many side effect. They're also really helpful in growing new body part outside of our bodies, like tissues and organs. This can give hopes to people waiting for organ transplants.

The technology behind BioMEMS relies heavily on microfabrication techniques, often involving silicon and other materials at the microscale. These devices are meticulously designed and engineered to ensure biocompatibility and durability within the human body. Furthermore, BioMEMS incorporates microsensors, microactuators, and microcontrollers to capture, process, and transmit big data while interacting with the surrounding biological environment. Fig. 8.1 below show some of the Implantable BioMEMS devices.

In this chapter, we will explore deeper into the multifaceted world of BioMEMS in medical implants, exploring their applications in detail, their fabrication methods, and their potential to shape a future where implantable devices are not just life saving but also adaptive, responsive, and individually tailored to the special needs of each patient Fig. 8.1.

8.2 Implantable hearing aids

Micro-Electro-Mechanical Systems (MEMS) acoustic sensors have played a transformative role in the evolution of hearing-aid technology, especially in the development of fully implantable hearing-aid systems. Their extremely small dimensions and high sensitivity make them ideal for device proposed to remain within the human body for long periods of time (Ko et al., 2009). Because these sensors can detect minute acoustic signals with precision, they serve as essential component in modern implantable designs that aim to deliver natural, stable, and high quality auditory perception. This section presents a full explanation of the characteristics, function, and operational principle of MEMS acoustic sensor within totally implantable hearing aid systems.

MEMS acoustic sensor technology is based on the integration of microelectronics with advanced microfabrication technique that allow engineers to creates small mechanical structures on silicon substrates. This manufacturing approach makes it possible to design sensor that are extremely compact, energy efficient, and finely tuned to respond perfectly to acoustic input. The accuracy of MEMS fabrication have main rule in producing sensor that satisfy the strict performance requirement of implantable hearing devices.

The major advantage of MEMS acoustic sensor is the miniature size, this allow them to be implanted in the ear canal or various region of the auditory system without causing discomfort or interfering with the ear natural anatomy. The compact form enable fully implantable hearing aids to remains discreet and almost invisible, this improve wearer comfort and eliminating the cosmetic concern associated with traditional external hearing aids. Also, MEMS acoustic sensor offer excellent sensitivity and

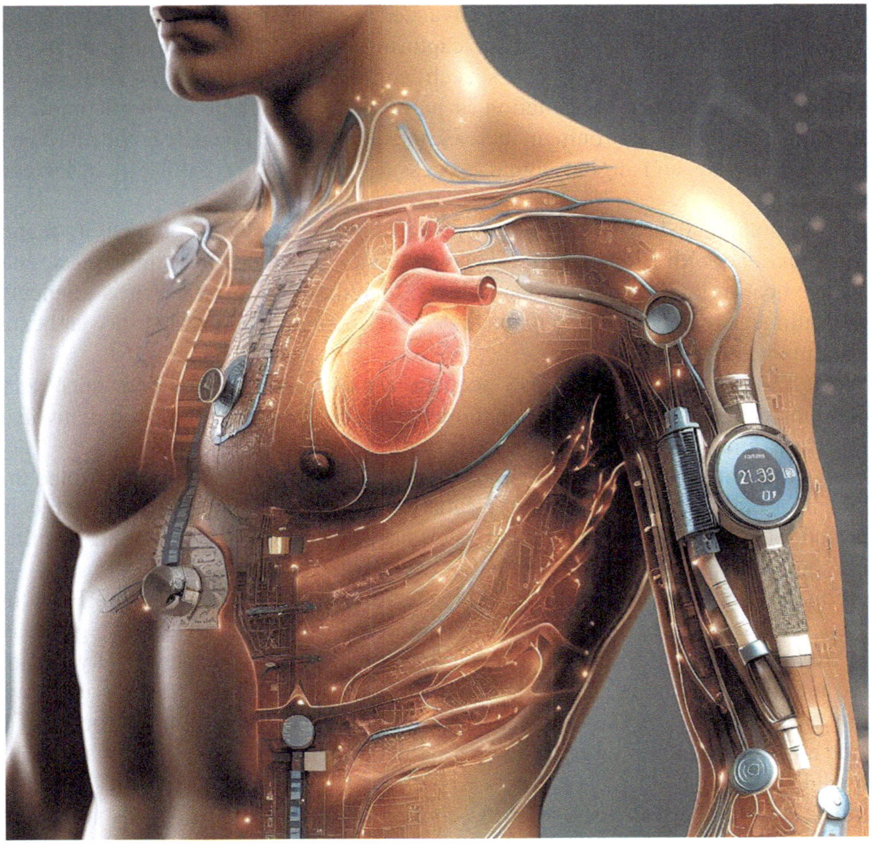

FIGURE 8.1

Implantable BioMEMS devices.

performance. They can detect a wide range of acoustic frequencies, and their sensitivity can be precisely calibrated according to the type and degree of hearing impairment experienced by the user. This ability to adapt sensor performance ensure a more natural and personalized auditory experience.

Low power consumptions is another critical attribute of MEMS acoustic sensor. Implanted device must work continuously without frequent battery replacement, making energy efficiency essential. MEMS sensor operate with minimal power, contributing to longer battery life and supporting the long term operation required in fully implantable hearing systems.

Biocompatibility is also a central consideration in the design of implantable hearing aid. All component must be compatible with human tissue to avoid irritation, inflammation, or long term biological complication. MEMS sensor can be fabricated from material that are well tolerated by the body, reducing the risk of adverse reactions. Their structure can be protected with biocompatible coatings to ensure safe and stable integrations within the moist, temperature variable environment of the ear.

MEMS acoustic sensor work closely with advanced signal processing algorithms to improve the quality of sound delivered to the user. After capturing sound, the sensor output is processed to remove unwanted background noise, clarify speech, and prevent feedback that may causes whistling or squealing. Effective feedback cancellation is particularly important for implantable hearing aid because the compact, enclosed environment increase the likelihood of acoustic feedback. The interaction between MEMS hardware and digital signal processing significantly enhance clarity and listening comfort.

Modern fully implantable hearing aid system also incorporate wireless connectivity feature. MEMS sensor can be integrated with Bluetooth or other wireless communication technologies, enabling user to connects their hearing aid to smartphone, tablets, or remote control. Wireless connectivity allow user to adjust settings, stream audio, and manage device function, enhance usability and expanding the capability of implantable hearing systems.

The longevity and reliability of MEMS acoustic sensors essential for their success in medical implant. These sensor designed to remains functional in challenging environment that involve moisture, fluctuating pressure, and continuous mechanical vibration.

Another advantage of MEMS based design is the ability to be customized and adapted to individual user. MEMS sensor optimized for different type and level of hearing loss by adjusting mechanical and electrical properties. This customization allow each user to receive tailored sound amplification that closely match their hearing profiles and improving overall effectiveness.

MEMS acoustic sensor have driven major advancement in hearing aid design. Fully implantable hearing aid become smaller, less noticeable, and more capables of delivering high quality sound. Compared to traditional device, MEMS enabled systems offers superior sound quality, greater comfort, and increased reliability. These improvement enhanced the quality of life for individual with hearing impairment by providing natural, stable, and personalized auditory support. Fig. 8.2 below (Chung et al., 2021), provide an in depth exploration of the operational principles behind MEMS acoustic sensors in these hearing aid systems.

Beyond their structural advantages, MEMS acoustic sensor operate through a series of functional principle that allow them to captures, convert, process, and transmit sound efficiently. The operational process begin with sound capture. Similar to a conventional microphone, the MEMS acoustic sensor contain a tiny diaphragm that vibrate when exposed to sound wave. These vibrations converts acoustic energy into mechanical motion, forming the first step in the signal acquisition process.

To transform these mechanical vibrations into electrical signal, MEMS acoustic sensors may use either piezoelectric or capacitive transduction. Piezoelectric sensor utilize diaphragms made from material that produce electrical charge when mechanically deformed. As the diaphragm vibrate in response to sound, it generate electrical signals that accurately represent the acoustic input. In contrast, capacitive MEMS sensor employ a diaphragm positioned very close to a fixed back plate. Sound induced diaphragm motion change the capacitance between the two surfaces, and these capacitance variations creates electrical voltage changes corresponding to the incoming sound.

The initial electrical signal produced by the sensors is in analog form. In order for the hearing aid to process, optimize, and manipulates this signal, it must be converted into digital data. An analog to digital converter (ADC) performs this task by sampling the continuous analog signal and translating it into discrete digital information. Once the signal is digitized, it can be processed by the hearing aid digital signal processing (DSP) unit.

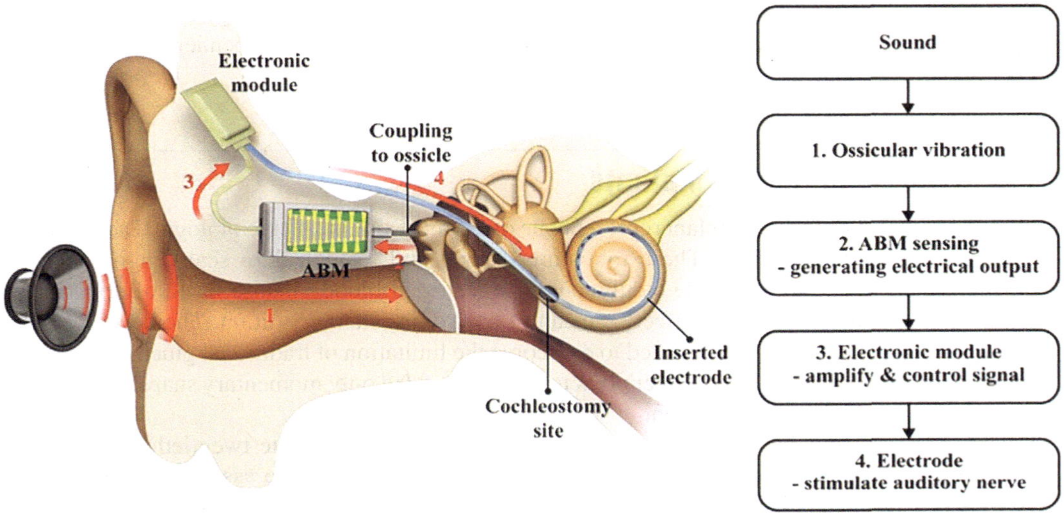

FIGURE 8.2

Implantable hearing aids system. A schematic drawing of the entire Implantable Hearing Aids system.

(Reprinted from Chung et al. (2021), open access.)

Digital signal processing plays a central role in refining the sound delivered to the user. The DSP algorithm analyze the digital audio signal and apply operation such as noise reduction, amplifications, equalization, and automatic feedback cancellation. These algorithms adjust the output according to the user hearing profile and the acoustic environment, ensuring a comfortable listening experience. DSP also enhance the signal to noise ratio by reducing unwanted noise and emphasizing speech or other important sounds.

After processing, the refined digital audio signal is amplified to a levels that meets the user specific hearing requirement. This amplification stage is customizable and ensure that the output remains clear, stable, and comfortable. In fully implantable systems, the amplified signal is then converted back into sound wave or mechanical vibration through component such as bone conduction transducer or miniature implantable speakers. These devices deliver sound directly to the inner ear, bypassing the outer and middle ear, and providing effective auditory support for user with various types of hearing loss.

In many modern implantable system, wireless connectivity enhance the sensor functionality by enabling remote adjustment and audio streaming. MEMS sensors can transmit processed sound or device data wirelessly to external devices. Users can control setting, adjust volume, or stream audio without the need for physical interaction with the implant. This capability significantly enhance flexibility, convenience, and overall user satisfaction.

In summary, MEMS acoustic sensor serve as the core building blocks of fully implantable hearing aid system. Their compact size, high sensitivity, low power consumption, biocompatibility, and reliability make them ideal for long term operation within the human body. By combining precises mechanical transduction with advanced digital processing and wireless communication, MEMS based hearing aid offer natural, personalized, and durable auditory enhancement. These technology continues to propel innovation in hearing rehabilitation, improving sound quality, comfort, and independence for individuals with hearing impairment

8.3 Glucose monitoring implants

BioMEMS glucose monitoring implant represent one of the most significant technological advancement in modern diabetes management. These miniature device, built using micro scale engineering and biological sensing technology, enable continuous monitoring of blood glucose level, offering patient a safer, more convenient, and more effective mean of controlling their condition. The development of these implant stems from the growing need to overcome the limitation of traditional glucose monitoring techniques, which required frequent finger prick tests and provided only momentary snapshot of glucose level rather than continuous information.

The origin of glucose monitoring implant can be traced back to the late twentieth century, when scientist began exploring ways to reduce the discomfort and inconvenience associated with repeated finger stick measurement. Traditional method required to obtain a drop of blood several times a day, which is both painful and insufficient for detecting sudden change in glucose level. This limitation inspired researcher to investigates the use of micro scale technology to create small and implantable device capable of continuously tracking blood sugar. A major milestone occurred in the early 2000s with the introduction of the first generation of implantable glucose sensor. These early system employed biochemical detection methods to measure glucose but faced challenges related to accuracy, longevity, and stability once placed inside the human body (Birkholz et al., 2013). Despite these limitation, they laid the groundwork for the highly advanced BioMEMS based systems used today.

BioMEMS glucose monitoring implant operate through a combination of biological sensing, electrical signal conversion, and wireless communication, Fig. 8.3 shows the Implantable Glucometer. And it's working principles.

At the core of the implants is a tiny biosensor that detects glucose concentrations in bodily fluid. This is usually achieved through the enzyme glucose oxidase, which reacts with glucose and produces by products that can be measured electrically. The biochemical reactions produces an electrical signal proportional to the glucose level present in the surrounding fluid. This signal is then converted into a measurable electrical output, such as a current or voltage, through a process known as signal transduction.

Once generated, the electrical signal undergoes processing inside the implant. A miniature processor evaluate the data, filter out unwanted noise, and ensure the accuracy of the measurement. This internal data processing stage is essential for maintaining consistent and reliable glucose reading, especially considering the dynamic environments within the human body. In many modern implant, the processed data is transmitted wirelessly to an external devices, such as a smartphone or medical receiver. The wireless connection enable patient and clinician to monitor glucose level in real time, rapid decision making and improved management of diabetes can be obtained. Continuous access to such data allows patient to regulate their diet, insulin dosage, and overall lifestyle more effectively. Healthcare provider can also review historical data to tailor treatment plan more precisely (Zafar et al., 2022).

BioMEMS glucose monitoring implants offers several key advantage over traditional glucose testing methods. Continuous data collection provide a more comprehensive understanding of glucose

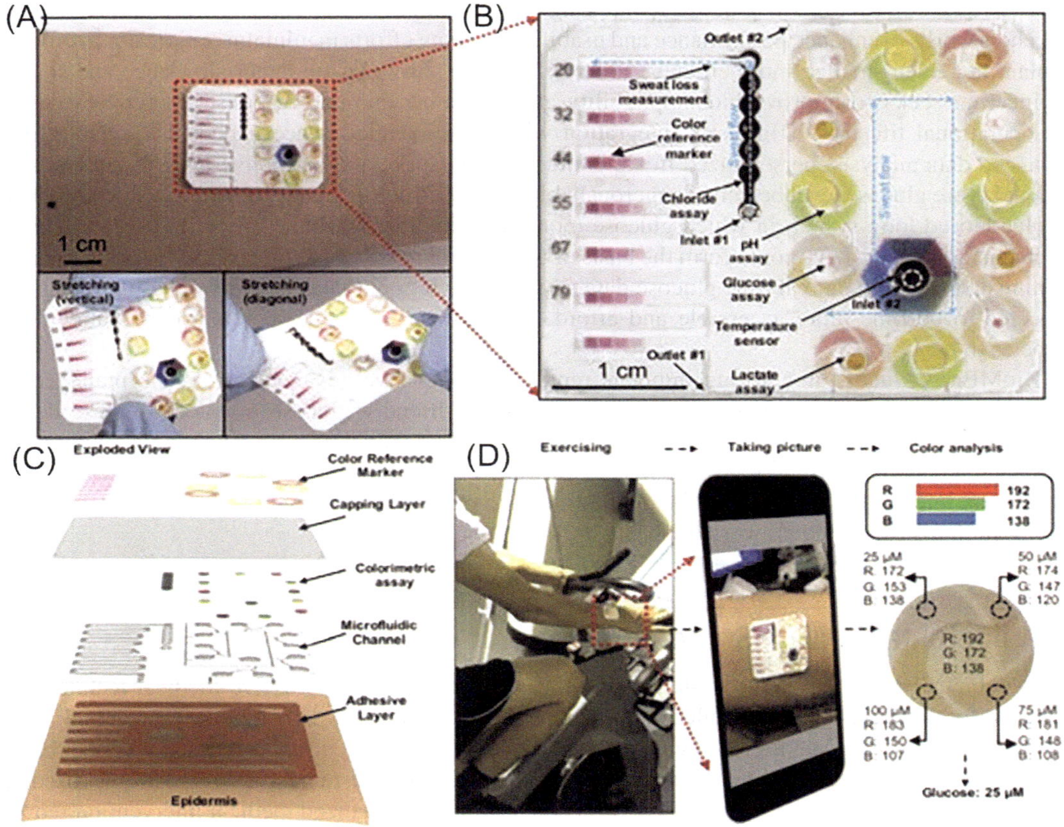

FIGURE 8.3

Implantable glucometer. (A) Optical images of the flexible and soft microfluidic device designed for colorimetric sweat analysis on the skin, showing its ability to undergo mechanical deformations such as bending (bottom left) and twisting (top left and bottom right); (B) microfluidic channels filled with blue-colored water, viewed from above; (C) an exploded view illustrating the device structure and its interface with the skin; (D) the process of sweat sample collection and the subsequent color-based analysis using digital images of the device.

(Reprinted from Zafar et al. (2022), open access.)

fluctuations during the day, allowing the detection of pattern that would otherwise go unnoticed. Eliminating the need for frequent finger stick test greatly reduce discomfort, particularly benefiting individuals with type 1 diabetes who requires multiple measurements daily. The accuracy of these implant generally higher than that of conventional techniques, reducing the risk of dangerous episode of hypoglycemia or hyperglycemia. Continuous automated monitoring also reduce the possibility of human error, as measurement taken without user intervention. And the remote data accessibility allow clinicians to monitor patient health and adjust treatment plan without requiring frequent in person visits.

The future of BioMEMS glucose monitoring implant filled with promising developments as researchers works to enhance performance and usability. ongoing efforts in miniaturization aim to produce implant that even smaller, more efficient, and less invasive. Advances in material science and surface engineering seeks to improve biocompatibility and minimize immune response, thereby extending the operational lifespan of implant. Integration with artificial intelligence expected to enable more advanced data analysis, personalized treatment recommendation, and predictive insight that can alert patient before glucose level become dangerously high or low. Another major possibility of development involve closed loop system in which glucose monitoring implant work directly with insulin delivery mechanism. Such systems could form the basis of an artificial pancreas, automatically adjusting insulin administration based on real time glucose data. as technologies continues to evolve, these implant also expected to become more accessible and affordable, allowing a broader population of patient with diabetes to benefit.

BioMEMS glucose monitoring implants represents innovative advancement in the management of diabetes. By providing continuous, accurate, and user friendly glucose monitoring, these device significantly improve disease control and reduces the risk of complication. As innovation in microfabrication, biotechnology, artificial intelligence, and wireless communication continue to advance, glucose monitoring implant hold the potential to revolutionize diabetes care and greatly enhances the quality of life for millions of individual worldwide

8.4 Implantable MEMS micropump

implantable MEMS micropump system represent one of the most progressive medical technologies designed for the precise and controlled administration of medication directly into the human body. This devices capable to deliver highly regulated doses of therapeutic agent and hold significant promise across a broad spectrums of clinical application, these include long term treatment for chronic diseases, targeted cancer therapy, pain management protocol, and other specialized medical intervention require continuous or programmable drug delivery. By combining microfabrication technique with biomedical engineering, MEMS based micropumps emerged as powerful tool capable of improving treatment accuracy and patient comfort (Kalra & Mashuq-Un-Nabi, 2018). This section provide a detail overview of the essential component of such system, their operational mechanisms, the benefit they offer, and the challenge involved in their design and implementation.

The core component of an implantable micropump system is the MEMS based pumping unit itself. its tiny mechanical device designed to move precise quantities of fluid, such as pharmaceuticals, through miniature channel or reservoirs. its fabricated from silicon through sophisticated microfabrication processes, the micropump may adopts several structural design, including diaphragm driven pumps, piston driven pumps, or peristaltic mechanism. Each design configuration offer distinct advantages depending on the therapeutic application and the required dosing precision. medication reservoir which serve as the storage unit for the delivered drug. The reservoir constructed from biocompatible material to ensure patient safety and prevent adverse reaction with human body fluid. Depending on clinical need, the reservoir designed as a refillable chamber or as a replaceable components that can be exchanged periodically.

Flow control valve have an essential role within the system by regulating the movement of medications from the reservoir through the pump and toward the targeted delivery sites. This microvalve, controlled electronically, enable highly accurates modulation of fluid flow, allow the system to dispense precise micro volumes of medication needed. In more advanced system, sensory component are integrated to monitor various physiological and environmental parameters. These sensors may track the patient health status, the concentration of the drug in the bloodstream, and other factor that could influence delivery efficiency. Information gathered from the sensor used in real time to adjusts dosing schedules or delivery rates.

Managing all these components requires a sophisticated electronic control architecture. The device contain miniaturized circuit responsible for operating the pump, valve, and sensors, along with a power source that consists of a battery or an energy harvesting unit. The electronics subsystem ensures seamless coordination among device component and support communication with external controller when necessary.

The operational sequence of an implantable MEMS micropump begin with loading the medication into the reservoir. This step performed at the time of surgical implantation or during later procedure designed to replenish the drug supply. When medication delivery required, the device control electronics activate the micropump and its associated valve, allowing the programmed amount of drug to be extracted from the reservoir and delivered to the intended physiological sites. The sensors continuously monitor the patient conditions and measure drug concentration level in real time. This data enable automatic adjustment of dosing parameters, ensuring precise and personalized therapy.

Implantable MEMS micropump systems offers several significant advantages in modern medicine. the ability to deliver medication with exceptional precision reduce the risks of both overdosing and underdosing, a crucial factor when administering potent or narrow therapeutic window drug. The ability to target delivery to specific tissue or organ minimize unwanted systemic side effect. implantable system eliminate the need for frequent injection or oral dosing, thereby improving patient comfort and treatment adherence. The inclusion of sensors and adaptive control mechanism enables personalized therapy that can dynamically respond to changing clinical condition. The overall minimally invasive nature of these system further enhance their appeal, it require only a single implantation procedure followed by remote or automated operation.

Implantable MEMS micropump system face several important challenge. Ensuring biocompatibility is critical, all device material must be safe for long term contact with bodily tissues and fluids. Device longevity also essential, as the system must remain functional for extended duration without frequent surgical replacements. Powering the device is another technical difficulty, the micropump and control electronics requires reliable energy sources. Possible solution include micro batteries, wireless power transfer, or energy harvesting method. Additionally, because all implantable device carry natural infection risk, strict design and sterilization protocols must be implemented to minimize the possibility of infection. Regulatory approval present another substantial challenges, these systems must undergo extensive testing to confirm biocompatibility, reliability, safety, and clinical effectiveness before they can be used in human patients.

The creation of an implantable MEMS micropump system involve a highly precise and multistage fabrication process. The journey begin with the d esign and configuration of the micropump architecture. Engineers define the arrangement of microcomponents such as the pump, valves, reservoir, and sensors, optimizing the structural layout to meet t he targeted drug delivery requirements and implantation constraints (Gensler et al., 2012). Selecting suitable material is critical to ensuring device performance and biocompatibility. Silicon remain the most widely used substrate due to its compatibility with microfabrication processe, although additional biocompatible coating or polymers may be incorporated to improve safety and durability.

Microfabrication techniques form the backbone of the manufacturing process. Photolithography is used to create patterns on silicon wafers coated with photosensitive material, defining the shapes and boundaries of pump component, valves, and flow channels. Etching processes, both wet and dry, remove selected areas of material to create three dimensional structures essential for pump operation. Deposition techniques such as chemical vapor deposition (CVD) and physical vapor deposition (PVD) apply thin films that form structural layer or functional coating. Multiple wafer layer may be joined through bonding method including adhesive and anodic bonding to produce integrated microfluidic structures.

Fabricating the micropump itself require shaping the diaphragm, designing the valve mechanisms, and implementing actuation strategy that enables controlled movement of the diaphragm or other mechanical component. Reservoir fabrication employ similar methods and often incorporate features allow refilling or controlled drug release. Sensors and control electronics integrated on the device, including the placement of miniature pressure or flow sensors and the bonding of electronic control circuits.

Rigorous testing and quality assurance procedure verify the performance of each component is nessecary after the device fabrication. This include assessing pumping accuracy, checking for f liquid leak, evaluating valve function, and confirming the reliability of the electronics. The entire device must be encapsulated within biocompatible material to protect it from bodily fluids and ensure long term operation. Before implantation, the device undergoe sterilization using method compatible with its materials and construction, such as ethylene oxide treatments, autoclaving, or gamma irradiation. Then the device include additional accessories such as catheters or external connectors to support clinical use. after sterilization process, the implantable MEMS micropump system undergo surgical implantation and subsequent clinical evaluation. these trial assess not only the technical performance of the device but also patient safety, therapeutic effectiveness, and long term reliability. The complexity of fabricating such system require expertise in microfabrication, material science, and biomedical engineering. However, the outcome is a sophisticated medical platform capables of delivering precise and targeted drug therapy with significant potential to improve treatment outcomes across numerous clinical applications (Au et al., 2011; Bußmann et al., 2021) (Figs. 8.4 and 8.5).

8.5 Fully implantable MEMS cochlea

Creating a fully implantable MEMS cochlea is specialized task require extensive knowledge in biomedical engineering, microfabrication, material science, and auditory physiology. MEMS cochlea is a miniature biomedical device designed to emulates the essential function of the human cochlea, a spiral shaped structure of the inner ear responsible for converting sound wave into electrical signal for auditory perception. Fully implantable MEMS cochlear system are primarily for patient with profound hearing impairment, offers a potential pathway to restore auditory function. The following section presents the detailed design considerations, fabrication steps, and testing procedure involved in the development of a fully implantable MEMS cochlear implant.

The design process begin with the careful selection of biocompatible material suitable for long term implantation within the human body. Common material used in MEMS structures include silicon, silicon

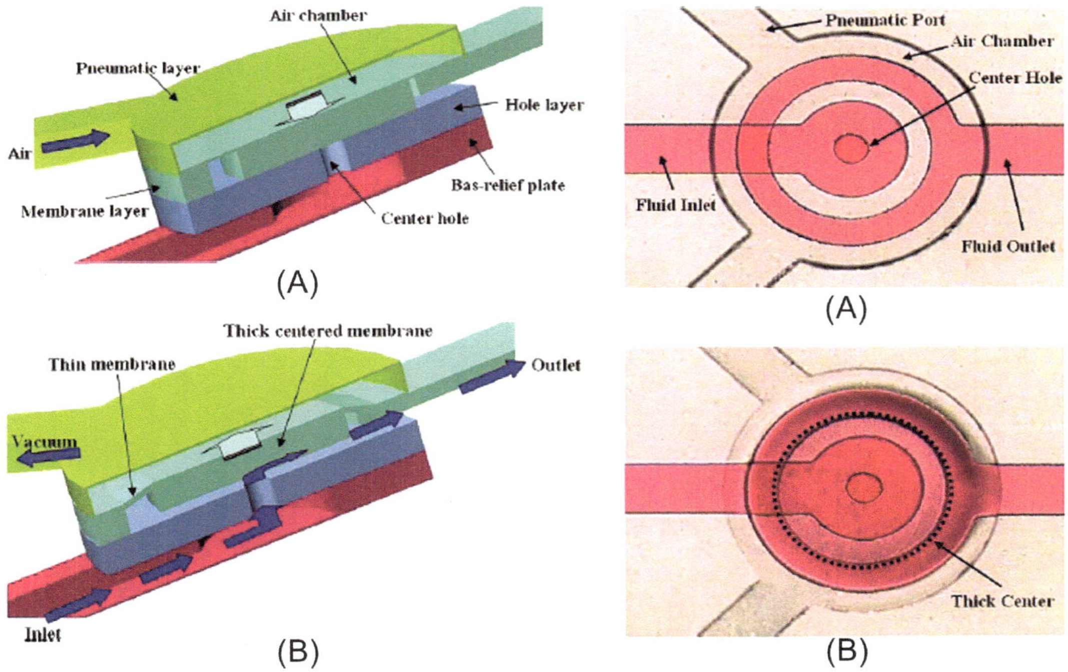

FIGURE 8.4

Microvalves Schematic diagram. 3D schematic representations (*left*) and microscopic images (*right*) of the plunger microvalve in its closed (a) and open (b) states.

(Reprinted from Au et al. (2011), open access.)

nitride, and specialized biocompatible coating designed to minimize adverse tissue responses. Because MEMS device are extremely small, they naturally exhibit low power consumption, making them suitable for operation with energy harvesting systems (İlik et al., 2018). The structural design center on recreating the compact, spiral form of the natural cochlea, which is essential for separating and analyzing different sound frequency. This geometry is translated into a MEMS based framework capables of supporting multiple microstructure that follow the mechanical filtering characteristics of the biological cochlea.

A network of microfluidic pathway incorporated into the device to guide sound induced vibration and fluid motion. These pathway are typically etched into silicon substrate and are essential for reproducing the hydromechanical behavior of the natural cochlear fluid. To replicate the biological function of the cochlea, the device integrate a set of micro electromechanical element that include miniature microphones, sensors, and actuator. These component detect incoming sound waves, converts mechanical vibrations to electrical signal, and stimulate the auditory nerve. The sensing elements consist of finely tuned piezoresistive diaphragm designed to operate within the audible frequency range. The number and arrangement of these diaphragms forms an electromechanical filter bank that identifies and separate primary acoustic frequency, closely approximating the frequency discrimination performed by the natural cochlea. The diaphragm driven by sound waves transmitted through a fluid medium, allowing the system to mirror the operation of the biological organ (Kulah et al., 2022).

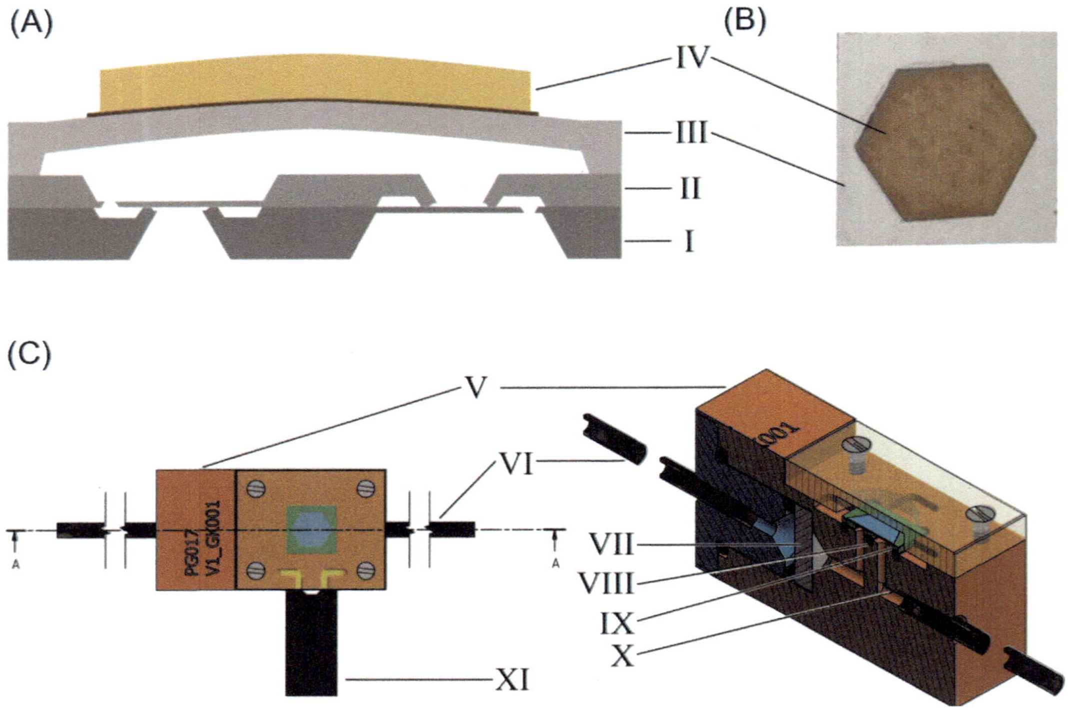

FIGURE 8.5

Micropump Schematic cross-section. (A) A 5 × 5 × 0.6 mm^3 silicon micropump, and (B) a pump chamber with a diameter of 4.5 mm. The three gray layers represent the silicon pump body, consisting of the lower valve unit (I), upper valve unit (II), and actuator diaphragm (III) with an attached piezocëramic element (IV). For fluidic characterization, the pumps (IX) are installed in a test housing (C) made of a polyether ether ketone (PEEK) body (V) containing drilled fluidic channels (X), inlet and outlet capillaries (VI), a stainless-steel filter (VII), and a soft sealing component (VIII).

(Reprinted from Bußmann et al. (2021), open access.)

Signal processing electronic incorporated to handle the electrical signal generated by the MEMS sensor. These electronics may include analog to digital conversion and digital signal processing module designed to amplify, filter, and interpret the faint electrical signal produced by the piezoelectric or piezoresistive component. Because of the extremely small voltage signals generated by these elements often far lower than typical acoustic microphone outputs—significant amplification is required. The electronic design must also operate with ultra-low power consumption, enabling the entire system to rely on energy-harvesting mechanisms rather than implanted batteries. Conventional power-saving techniques, such as sleep modes, are not suitable because they may cause the system to miss brief acoustic events. Therefore, the signal processing electronics are optimized for continuous, efficient charge transfer from the energy-harvesting subsystem directly to the implanted cochlear electrodes.

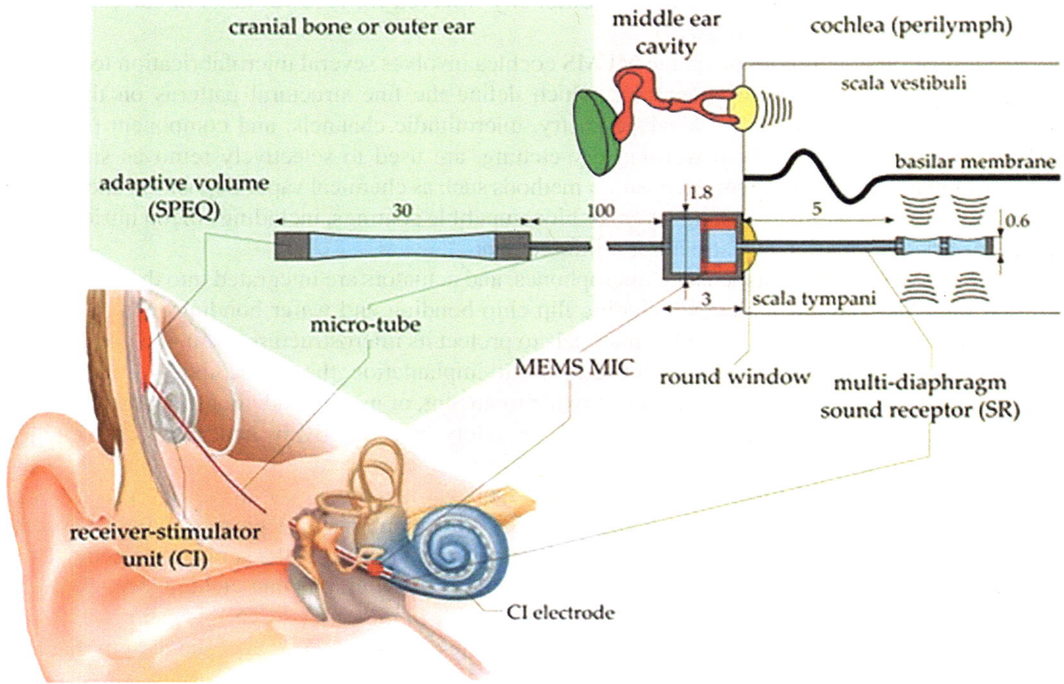

FIGURE 8.6

MEMS microphone. The positions of the individual structural components within the human body are depicted both schematically and through an anatomical illustration of the human ear. In the anatomical representation, the intracochlear acoustic receiver (ICAR) is illustrated as an integral component of a cochlear implant system.

(Reprinted from Prochazka et al. (2019), open access.)

Ideally, nearly all harvested electrical charge is redirected toward stimulating the auditory nerve, with minimal leakage currents to ground.

Powering the implant requires a highly efficient energy source. Fully implantable MEMS cochlear systems may use batteries or, more commonly, energy-harvesting mechanisms integrated into the skull behind the ear. Thermoelectric harvesting can exploits temperature differences of around 5°C, which occurs naturally due to metabolic heat productions. Additional energy may also be captured from body motions through electromagnetic or piezoelectric harvester. The MEMS device compact size allow it to function at extremely low power level, making these form of energy harvesting both feasible and practical. To prevent tissue irritation or adverse reactions, all external surface of the cochlear device are coated with biocompatible materials. The system also include communication and control interfaces, often using wireless technologies, to allow clinician to adjust device parameter or perform software updates without additional surgery.

Fig. 8.6 illustrate the internal configuration of a fully implantable MEMS cochlear system. The cochlear implant is placed in the middle ear cavity, while the energy harvesting module is positioned within the skull behind the ear, a location traditionally reserved for receiver units in current cochlear implant design. (Prochazka et al., 2019).

The fabrication of a fully implantable MEMS cochlea involves several microfabrication techniques. The process begins with photolithography, which define the fine structural patterns on the silicon substrates, including the spiral cochlear geometry, microfluidic channels, and component placement regions. Etching techniques, both wet and dry etching, are used to selectively removes silicon and sculpt the channels and diaphragm. Deposition methods such as chemical vapor deposition and physical vapor deposition apply thin insulating layers or biocompatible coatings, including silicon nitride, which provide electrical insulation and structural reinforcement.

MEMS components such as sensors, microphones, and actuators are integrated into the device using precision microassembly techniques including flip chip bonding and wafer bonding. After assembly, the device is covered with biocompatible materials to protect its microstructures from the physiological environment and to prevent immune responses. Prior to implantation, the device undergoe sterilization using technique such as autoclaving, ethylene oxide treatment, or gamma radiation. Finally, the system undergoes clinical testing to ensure that it operate safely and effectively under real physiological conditions, followed by surgical implantation into the patient inner ear.

Figs. 8.7 and 8.8 further illustrates the fabrication sequence and the sensor model. The process begin with applying an isolation layer such as SiXNY to the silicon substrate. A piezoresistive layer, often NiCr, is deposited using reactive sputtering or RF sputtering and serve as the sensing element. A thin, conductive gold layer is deposited to form electrical contact. Lithography and etching steps define the gold pattern and piezoresistive elements. On the opposite side of the wafer, SiXNY is patterned to act as a protective mask for KOH wet etching, which releases the diaphragm. The final diaphragm thickness is determined by the SiXNY layer, offering strong etch selectivity and precise control over structural dimension (Latifet al., 2021; Yang et al., 2022).

The testing of fully implantable MEMS cochlear device is a critical step to ensure safety, functional accuracy, and long term reliability. Functional testing evaluates the device ability to perceive and analyze sound using simulated acoustic environment. Frequency response testing verify the implant ability to discriminate between different sound frequencies, reflecting the natural cochlea frequency based separation of signals. Dynamic range test assess whether the implant can handle soft and loud sounds without distortion, while sound localization test determine whether the device can reproduce directional hearing cues.

Electrical testing includes impedance measurements to confirm proper electrode interaction with auditory nerve fibers. Each electrode in the array is tested individually to ensure that it can stimulates neural tissue effectively. Signal processing electronic are evaluated to confirm their ability to convert acoustic signal into electrical impulses suitable for nerve stimulation.

Biocompatibility testing ensure that device materials do not trigger toxic or inflammatory response. Cytotoxicity evaluations assess the impact of device materials on surrounding cells, while tissue response studies examine real biological tissue for inflammation or fibrosis. Durability testing subject the device to mechanical stress, including bending, torsion, and compression, to ensure that it can withstand the mechanical environment of the human body. Environmental tests simulate internal body conditions, such as temperature fluctuations, moisture, and pH, to ensure device stability.

Electromagnetic compatibility testing determine whether the implant can function without interference from external electromagnetic fields and ensures that it does not cause interference to other device.

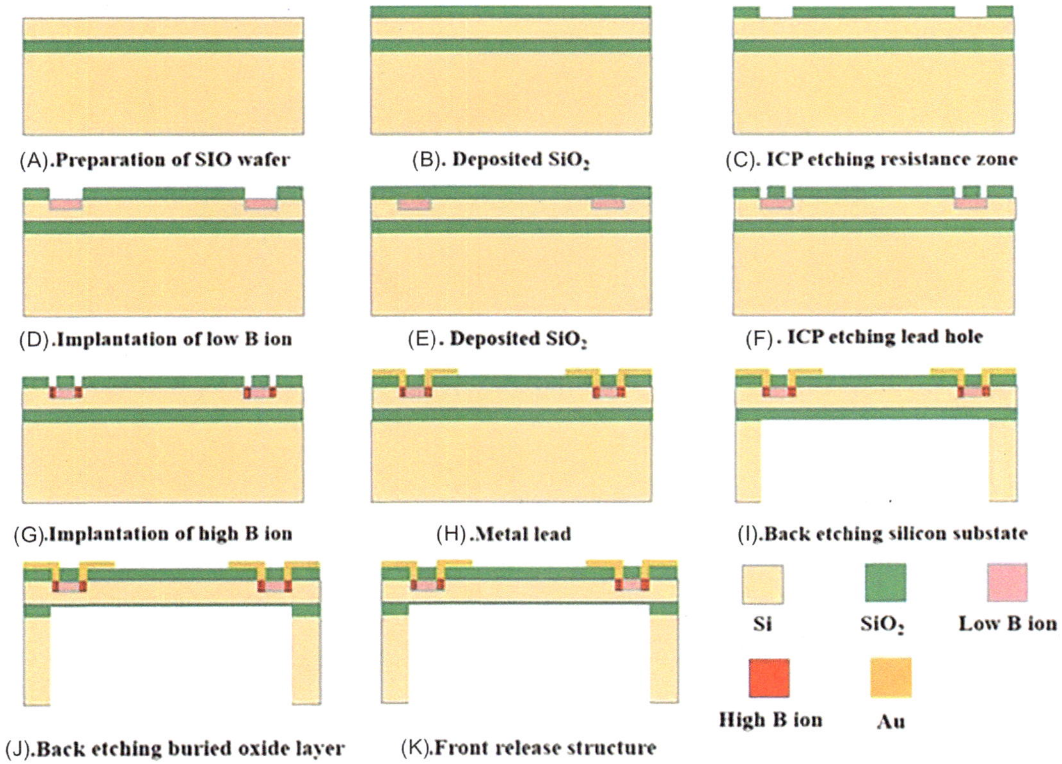

FIGURE 8.7

The process flow of the MEMS microstructure.

(Reprinted from Yang et al. (2022), open access.)

Wireless communication and control function are tested to verify the reliability of remote adjustment, diagnostics, and software updates.

In vivo testing begins with animal studies that assess the implant performance in living organism, followed by phased human clinical trials to evaluate safety, effectiveness, and real world functionality. Safety testing extends to the implant power system, evaluating the integrity of batteries or energy harvesting systems, checking for overheating or leakage, and ensuring compliance with regulatory standard. Infection control evaluations confirms that the implant design reduce the likelihood of postoperative infection. Long term monitoring of implanted patients provide data on performance over years and helps identify any late emerging issues.

In conclusion, the design, fabrication, and testing of a fully implantable MEMS cochlea constitute a sophisticated engineering process that integrate microfabrication, biocompatibility, electronics, and auditory science. This technology hold transformative potential for individuals with severe hearing loss, offering them the possibility of restored auditory function through a highly advanced biomedical implant capable of mimicking the complex behavior of the natural cochlea.

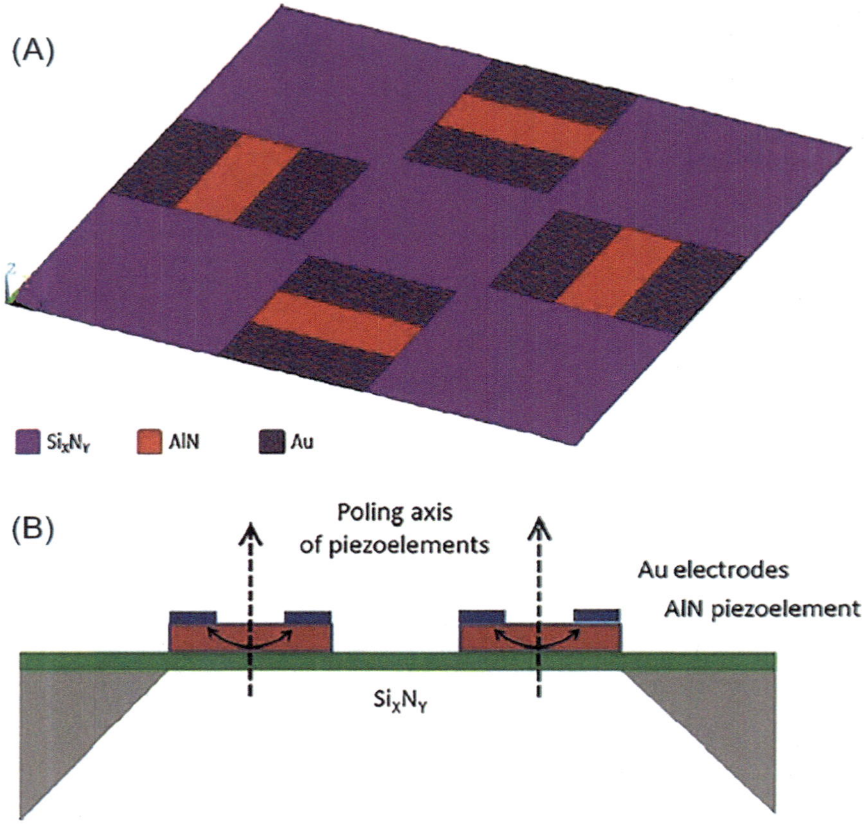

FIGURE 8.8

MEMS sensor model: A membrane equipped with a piezoelectric component and gold electrodes functioning in mode 33.

(Reprinted from Latif et al. (2021), open access.)

8.6 Artificial cardiac pacemaker

An artificial cardiac pacemaker is a medical device that is surgically implanted inside a patient's body to reinstate the normal rhythmic function of the heart. It stands as one of the earliest electronic devices successfully introduced into the human chest cavity, yielding positive outcomes. Nevertheless, the initial pacemakers were quite cumbersome due to the size of their components. However, the advent of MEMS technology has facilitated a significant reduction in the size of pacemaker components.

In the past, pacemaker used to be relatively large, single chambered, and operated at a fixed rates. However, with the advancements in miniaturization technology, modern pacemakers have transformed into compact, dual chambered, rate adaptive, and programmable devices that also offer diagnostic capabilities. These state of the art pacemakers, designed using MEMS technology, prioritize precision,

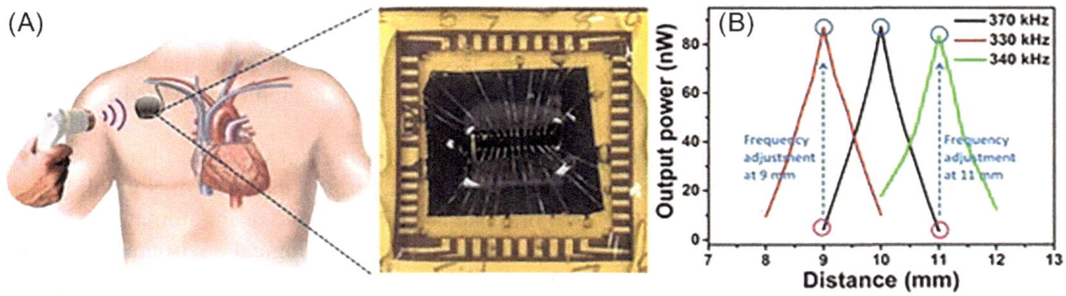

FIGURE 8.9

Ultrasonic energy harvester integrated with a pacemaker. (A) The fabricated chip is displayed on the right side. (B) Frequency tuning is performed to maintain high power transfer and efficiency, compensating for a 1 mm distance variation caused by breathing and muscle movement.

(Reprinted from Shuvo et al. (2022), open access.)

reliability, stability, and energy efficiency (Alrashdan et al., 2015). As a result, they can accurately adjusts the patient heart rate to accommodate the body varying needs.

Contemporary artificial pacemakers now serves as actuator capable of rectifying irregular heartbeat. These device interact with the patient heart, identifying any irregularities that lead to abnormal heart rhythm. Subsequently, they respond by delivering electrical impulse to restore the heart normal rhythmic activity.

Rate adaptive pacemakers, which adjust cardiac output to match metabolic needs, are preferred over fixed rate pacemakers. They prove especially valuable for patient whose heart rate remain steady even during physical activity. Rate adaptive pacing relies on sensors, which should match the responsiveness of the heart natural pacemaker (the SA node). These sensors need to exhibit exceptional sensitivity when detecting irregular heart rate, offer rate adjustments that align with exercise and metabolic demand, and remain unaffected by non cardiac stimuli. Additionally, it's crucial that they refrain from causing excessive pacing (Jeyalakshmi Devi, 2017).

MEMS based inertial accelerometer based sensors are among the most prized sensors used in pacemakers, playing a significant role in this technology. These sensor are part of the vital category of silicon based sensors, and microaccelerometer based sensors rank second only to pressure sensors in terms of sales.

Currently, a growing area of research focused on the development of smart pacemaker with automatic functionality. The goal is to develop pacemaker with versatile capabilities, encompassing diagnostic functions and therapeutic interventions. New algorithms to enable pacemakers to automatically detect capture, sense abnormality, measure impedance, and determines longevity. as pacemaker functionalities become more intricates, its size continues to diminish, a feat made possible by advancements in miniaturization technologies (Shuvoet al., 2022) Fig. 8.9.

8.7 Stents

Completely implantable BioMEMS stents represents an innovative class of medical devices designed to address a range of clinical problems, particularly those involving the cardiovascular and gastrointestinal systems (Cros, 2009). These stents integrate MEMS technology with biocompatible materials to create intelligent implant capable not only of providing mechanical support, but also of monitoring physiological condition and, in some cases, delivering targeted therapy. In what follows, the design principle, major applications, and fabrication processes of fully implantable BioMEMS stents are discussed in the following discussion.

The design of fully implantable BioMEMS stents begin with the careful selection of biocompatible materials. Common choices include medical grade stainless steel, nitinol (a nickel titanium alloy), and biodegradable polymers. These material must not provoke adverse immune or inflammatory response and must maintain their mechanical and chemical stability in the physiological environment. In some systems, biodegradable material are intentionally selected so that the stent provide only temporary supports and gradually resorbs as healing progresses, eliminating the need for long term permanent implants.

Once an appropriate material system has been identified, attention turns to the microstructure design of the stent. Using MEMS based methodologies, engineers defines precise strut or lattice geometries with controlled thickness, width, and pattern. The specific dimension and configuration depend on the target application, for example, vascular stents requires designs that can withstand pulsatile blood flow and vessel motion, whereas gastrointestinal stent may prioritize flexibility and resistance to compression. In many cases, the surface of the stent is modified with biocompatible or bioactive coating. These may serve to reduce inflammation, inhibit thrombosis, or provide a vehicle for local drug delivery, thereby combining mechanical scaffolding with therapeutic functionality.

A distinguishing feature of BioMEMS stents is the potential incorporation of microsensor and, in some designs, microactuators. Depending on the clinical objective, integrated sensors may monitor parameter such as blood flow, intraluminal pressure, local temperature, or chemical markers. This real time information can be used to detect restenosis, occlusion, or other complication at an early stage. Actuators, when included, may enables dynamic function such as controlled expansion, shape adaptation, or modulation of drug release. To support such active functions, miniaturized power and control subsystems are required. These typically consists of compact power sources, e.g., microbatteries or energy harvesting elements, and low power control electronic designed to operates reliably in the in vivo environment with high energy efficiency.

Fully implantable BioMEMS stent have several important application domains. In the vascular system, they are used to treat condition such as atherosclerosis by being deployed into arterie or vein to maintain or restore adequate blood flow. When equipped with integrated sensors, vascular BioMEMS stent can provide continuous monitoring of blood pressure or detect early signs of re occlusion. In the gastrointestinal tract, BioMEMS stents may be employed to treat esophageal or intestinal strictures. self-expanding designs are often used to ensure proper positioning and to maintain lumen patency, and some model incorporate localized drug release to reduce inflammation at the stenotic site.

The urinary tract represent another important application area, where stent are implanted to relieve obstructions associated with ureteral narrowing or kidney stones. BioMEMS urinary stent designed to monitor urinary flow rate and pressure, providing clinician with valuable data on the progression or resolutions of obstruction. Drug eluting BioMEMS stent, particularly in the cardiovascular field, specifically designed to releases therapeutic agents in a controlled manner over time. Such system widely used to reduces the risk of restenosis following angioplasty by inhibiting smooth muscle cell proliferation. biodegradable BioMEMS stents are increasingly explored as temporary scaffold that provide support during the critical healing phase and then safely degrade, removing the need for long term implants or surgical retrieval.

The fabrication of fully implantable BioMEMS stents is a highly specialized process that relie on advanced microfabrication techniques. The microscopic lattice structure of the stent is typically defined using methods such as photolithography to pattern features on a substrate, followed by etching processes to create the three dimensional geometry. Depending on the material system, both wet and dry etching may be used to achieve the desired precision and surface quality. Thin film deposition techniques, including CVD and PVD, may be employed to add structural layer, protective coatings, or functional layer that support sensing or drugdelivery functionalities.

When stent are designed to be drug eluting, an additional coating and drug loading steps is required. Biocompatible polymer layers can be applied to the stent surface, into which therapeutic agent are either physically embedded or chemically bound. The release profile is controlled through the choice of polymer, layer thickness, and drug formulation. If sensors and actuators are part of the design, these microcomponent must be integrated with the stent body, often by attaching or embedding them at predetermined locations. This integration step require precise alignment and robust bonding techniques to ensure long term stability in the physiological environment.

Active BioMEMS stents that require on board intelligence also incorporate power and control electronics. These miniaturized circuit are designed to be compact, energy efficient, and robust under mechanical and chemical stresses. Once all component have been assembled, the entire device undergoe thorough sterilization to eliminate microbial contamination before implantation. Common sterilization method include ethylene oxide gas, gamma irradiation, or other techniques compatible with the stent materials and electronics.

Before reaching clinical use, fully implantable Bio MEMS stents must pass difficult biocompatibility testing. These assessments evaluate whether the stent material and coatings provoke cytotoxicity, excessive inflammation, thrombosis, or other adverse tissue responses. Functional testing is also performed to confirms mechanical integrity, expansion characteristics, and, where applicable, sensor accuracy and reliability. Subsequent preclinical and clinical study involve evaluating safety and performance in animal model and human subject, respectively, to demonstrate that the device meets regulatory requirements for efficacy and safety. completely implantable BioMEMS stents represent a significant advance in medical technology by combining MEMSbasedmicroengineering with biocompatible material and, in some cases, intelligent sensing and actuation. These device provide precise mechanical support to diseased vessels and duct, enable localized and controlled drug administration, and supports continuous in situ monitoring of physiological parameters. By offering minimally invasive, highly tailored therapeutic options, fully implantable BioMEMS stents have the potential to substantially improve clinical outcome and quality of life for patient with a wide range of vascular, gastrointestinal, and urological conditions Fig. 8.10.

8.8 Neurostimulators

The development of BioMEMS neurostimulator implants reflect a notable convergence of microscale engineering and modern neuroscience. Although neurostimulation can be traced back to ancient civilizations, where physicians used electric fish or electric eels to alleviate pain, scientific progress accelerated

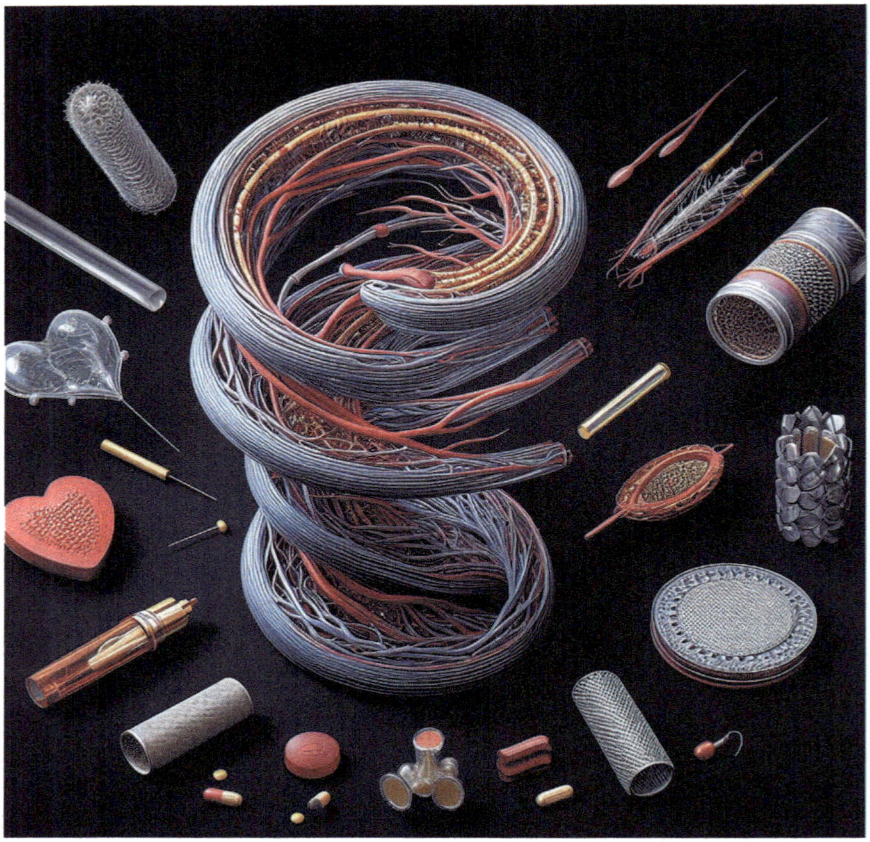

FIGURE 8.10

Different shape and configurations of stents.

significantly during the twentieth century. The development of early implantable device such as cardiac pacemakers and spinal cord stimulator marked the beginning of clinically regulated neurostimulation. Over the last few decades, BioMEMS technology has transformed this field by enabling device with unprecedented levels of precision, miniaturization, and functional integration. Early neurostimulators were relatively large and offered only limited control, but advances in microelectronic, biosensors, and biocompatible materials have made it possible to design highly sophisticated implants capable of finely regulating neural circuits.

The therapeutic use of electricity for disease managements has been documented for more than two thousand year. Ancient Roman physicians, placed electric fish on specific part of the body to treat pain caused by rheumatism or gout. However, it took centuries before scientist understood the bioelectric phenomena underlying human sensory perception, movement, and neurological disorder. Even today, the physiology of the nervous system and its pathological transformation such as those following spinal cord injury or sensory degeneration remain subjects of ongoing research. The main challenges lies not in the absence of scientific knowledge, but rather in the difficulty of translating that knowledge into effective clinical therapies and rehabilitation device.

Spinal cord injury continue to impose a heavy biological, social, and financial burden. In Germany, approximately 80,000 people live with spinal cord injury, and an estimated 1800 new cases occur each year. In the United States, nearly 40 million individual are affected, with more than 10,000 new incidents annually. Historically, the prognosis for spinal cord injury was extremely poor, many patient did not survive more than six months after injury. Early rehabilitation pioneer, such as Sir Ludwig Guttmann in the mid twentieth century, radically changed these outcome, laying the foundation for modern neurorehabilitation. Today, the life expectancy of individual with spinal cord injuries approache that of the general population.

Beyond trauma, sensory related disorder are widespread. In Europe, around 600,000 individual suffer from incurable retinal disease such as retinitis pigmentosa or macular degeneration, ultimately leading to blindness. Many other sensory deficits are even more common. Despite advances in prosthetics and neural rehabilitation, developing device that restore lost function remains a significant challenge. Effective neural prostheses must bridge the gap between biological signals and technical system, enabling the translation of human intention into machine action while allowing artificial device to deliver sensory feedback to the user. Over the past three decade, implantable neural prostheses have made remarkable progress, yet they often remain under recognized by clinicians, patients, and insurance provider.

Neural prostheses, including BioMEMS neurostimulator, share a common functional foundation, electrodes that serves as interfaces between the biological nervous system and engineered technologies. These electrode enable bidirectional communication, capturing neural signals for analysis and delivering electrical stimuli to modulate neural activity. Although clinically approved device have demonstrated long term reliability, their miniaturization has historically imposed functional limitation. Microsystems technologies now offer new opportunities for designing complex neural interface equipped with large numbers of electrode capable of performing highly specific neural interactions.

The magnitude of recent advancement is evident in clinical statistic. More than 200,000 neural device have been implanted worldwide. Over 130,000 spinal cord stimulator have been used to treat chronic pain and spasticity. Cochlear implants, which restore hearing for individual with damaged outer hair cell but intact auditory nerves, have been implanted in more than 55,000 patient, including infants as young as six months old. More than 20,000 deep brain stimulation system have been implanted to manage Parkinsonian tremor and dyskinesia. Similarly, over 17,000 vagal nerve stimulator have been deployed to treat intractable epilepsy. Additional neurostimulators targets urinary urge incontinence, a condition affecting roughly 10% of individuals over sixty in Western countries, while sacral nerve stimulator support bladder control and social reintegration for patient with spinal cord injuries. Phrenic nerve pacemakers provide life changing breathing assistance for individual with severe cervical spinal cord injuries by electrically stimulating the diaphragm via the phrenic nerve.

Despite these achievements, the development of a truly ideal neural interface remain an unsolved challenge. Neural implant must be safe, minimally invasive, and capable of long term operation without damaging surrounding tissues. Existing device lack the ability to monitor nerve health in real time or adjusts stimulation based on detailed biological feedback. Future systems will require electrode capable of providing new type of sensory information such as tissue growth patterns, local inflammation, or real time nerve condition allowing clinician to refine stimulation parameters and enhance therapeutic outcomes (Cho et al., 2010).

BioMEMSneurostimulators operate through a sequence of finely tuned microscale mechanisms. The process begins with sensing, where small sensor detect neural activity or biomarkers associated with neurological processes. These sensors may includes electrodes that record action potentials, microfluidic channels that capture biochemical signal, or optical component that detect fluorescence or light based signals. Once data are collected, the implant onboard electronics process the signal through amplification, noise filtering, and feature extraction. This processing ensure that the device interprets neural activity accurately and efficiently, even when signal are faint or distorted by surrounding tissues.

Based on the processed informations, the neurostimulator generate targeted electrical, chemical, or optical stimulation. The optical stimulation modulate neural circuit to achieve therapeutic effect, such as reducing chronic pain, interrupting epileptic seizure, improving motor function, or regulating abnormal neural activity. More advanced system employ closed loop operation, to continuously monitor neural activity and adjust stimulation parameter in real time. This adaptive process ensure that the implant maintain optimal therapeutic condition without requiring manual recalibration.

Modern neurostimulators incorporates wireless communication system allows clinicians to adjust settings, download performance data, and perform diagnostics remotely. It reduce the need for invasive procedure and improves overall patient care.

BioMEMSneurostimulators offers several advantage in modern neuroscience and clinical practices. high precision enable highly targeted neural modulation with negligible side effect. Miniaturization reduced surgical invasiveness lead for shorter recovery times and increased patient comfort. Customizable stimulation parameters allows treatment tailored to each patient unique neural profile. Advances in energy efficient electronics and durable material improved device longevity, reducing the frequency of surgical replacement. These device play a central role in managing chronic pain, Parkinson disease, epilepsy, depression, and other neurological disorder.

Research effort focused on developing advanced sensors provide richer and more accurates neural information, enabling deeper insight to neurological disorder which make the future of BioMEMSneurostimulator hold wonderful potential. closed loop systems augmented with artificial intelligence and machine learning allows implants to autonomously adapt stimulation pattern in responses to real time physiological feedback. wireless power transfer and data communication enhances convenience and device lifespan. BioMEMSneurostimulators are also expected to plays a pivotal role in next generation neuroprosthetics aimed at restoring lost motor or sensory function for individuals with paralysis. As our understanding of brain function extend, the scope of neurostimulator applications may expand to include condition such as Alzheimer disease, mood disorders, and other complex neurological condition.

BioMEMSneurostimulator implant represent a transformative frontier in neural modulation and rehabilitation. Their precision, flexibility, and compatibility with advanced microsystem technologies position them as key contributor to the future of neurological medicine. As these technologies continues to evolves, they offer profound hope for improving the lives of individual affected by neurological disorders, enhancing functional recovery, and restoring independence Fig. 8.11.

8.9 BioMEMS-based drug delivery implants

Totally implantable BioMEMS based drug delivery implant represent an advanced approachs to controlled and targeted drug administration within the human body. These device rely on BioMEMS that integrate microscale designs with biomedical science to achieve precise control over medication release

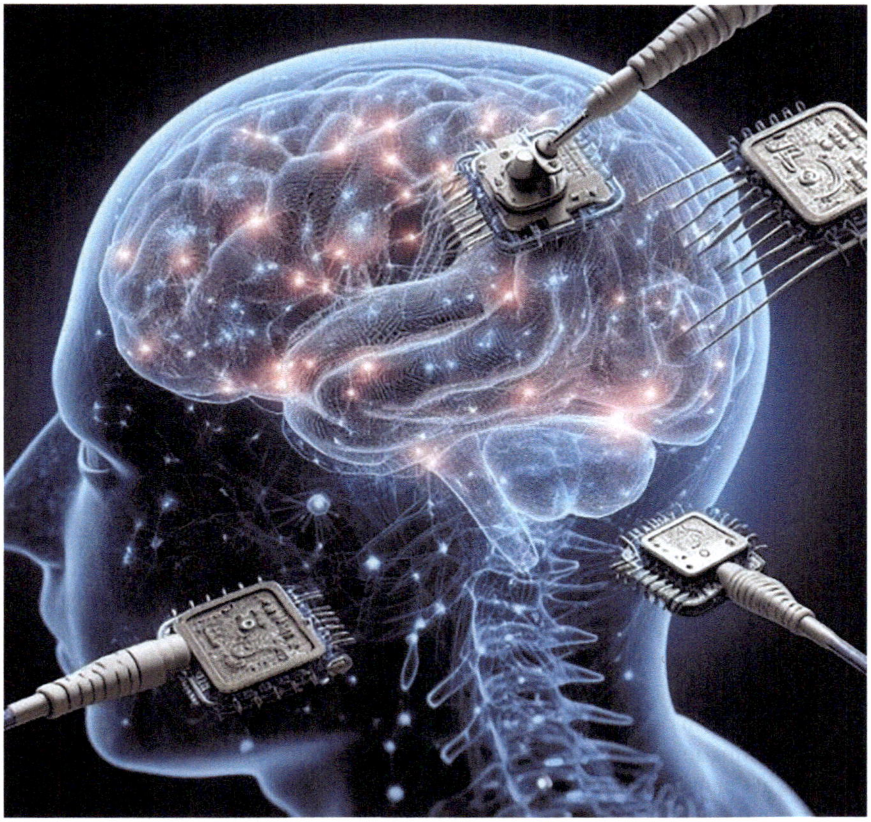

FIGURE 8.11

Illustration of the BioMEMS neurostimulator implants.

(Nisaret al., 2008). The use of MEMS technology allow these implants to deliver therapeutic agent at highly regulated rates, respond dynamically to patient physiology, and sustain long term operation with minimal intervention. The following section provides a detailed explanation of how these implant function, the advantages they offer, and the MEMS technologie that enable their performance.

The working principles of totally implantable BioMEMS drug delivery implant begin with the fabrication of a micro scale drug reservoir. This reservoir designed using microfabrication technique to hold a specific amount of medication or therapeutic compound within a confined structure. It is sealed to prevents leakage or contamination, this will ensure the drug remain stable and sterile during the implant lifespan. a network of microfluidic component located around the reservoir, including microchannel, microvalves, and micropump, all fabricated using MEMS technique. These microfluidic structures regulates the movement, direction, and quantity of drug released from the reservoir. Microchannels guide the drug to its targets site, and microvalves control the rate and timing of releases with exceptional precision.

Many design of BioMEMS drug delivery implants need sensing and feedback mechanisms. Embedded microsensor continuously monitor physiological parameters such as glucose levels, pH, or the concentration of specific biomarker in bodily fluids. the real time data produced by these sensor used to adjust drug delivery rate automatically, ensuring that patients receives the correct dosage at the most appropriate moment. This approach enhance treatment accuracy and reduces the likelihood of over or under medication.

Actuation mechanism also play essential role in the function of these implants. MEMS based microactuators including piezoelectric and electrostatic mechanism enable precise control over microvalves and microfluidic structure. The actuation system respond either to internal sensor data or to external signal, ensuring that the medication dispensed accurately and consistently. The entire operation coordinated by embedded control electronic. These circuit interpret sensor data, compute the appropriate response, and send instructions to microactuators to regulate medication release. The electronics must be energy efficient and reliable to support long term implanted operation.

Powering these implant require careful design. MEMS based drug delivery system use internal batteries, and many design come with energy harvesting mechanisms that convert body heat, motion, or external electromagnetic fields to usable power. Other system use wireless power transfer which allow the implant to operates indefinitely without battery replacement. Effective power management essential to ensure long term operation and maintain patient safety.

Implantation of BioMEMS drug delivery system require robust packaging to protect the internal component from bodily fluids and immune responses. The implant encased in biocompatible materials to minimize tissue irritation, reduces the likelihood of rejection, and prevent microbial contamination. proper packaging design also minimize infection risks and ensure long term biocompatibility.

These implantable device offer several important advantage. its able to deliver medication with extraordinary precision. This capability allow clinicians to tailor treatments to the individual need of each patient and minimizing the risk of adverse effects. The integration of sensor and closed loop feedback mechanisms enable continuous monitoring of physiological parameter, this allow the drug dosages to be adjusted dynamically in response to the patient real time condition. Such implant also improve patient compliance by eliminating the need for frequent injection or complex medications schedule. By maintaining drug concentration within a tightly controlled therapeutic window, BioMEMS drug delivery implants can reduces side effects associated with fluctuating medication level. Their long operational lifespan, supported by MEMS based miniaturization and energy efficient electronics, further reduce the frequency of surgical intervention. Additionally, the targeted nature of drug releases minimizes exposure to healthy tissue, improving therapeutic outcome and reducing systemic toxicity.

The MEMS technologies that makes these implant possible rely on advanced microfabrication method. Photolithography, etching, and thin film deposition techniques are used to create microscale drug reservoirs, microchannels, and microvalves. Microactuator fabricated with piezoelectric or electrostatic principle enable precise control of valve opening and medication flow. Integrated microsensor, including microelectrode and biochemical sensing elements monitor physiological condition and provide feedback signal to regulate dosing. Microfluidic technology guide the drug from reservoir to the target site essential to the proper functioning of these system, these component create microsystem can deliver medication with exceptional precision and adaptability.

Fully implantable BioMEMS based drug delivery implant show a transformative approach to personalized and tightly regulated drug administrations. it combine microfabrication, sensing, actuation, and control technologies to achieve excellent accuracy in medication dosing, improve patient adherence, minimize side effect, and enable targeted therapy. BioMEMS technologies continue to evolve, these implants hold significant promise to enhance the effectiveness, efficiency, and safety of medical treatment across a wide range of clinical conditions.

8.10 MEMS devices for artificial lung

For many years, the medical community require artificial organ to address organ shortages and improve the quality of life for patient with organ diseases. building artificial lungs is one of the most complex and challenging question. The human lung is a highly sophisticated organ with delicate structure and efficient gas exchange mechanisms are difficult to replicate artificially. However, recent advancements in MEMS technology introduced new possibilities to produce micro scale devices capables of functioning as artificial lungs. These development have generated hope for patients with severe respiratory conditions and for researcher exploring broader biomedical and technological applications. This section examine how MEMS technology utilized to create microfabricated device designed to replicate the essential functions of the natural lung.

MEMS technology ranging from a few micrometer to several millimeters. MEMS offers several unique advantages in artificial lung engineering. One of the most significant benefit is miniaturization. MEMS make it possible to develops artificial lung systems that are much smaller and lighter than traditional extracorporeal life support device. the reduced size enhance the portability of artificial lung system and decreases invasiveness when implanted in the human body. MEMS fabrication technique is also provide exceptional precision, this enable the creation of microscale channels, structure, and surfaces mimic the highly intricate architecture of natural lung tissues. This precision essential for replicating the complex gas exchange processes occurring in the alveoli.

In addition, MEMS allow engineers to select materials based on their biocompatibility, so the artificial lung component are safe when contact with blood, tissue, and bodily fluids. Biocompatible material such as silicon, polymers, or advanced coatings can help prevent clot formation, inflammation, and immune reactions. MEMS devices can also integrate sensors directly to the artificial lung structure. sensor monitor key operational parameters such as oxygen and carbon dioxide concentration, pressure level, and airflow rates. Real time feedback from the sensor allow clinician to track device performance and patient status continuously. MEMS based artificial lungs often incorporates microfluidic systems regulates the movement of gases and fluids. The microfluidic structures control airflow, distribute gases, and facilitate the exchange of oxygen and carbon dioxide in a manner closely resembling natural pulmonary function.

The manufacturing process of microfabricated artificial lung using MEMS technology normally need several steps. design and simulation is first step, engineers determines the device geometry, dimensions, and gas exchange characteristics. Computational simulation tools to refines the designs and optimize parameter such as flow resistance, surface area, and gas permeability. When the final design validated, MEMS fabrication techniques used to construct the device. These techniques involves creating complex microstructures such as microchannels and gas exchange membrane, using materials selected based on the biocompatibility. Then, the sensor for pressure, flow rate, and gas concentration integrated into the device to ensure proper monitoring and operational control. These sensor help maintain safe and efficient functioning in the real time input to regulate airflow and gas exchange processes. Next, microfluidic element incorporated to manage oxygen and carbon dioxide transport within the device, simulating the natural circulation of gas within the lung. After assembly, the artificial lung undergoe extensive testing and optimization to validate its gas exchange performance, and to ensure structural integrity, and minimize complications such as leakage, clotting, or flow imbalance. This stage important to ensure the device can operates safely and efficiently under physiological conditions.

Microfabricated artificial lung fabricated using MEMS technology can be used with patients with acute respiratory failure as temporary life support systems, providing critical oxygenation while the patient awaits lung transplantation or undergoes recovery. the patients with chronic lung diseases such as COPD or pulmonary fibrosis, MEMS based artificial lungs offer long term respiratory support, improving quality of life and reducing the dependence on donor lungs. These device also hold considerable values in research environments. it can serve as platforms for drug testing, disease modeling, and studies of lung physiology, allow researcher to control microscale environment that mimics key aspect of natural pulmonary function.

MEMS based artificial lungs face several challenges, where the long term biocompatibility remain a major concern, implanted device must withstand continuous interaction with biological fluids without inflammation or clotting. Power supply is another challenge, especially for portable or implantable device that must operate without frequent battery replacement. and it must undergo rigorous regulatory evaluations to ensure safety and effectiveness before it can be used inside human body. Overcoming these challenges requires continues innovation in materials science, microengineering, and biomedical research.

Microfabricated artificial lung important role in the future of respiratory medicine increased. Since the MEMS technology and lung physiology knowledge continue to be improved. These device hold the potential to transform treatment strategies, allow earlier intervention in respiratory diseases, and enhance patient care across a range of clinical scenarios (Gharibet al., 2022) Fig. 8.12.

8.11 Implantable BioMEMS and tissue engineering

Micro Electro Mechanical Systems improved numerous scientific and medical field. Tissue engineering is the field concerned with the regeneration, restorations, or creation of functional human tissues for clinical treatment, transplantation, and research application (Farahani et al., 2023). MEMS technology accelerated the improvement of this field by founding high precise tools for microscale control, monitoring, and manipulation. These technologies allows researchers to emulate physiological environment, regulate biological processes, and engineer tissues with exceptional accuracy.

MEMS enabled bioprinters which allow researchers to position cell, biomaterials, and growth factor with extreme precision, enabling the construction of intricate three dimensional tissue structure, this enhance the bioprinting field itself. Depositing living cells and scaffold material in controlled pattern with high accuracy, bioprinting enhance the development of functional tissues and organs. MEMS device played a key role in replicating the cell microenvironment. MEMS enable the creation of microenvironments closely mimic natural biological conditions, including fluid flow, mechanical stresses, and chemical gradient. These parameters essential for proper tissue maturation. MEMS based microfluidic systems allow scientist to recreate physiological cues that guide tissue growth.

Cell culture systems benefited from MEMS innovation. Microscale culture platform allow cells to grow on specialized substrates where parameter such as temperature, pH, nutrient levels, and oxygen

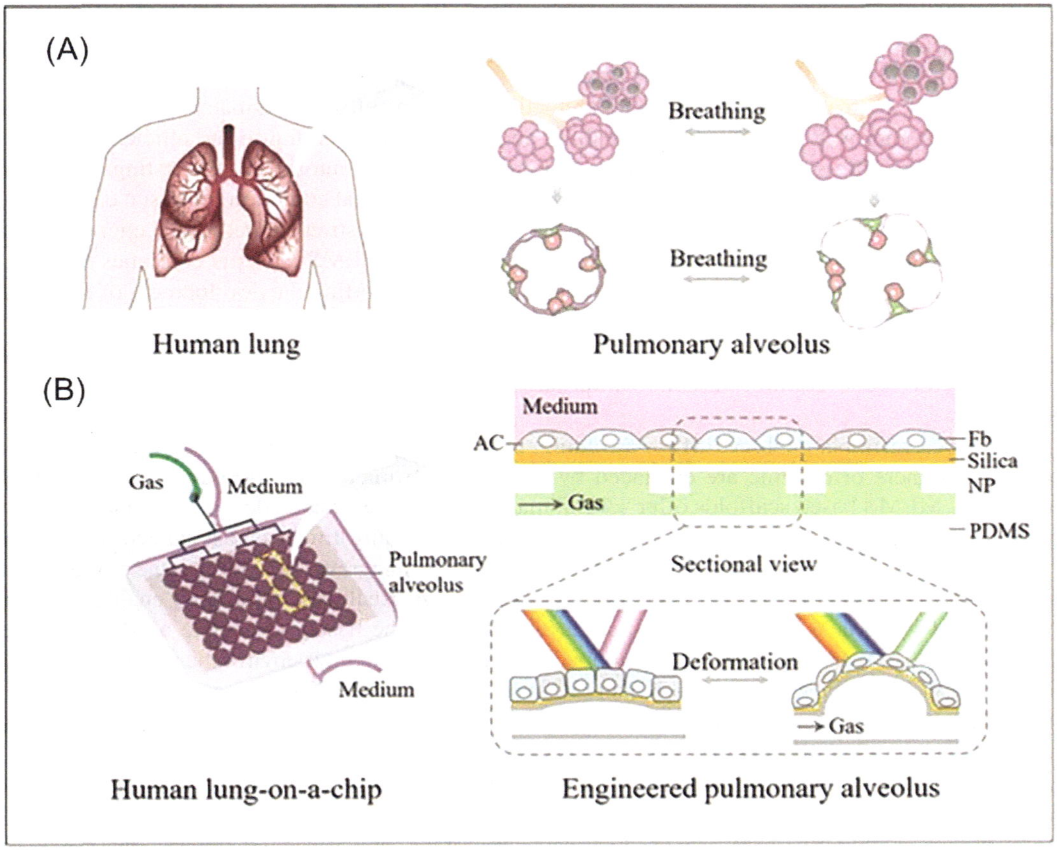

FIGURE 8.12

Microfluidic artificial lung technology.

(Reprinted from Gharib et al. (2022), open access.)

concentration precisely regulated. MEMS technology enabled the development of organ on a chip platforms, which is miniature systems follow the functions of specific organs. These devices provides an invaluable tool for drug testing, toxicity studies, and diseases modeling, this allow researchers to study tissue responses with high controlled conditions.

MEMS sensors enhances tissue engineering with real time monitoring of designed tissues. These embedded sensor measure pH, oxygen levels, mechanical forces, and other critical variables, it provide continuous feedback on tissue health and function. MEMS based drug delivery system also contribute to tissue engineering by controlling the release of therapeutic agents. The precise delivery of growth factors, hormones, or bioactive compound essential for guiding tissue regeneration. some MEMS devices implanted directly into the body to monitor tissue regeneration or support healing processes. Implanted MEMS sensors can track oxygen concentration, pH, or local mechanical stress, enabling clinician to adjust treatments accordingly.

Also, MEMS technology used to sort and analyze cells within engineered tissues. Microfluidic devices separates specific cell types, ensuring that only desired cells used in constructing tissue structure. This capability is vital for producing uniform and functional engineered tissue. MEMS frequently integrate nanoscale features, enhancing the interaction between cells and biomaterial. Such nanoscale enhancements improves cell adhesion, signaling, and growth, contributing to the efficiency of tissue regeneration. MEMS device support high throughput screening by automating the testing of multiple material and methodologies, accelerating the identification of optimal condition for tissue engineering. these advances demonstrates how MEMS technology significantly strengthened tissue engineering by enabling precise control and monitoring at the microscale. As MEMS systems continues to evolve, they are expected to further enhances regenerative medicine, supporting the development of functional tissues and organ for transplantation, research, and therapeutic application.

A number of MEMS application in tissue engineering warrant deeper examination, including scaffolding for tissue culture, vascularization strategies, liver tissue engineering, and microfluidic bioreactors.

In the area of implantable MEMS scaffolding, traditional tissue engineering scaffolds commonly made of polymers or ceramic are enhanced by incorporating microscale mechanical and electronic component. MEMS based scaffolds offer several major benefit, they can be designed to release growth factors or bioactive substances in a controlled manner, providing timed and localized signals that promotes tissue regeneration. These scaffolds may also include integrated sensors capable of monitoring tissue condition such as oxygenation, pH, or mechanical stress. Real time monitoring enable clinicians or automated systems to adjust treatment strategies based on the tissue immediate needs. Moreover, MEMS scaffold can respond dynamically to mechanical changes in their environment, modifying their stiffness or structure in ways that copy natural tissue behavior. The integration of electronics allow wireless communication and data transmission, enabling continuous tracking of scaffold performance and tissue development (Borenstein & Vunjak-Novakovic, 2011).

MEMS technology play critical role in promoting vascularization, which is essential requirement for engineered tissue to survive and function. MEMS devices replicates the circulatory pathways needed to transport oxygen, nutrients, and waste products using microfabricated vascular networks. These network follow natural blood vessel structures and provide the structure for healthy tissue growth. MEMS systems releases pro angiogenic factor or generate mechanical cues that stimulates new blood vessel formation. Microfluidic perfusion systems further emulate the natural flow of blood, its also support the development of functional vascular network. vasculature on a chip devices simulates real blood vessel behavior in a controlled environment, and offer platform for drug testing, disease modeling, and vascular biology research.

MEMS technology enable the creation of highly detailed scaffold that replicate the liver intricate architecture. These microscale scaffolds provides the structural support needed for hepatocytes the primary liver cells to grow and function properly. MEMS devices allows precise control over the hepatocyte microenvironment, ensuring optimal nutrient flow, oxygen supply, and biochemical signaling. This control is essential for maintaining proper metabolic activity. MEMS based liver models are invaluables for drug metabolism studies, they allows researchers to evaluate how medications are processed by liver cell and predict human physiological responses. Moreover, MEMS contribute to the development of bioartificial liver devices designed to provide temporary supports for patients experiencing liver failure.

To construct liver tissue using MEMS technology, several step are required. The process begin with isolating hepatocytes from donor tissue or deriving them from stem cells, and culturing them in controlled laboratory conditions. MEMS based scaffold are then designed to follow the liver natural structure, including microfluidic channels that regulates the flow of nutrients and waste products. Cell patterning technique are used to position hepatocyte and other liver cell types in precise arrangements, forming functional hepatic units. The engineered tissue is then placed in a bioreactor that replicates physiological condition to support maturation. Over time, the tissue develops capability such as detoxification, protein synthesis, and bile production. Functional testing ensures that the engineered tissue performs as expected. While highly promising, MEMS based liver engineering still faces challenge such as achieving long term function, establishing robust vascularization, and ensuring immune compatibility. However, continued technological advance and a deeper understanding of liver biology point toward a promising future in which MEMS engineered liver tissue may eventually serve as a substitute for organ transplantation (Borenstein & Vunjak-Novakovic, 2011).

Microfluidic bioreactor represent another important category of MEMS devices in tissue engineering. These systems provide finely controlled microenvironment for cell growth and tissue development. They make it possible to cultivate cells in three dimensional structures and offer highly controlled condition for fluid flow, nutrient delivery, and waste removal. Microfluidic bioreactor also support high throughput screening, enabling rapid evaluation of different tissue engineering strategies and they serves as foundational technology for organ on a chip systems, which model human organ function for drug testing or disease research. By reproducing natural nutrient and oxygen supply pattern, microfluidic bioreactors creates highly realistic conditions that enhances tissue maturation and viability.

Overall, MEMS technology has dramatically expanded the potential of tissue engineering by introducing tools for fabrication, monitoring, and optimization of tissue constructs. Technologies such as implantable MEMS scaffold, microfabricated vascular systems, liver on chip model, and microfluidic bioreactor demonstrates the transformative impact of MEMS on regenerative medicine. when technologies continue to advance, they promises to improve the effectiveness of tissue regeneration and contributes to new solutions for transplantation, therapy, and biomedical research.

8.12 Biocompatibility and biodegradability of implantable BioMEMS devices/components

Biocompatibility and biodegradability are fundamental considerations in the design and development of implantable BioMEMS devices. These systems intended to function within the human body for extended periods, and their long term success depend on their ability to coexist harmoniously with biological tissues (Chen et al., 2018). In this section, the importance of biocompatibility and biodegradability, review established standards and testing procedure, and examine materials commonly used in BioMEMS devices that fulfill these requirements will be explored. biocompatibility is essential for patient safety and therapeutic effectiveness during the design of implantable BioMEMS devices. Biocompatible materials critical because they minimize the risk of immune responses, inflammation, or other adverse biological reactions. A device that provokes irritation or immune activation compromise patient health and device performance. Biocompatibility is important in achieving long term functionality. stable and non reactive materials within the body is less likely to corrode, degrade prematurely, or lose mechanical integrity, it maintain the functionality of the device during its intended lifespan. materials that integrate well with surrounding tissues reduce the risk of infection or rejection and support proper healing. Temporary implants, such as scaffolds or controlled drug delivery system require materials gradually break down and are absorbed naturally by the body. The biodegradable components eliminate needs for surgical removal, reduce patient burden and healthcare costs.

Implantable BioMEMS devices must follow the biocompatibility standards for safety and reliability purposes. ISO 10993, is series of standards developed by the International Organization for Standardization. it provide a comprehensive framework to evaluate the biological effect of medical material, including test for cytotoxicity, sensitization, irritation, systemic toxicity, and chronic exposure effect. The ISO 10993 standards help ensure that implantable device undergo rigorous and consistent evaluation. The United States Pharmacopeia (USP) provide biological reactivity testing protocol specifically designed to assess material used in medical devices and pharmaceutical. It help to confirm that materials do not trigger harmful biological responses when placed in contact with tissues or fluids.

Biocompatibility standards involve a wide array of in vitro and in vivo test. Cytotoxicity testing is the first step, it measure how material affect cell survival and its functions. Common standard, such as those described in ISO 10993-5, expose cultured cell to material extracts to detect toxic effects. Irritation and sensitization testing examines the ability of device or material to cause tissue irritation or allergic reaction when in contact with skin or internal tissues. Hemocompatibility testing required for device interact with blood, evaluating factor such as hemolysis, clot formation, coagulation, and platelet activation. Implantation studies to evaluate the device in animal models or in later phases, in human clinical trials to observe tissue responses over time, such as inflammation, fibrosis, or encapsulation. additional studies are required to evaluate the degradation rate of the biodegradable materials, the nature of degradation byproducts, and their impact on the body. These tests ensures that degradation occurs predictably and safely.

Several materials commonly used in BioMEMS devices because of the favorable biocompatibility and biodegradability profiles. Silicon most widely used materials in MEMS technology because of its excellent mechanical properties and compatibility with microfabrication techniques. Silicon nitride similarly valued for its biocompatibility, mechanical strength, and insulating properties. Polysilicon frequently used in MEMS thin films and structural components due to its stability and compatibility with biological environment. Biodegradable polymers such as poly(lactic-co-glycolic acid) (PLGA) widely used because they degrade into lactic acid and glycolic acid naturally occurring metabolites in the body. PLGA and related polymers like poly(lactic acid) (PLA) and poly(caprolactone) (PCL) important materials for temporary implants, scaffolds, and controlled drug delivery system. Titanium and titanium alloys renowned for their exceptional biocompatibility and commonly employed in orthopedic, dental, and MEMS based implantable components. Biodegradable magnesium alloys gaining interest for temporary BioMEMS applications because they degrade slowly and safely within the body.

Biocompatibility and biodegradability are critical attribute in the design of implantable BioMEMS devices to ensure that devices function safely within the body, support long term therapeutic applications, and eliminate the need for surgical removal. following international testing standards and employing well characterized materials, engineers and researcher develop BioMEMS technologies that meet stringent safety requirements while enabling advanced medical functionalities. Since the BioMEMS continues to expand into new therapeutic domain, the importance of carefully selecting and evaluating materials remain central to ensuring patient safety, device reliability, and clinical success.

8.13 Chapter summary

This chapter explored the rapidly developing field of BioMEMS and its important role in medical implants. It began by outlining the historical development of BioMEMS, highlighting how advances in microfabrication, materials science, and biotechnology enable the creation of miniature, highly functional implantable devices that interact with biological systems. The chapter emphasized the growing importance of BioMEMS across healthcare system due to their precision, biocompatibility, and ability to support patient medicine.

A series of major BioMEMS based implantable systems examined in depth. Implantable hearing aids demonstrated how MEMS acoustic sensor provide miniaturization, sensitivity, low power performance, and wireless connectivity, enabling fully implantable auditory solutions. Glucose monitoring implants shown to transform diabetes management through continuous sensing, feedback mechanisms, and real time data transmission. The chapter also detailed the architecture and fabrication of implantable MEMS micropumps, which offer accurate, programmable drug delivery for chronic diseases, pain management, and oncology.

The discussion expanded into more complex systems, such as the fully implantable MEMS cochlea, outlining its intricate microfabricated structure, piezoelectric sensing elements, low power signal processing electronics, and stringent testing requirements to validate function and biocompatibility. Additional implantable technologies, including cardiac pacemakers, BioMEMS-based stents, neurostimulators, drug delivery microdevices, and MEMS based artificial lung components, further illustrated how microscale engineering reshaping therapeutic interventions across cardiology, neurology, endocrinology, and respiratory care.

The chapter highlighted the transformative potential of BioMEMS systems: higher precision, continuous physiological monitoring, targeted therapy, reduced invasiveness, and increased patient comforts. Ongoing advancements in microfabrication, wireless power, energy harvesting, AI-driven closed loop control, and biocompatible materials promise to make future BioMEMS implants even more intelligents, adaptive, and personalized. The chapter demonstrates that BioMEMS technologies not only redefining modern medical implants but also paving the way toward new generation of responsive, lifelong, and patient specific healthcare solutions.

References

Alrashdan, M. H. S., Hamzah, A. A., & Majlis, B. Y. (2015). Design and optimization of cantilever based piezoelectric micro power generator for cardiac pacemaker. *Microsystem Technologies, 21*(8), 1607–1617. https://doi.org/10.1007/s00542-014-2334-1.

Au, A. K., Lai, H., Utela, B. R., & Folch, A. (2011). Microvalves and micropumps for BioMEMS. *Micromachines., 2*(2), 179–220. https://doi.org/10.3390/mi2020179.

Birkholz, M., Ehwald, K. E., Basmer, T., Kulse, P., Reich, C., Drews, J., Genschow, D., Haak, U., Marschmeyer, S., Matthus, E., Schulz, K., Wolansky, D., Winkler, W., Guschauski, T., & Ehwald, R. (2013). Sensing glucose concentrations at GHz frequencies with a fully embedded Biomicro-electromechanical system (BioMEMS). *Journal of Applied Physics, 113*(24). https://doi.org/10.1063/1.4811351.

Borenstein, J. T., & Vunjak-Novakovic, G. (2011). Engineering tissue with BioMEMS. *IEEE Pulse, 2*(6), 28–34. https://doi.org/10.1109/MPUL.2011.942764.

Bußmann, A., Leistner, H., Zhou, D., Wackerle, M., Congar, Y. ü, Richter, M., & Hubbuch, J. (2021). Piezoelectric silicon micropump for drug delivery applications. *Applied Sciences., 11*(17), 8008. https://doi.org/10.3390/app11178008.

Chen, Y., Kim, Y.-S., Tillman, B., Yeo, W.-H., & Chun, Y. (2018). Advances in materials for recent low-profile implantable bioelectronics. *Materials., 11*(4), 522. https://doi.org/10.3390/ma11040522.

Cho, S. H., Cauller, L., Rosellini, W., & Lee, J. B. (2010). A MEMS-based fully-integrated wireless eurostimulator. In *Proceedings of the IEEE International Conference on Micro Electro Mechanical Systems (MEMS)* (pp. 300–303). https://doi.org/10.1109/MEMSYS.2010.5442507.

Chung, J., Jung, Y., Hur, S., Kim, J. H., Kim, S. J., Kim, W. D., Choung, Y. H., & Oh, S. H. (2021). Development and characterization of a biomimetic totally implantable artificial basilar membrane system. *Frontiers in Bioengineering and Biotechnology, 9*. https://doi.org/10.3389/fbioe.2021.693849.

Cros, F. (2009). *BioMEMS biomedical materials* (pp. 443–475). United States: Springer US. https://doi.org/10.1007/978-0-387-84872-3_16.

Farahani, A., Zarei-Hanzaki, A., Abedi, H. R., Daryoush, S., Ragheb, Z. D., Mianabadi, F., Shahparvar, S., Akrami, M., Mostafavi, E., Khanbareh, H., & Nezami, F. R. (2023). Silk-based biopolymers promise extensive biomedical applications in tissue engineering, drug delivery, and BioMEMS. *Journal of Polymers and the Environment, 31*(11), 4559–4582. https://doi.org/10.1007/s10924-023-02906-x.

Gensler, H., Sheybani, R., Li, P. Y., Mann, R. L., & Meng, E. (2012). An implantable MEMS micropump system for drug delivery in small animals. *Biomedical Microdevices, 14*(3), 483–496. https://doi.org/10.1007/s10544-011-9625-4.

Gharib, G., Bütün, İ., Muganlı, Z., Kozalak, G., Namlı, İ., Sarraf, S. S., Ahmadi, V. E., Toyran, E., van Wijnen, A. J., & Koşar, A. (2022). Biomedical applications of microfluidic devices: A review. *Biosensors., 12*(11). https://doi.org/10.3390/bios12111023.

Grayson, A. C. R., Shawgo, R. S., Johnson, A. M., Flynn, N. T., Li, Y., Cima, M. J., & Langer, R. (2004). A BioMEMS review: MEMS technology for physiologically integrated devices. In *Proceedings of the IEEE: 92* (pp. 16–21). United States: Institute of Electrical and Electronics Engineers Inc. https://doi.org/10.1109/JPROC.2003.820534.

İlik, B., Koyuncuoğlu, A., Şardan-Sukas, Ö., & Külah, H. (2018). Thin film piezoelectric acoustic transducer for fully implantable cochlear implants. *Sensors and Actuators, A: Physical, 280*, 38–46. https://doi.org/10.1016/j.sna.2018.07.020.

Jeyalakshmi Devi, V. (2017). Artificial cardiac pacemaker 2018-. In *ICONSTEM 2017 - Proceedings: 3rd IEEE International Conference on Science Technology, Engineering and Management* (pp. 1015–1017). India: Institute of Electrical and Electronics Engineers Inc. https://doi.org/10.1109/ICONSTEM.2017.8261320.

Kalra, S., & Mashuq-Un-Nabi (2018). Implantable Bio-MEMS applications: A review. In *2017 Recent Developments in Control, Automation and Power Engineering, RDCAPE* (pp. 131–136). India: Institute of Electrical and Electronics Engineers Inc. https://doi.org/10.1109/RDCAPE.2017.8358254.

Ko, W. H., Zhang, R., Huang, P., Guo, J., Ye, X., Young, D. J., & Megerian, C. A. (2009). Studies of MEMS acoustic sensors as implantable microphones for totally implantable hearing-aid systems. *IEEE Transactions on Biomedical Circuits and Systems, 3*(5), 277–285. https://doi.org/10.1109/TBCAS.2009.2032267.

Kulah, H., Ulusah, H., Chamanian, S., Batu, A., Ugur, M. B., Yuksel, M. B., Yilmaz, A. M., Yigit, H. A., Koyuncuoglu, A., Topcu, O., & Soydan, A. K. (2022). A Fully-Implantable Mems-Based Autonomous Cochlear Implant 2022. In *Proceedings of the IEEE International Conference on Micro Electro Mechanical Systems (MEMS)* (pp. 396–399). Turkey: Institute of Electrical and Electronics Engineers Inc. https://doi.org/10.1109/MEMS51670.2022.9699689.

Latif, R., Noor, M. M., Yunas, J., & Hamzah, A. A. (2021). Mechanical energy sensing and harvesting in micromachined polymer-based piezoelectric transducers for fully implanted hearing systems: A review. *Polymers, 13*(14). https://doi.org/10.3390/polym13142276.

Nisar, A., Afzulpurkar, N., Mahaisavariya, B., & Tuantranont, A. (2008). MEMS-based micropumps in drug delivery and biomedical applications. *Sensors and Actuators, B: Chemical, 130*(2), 917–942. https://doi.org/10.1016/j.snb.2007.10.064.

Prochazka, L., Huber, A., Dobrev, I., Harris, F., Dalbert, A., Röösli, C., Obrist, D., & Pfiffner, F. (2019). Packaging technology for an implantable inner ear MEMS microphone. *Sensors, 19*(20), 4487. https://doi.org/10.3390/s19204487.

Shuvo, M. dM. H, Titirsha, T., Amin, N., & Islam, S. K. (2022). Energy harvesting in implantable and wearable medical devices for enduring precision healthcare. *Energies., 15*(20), 7495. https://doi.org/10.3390/en15207495.
Yang, Y., Wang, B., Cui, J., Zhang, G., Wang, R., Zhang, W., He, C., Li, Y., Shi, P., & Wang, S. (2022). Design and realization of MEMS heart sound sensor with concave, racket-shaped cilium. *Biosensors, 12*(7), 534. https://doi.org/10.3390/bios12070534.
Zafar, H., Channa, A., Jeoti, V., & Stojanović, G. M. (2022). Comprehensive review on wearable sweat-glucose sensors for continuous glucose monitoring. *Sensors., 22*(2), 638. https://doi.org/10.3390/s22020638.

CHAPTER

Biological microelectromechanical systems applications

9

Nashrul Fazli Mohd Nasir[a], Anas Mohd Noor[a], Mazlee Mazalan[b], Mohd Hafizuddin Junaidi[b] and Ruslinda A Rahim[c]

[a]*Biomedical Electronic Engineering Program, Faculty of Electronic Engineering and Technology, Universiti Malaysia Perlis, Arau, Malaysia,* [b]*Faculty of Electronic Engineering and Technology, Universiti Malaysia Perlis, Arau, Malaysia,* [c]*Lynas Rare Earths, Kuantan, Pahang, Malaysia*

9.1 Biological microelectromechanical systems for wearable health monitoring

This section explores the application of biological microelectromechanical systems (BioMEMS) in the field of wearable health technology. It details the transition toward continuous, non-invasive physiological assessment by examining the flexible substrate materials required for skin integration, various sensing modalities including electrochemical and electrophysiological methods and the incorporation of microfluidics and wireless data processing for real-time monitoring.

9.1.1 Overview and clinical motivation

Wearable biological microelectromechanical systems (BioMEMS) represent a paradigm shift from periodic clinical diagnostics toward continuous, non-invasive physiological assessment. These microsystems provide real-time data streams essential for detecting temporal patterns and early physiological indicators (Kharb, 2025; Marinescu et al., 2025). The key advantage of MEMS technology in this domain is the ability to integrate sensing, fluid handling, and signal transduction into a unified, skin-mountable platform.

9.1.2 Substrate materials and mechanical integration

Long-term wearable monitoring requires substrate materials that accommodate mechanical mismatch between rigid silicon components and human skin (modulus ~0.1–1 MPa). Researchers have adopted ultrathin, flexible materials, including polyimide (PI) and parylene-C, as structural backbones for these microsystems. Polyimide offers compatibility with standard photolithography and metal deposition processes, enabling precise patterning of microelectrodes and interconnects (Rogers et al., 2016).

The flexible and wearable rGO-GOx/LIG-IEs biosensor on polyimide, fabricated by Sun et al. (2023), demonstrates this approach effectively (Fig. 9.1). Mechanical durability is achieved through serpentine interconnects and fractal microstructures that distribute mechanical strain across

BioMEMS Devices. DOI: https://doi.org/10.1016/B978-0-443-24015-7.00003-6

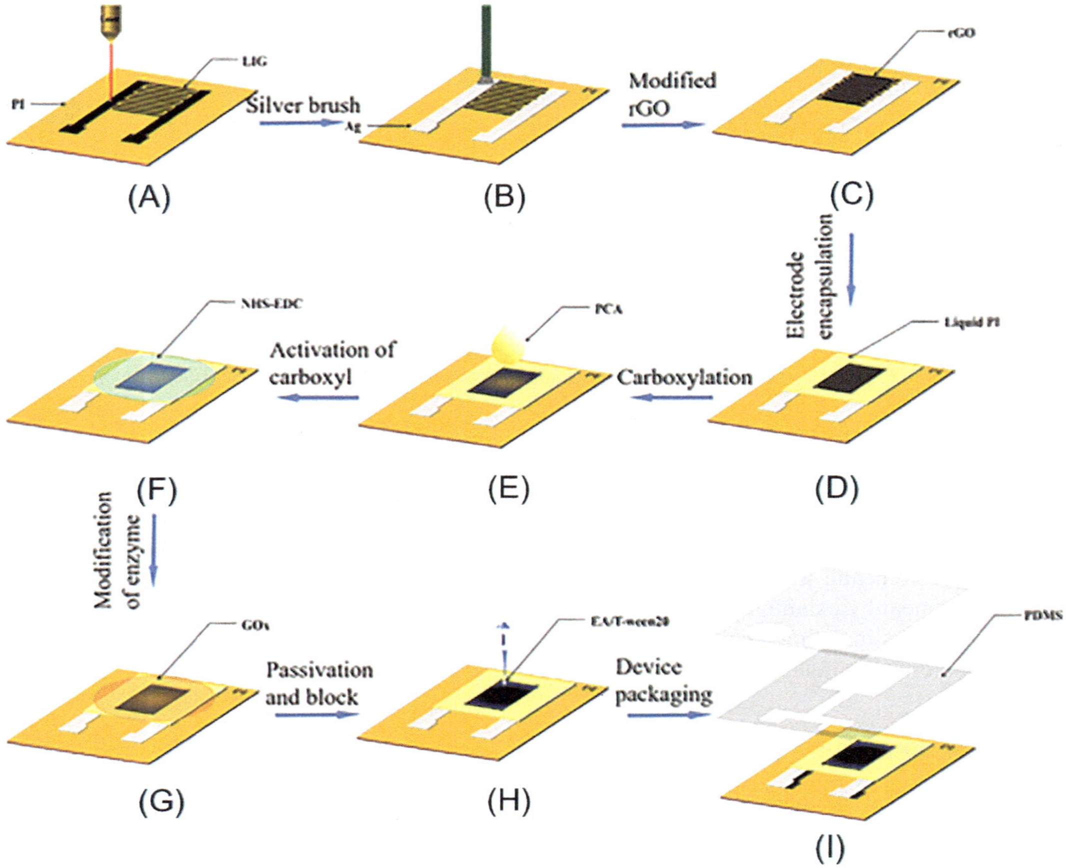

FIGURE 9.1

Fabrication process of the flexible and wearable rGO-GOx/LIG-IEs biosensor. Sequential fabrication of the flexible biosensor, from (A) laser-induced graphene (LIG) patterning and (B) silver coating, to (C) rGO deposition, (D) PI encapsulation, (E–F) surface carboxylation and activation, (G) GOx immobilization, and (H–I) final stabilization and PDMS packaging (Sun et al., 2023).

the device, preventing stress concentration at sensor interfaces. Such designs accommodate strains exceeding 20%–30% while maintaining electrical continuity (Marinescu et al., 2025).

9.1.3 Electrochemical and electrophysiological sensing

Electrochemical and electrophysiological sensing represent complementary approaches in wearable BioMEMS, enabling real-time monitoring of biochemical and bioelectrical states (Luo et al., 2021; Rogers et al., 2016). These platforms achieve high sensitivity at the skin interface while maintaining user comfort through advances in microfabrication and materials engineering.

Micro-electrochemical sensors typically employ screen-printed or sputtered microelectrodes functionalized with specific enzymes (e.g., glucose oxidase) or ion-selective membranes. These devices utilize potentiometric and amperometric transduction to quantify electrolytes and metabolites in nanoliter biofluid volumes (Marinescu et al., 2025).

Conversely, micro-electrophysiological sensors use flexible microelectrode arrays fabricated from conducting polymers (e.g., PEDOT: PSS) or nanomaterials to record bioelectrical signals, including ECG, EMG, and EEG. The micro-scale electrode geometry improves signal-to-noise ratios by reducing impedance and conforming closely to the skin surface (Rogers et al., 2016).

9.1.4 Microfluidic sweat and interstitial fluid sensors

Non-invasive biofluid collection and transport often require integrated microfluidic channels. These channels, fabricated from soft polymers (PDMS) or laminated thermoplastics, are incorporated directly into wearable patches. Capillary action passively transports sweat from the skin surface to sensor arrays without external pumps (Kharb, 2025; Marinescu et al., 2025).

Microfluidic architecture provides multiple MEMS-specific functions: defining precise sample volumes (nanoliter to microliter ranges), reducing evaporation, and enabling fresh sample delivery to sensing sites. Recent designs incorporate passive capillary valves for sequential fluid routing, allowing time-resolved biomarker analysis.

Noor et al. (2025) demonstrated a notable wearable platform for continuous, real-time monitoring of human sweat sodium. Their skin-interfaced microfluidic layer continuously transports fresh sweat across ion-selective electrode arrays, enabling stable on-body sodium readout during exercise without external pumping. This design illustrates how carefully engineered microfluidic geometries sustain unidirectional fresh sweat flow, mitigating signal drift from stagnation and evaporation while supporting long-term electrolyte tracking (Fig. 9.2).

Schematic drawing and 3D print of the microfluidic chip and sodium sensor (Noor et al., 2025).

9.1.5 Surface acoustic wave devices in wearable biosensing

SAW devices offer a unique MEMS-based sensing mechanism distinct from electrochemical approaches. Unlike electrochemical sensors that rely on chemical reactions, SAW resonators detect mass or viscosity changes on device surfaces. Flexible SAW sensors on piezoelectric polymer substrates (e.g., PVDF) or thin-film piezoelectric layers (ZnO) respond to mechanical strain and mass loading, enabling monitoring of physiological parameters such as respiration rate and skin hydration (Collins et al., 2020; Huang et al., 2024).

Furthermore, SAW devices induce acoustic streaming within microfluidic channels, enhancing mixing and reaction kinetics in diffusion-limited sensing regimes. Wang et al. (2021) provided design guidelines for flexible acoustofluidic SAW devices through characterization of ZnO/Al sheet configurations versus conventional ZnO/Si SAWs, emphasizing thickness effects on wave modes and microfluidic actuation. A 200 μm Al sheet thickness optimized fluid transport across bent/deformed surfaces, enabling robust wearable biosensing performance.

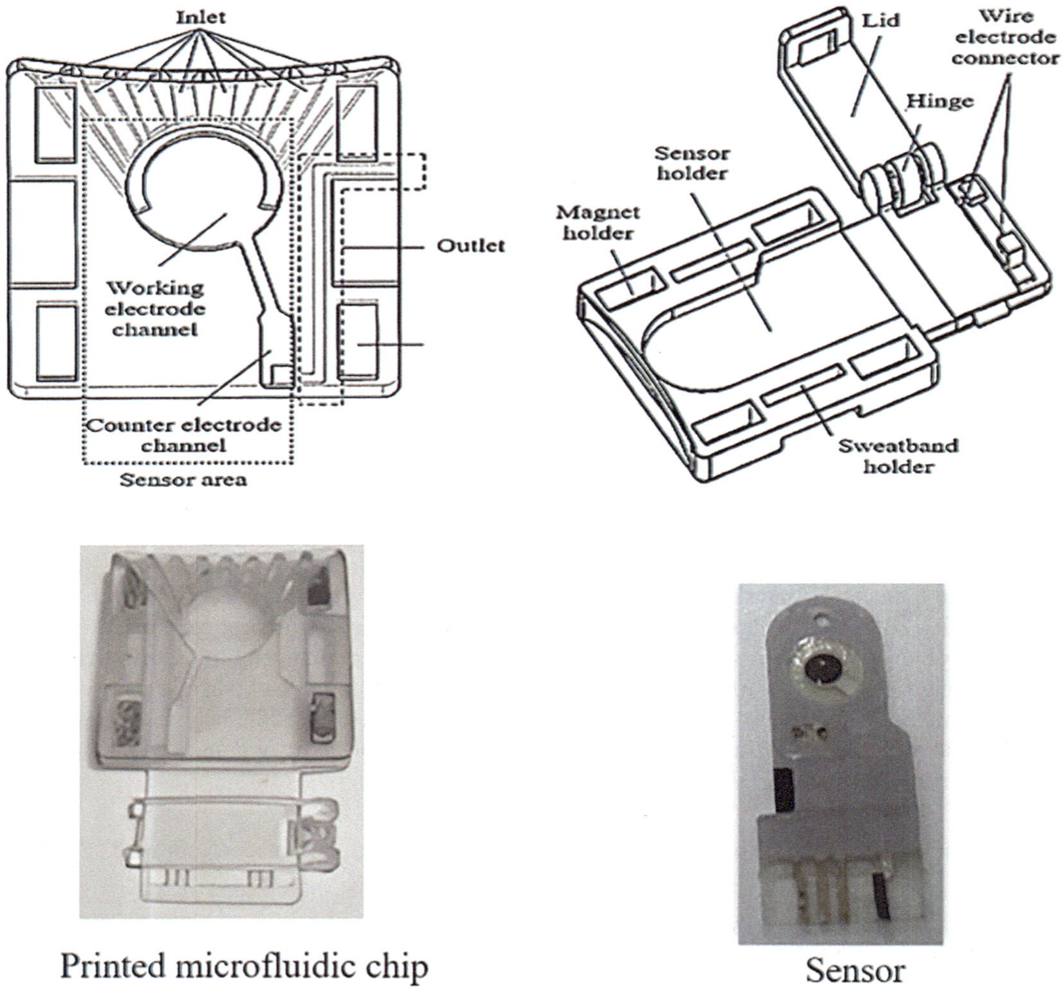

FIGURE 9.2

Schematic drawing and 3D print of the microfluidic chip and sodium sensor for wearable sweat sensing monitoring application (Noor et al., 2025).

9.1.6 Signal processing, wireless communication, and data analysis

Clinical utility of wearable BioMEMS depends on miniaturized electronics for signal conditioning. Application-Specific Integrated Circuits (ASICs) or microcontrollers perform analog-to-digital conversion (ADC) and digital signal processing (DSP) directly on the wearable node. Low-energy wireless protocols such as Bluetooth Low Energy (BLE) minimize power consumption (Kharb, 2025; Marinescu et al., 2025).

Recent advancements employ edge-computing architectures where machine learning algorithms execute on the device's microcontroller. This enables real-time anomaly detection without continuous raw data transmission, supporting autonomous operation (Kharb, 2025; Marinescu et al., 2025).

9.2 Biological microelectromechanical systems as surgical tools and micromanipulation platforms

This section examines the transition of BioMEMS from passive monitoring devices into active surgical tools and micromanipulation platforms. It highlights the use of traditional microactuation mechanisms for precise, cellular-level interventions, the implementation of surface acoustic waves (SAW) for contactless particle and cell manipulation, and the integration of microfluidics to enable targeted therapeutic delivery during microsurgery.

9.2.1 Context and clinical motivation

Surgical BioMEMS facilitates active microscale manipulation, transitioning from passive monitoring to active intervention. These tools employ microactuators for enhanced precision in procedures requiring cellular-level manipulation, including single-cell biopsy and targeted drug delivery. Generating piconewton to nanonewton forces is critical for interacting with biological structures without mechanical damage (Kim et al., 2010; Menciassi & Dario, 2013).

9.2.2 Traditional biological microelectromechanical systems surgical tools

Conventional surgical MEMS employ diverse microactuation mechanisms for precise micromanipulation. Electrothermal microactuators configured as V-shaped beams or bimorphs (e.g., Al–SiO_2) leverage differential thermal expansion from Joule heating to generate displacements and gripping forces suitable for handling individual cells or organelles (Kim et al., 2010).

Shape-memory alloy (SMA) microgrippers utilize thin-film phase transitions to achieve large displacements and high force-to-weight ratios, enabling autonomous cyclic motion in biomedical gripping tasks (Menciassi & Dario 2013). Piezoelectric microactuators provide complementary high-frequency, sub-micron precision for cellular tissue cutting or penetration, enhancing system versatility.

9.2.3 Surface acoustic wave-enhanced micromanipulation

SAW-based micromanipulation offers contactless alternatives to mechanical grippers. Interdigital transducers (IDTs) on piezoelectric substrates generate standing or traveling waves that create localized acoustic radiation forces. These forces trap and translate particles or cells within microfluidic chambers without physical contact, eliminating mechanical trauma and contamination risks (Collins et al., 2020; Huang et al., 2024).

Force magnitude is precisely tunable via input voltage amplitude. Additionally, SAW-induced acoustic streaming manipulates fluids, assisting debris removal and reagent mixing in microsurgical fields. Zhang et al. (2021) developed acoustoelectronic nanotweezers that generate dynamic acoustic

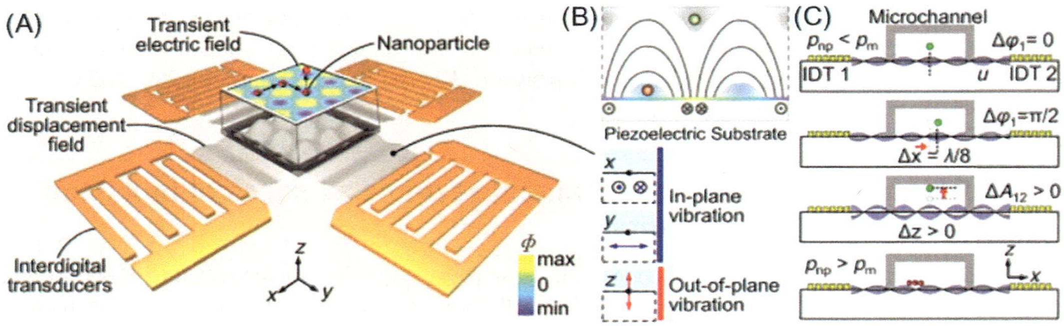

FIGURE 9.3

(A) Acoustoelectronic fields are generated via dynamic acoustic wave interactions. (B) Schematic side-view of the electric field distribution and trapping positions for particles with different polarizabilities relative to the medium. (C) Schematic mechanism of manipulating nanoparticles with lower or higher polarizability (Zhang et al., 2021).

fields via interdigital electrodes on optimized substrates like Y-41° $LiNbO_3$, enabling precise sub-100 nm nanoparticle trapping and patterning (Fig. 9.3).

9.2.4 Integration of microfluidics with surgical biological microelectromechanical systems

Advanced surgical platforms increasingly integrate microfluidic components for complex procedures. Microchannels embedded within surgical tools deliver therapeutic agents or saline irrigation directly to device tips. Recent integrated systems combine hollow microneedles with acoustic actuation: microneedles penetrate tissue barriers while acoustic energy enhances genetic material and drug diffusion into target cells (Kim et al., 2010; Menciassi & Dario 2013). This convergence represents a fully MEMS-based surgical intervention approach.

9.3 Biological microelectromechanical systems for neural interfaces

This section explores the development of BioMEMS for neural interfaces, emphasizing the critical need to overcome mechanical mismatches between rigid probes and soft neural tissue. It discusses the design and fabrication of flexible neural probes, multimodal sensing and stimulation capabilities, and the application of SAW technology and soft neural cuffs for both central and peripheral nervous system monitoring and modulation.

9.3.1 Overview and clinical context

Neural interfaces require high-fidelity recording and stimulation for applications including brain-computer interfaces (BCIs) and deep brain stimulation (DBS). The primary challenge is a mechanical

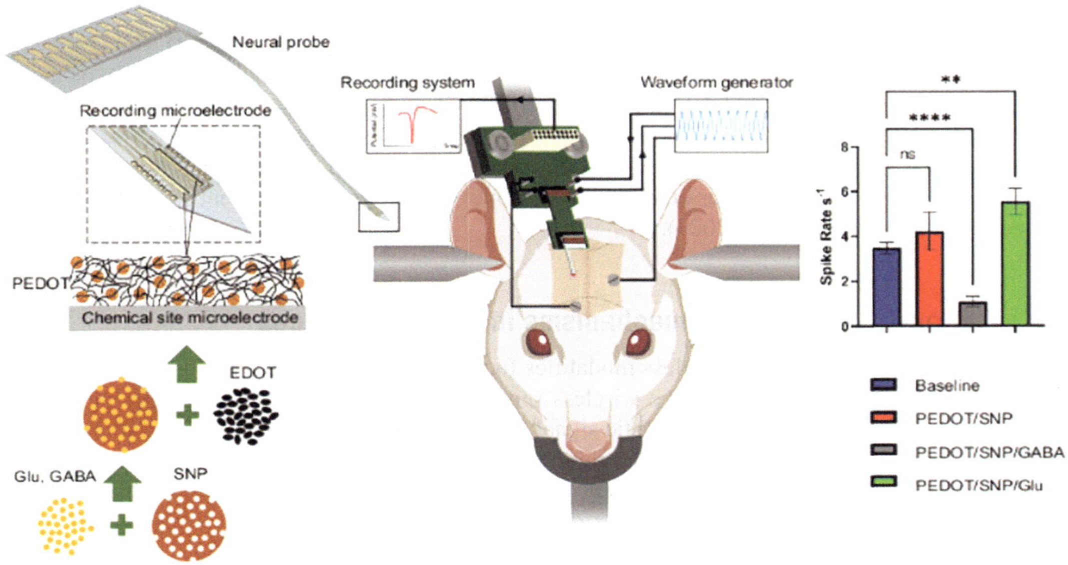

FIGURE 9.4

Fully flexible implantable neural probes for electrophysiology recording and controlled neurochemical modulation (Malekoshoaraie et al., 2024).

mismatch between rigid silicon microprobes (~200 GPa) and soft neural tissue (~3 kPa), leading to chronic inflammation and device failure (Lv et al., 2025; Zhao et al., 2022).

9.3.2 Flexible neural probes: Materials and design

Next-generation neural probes improve biocompatibility through microfabrication techniques adapted for flexible polymers. Ultrathin substrates (polyimide, parylene-C, SU-8) serve as structural bases, reduced to micro-scales (10–50 μm) to minimize flexural stiffness (Lv et al., 2025; Rogers et al., 2016).

Microelectrodes are patterned onto substrates using photolithography and thin-film deposition. Temporary stiffeners made from bioresorbable polymers (e.g., PEG, PLGA) or microfluidic insertion guides facilitate insertion of these compliant structures (Zhao et al., 2022).

Malekoshoaraie et al. (2024) developed a fully flexible implantable neural probe on a biocompatible Parylene C substrate that integrates 16 small recording electrodes with two large chemical sites loaded with porous silica nanoparticles for localized drug delivery. This design enables simultaneous high-density electrophysiology recording and electrically controlled neurochemical release from the compliant shank. In vivo rat cortex tests demonstrated the probe's ability to conform to brain tissue, minimizing mismatch while maintaining stable chronic neuronal monitoring performance (Fig. 9.4).

9.3.3 Multimodal sensing and stimulation

Neural BioMEMS increasingly integrate multimodal capabilities on single micro-devices:

Microfluidic delivery channels: Probe shank channels enable precise, localized delivery of neuromodulators or anti-inflammatory drugs directly to recording sites (Cai et al., 2024).
Integrated optical waveguides: For optogenetic applications, micro-scale waveguides or micro-LEDs integrated into probes deliver light to specific neural populations with high spatial resolution (Zhao et al., 2022; Lee et al., 2025).
Biochemical sensors: Organic electrochemical transistors (OECTs) integrated onto probes enable detection of specific neurotransmitters (e.g., dopamine) by transducing chemical concentrations into electrical currents (Liu et al., 2024).

9.3.4 Surface acoustic wave mechanisms in neural interfaces

SAW technology provides novel wireless modalities for neural interfacing. SAW resonators integrated into neural implants function as passive wireless sensors, detecting tissue viscoelasticity or pressure changes by modulating the reflected acoustic signal frequency. Acoustic actuation offers non-electrical neuromodulation methods: mechanical stimulation via acoustic waves modulates ion channel activity (mechanotransduction), providing therapeutic alternatives to electrical stimulation while avoiding charge-injection tissue damage (Collins et al., 2020; Huang et al., 2024).

9.3.5 Peripheral neural interfaces and soft neural cuffs

Peripheral neural interfaces utilize MEMS fabrication to create soft, conformal cuff electrodes. These devices employ flexible substrates with stretchable, serpentine metal interconnects that wrap around peripheral nerves without compression (Lv et al., 2025; Zhao et al., 2022). Recent designs incorporate micro-patterned electrode arrays for selective stimulation of specific nerve fascicles. SAW mechanisms are being explored for peripheral nerve regeneration, where acoustic stimulation may enhance axonal outgrowth through mechano-activation of growth pathways (Collins et al., 2020; Huang et al., 2024).

9.4 Future directions, challenges, and clinical translation

This concluding section outlines the future trajectory of BioMEMS technologies. It addresses the critical hurdles that must be overcome for widespread clinical adoption, such as complex system-level integration, long-term biocompatibility, and stringent regulatory validation. Additionally, it highlights emerging opportunities driven by the convergence of MEMS with artificial intelligence and closed-loop systems for personalized medicine.

9.4.1 Integration and system-level design

Achieving cohesive system-level integration remains a significant challenge. Future BioMEMS must integrate sensing, actuation, power management, and communication into highly miniaturized form factors. Advances in 3D micro-integration (e.g., stacking MEMS layers) and hierarchical substrate architectures provide pathways toward more sophisticated, multi-functional microsystems (Kharb, 2025; Marinescu et al., 2025).

9.4.2 Biocompatibility and long-term reliability

A critical barrier for implantable BioMEMS is the foreign body response. Biofouling and encapsulation can isolate micro-sensors from the biological environment. Mitigation strategies include biomimetic surface coatings (e.g., hydrogels) and microfluidic channels for controlled anti-inflammatory agent release (Gao et al., 2025; Lv et al., 2025).

9.4.3 Clinical validation and regulatory pathways

Transitioning these technologies to market requires rigorous validation. Regulatory frameworks continue adapting to BioMEMS-specific challenges, including nanomaterial safety and wireless micro-component reliability. Standardized testing protocols for MEMS-tissue interactions are essential for ensuring patient safety (Kharb, 2025; Marinescu et al., 2025).

9.4.4 Convergence of technologies and emerging opportunities

Future opportunities emerge from MEMS modality convergence. Wearable SAW sensors could combine with closed-loop micro-stimulation systems for personalized medicine. On-chip machine learning will enable these integrated microsystems to function autonomously, analyzing multimodal data streams to predict clinical events and adjust therapy in real-time (Kharb, 2025; Marinescu et al., 2025).

9.5 Chapter summary

BioMEMS technologies have established themselves as transformative platforms for modern biomedical engineering. By leveraging micro- and nanoscale phenomena, these devices enable capabilities impossible with macroscopic tools: single-cell manipulation, wireless passive sensing, and conformal neural recording. Continued research into materials, microfabrication techniques, and system integration will accelerate clinical translation, ultimately delivering significant improvements in patient outcomes.

AI disclosure

During the preparation of this work, the author(s) used AI Scopus and Gemini in order to list down references related to the topic and to summarize certain articles downloaded. After using this tool/service, the author(s) reviewed and edited the content as needed and take(s) full responsibility for the content of the publication.

References

Cai, X., Liu, J., & Chen, B. (2024). Recent advances in intracortical neural interfaces for brain-computer interfaces. *Progress in Neurobiology, 238*, 102578. https://doi.org/10.1016/j.pneurobio.2024.102578.

Collins, D. J., Neild, A., & Friend, R. (2020). Surface acoustic wave microfluidics and its applications. *Lab on a Chip, 20*(6), 1185–1204. https://doi.org/10.1039/D0LC00150C.

Gao, W., Yan, Z., Zhou, H., Xie, Y., Wang, H., Yang, J., … Ye, Z. (2025). Revolutionizing brain–computer interfaces: Overcoming biocompatibility challenges in implantable neural interfaces. *Journal of Nanobiotechnology, 23*(1), 498. https://doi.org/10.1186/s12951-025-03573-x.

Huang, Y., Ho, J., Xu, X., & Zhang, Y. (2024). Surface acoustic wave microfluidics: Fundamentals, design paradigms, and emerging biomedical applications. *Advanced Materials Interfaces, 11*(9), 2302238. https://doi.org/10.1002/admi.202302238.

Kharb, S. (2025). Future directions: what LIES ahead for smart biochemical wearables in health monitoring? *Frontiers in Analytical Science, 4*, 1509815. doi:10.3389/frans.2024.1509815.

Kim, K., Liu, X., Zhang, Y., & Sun, Y. (2010). Nanotweezers: Biomedical applications and technology. In *MEMS for Biomedical Applications* (pp. 245–268). Woodhead Publishing. https://doi.org/10.1533/9780857090743.2.245.

Lee, J. Y., Kim, T., Cho, S., Shin, J., Yeo, W. H., Kim, T. S., & Yu, K. J. (2025). Design considerations for optogenetic applications of soft micro-LED-based device systems across diverse nervous systems. *Bioactive Materials, 48*, 217–241.

Liu, H., Song, J., Zhao, Z., Zhao, S., Tian, Z., & Yan, F. (2024). Organic electrochemical transistors for biomarker detections. *Advanced Science, 11*(27), e2305347. https://doi.org/10.1002/advs.202305347.

Luo, Y., Abdalrahman, I., & Wang, Z. L. (2021). Wearable electrochemical biosensors for health monitoring: Progress and perspectives. *Advanced Functional Materials, 31*(48), 2106339. https://doi.org/10.1002/adfm.202106339.

Lv, S., Wu, B., & Zhang, X. (2025). Long-term stability strategies of deep brain flexible neural interfaces: A systematic review. *Nature Biomedical Engineering, 9*(4), 451–478. https://doi.org/10.1038/s41528-025-00410-x.

Malekoshoaraie, M. H., Wu, B., Krahe, D. D., Ahmed, Z., Pupa, S., Jain, V., … Chamanzar, M. (2024). Fully flexible implantable neural probes for electrophysiology recording and controlled neurochemical modulation. *Microsystems and Nanoengineering, 10*(1). https://doi.org/10.1038/s41378-024-00685-6.

Marinescu, M. R., Cristea, I. A., & Oancea, R. (2025). Next-generation healthcare devices: Evolution of MEMS and BioMEMS technologies. *IEEE Reviews in Biomedical Engineering, 18*, 45–78. https://doi.org/10.1109/RBME.2024.3478901.

Menciassi, A., & Dario, P. (2013). Surgical microrobots: Advances and perspectives. In *Surgical Robotics* (pp. 139–165). Springer. https://doi.org/10.1007/978-1-4614-4693-2_8.

Noor, A. M., Al Farisi, M. S., Mazalan, M., Ibrahim, N. F. A., Wahab, A. A., Zakaria, Z., Rusli, N. I., Sabani, N., & Manaf, A. A. (2025). Wearable device for continuous and real-time monitoring of human sweat sodium. *Sensors, 25*(11). https://doi.org/10.3390/s25113467.

Rogers, J. A., Ghaffari, R., & Kim, D. (2016). Stretchable and flexible electronics for biomedical applications. *Advanced Materials, 28*(22), 4553–4570. https://doi.org/10.1002/adma.201504523.

Sun, H. X, Song, S., Zhao, G., Wang, X., & Liu, G. (2023). A flexible and wearable chemiresistive biosensor fabricated by laser inducing for real-time glucose analysis of sweat. *Advanced Materials Interfaces, 10*(22). https://doi.org/10.1002/admi.202300281.

Wang, Z., Chen, Y., & Xu, X. (2021). Design optimization of flexible acoustofluidic SAW devices for biomedical applications. *Ultrasonics, 112*, 106351. https://doi.org/10.1016/j.ultras.2020.106351.

Zhang, P., Rufo, J., Chen, C., Xia, J., Tian, Z., Zhang, L., Hao, N., Zhong, Z., Gu, Y., Chakrabarty, K., & Huang, T. J (2021). Acoustoelectronic nanotweezers enable dynamic and large-scale control of nanomaterials. *Nature Communications, 12*(1). https://doi.org/10.1038/s41467-021-24101-z.

Zhao, H., Liu, R., Zhang, H., Cao, P., Liu, Z., & Li, Y. (2022). Research progress on the flexibility of an implantable neural microelectrode. *Micromachines, 13*(3), 386. https://doi.org/10.3390/mi13030386.

CHAPTER

Lab-on-a-chip devices

10

Ahmad Fairuzabadi Mohd Mansor[a] and Ioana Voiculescu[b]

[a]*Department of Electrical and Computer Engineering, Kulliyyah of Engineering, International Islamic University Malaysia, Kuala Lumpur, Malaysia,* [b]*Mechanical Engineering Department, The City College of New York, The City University of New York, New York, NY, United States*

10.1 Introduction

Life science testing and experimentation often involve extensive manual handling of tools, chemicals, and equipment in laboratories by researchers and students. This could impact the productivity and increase costs in research. To address this issue, automated and miniaturized technologies have been developed, such as Lab-on-Chip (LoC) technology. LoC is a class of miniaturized devices that integrate manual laboratory procedures on a single chip, allowing for faster, more efficient, and higher-throughput experimentation and analysis. The core of LoC devices is miniaturization, where complex laboratory processes are scaled down and integrated onto chip-sized platforms. This downsizing brings advantages such as reduced sample and reagent volumes, faster reaction kinetics, increased automation, portability, and cost-effectiveness. Fig. 10.1 shows an example of conventional methods vs LoC platforms for performing several processes such as mixing, heating, and testing. These benefits enhance experimental throughput, sensitivity, and enable point-of-care testing (Oliva-Cárdenas et al., 2021; Subudhi & Das, 2023).

LOC devices are versatile, capable of performing various functions like sample preparation (Poddar et al., 2022), high-throughput screening (Vecchio et al., 2014), cell culture (Mansor et al., 2018; Podwin et al., 2020), and organ-on-a-chip systems (Ingber, 2022). Their modular nature allows seamless integration of components like microfluidic channels, sensors, valves, and actuators, enabling customization to specific applications and experimental needs (Cunha et al., 2022). In this chapter, we explore the capabilities and applications of LOC devices, covering methodologies and technologies for sample preparation, high-throughput screening, cell culture, tissue engineering and organ-on-a-chip systems. Our aim is to provide readers with a basic understanding of LoC technology's principles, methodologies, and emerging trends.

10.2 Sample preparation using lab-on-chip devices

Sample preparation is a laborious process, and difficult to implement on a chip format, due to the requirement for multiple steps, sample manipulation, and purification of samples. The introduction of lab on a chip device for sample preparation, eliminates the need for skilful professionals, due to their simplicity. The miniature dimensions of the lab on a chip system allows for device portability to

BioMEMS Devices. DOI: https://doi.org/10.1016/B978-0-443-24015-7.00011-5

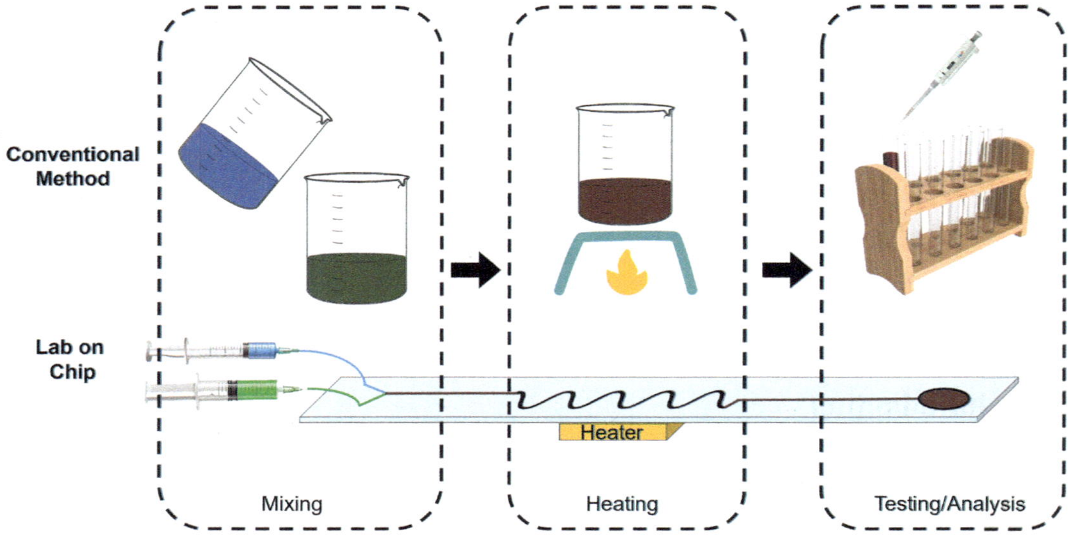

FIGURE 10.1

Example comparison of conventional method with lab-on-chip platform.

resource-limited environments. The reduced device size allows for lower reagent consumption, faster results, and allow automation via microfluidics (Cunha et al., 2022). Since the user intervention is reduced, the contamination is minimal, compared to traditional methodologies.

The majority of lab on chip devices for sample preparations are used for nucleic acid extraction, amplification and detection. For example, one study reports research for the detection of Plasmodium falciparum in plasma samples and Neisseria meningitides in cerebrospinal fluid (CSF) samples using a loop-mediated, isothermal amplification (LAMP) system that can produce results within 50 min (Kadimisetty et al., 2018). This rapid and inexpensive 3D printed device has great potential for point-of-care molecular diagnosis of infectious disease in resource-limited settings, see Fig. 10.2A Isolation of nucleic acids inside the 3D printed reactor array was performed by pipetting the sample into the inlet port and forcing the samples to pass through a silica membrane to bind the nucleic acids, while the lysed cellular debris flowed through. The nucleic acid extraction and amplification was also performed by this lab on chip. Detection of the Plasmodium falciparum parasite was done using both the colorimetric method and real-time fluorescence analysis (Kadimisetty et al., 2018).

Another interesting device was demonstrated to detect 1 pg mass of both Phytophthora infestant DNA and tomato spotted wilt virus (TSWV) RNA (Paul et al., 2021). This device is shown in Fig. 10.2B. A microneedle patch for nucleic acid amplification was integrated with a smartphone and the diagnosis of multiplexed plant pathogens was generated within 30 min. This system is portable and contains polymeric microneedles for fast and complete separation of nucleic acid.

Another notable detection method is based on magnetic beads, which were used to assist the extraction of nucleic acids. These nucleic acids bind to magnetic beads in the presence of high salt solution. An external magnetic field was used to separate the genetic material from waste compounds,

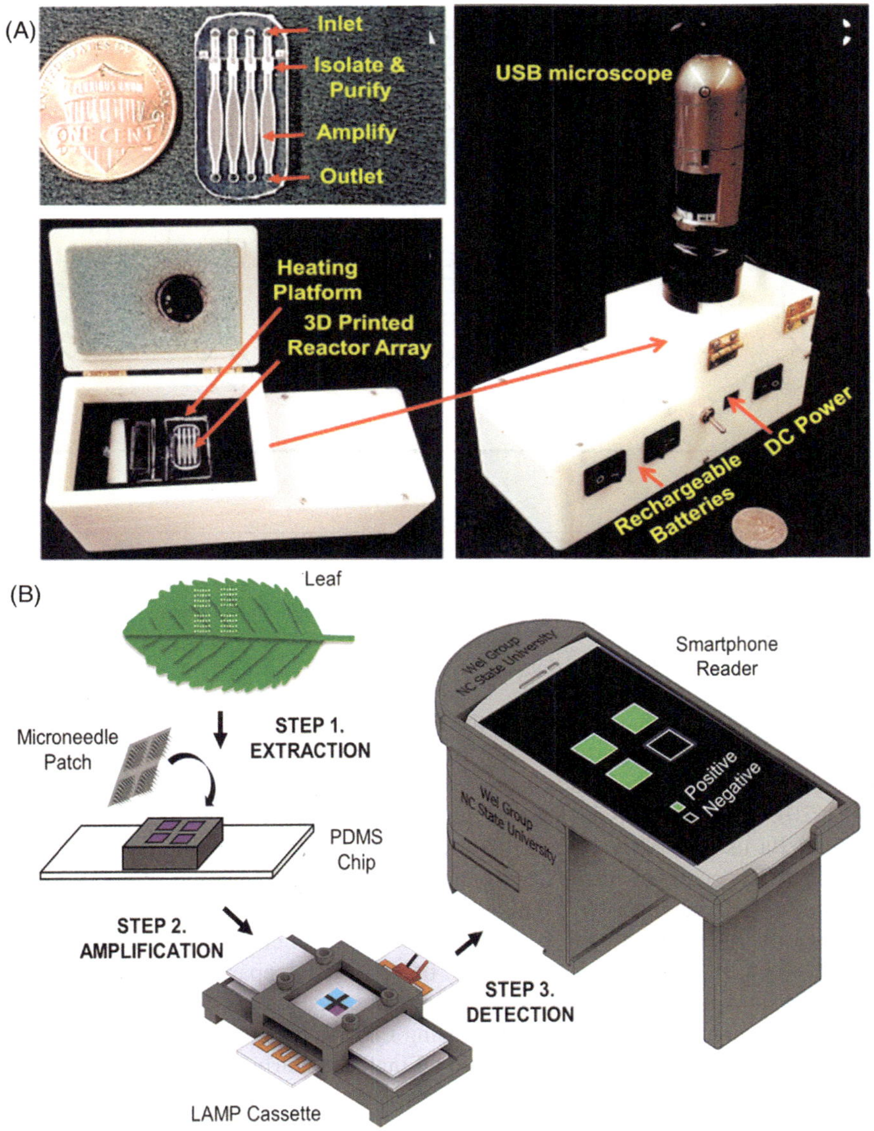

FIGURE 10.2

(A) Portable heater module with USB microscope and heat platform for 3D printed array to process LAMP amplification. (B) Smart phone with microneedle patch for detection of Phytophthora infestant DNA and tomato spotted wilt virus (TSWV) RNA.

(A) From Kadimisetty, K., Song, J., Doto, A. M., Hwang, Y., Peng, J., Mauk, M. G., Bushman, F. D., Gross, R., Jarvis, J. N., & Liu, C. (2018). Fully 3D printed integrated reactor array for point-of-care molecular diagnostics. Biosensors and Bioelectronics, 109, *156–163. https://doi.org/10.1016/j.bios.2018.03.009; (B) from Paul, R., Ostermann, E., Chen, Y., Saville, A. C., Yang, Y., Gu, Z., Whitfield, A.E., & Ristaino, J. B., & Wei, Q. (2021). Integrated microneedle-smartphone nucleic acid amplification platform for in-field diagnosis of plant diseases.* Biosensors and Bioelectronics, 187, *113312. https://doi.org/10.1016/j.bios.2021.113312.*

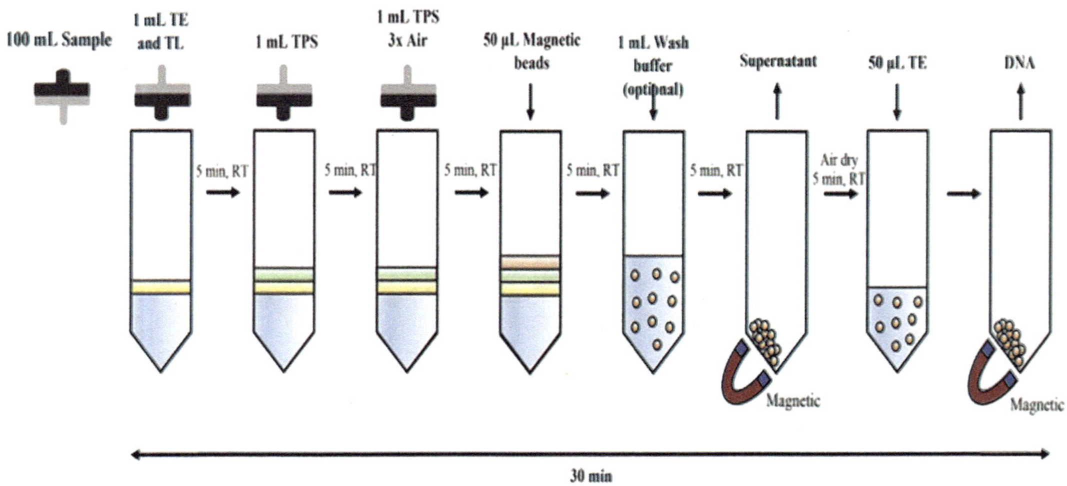

FIGURE 10.3

A diagram of syringe-filter based DNA extraction method.

From Lee, S., Khoo, V. S. L., Medriano, C. A. D., Lee, T., Park, S. Y., & Bae, S. (2019). Rapid and in-situ detection of fecal indicator bacteria in water using simple DNA extraction and portable loop-mediated isothermal amplification (LAMP)PCR methods. Water Research, 160, *371–379. https://doi.org/10.1016/j.watres.2019.05.049.*

see Fig. 10.3 (Lee et al., 2019) In this research the detection is focused on indicator of fecal bacteria in water, using DNA extraction and portable LAMP-polymerase chain reaction (PCR) device.

In another example, magnetic beads were used for a Salmonella typhimurium detection study (Wang et al., 2020). A lab-on-chip device was developed for rapid, automatic and sensitive detection of viable Salmonella using LAMP, a smart phone and magnetic nanoparticles (MNPs) as shown in Fig. 10.4. First, MNPs coated with anti-Salmonella were used to capture antibodies in propidium monoazide (PMA). The beads were fully mixed with bacterial samples, and incubated in a serpentine channel, allowing specific formation of magnetic bacteria and sufficient PMA pretreatment of the DNA of dead bacteria. Then, the PMA-treated magnetic bacteria were separated in the separation chamber using the magnetic field and their genomic DNA templates were extracted using lysis buffer at 70°C. Finally, the viable bacteria's DNA was amplified using LAMP in the detection chamber after blocking with paraffin oil to avoid aerosol cross contamination. Finally, the turbidity of the LAMP reaction system was monitored in a real-time manner for the quantitative detection of viable bacteria. The experimental results demonstrated that this device was able to automatically detect viable Salmonella as low as 14 CFU mL^{-1} in spiked chicken meat supernatants within 1.5 h. In another device magnetic beads coated with vancomycin were utilized to extract the genetic material of *Staphylococcus aureus*; (Shu et al., 2017).

In this case, the magnetic field was combined with a laser diode, which resulted in bacterial capture, concentration, and photothermolysis. The nucleic acids were extracted from bacterial samples and utilized for a PCR assay (Shu et al., 2017). In another technique for detection of the *S. aureus*, the bead were coated with a specific nucleotide, wherein genetic material from methicillin-resistant *S. aureus*

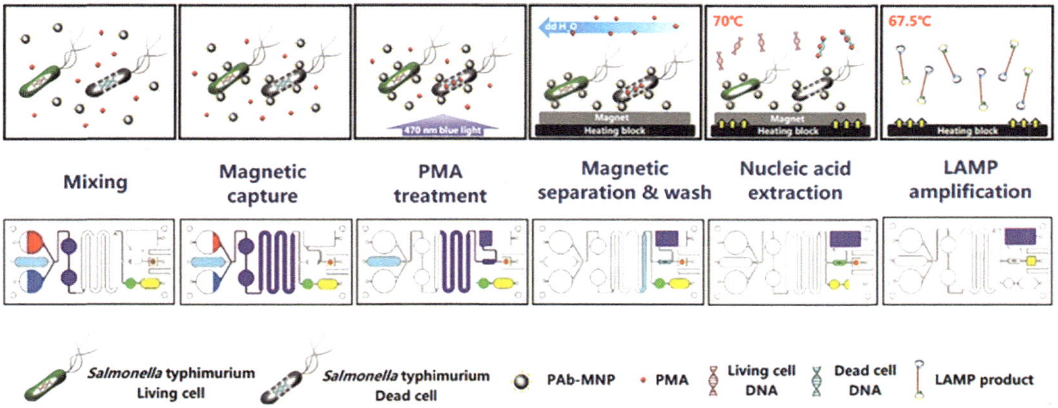

FIGURE 10.4

Schematic of the lab-on-chip device for automatic and sample-in-result-out detection of viable *Salmonella typhimurium*.

From Wang, S., Liu, N., Zheng, L., Cai, G., & Lin, J. (2020). A lab-on-chip device for the sample-in-result-out detection of viable Salmonella using loop-mediated isothermal amplification and real-time turbidity monitoring. Lab on a Chip, 20*(13), 2296–2305. https://doi.org/10.1039/d0lc00290a.*

was extracted from spiked milk, sputum, and serum for its detection via LAMP assay (Wang et al., 2011).

Magnetic beads can also be coupled with capillaries and droplet formation techniques, forming a water-in-oil droplet that prevents evaporation of reagents during amplification (Liu et al., 2013a). This particular method was used for the extraction of genetic material of Mycobacterium tuberculosis from the Bacillus Calmette Guerin (BCG) vaccine and sputum saliva samples, which were later identified using LAMP reaction.

Centrifugal microfluidics (Ducrée et al., 2007; Madou et al., 2006; Mark, et al., 2010b) are one of the most remarkable microfluidic platforms. The rotational speed control of the platform facilitates a set of microfluidic unit operations such as liquid pumping, valving, metering, aliquoting (Gorkin et al., 2010; Strohmeier et al., 2015). Unlike other microfluidic platforms that require external pumps and tubing systems, only a single motor is the necessary accessory for operating the centrifugal microfluidic platform (Gorkin et al., 2010). Owing to the simplicity of operation and the portability for point-of-care diagnostics, such devices have been widely applied for many areas including immunoassay (Park et al., 2012b), cell analysis (Burger et al., 2015), nanoparticle synthesis (Park et al., 2015a, 2015b), and molecular diagnostics (Focke et al., 2010; Jung et al., 2012; Keller et al., 2015; Park et al., 2012a; Strohmeier et al., 2014). Point-of-care molecular diagnostics have been advanced by the aid of centrifugal microfluidics and isothermal DNA amplification techniques and provide advantages of miniaturization, operational simplicity, and user-friendliness.

One interesting application of the centrifugal microfluidic device is for detection of food borne pathogens (*Escherichia coli, Salmonella typhimurium, Vibrio parahaemolyticus* and *Listeria monocytogenes*), see Fig. 10.5A (Oh & Seo, 2019). In this research it was demonstrated that the centrifugal microdevice could be combined with a 3D printed solution-loading cartridge to achieve a fully

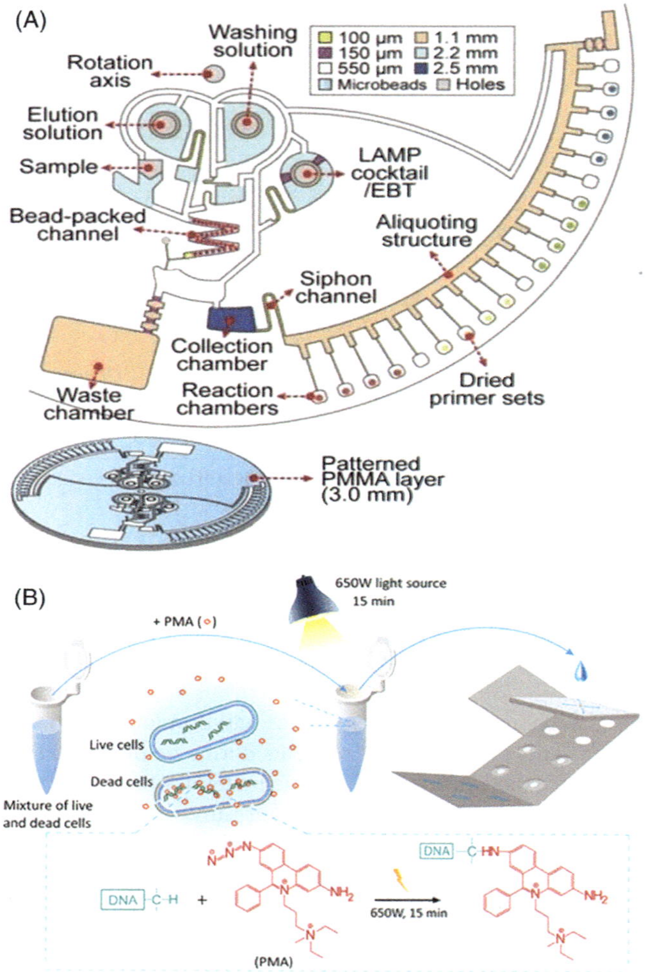

FIGURE 10.5

(A) Illustration of the design and components of a centrifugal microfluidic device with a PMMA layer disk. The gray circles represent the places where the sample, washing solution, elution solution, and LAMP cocktail are added. The bead-packed channel is where DNA extraction occurs via silica-coated microbeads. In this disk, there are 20 reaction chambers where multiplex LAMP is performed.(B) Schematic representation of the concept of sample preparation for the performance of the PMA–LAMP assay on the origami paper microdevice.

(A) From Oh, S. J., & Seo, T. S. (2019). Combination of a centrifugal microfluidic device with a solution-loading cartridge for fully automatic molecular diagnostics. Analyst, *144(19), 5766–5774. https://doi.org/10.1039/c9an00900k; (B) from Trieu, P. T., & Lee, N. Y. (2019). Paper-based all-in-one origami microdevice for nucleic acid amplification testing for rapid colorimetric identification of live cells for point-of-care testing.* Analytical Chemistry, *91(17), 11013–11022. https://doi.org/10.1021/acs.analchem.9b01263.*

automated lab-on-a-chip (LoC)-based molecular diagnostics. The entire detection process was completed within 65 min, and the limit of detection of the assay was 100 bacterial cells. The usage of the solution-loading cartridge successfully replaced the laborious and error-prone manual loading processes and realized true automation of molecular diagnostics (Oh & Seo, 2019).

An innovative paper-based microdevice was used for the colorimetric identification of live cells based on a nucleic acid amplification testing (NAAT). This all-in-one origami paper microdevice shown in Fig. 10.5B integrated with DNA purification, LAMP, and on-site colorimetric detection (Trieu & Lee, 2019). The origami paper microdevice was fabricated with low-cost paper with an embossing technique and can be used for remote and resource-limited regions. This disposable and environmentally friendly device could be used for viability tests of harmful pathogens, to control and monitor foodborne disease outbreaks, to ensure the health of ecosystems in ballast-water exchanges, and to conduct studies on antibiotic resistance, as a replacement for culture-based methods (Trieu & Lee, 2019).

In conclusion all these applications demonstrate the potential of microfluidic devices over conventional benchtop testing. They show that simple and rapid DNA extraction does not require costly laboratory equipment or highly trained laboratory technicians. Embedding an automated nucleic acid extraction method into these devices has been a challenge, because it involves multiple steps (lysis, extraction, washing, recovery). Nevertheless, various methods have been invented to simplify the preparation steps such as: the use of magnetic bead, microdevice with origami patterns, centrifugal microfluidics device. LoC devices can be fabricated using inexpensive and disposable material, the facile and robust fabrication processes can be tailored to suit limited-resource regions. For rapid screening of diseases and control of disease outbreaks, lab and equipment intensive testing methods such as PCR can be replaced with isothermal DNA amplification methods, and colorimetric detection. Disease screening can be extended to different targets by simply changing biomarkers and microfluidic design to increase the throughput for commercialization purposes.

10.3 High throughput screening (HTS) using lab-on-chip devices

High-throughput screening (HTS) using lab-on-chip devices represents an important advancement in modern biomedical research, diagnostics, drug discovery, and environmental monitoring. HTS allows a large number of samples, compounds, or experimental conditions to be tested rapidly and systematically, making it useful for identifying effective drugs, detecting diseases, studying cell responses, and evaluating biological or chemical reactions. However, conventional HTS platforms often require large reagent volumes, expensive equipment, complex laboratory procedures, and longer processing times. To overcome these limitations, lab-on-chip technology offers a miniaturized and integrated platform that can perform multiple laboratory operations on a small chip. By combining microfluidics, automation, precise sample handling, and sensitive detection methods, lab-on-chip devices enable faster, cheaper, and more efficient screening processes. Therefore, this chapter discusses the importance of HTS, the advantages of lab-on-chip devices, key design considerations, various applications, and the challenges and future directions of this technology.

10.3.1 Introduction to HTS

High-throughput screening (HTS) is a laboratory method that rapidly tests vast numbers of chemical or biological compounds to determine their specific activities, such as their ability to inhibit or activate particular biological targets (Payne et al., 2020). This process involves screening thousands to millions

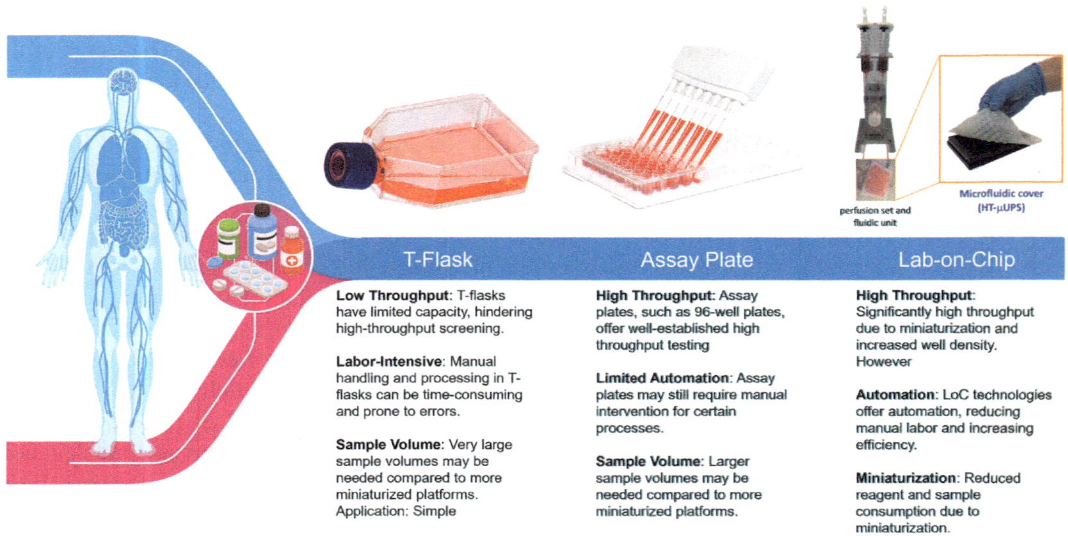

FIGURE 10.6

Comparison of HTS techniques (T-Flask, Assay Plate, and LoC-based) platforms in cell culture applications (Wei et al., 2020).

From Wei, L., Li, W., Entcheva, E., & Li, Z. (2020). Microfluidics-enabled 96-well perfusion system for high-throughput tissue engineering and long-term all-optical electrophysiology. Lab on a Chip, 20*(21), 4031–4042. https://doi.org/10.1039/d0lc00615g.*

of substances in a brief period, allowing researchers to identify potential drug candidates, study biological pathways, or investigate molecular interactions on a large scale. In simpler terms, HTS is like running many small experiments simultaneously to find out which substances might be useful for a particular purpose, like developing new medicines or understanding how cells work. It's a way for scientists to test lots of different substances quickly and efficiently. An example of progress on high throughput testing for cell culture applications is shown in Fig. 10.6.

10.3.1(a) *Importance of high throughput screening and limitations in traditional methods*

HTS has fundamentally reshaped the scientific discovery landscape, transforming the laborious task of identifying promising candidates from vast datasets into a streamlined and efficient process. For instance, HTS has become a cornerstone of modern drug discovery, enabling the rapid screening of extensive compound libraries to efficiently identify potential drug candidates (Mansor et al., 2023). By automating the screening of compounds against biological targets, HTS has significantly streamlined the drug discovery process, leading to the identification of promising leads for further development.

The use of HTS technologies has significantly improved the efficiency and speed of drug discovery processes. Traditional methods for drug discovery, such as manual screening and biochemical assays, were limited by their time-consuming nature and the need for large amounts of resources and labor (Wang, 2023). In the context of LoC applications, the limitations of traditional methods are further pronounced due to the need for miniaturization, automation, and high-throughput capabilities. LoC platforms aim to integrate multiple laboratory functions on a single chip, enabling rapid analysis of compounds with minimal sample volumes. Traditional methods struggle to meet the demands of LoC applications due to their manual and labor-intensive nature, making them impractical for high-throughput screening on a miniaturized scale.

10.3.1(b) ***Advantages of using Lab-on-Chip devices for high throughput screening***

While HTS has revolutionized scientific discovery, traditional methods remain hindered by time-consuming execution and resource-intensive requirements. LoC devices, miniaturized laboratories integrated onto a single chip, offer significant advantages for HTS applications. Firstly, LoC platforms contribute to reduced assay variability and error, ensuring greater consistency and reliability in screening results. Their miniaturized format allows for the screening of a vast number of compounds within a smaller footprint, leading to enhanced efficiency in terms of time and resource utilization. Integration with robotic systems and scheduling software further streamlines the process by enabling automated and continuous screening, minimizing human error and inconsistencies. Furthermore, the cost-effectiveness achieved through miniaturization and automation renders LoC devices a financially viable option for screening processes. Their inherent flexibility allows for the accommodation of diverse assays, facilitating tailored experimentation based on specific biological parameters. These advancements position LoC devices as invaluable tools for HTS, offering significant improvements in efficiency, accuracy, and cost-effectiveness within the drug discovery pipeline. Many people are interested to move towards medical testing on human cells cultured on LoC, rather than to do testing on live animals, for legal, ethical and moral reasons.

10.3.1(c) ***The diverse applications of lab-on-chip for high-throughput screening***

Beyond revolutionizing HTS, LoC devices have become established as critical tools across a broad spectrum of scientific disciplines. Their unique functionalities of miniaturization, automation, and the ability to integrate complex processes have unlocked a vast potential for scientific discovery, propelling research advancements at an accelerated rate.

In the field of drug discovery, LoC platforms have emerged as indispensable instruments. These devices facilitate the rapid screening of extensive compound libraries for interactions with targeted molecules, significantly expediting the identification of promising drug candidates. This translates to condensed development timelines, reduced costs, and ultimately, the faster delivery of life-saving therapeutics. The ability of LoC platforms to explore numerous possibilities simultaneously paves the way for the development of more targeted and efficacious treatments.

Similarly, LoC devices are transforming the landscape of materials science. The quest for new materials with specific properties often involves testing countless combinations. LoC technology streamlines this process by enabling parallel evaluation of various material compositions and properties. This expedites the development of advanced materials for diverse applications, ranging from energy storage to lightweight aerospace components, ultimately impacting fields as varied as construction and transportation.

In the environmental sensing, LoC devices serve as powerful tools with their ability to be deployed in remote locations or integrated into wearable devices allows for real-time detection of pollutants, pathogens, and other environmental hazards. This empowers proactive responses to environmental threats, safeguarding public health and contributing to a cleaner, healthier planet. Their miniature size and portability make them ideal for monitoring air and water quality, providing invaluable data for policymakers and environmental scientists.

Furthermore, LoC technology plays a vital role in ensuring food safety. These devices can rapidly detect foodborne pathogens, allergens, and toxins, providing accurate results that safeguard public health. This contributes to a safer and more reliable food supply chain for everyone, protecting consumers and ensuring the integrity of the food system. Their rapid and precise detection capabilities offer significant advantages over traditional methods, allowing for preventative measures and targeted interventions.

The future of healthcare centers on personalized medicine, and LoC devices hold immense potential in this arena. By analyzing individual patient samples on a microfluidic chip, they can provide insights into genetic predispositions, disease susceptibility, and drug response. This empowers tailored treatment strategies, leading to improved patient outcomes and a more holistic approach to healthcare. LoC technology offers the potential to revolutionize diagnostics and treatment, paving the way for a future of precision medicine.

These represent just a fraction of the vast potential of LoC technology. As research and development continue, we can expect even more groundbreaking applications to emerge, pushing the boundaries of scientific exploration and impacting fields ranging from agriculture to diagnostics. From miniature labs on a chip to a universe of possibilities, LoC technology is poised to revolutionize various sectors, shaping a brighter future for generations to come.

10.3.2 Design considerations for high throughput screening

LoC platforms have become indispensable tools for HTS, offering unparalleled miniaturization, automation, and integration. However, unleashing their full potential necessitates meticulous design considerations. These microfluidic marvels, aptly termed miniature laboratories, demand careful engineering to ensure efficient, reliable, and accurate operation within demanding HTS environments.

Miniaturization is a key technology in transforming low throughput screening to HTS or ultra-high throughput screening (uHTS) formats (Achyuthan & Whitten, 2007). It allows for carrying out testing with less reagents, resulting in significant cost savings and less waste. For instance, using 3456-well plates instead of 96-well plates can reduce reagent consumption by 100-fold and shorten screening time from weeks to a week.

Microfluidic Architecture: The success of a LoC device heavily relies on the design of its microfluidic channels. These channels, defined by their size and shape, control how fluids flow, mix, and react within the device. Finding the right balance between channel dimensions, desired sample volumes, and specific assay needs is crucial. Additionally, incorporating microfluidic valves and pumps enables precise control over the movement of samples and reagents, similar to how a conductor directs an orchestra to create a complex performance with individual instruments. (Sanka et al., 2019).

Sensor Integration and Data Acquisition: Accurate detection is essential for HTS. The choice of sensor technology, either optical or electrochemical, depends on the specific molecule being analyzed (target analyte) and the level of detail needed (sensitivity). Smooth integration of the sensor with the microfluidic channels is crucial for obtaining reliable detection results. Additionally, robust data collection systems and powerful analysis tools are necessary to convert the raw data into usable information.

Automation and Optimization: Efficiency and reproducibility are important for successful HTS. User-controlled software acts as the central control unit for the system, managing microfluidic movement, sensor function, and data collection. The software incorporates feedback mechanisms and optimization algorithms, allowing it to automatically adjust test parameters to ensure the highest quality data is obtained.

Biocompatibility and Sample Handling: Biocompatible materials and surface treatments are crucial for protecting samples and minimizing unintended interactions with the microfluidic channels. This is especially important when working with living cells or tissues, where biocompatibility becomes a top priority. Selecting appropriate biocompatible materials and optimizing the surface properties of the channels are essential to ensure that cells remain alive (maintain viability), receive necessary nutrients, and can effectively remove waste products. This careful balancing act guarantees reliable results and prevents unwanted cellular responses within the microfluidic environment (Yang et al., 2021).

Cost and Scalability: Cost-effectiveness is essential for HTS system. Material selection and fabrication techniques must strike a balance between affordability and desired performance. The design should readily adapt to handle larger sample volumes and higher throughput, catering to the demands of large-scale HTS (Szymański et al., 2012).

10.3.3 Examples of high throughput screening applications

LoC devices have evolved beyond simply miniaturized laboratories. They now serve as potent instruments for HTS across a wide range of scientific disciplines. The confluence of miniaturization, automation, and integration capabilities within LoC devices empowers researchers to explore a vast spectrum of possibilities in various applications. Examples of applications are shown in Fig. 10.7:

Seidael et al. (2024) in their work introduces a novel thermal device designed for precise temperature control of 96-well plates within automated liquid handling systems. This device leverages Smart Sensor and Sensor 4.0 principles, allowing for integration with standard plates. It features a modular design with three subunits: a thermal unit with individually controlled heating elements, a control unit with advanced temperature regulation algorithms, and a cooling unit. The control unit also provides data visualization and monitoring capabilities. The device seamlessly integrates with existing automation systems through its multiple communication interfaces. In a separate study, the researchers demonstrate the effectiveness of this thermal device in a high-throughput enzyme assay. Focusing on the temperature dependence of enzyme activity, the study utilizes a luminescence-based assay to measure reaction rates. This fully automated setup allows for large-scale screening through model-based analysis of reaction parameters and enzyme kinetics. Overall, this research demonstrates the potential of the thermal device for high-throughput applications within automated liquid handling systems such as small-scale fermentation and investigation of heat shock responses.

Another work done by Cliffe et al. (2023) addresses the need for improved Microphysiological Systems (MPSs) to enhance drug development by better predicting efficacy and toxicity during preclinical stages. The high throughput test in this work is the use of Mera, an automated, modular and scalable system for culturing and assaying microtissues with interconnected fluidics, inbuilt environmental control and automated image capture. The system can culture up to 40 microtissues on a single Bioplate and perform various biological assays such as viability, toxicity, albumin, urea and Cytochrome P450 (CYP) expression. The system can also switch the fluid flow direction to create a recirculating circuit of

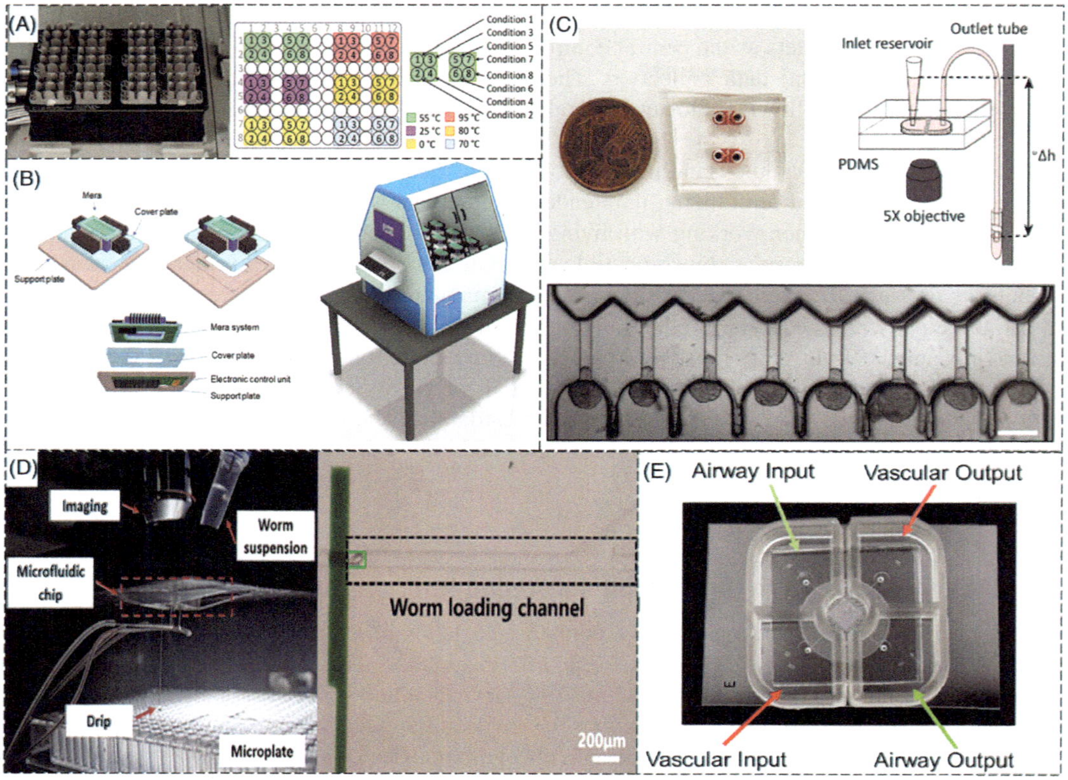

FIGURE 10.7

Example applications of HTS system. (A) Automated liquid handling systems for high-throughput enzyme assay (Seidel et al., 2024). (B) An automated, modular and scalable system for culturing and assaying microtissues with interconnected fluidics, inbuilt environmental control and automated image capture. (C) Microfluidic chip that allows for high-throughput testing of cell spheroids (Boot et al., 2023). (D) High-throughput toxicity testing using Caenorhabditis for detection of environmental contaminants. (E) Human airway chip specifically designed for BSL-3 environments and high-throughput analysis for SARS-CoV-2 infection.

(B) From Cliffe, F. E., Madden, C., Costello, P., Devitt, S., Mukkunda, S. R., Keshava, B. B., Fearnhead, H.O., Vitkauskaite, A., Dehkordi, M. H., Chingwaru, W., Przyjalgowski, M., Rebrova, N., & Lyons, M. (2023). Mera: A scalable high throughput automated micro-physiological system. SLAS Technology, 28*(4), 230–242. https://doi.org/10.1016/j.slast.2023.01.004; (D) from Wu, J., Gao, Y., Xi, J., You, X., Zhang, X., Zhang, X., Cao, Y., Liu, P., Chen, X., & Luan Y. (2022). A high-throughput microplate toxicity screening platform based on* Caenorhabditis elegans. Ecotoxicology and Environmental Safety, 245, *114089. https://doi.org/10.1016/j.ecoenv.2022.114089; (F) from Faley, S. L., Boghdeh, N. A., Schaffer, D. K., Spivey, E. C., Alem, F., Narayanan, A., Wikswo, J. P., & Brown, J. A. (2025). Gravity-perfused airway-on-a-chip optimized for quantitative BSL-3 studies of SARS-CoV-2 infection: Barrier permeability, cytokine production, immunohistochemistry, and viral load assays.* Lab on a Chip, 25*(23), 6231–6247. https://doi.org/10.1039/D5LC00510H.*

inter-connected channels to allow inter-organ communication and drug metabolism. The system aims to provide a more accurate and ethical model for preclinical drug screening.

Traditional methods for studying cell spheroids, which are models that mimic biological tissues, are slow and can only analyze one spheroid at a time. This makes it difficult to collect statistically significant data. Boot et al. (2023) developed a microfluidic chip that allows for high-throughput testing of cell spheroids. This chip can analyze ten spheroids simultaneously using a design inspired by micropipette aspiration. The chip operates by strategically loading spheroids into designated wells. Subsequently, pressure is applied to draw the spheroids into designated channels. This efficient process facilitates the rapid testing of a large number of spheroids, while still providing accurate data on their viscoelastic properties. This novel technology offers significant advantages to researchers investigating the mechanics of various tissues and how these mechanics influence physiological processes such as tissue remodelling and cancer metastasis. The high-throughput nature of the chip allows researchers to collect statistically significant data in a much shorter timeframe compared to traditional methods. This paves the way for a more comprehensive understanding of tissue mechanics and its role in health and disease.

A novel automated platform has been developed by Wu et al. (2022) for high-throughput toxicity testing using *Caenorhabditis elegans* (*C. elegans*) as a model organism, targeting both general toxicology and neurotoxicology endpoints relevant to environmental contaminants. This integrated system combines microplates for culturing worms, a microfluidic-based module for dispensing worms, an image acquisition/analysis module for monitoring toxicity endpoints, and a commercial pipetting workstation for automated food/drug delivery. Additionally, a high-content analysis system enables fluorescence detection, while connectivity to the web allows for remote operation and data analysis. The platform facilitates efficient dispensing, long-term maintenance, monitoring, survival assays, behavioral analyses, and gene reporter assays. Demonstrated with cadmium treatments, the platform effectively assesses longevity, neurotoxicity, developmental toxicity, and gene expression. This automated platform alleviates manual tasks associated with worm handling and analysis, promising rapid toxicity assessment of emerging compounds and ecological risk assessment of environmental samples in high-throughput screening applications.

Faley et al. (2025) addresses limitations in existing microphysiological systems for high-throughput infectious disease testing. The authors introduce a new human airway chip specifically designed for Biosafety Level 3 (BSL-3) environments and high-throughput analysis. This chip mimics human airway structure and function, allowing researchers to study SARS-CoV-2 infection and immune response. The chip's compatibility with BSL-3 environments and ease of use in other settings make it a valuable tool for rapid drug and intervention testing against COVID-19 and future respiratory pathogens.

Primary cancer cells hold greater predictive value for in vivo responses compared to immortalized cell lines. This is due to their closer resemblance to the original tumor microenvironment. Isolation and subsequent culture of primary cells are essential for such studies. A study by Mansor et al. (2023) explored the concept of HTS using a commercially available biosensor on primary cell cultures as shown in Fig. 10.8. The study employed the xCELLigence Real-Time Cell Analysis (RTCA) device to monitor the response of primary lung cancer cells to varying concentrations (1–120 μM) of cisplatin. This approach revealed distinct cisplatin effects across four patient-derived primary lung cancer cell lines. The findings suggest that biosensor assays offer a significant advantage over conventional optical assays in the context of personalized medicine. The portability, real-time analysis capabilities, and label-free nature of this technique make it a promising candidate for cost-effective patient care. While the current system utilizes

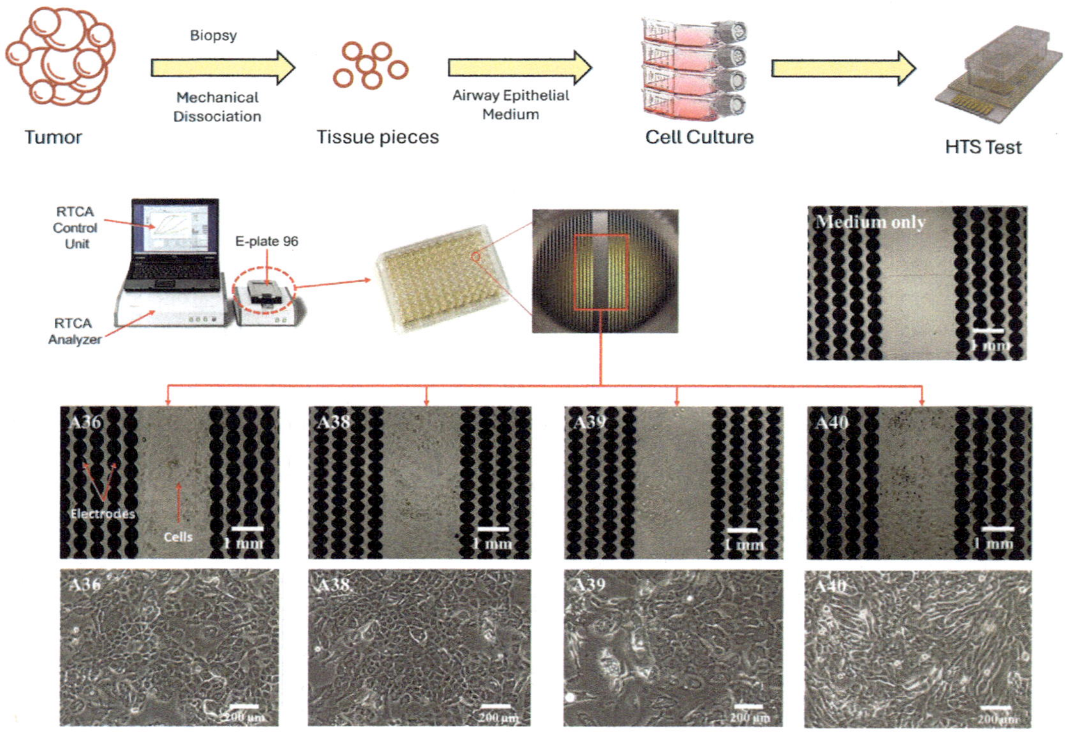

FIGURE 10.8

(A) The concept of HTS for chemotherapy prediction on primary cell culture using biosensors. (B) Overview of the HTS test using xCELLigence system on four patient-derived primary lung cancer.

(B) From Mansor, A. F. M., Sedek, S., Nordin, A. N, Mohammad, M., Hashim, Y. Z. H. -Y., Nordin, A. F. W., Goh, K. L., Azih, M. N. A., Abdul Razak, M.R., Ibrahim, A., & Hin How, S. (2023). Application of real-time cell analysis biosensor technology for drug cytotoxicity studies in primary lung cancer cells. Biomedical Materials & Devices, 1*(2), 825–835. https://doi.org/10.1007/s44174-022-00045-w.*

electrodes, exploring cheaper alternatives for their fabrication could further enhance affordability. Additionally, the minimal sample requirement combined with the high-throughput potential allows practitioners to analyze multiple drug combinations simultaneously.

Comprehensive identification of chemical contaminants in the field water supplies can be a lengthy process, and rapid analytical methods suitable for field use are limited. A field-portable toxicity device based on live mammalian cells that was highly sensitive to the toxicants in the water was presented in the literature (Liu et al., 2013b, 2014). The cell-based device provided rapid assessments of water quality and contributed to increase the security of drinking water. This device included a combination of two independent sensors, that were able to monitor the mammalian cell attachment and viability when exposed to toxicants. One of the sensors was a quartz crystal microbalance (QCM) resonator that was integrated in a single device with the second sensor that was performing electric cell-substrate

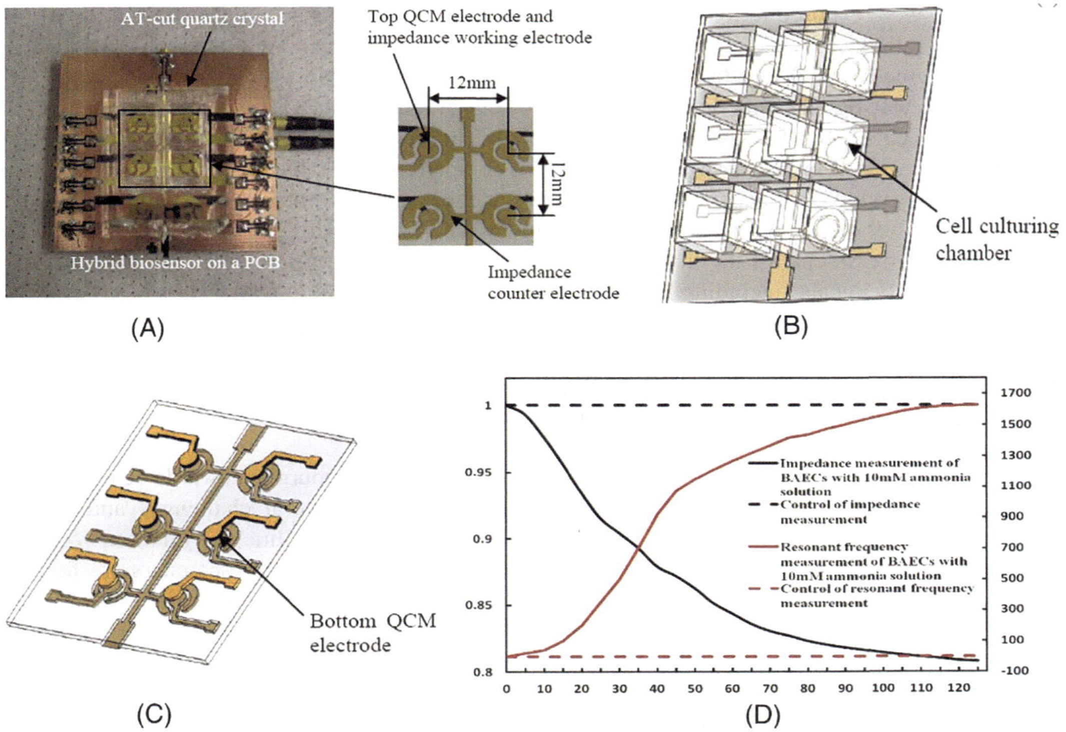

FIGURE 10.9

The hybrid sensors configuration (2 × 3 array). (A) Picture of the hybrid sensor array on PCB; (B) Top view of the sensor array assembled with cells culturing chamber; (C) Bottom view of the sensor showing bottom QCM electrodes. (D) Simultaneous impedance measurements and resonant frequency measurements for testing the BAECs with 10 mM of ammonia (Voiculescu et al., 2018).

(C) Liu, F., Li, F., Nordin, A., & Voiculescu, I. (2013). A novel cell-based hybrid acoustic wave biosensor with impedimetric sensing capabilities. Sensors, *13(3), 3039–3055. https://doi.org/10.3390/s130303039; (D) from Voiculescu, I., Toda, M., Inomata, N., Ono, T., & Li, F. (2018). Nano and microsensors for mammalian cell studies.* Micromachines, *9(9), 439. https://doi.org/10.3390/mi9090439.*

impedance sensing (ECIS) on mammalian endothelial cell monolayer. Both techniques; the resonant frequency measurements with QCM and ECIS measurements were demonstrated to characterize in real time the attachment and viability of the endothelial cells exposed to different types of toxicants. Bovine aortic endothelial cells (BAECs) were used in this research, because of their ability for long-term survival (Voiculescu et al., 2013). The device, shown in Fig. 10.9A–C, directly indicates the toxicity effect on live cells, instead of detecting the individual chemical constituents in the water (Liu et al., 2013b, 2014). Impedance values and resonant frequency values were monitored and recorded in real-timely for several hours after introducing ammonia toxicant. When exposed to ammonia with the concentration of 10 Mm the BAECs become apoptotic, see Fig. 10.9D. When the cells cultured on this hybrid sensor were exposed to toxicant and were apoptotic the impedance values were minimal since the apoptotic cells lost their dielectric properties of the cell membrane. Apoptotic cells also become detached from the QCM electrode, and the resonant frequency increased.

The device was based on the innovative employment of the upper QCM electrode as the working electrode for ECIS technique as shown in Fig. 10.9. The device had two water toxicity detection methods, and provided higher accuracy of detection, compared to a device with only a single sensing method. For rapid studies of water toxicity, the simultaneous responses from two different sensors will confirm the quality of the water and will offer higher detection security. This device is portable and provides rapid assessments of the field water toxicity. Living cell sensors indicate the presence of a broad range of chemicals (including unknown agents) that cause a toxic response, whereas analyte-specific sensors can only quantify and identify specific chemicals.

10.3.4 Challenges and future directions

LoC devices have emerged as powerful tools for HTS, yet their journey towards realizing their full potential remains ongoing. Researchers are confronted with several challenges and exciting opportunities that drive them to further innovate and explore the possibilities of this transformative technology. In the quest for even smaller and faster screening capabilities, researchers are focusing on scaling down microfluidic channels to process a larger number of samples simultaneously, thereby accelerating screening speed and data generation. Integrating multiple functionalities such as sample preparation, mixing, and detection onto a single chip hold promise for streamlining workflows and maximizing efficiency. Additionally, the fabrication of microfluidic arrays on a single chip offers the potential for massively parallel screening, significantly enhancing throughput and accelerating scientific progress.

To unlock deeper insights, researchers are developing ultra-sensitive sensors capable of detecting faint signals, enabling the analysis of smaller samples and reducing reagent consumption. Minimizing background noise through optimized chip design and surface treatments ensures clearer and more reliable results, leading to more accurate data interpretation. Multiplexed detection techniques provide a comprehensive understanding of sample properties, facilitating deeper insights and accelerating scientific breakthroughs.

Efforts to bridge the gap between chips and organisms involve creating biomimetic microenvironments that replicate tissues and Organs-on-chips (OOC), offering more relevant and predictive biological assays. Integration of living cells and tissues on LoC devices allows for real-time analysis and holds immense potential for unlocking biological secrets. Automation of complex biological workflows, including sample preparation, cell handling, and data analysis, streamlines assays and enables researchers to focus on interpretation and discovery.

In addition to technical challenges, there are also considerations for making LoC devices more accessible and cost-effective. This involves optimizing manufacturing processes, standardizing chip designs, and promoting collaboration and knowledge sharing within the scientific community. By addressing these challenges and exploring these opportunities, LoC technology holds immense promise for the future of HTS, with potential applications ranging from personalized medicine to environmental monitoring. With continued advancements, LoC devices have the potential to unlock a universe of possibilities, shaping a brighter future for generations to come.

10.4 Cell culture and organ-on-a-chip using lab on chip devices

Cell culture and organ-on-a-chip technologies using lab-on-chip devices have transformed the way biological systems are studied in modern biomedical research. Traditional cell culture methods are commonly performed in static environments, such as culture plates or flasks, which do not fully represent the dynamic conditions found inside the human body. In contrast, lab-on-chip devices provide a miniaturized and controlled platform where cells can be cultured under more realistic physiological conditions, including continuous fluid flow, chemical gradients, mechanical stimulation, and controlled microenvironments. These features allow researchers to better mimic tissue and organ functions, making organ-on-a-chip systems valuable for drug testing, disease modelling, toxicity assessment, and personalized medicine. This chapter discusses how lab-on-chip devices improve cell culture and tissue engineering, the key design considerations such as cell type, extracellular matrix, material selection, and microenvironment control, as well as important applications including multiple-organ-on-chip and stretchable lab-on-chip devices.

10.4.1 Introduction: How LoC devices revolutionize cell culture and tissue engineering

While traditional cell culture and animal models have laid the groundwork for countless discoveries, their limitations are becoming increasingly apparent. The artificial environment of plastic dishes fails to capture the interactions and microenvironments of living tissues, leading to inaccurate and misleading results. Similarly, animal models raise ethical concerns and often fall short in mirroring human physiology, limiting their translational value. These limitations highlight the need for alternative approaches that can better mimic the complexity of tissues and organs while providing more ethical and cost-effective solutions. This is where the LoC devices present a promising solution to these challenges. Fig. 10.10 summarizes the improvement and advantages of using LoC compared to the conventional method.

Firstly, the flat, LoC devices improve on the flat, artificial environment typically found in petri dishes (Damiati et al., 2018). The complex architecture and dynamic cues of natural tissues are mimicked by creating complex 3D microfluidic channels. A more physiologically relevant manner for cells to interact with each other and their surroundings is enabled, leading to more accurate and reliable data. Precise manipulation of factors like flow and nutrient delivery is allowed by microfluidic control, further mimicking the in vivo environment (Damiati et al., 2018). Additionally, scaffolds supporting cell growth and differentiation, resembling the natural extracellular matrix, can be created by integrating biocompatible materials into these chips.

Secondly, LoC devices provide platform for studying multiple cells interaction (Huh et al., 2011). Unlike the conventional method, these devices can hold various cell types on one chip. This helps researcher to study how cells interact and communicate with each other in a more realistic way. It also lets them learn how different groups of cells affect each other, which gives important insights into how tissues work and what goes wrong in diseases. Also, these devices create spaces that mimic how tissues are organized in the body. This helps scientists' study complicated things like how wounds heal and how tumors grow with more detail than ever before. Importantly, scientists can use cells taken from patients in these models. This helps develop personalized medicine because drugs can be tested to see how well they work and how safe they are for each individual patient.

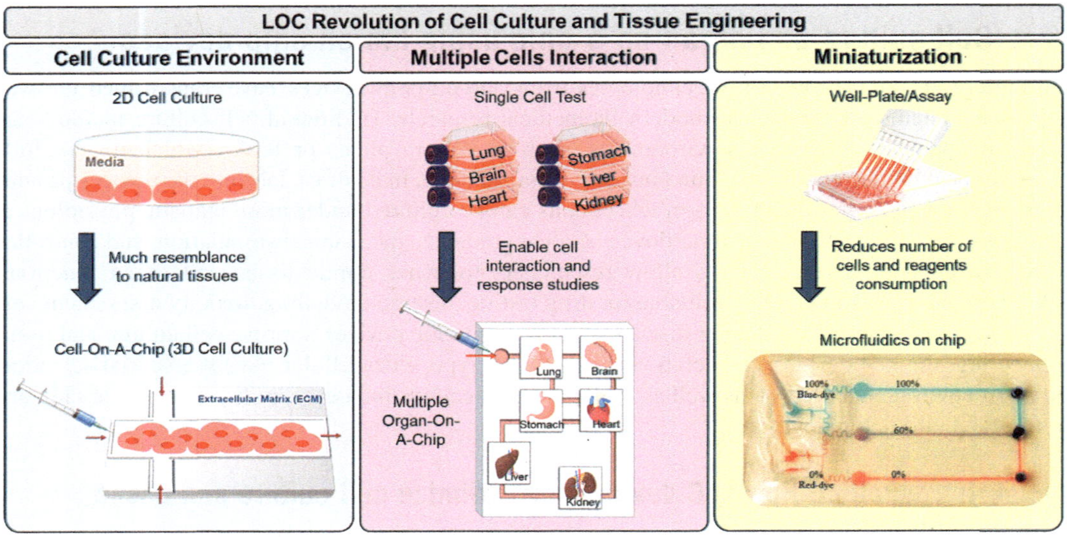

FIGURE 10.10

Advantages of using LoC method compared to conventional practices. Figures were adapted from Damiati et al. (2018) and Zaidon et al. (2016).

Finally, practical advantages are offered by LoC devices that optimize research processes. Their miniaturization reduces the number of cells and reagents needed (Mark et al., 2010a), making them cost-effective and sustainable. High-throughput experimentation and precise manipulation of experimental conditions are allowed by the controlled and automated nature of microfluidic environments, accelerating research progress. Additionally, real-time monitoring of cell behavior and generation of rich datasets are enabled by integration with sensors and analytical tools, facilitating a deeper understanding of cellular responses and development of predictive models for drug discovery and disease diagnosis.

In conclusion, microfabrication technologies and microfluidics have revolutionized cell culture and tissue engineering by enabling the development LoC devices. These innovative platforms provide precise control over the cellular microenvironment, allowing for the creation of sophisticated 3D cell-culture models that closely mimic the structure and function of living organs. By integrating living cells within microengineered structures, LoC devices offer a unique opportunity to study human physiology in an organ-specific context and to develop specialized in vitro disease models. Furthermore, these devices have the potential to replace traditional animal models in drug development and toxin testing, offering a more ethical, cost-effective, and predictive approach to understanding human health and disease. In this rapidly evolving field, LoC devices are paving the way for groundbreaking advancements in cell biology, tissue engineering, and personalized medicine.

10.4.2 Design considerations for cell culture and tissue engineering

LoC technology offers a powerful platform for cell culture and tissue engineering due to its ability to mimic the complex microenvironments of living tissues. The main use of LoC and OOC in biomedical engineering research, specifically in bio-MEMS, is as an alternative to animal models for drug development and toxin testing. OOC have been designed to simulate various organs including the brain, lung, heart, kidney, liver, prostate, blood vessels, skin, bone, and cartilage. Design considerations for cell culture and tissue engineering involve various factors that are crucial for creating successful and effective systems. Some key considerations include:

10.4.2(a) Cell type and extracellular matrix

The fundamental goal of tissue engineering in LoC devices is to recreate the biological complexity and functionality of actual human tissues. This requires a particular focus on replicating the cellular components and the environment in which they interact. There are a lot of considerations for the cell's application. This section will focus on the cell type and extracellular matrix.

Choosing the appropriate cell type (primary, stem cell, or immortalized cell lines) is crucial for replicating the desired tissue functionality. Primary cells are directly isolated from tissues and possess a finite lifespan in culture. They are valued for their in vivo cellular functions and genetic stability, making them more representative of natural responses compared to immortalized cell lines. However, primary cells face challenges such as limited longevity in vitro and potential variability between donors. To address these limitations, culturing primary cells under flow perfusion in OOC systems has been instrumental in extending their lifespan, enabling a more accurate emulation of human in vivo responses. Stem cells, on the other hand, are undifferentiated cells with the ability to differentiate into various cell types. Induced pluripotent stem cells (iPSCs) are artificially reprogrammed cells derived from adult cells. While stem cells offer differentiation potential and improved human relevance over engineered cell lines, they may not fully replicate the characteristics of mature primary human cells. Immortalized cell lines, characterized by indefinite proliferation through mutations or gene expression alterations, are simple to handle and cost-effective. They are commonly used in research settings for high-throughput screening but may lack key cellular components and fail to fully replicate human physiology when compared to primary cells or stem cells. Each cell type brings unique advantages and challenges to LoC and OOC systems, influencing their suitability for different experimental goals and applications.

On the other hand, the extracellular matrix (ECM) is essential for the development of functional tissue and organ models. First, the ECM provides structural support, serving as a scaffold for cells and mimicking the native tissue environment. Second, it regulates cell behavior, influencing processes like differentiation, proliferation, and migration. Third, the mechanical cues from the ECM, such as matrix elasticity, impact cellular responses such as proliferation, differentiation, migration, and adhesion. The ECM's tissue-specific composition ensures that OOC models accurately represent different organs. The ECM in both OOC and LoC systems can be manipulated to study specific aspects of tissue and organ function, such as mechanical properties, cell-cell interactions, and cell-ECM interactions. This allows researchers to better understand the complex processes that occur within tissues and organs, and to develop more accurate models for drug testing and disease research.

The last part is the microfluidic design to incorporating channels and chambers that replicate the flow patterns, shear stress, and nutrient gradients found in the targeted tissue. This layer is fundamental in both LoC and OOC devices. It consists of channels with dimensions typically in the range of less than 100 μm, allowing precise control and manipulation of small amounts of fluids. The microfluidic enables the transport of fluids, cells, and reagents within the chip, facilitating various biological and chemical processes.

10.4.2(b) Material selection

Organ-on-chip devices rely on materials to support tissue attachment and organization, as well as to form the structure of device channels and features. The choice of material determines its application in specific devices. Many commercial OOC devices use a process called replica molding, where a microfabricated design is developed using soft lithography and replicas are generated using polydimethylsiloxane (PDMS). The success of using LoC devices in cell culture and tissue engineering hinges on the thoughtful selection and design of biomaterials. These materials play a critical role in crafting a microenvironment that closely mirrors the natural cellular niche, influencing various cellular processes. Several key considerations guide the selection and design of these biomaterials.

Firstly, the chosen material must be both biocompatible and cytocompatible, meaning it is non-toxic, non-immunogenic, and supportive of cell adhesion, growth, and function without compromising their viability (Huyer et al., 2015; Kutluk et al., 2023). Additionally, tailoring surface properties like topography, stiffness, and wettability is necessary. By mimicking the target tissue's ECM, these properties promote optimal cellular interactions within the microenvironment.

Secondly, the mechanical properties of the biomaterial, such as stiffness, porosity and elasticity should ideally match the target tissue (Ahadian et al., 2018; Huyer et al., 2015). This provides cells with the necessary mechanical cues that significantly influence their morphology, differentiation, and function. Additionally, compatibility with microfabrication techniques used in LoC device construction is essential. The chosen material must allow for the creation of desired microfluidic features and structures within the chip.

Thirdly, the biodegradation and biosorption of the material need to be considered (Ahadian et al., 2018). For applications like studying tissue regeneration or wound healing, biomaterials that degrade at a controlled rate may be preferred, allowing for the gradual replacement of the material by newly formed tissue. In specific cases, biomaterials that are gradually absorbed by the body (bioresorption) might be desirable, particularly for temporary implants or scaffolds in tissue engineering.

Different materials are used to develop OOC depending on the specific features required in mimicking the in vivo structural and/or biochemical environment. Glass is an older material employed in microfluidics, known for its transparency, resistance to mechanical stress, hydrophilicity, biocompatibility, and low drug absorptivity. However, its disadvantages include low gas permeability, which limits long-term cell studies, and high cost of fabrication (Beißner et al., 2016; Monteduro et al., 2023).

The experimentation of other materials led to the mainstream use of polymeric materials, which can be divided into three major groups: elastomers, thermoplastics, and thermosets. Between elastomers, polydimethylsiloxane (PDMS) is the most employed for the realization of devices for life science applications. It is flexible, optical transparent, gas permeable, and biocompatible. Its mechanical properties and hydrophobicity are tunable, allowing for easy microfluidic connections and prototyping of new devices. PDMS is also useful in mimicking biological processes involving mechanical deformation and measuring mechanical properties of monolayer cells (Ahadian et al., 2018; Ding et al., 2020).

Thermoplastics are a type of polymers that can be melted and reshaped multiple times through heating, typically using thermomolding. These materials are often optically transparent, more rigid than elastomers, and resistant to the diffusion of small molecules. While they are moderately resistant to alcohol, they are soluble in most organic solvents. However, due to their low gas permeability, thermoplastic sealed microchannels and microchambers are not suitable for long-term cell studies. Their stiffness also prevents the realization of diaphragm valves. Depending on their application, thermoplastics can have their surface functionalized using surface grafting techniques or dynamic coating.

Common thermoplastic polymers like polystyrene (PS), polycarbonate (PC), polyurethane (PU), polymethylmethacrylate (PMMA), and polyethylene glycol diacrylate (PEGDA) are widely used in microfluidic systems. For instance, PS is highly biocompatible and used for cell growth and adhesion, while PC membranes are integrated among microchannels in OOC structures to model tissue-tissue interfaces. PU is known for its high mechanical strength and is used in various biomedical devices like heart valves and pacemakers. PMMA is chosen as a substrate for OOC due to its stiffness, optical transparency, and low autofluorescence background. PEGDA thermoplastic is utilized to create different types of microfluidic valves and pumps (Monteduro et al., 2023).

On the other side, biomaterials used in tissue engineering can be broadly categorized into natural and synthetic types. Natural biomaterials, such as collagen, fibrin, and hyaluronic acid, are derived from ECM components and offer biocompatibility and enzymatic degradability. However, they often present challenges such as batch-to-batch variability and potential denaturation during processing. Synthetic biomaterials, on the other hand, offer tuneable mechanical and degradation properties, and minimal batch variability. Yet, they lack inherent cell adhesion ligands, which can be mitigated through surface modification with molecules like laminin or fibronectin. Summary of strengths and limitation of different biomaterials used in the LoC and OOC (Ahadian et al., 2018; Monteduro et al., 2023) were shown in Table 10.1. Hybrid biomaterials may combine the advantages of both natural and synthetic materials, providing a wide range of chemical and mechanical properties in a controllable and reproducible manner.

10.4.2(c) *Microenvironment control*

While replicating the 3D architecture and incorporating biocompatible materials are crucial steps in establishing physiologically relevant microenvironments within LoC devices, the ability to control microenvironment is equally important. These cues significantly influence cellular behavior and function, similar to environmental factors shaping biological responses in vivo. In the design of OOC devices, controlling the microenvironment is a critical consideration to mimic the tissue microenvironment of cells in intact organs accurately. For example, current methods for controlling microfluidic flow in OOC systems often rely on gravity, rocker systems, or external pressure pumps. While pressure pumps offer consistent flow profiles, they may not be ideal for high-throughput applications such as testing drug responses. To address this, a novel method has been developed that allows for the multiplexing of microfluidic channels in OOC devices and enables the recirculation of culture medium. This recirculation feature is particularly valuable when using expensive culture reagents that are not fully utilized by cells during transient unidirectional flow, thus optimizing cost considerations in OOC experiments (Azimzadeh et al., 2022).

The surface chemistry of microfluidic channels plays a pivotal role in dictating cellular adhesion, proliferation, and differentiation. Chemical modification techniques can be employed to introduce specific functional groups onto the surface, promoting the adsorption of proteins or ECM components. This creates a surface favorable for cell attachment and migration, mimicking the natural cellular interactions observed in tissues. Furthermore, immobilization of specific biomolecules like adhesion proteins or growth factors directly onto the surface can further enhance these interactions. Additionally, micropatterning techniques can be utilized to create specific topographical features on the surface,

Table 10.1 Summary of biomaterials used in the LoC and OOC platforms (Ahadian et al., 2018; Monteduro et al., 2023).

Materials	Strengths	Limitation
Glass	Transparency Resistance to mechanical stress Hydrophilicity Biocompatibility low drug absorptivity	Low gas permeability
PDMS	Flexible Optical transparent Gas permeable Biocompatible Low autofluorescent Deformable Ease of use Low cost	Hydrophobicity Incompatibility with organic solvents Tendency to adsorb protein and small analytes Leaching un-crosslinked oligomers
Thermoplastics	Optically transparent, More rigid than elastomers Resistant to the diffusion of small molecules	Low gas-permeability Inflexible Dissoluble in most organic solvents
Collagen	Biocompatible Low antigenicity/immunogenicity Degraded enzymatically Contains cell adhesive domains Can be formed into specific geometries Major component of native ECM Cells able to remodel and contract gel matrix Culture media, proteins, and growth factors can be transported across collagen gel Orientation of collagen fibers can be achieved Cells can penetrate, remodel, and contract gel matrix	Chemical crosslinking used for increased stability Without mechanical support, collagen-based cell models remain intact for short time
Fibrin	Biocompatible Noninflammatory Biodegradable Gel formation at room temperature through enzymatic polymerization of fibrinogen by thrombin Bioadhesive properties Delivery of proteins and growth factors	Weak mechanical properties
Hyaluronic acid	Biocompatible Natural ECM component Structural component of tissue and joints Degradable with hyaluronidase Tunable elastic modulus capability	Weak mechanical properties

(*continued on next page*)

Table 10.1 Summary of biomaterials used in the LoC and OOC platforms (Ahadian et al., 2018; Monteduro et al., 2023)—cont'd

Materials	Strengths	Limitation
Chitosan	Biocompatible Biodegradable Similar in structure to glycosaminoglycans Flexible and porous Minimal foreign-body resp	Mechanical weakness Instability
Alginate	Biocompatible Degradable Immediate gelation upon exposure to divalent cations Use as sacrificial material and gel dissolves culture medium	Uncontrollable degradation Limited protein adsorption Lack of cell binding
Gelatin	Biocompatible Biodegradable Similar in composition to collagen Contains cell adhesion sites Less antigenic than collagen Tunable and physiologically relevant elastic modulus	Chemical crosslinking for stability
Synthetic biomaterials	Tunable mechanical properties Tunable degradation properties Less batch-to-batch variability than natural biomaterials Chemical modification to incorporate bioactive molecules Polyesters degraded through hydrolysis Moldable	Lack of cell adhesion ligands prior to modification Degradation products could have cytotoxic effects Immune response must be evaluated

influencing cell alignment, migration, and morphology. These features further contribute to mimicking the natural tissue architecture and guiding cellular responses within the microfluidic environment.

Temperature and pH are critical factors influencing cellular activity. LoC devices offer the ability to precisely control these parameters through the integration of microfluidic heaters, coolers, buffers, and reservoirs. Microfluidic heaters and coolers maintain desired temperature ranges, mimicking the specific thermal environment of various tissues. Buffers and reservoirs, on the other hand, enable the maintenance of constant pH or the generation of controlled pH gradients, reflecting the diverse microenvironments found within different tissues and organs (Leung et al., 2022).

Tissues exhibit non-uniform distribution of oxygen and nutrients. LoC devices can be designed to create these gradients by incorporating microfluidic diffusion barriers into the channel network. These barriers selectively control the diffusion of specific molecules across different compartments within the chip, mimicking the diverse microenvironments encountered by cells in vivo. Additionally, integration of oxygen-permeable membranes allows for controlled exchange of oxygen between the microfluidic channels and the external environment. This enables the creation of hypoxic or normoxic conditions within specific regions of the chip, further mimicking the varying oxygen levels found in different tissues.

By carefully tailoring these physicochemical cues, LoC devices can be designed to offer precise control over the microenvironment and enable the study of complex biological processes with unprecedented detail. This refined control paves the way for developing more accurate and relevant models for disease modeling, drug discovery, and personalized medicine approaches.

10.4.3 Applications of organ-on-chip

The pharmaceutical industry faces challenges with the efficiency of drug research and development. LoC systems can play a pivotal role by providing microengineered cell culture models with miniaturized and automated assays for increased resolution and precision. The process of developing new drugs, from initial discovery in candidate libraries to preclinical and clinical phases, is known to be inefficient, lengthy, and expensive. It typically takes several years of research and requires significant financial investment, specialized personnel, and rigorous validation procedures through regulated clinical phases. The success rate of new drug compounds is low, around 5%, leading to substantial resource wastage annually. Improving therapeutic efficacy also involves finding effective ways to deliver pharmaceuticals to patients through various routes like oral, rectal, inhalation, transdermal, or intravenous administration. OOC devices offer valuable support by replicating organ/tissue features, microenvironments, and biological barriers for testing drug efficacy, solubility, permeability, targeted delivery, and toxicity in a more accurate and reliable manner compared to traditional methods like 2D cell cultures and animal experiments. These technologies are gaining attention globally for their potential in high-throughput screening and more efficient drug development processes. Many researchers have published and discussed the application of single OOC in their review articles (Ahadian et al., 2018; Huh et al., 2011; Monteduro et al., 2023). Therefore, this section will focus more on the applications of Multi-OoC and stretchable LoC as shown in Fig. 10.11.

10.4.3(a) *Multiple organ-on-chip*

A notable trend involves coupling multiple organ modules through vascular perfusion to recreate organ-level structures, simulate multi-organ interactions, and evaluate drug metabolism, clearance, and toxicity within a comprehensive 'body-on-a-chip' platform. Microfluidic technologies enable combinatorial approaches to assess libraries of drug candidates and evaluate toxicity across multiple tissues more effectively than before. Integration of miniaturized sensors further enhances automation and scalability of the assays.

A study by Maschmeyer et al. (2015) reports a novel microfluidic four-organ-chip system that successfully maintains viable human gut, liver, skin, and kidney tissues in co-culture for an extended 28-day period. All tissues exhibited high cell viability and retained their distinct physiological architectures throughout the experiment. Notably, the intestinal epithelium formed 3D villi-like structures, the skin developed a stratified barrier, and the kidney epithelium maintained a functional polarized barrier. Importantly, the system established a stable glucose gradient between compartments, mimicking physiological interactions. Further analysis revealed a reproducible metabolic and genetic homeostasis across all four tissues, independent of donor background. This innovative platform is designed for ADME (absorption, distribution, metabolism, and excretion) profiling of drugs and allows for repeated-dose systemic toxicity testing. The system effectively recapitulates the integrity and function of the intestinal and kidney barriers at a relevant scale. By enabling "oral" drug administration and mimicking

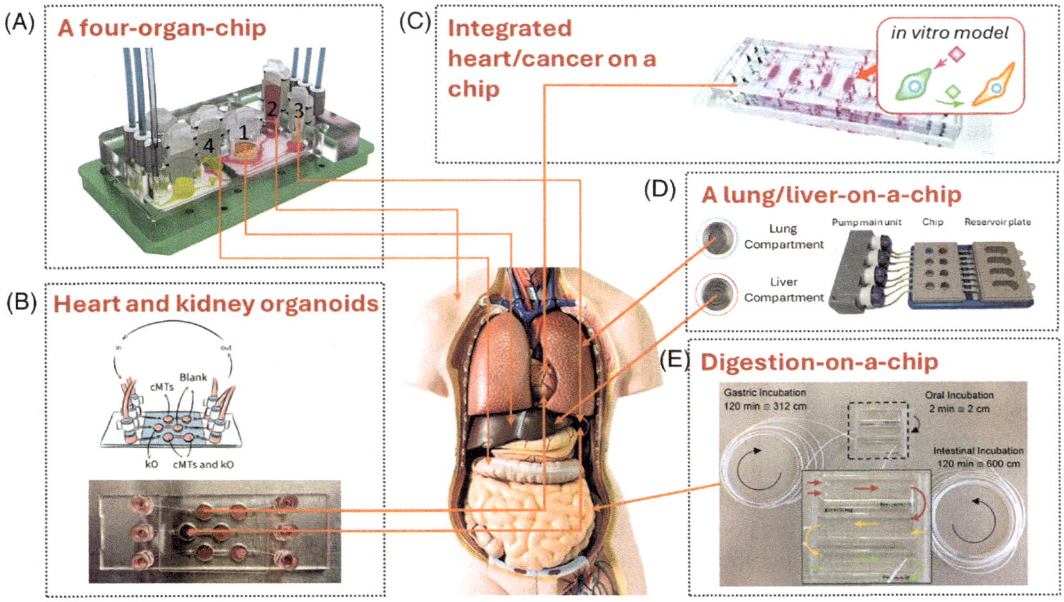

FIGURE 10.11

Example application of Multi-organ on a chip. (A) Four organ chip (Intestine, liver, skin and kidney) (Maschmeyer et al., 2015). (B) Heart and kidney on chip (Gabbin et al., 2023). (C) Heart and cancer on chip (Kamei et al., 2017). (D) Lung and liver on chip (Bovard et al., 2018) and (E) Digestion/intestines on a chip.

(E) From De Haan, P., Ianovska, M. A., Mathwig, K., Van Lieshout, G. A. A., Triantis, V., Bouwmeester, H., & Verpoorte, E. (2019). Digestion-on-a-chip: A continuous-flow modular microsystem recreating enzymatic digestion in the gastrointestinal tract. Lab on a Chip, 19*(9), 1599–1609. https://doi.org/10.1039/c8lc01080c.*

blood flow and organ interactions, the chip facilitates the study of a drug's absorption, metabolism (first-pass in liver and secondary elsewhere), and excretion through the kidney model. The accessible compartments and organ equivalents enable the evaluation of pharmacokinetic (drug movement) and pharmacodynamic (drug effects) parameters, providing valuable data for drug development.

Another study by Gabbin et al. (2023) developed a microfluidic system to study the interaction between heart and kidney organoids in vitro. The system allowed for the maintenance of tissue viability and organ-specific function for at least 72 h under both static and dynamic culture conditions. The cardiac microtissues (cMTs) displayed sarcomeric structures coupled with beating activity, while the kidney organoids (kOs) exhibited the presence of nephron structures and albumin uptake. The combination of cardiac and kidney organoids in the microfluidic platform provides a new in vitro model to study the cardiorenal axis and investigate diseases involving both organs and their potential treatments. Additionally, the study highlighted the potential for studying the dialogue between the heart and kidney, as well as the complex responses that characterize dynamic crosstalk. The microfluidic platform represents an important step in the development of in vitro models for understanding the interaction and function of multiple organs, paving the way for further research in disease modeling and therapeutic discovery.

While OOC technology holds promise for mimicking human physiological conditions, it's crucial to incorporate multiple tissue types and a circulatory system to accurately study drug side effects. Kamei et al. (2017) created a microfluidic device called Integrated Heart/Cancer on a Chip (iHCC), using human heart cells (hCMs) and liver cancer cells (HepG2), to simulate the side effects of the anti-cancer drug doxorubicin (DXR). This device features individual cultures of cells from different tissues, three artificial blood circulation loops, and precise fluid control through microfabrication technology. The advanced fabrication methods, including numerical optimization simulation and on-chip integration of valves and a micropump, ensure precise fluid flow. The iHCC enables modeling of DXR's side effects on heart cells, including the production of toxic metabolites (doxorubicinol; DXRol) by HepG2 cells and their delivery to heart cells via the circulation loop. These findings pave the way for the development of a comprehensive "Body-on-a-Chip" system (Wang et al., 2018).

Another work by Bovard et al. (2018) focuses on the development of a lung/liver-on-a-chip, connecting normal human bronchial epithelial cells cultured at the air–liquid interface with HepaRG™ liver spheroids. The individual tissues in the chip demonstrated improved viability and functionality when maintained together, with promising results observed over 28 days. The platform's capacity to metabolize compounds and mitigate toxicity was demonstrated using aflatoxin B1 (AFB1) as a model compound. The presence of HepaRG™ spheroids in the chip circuit reduced AFB1-induced cytotoxicity in lung tissues, indicating the potential of this platform to assess inhaled aerosol toxicity and evaluate new drug candidates targeting the human lung's safety and efficacy.

The research by Haan et al. (2019) focuses on developing a cell-free, miniaturized enzymatic digestive system crucial for assessing the bioavailability of compounds like nutrients. This system employs three micromixers connected in series to replicate the functions of the mouth, stomach, and small intestine, ensuring continuous processing of samples containing nutrients. The digested materials are then fed into a subsequent gut-on-a-chip absorption module containing living human intestinal cells. The system, featuring three-compartment enzymatic digestion, effectively mimics key gastrointestinal functions by modifying the chemical environment in each compartment to resemble physiological conditions. This innovation complements existing gut-on-a-chip systems, advancing the determination of orally administered compound bioavailability in a fast, continuous-flow ex vivo setup. Through experiments with model nutrients like starch and casein, successful enzymatic digestion is demonstrated in each compartment. The use of transparent microfluidic micromixers enables real-time monitoring of enzyme kinetics, providing insights from the onset of the reaction. Additionally, the system efficiently digests lactoferrin, a milk protein, in significantly shorter durations compared to conventional in vitro digestions using batch reactors.

Contemporary drug development protocols increasingly rely on predictive approaches utilizing accurate in vitro replicas of human tissues, specific diseases, and multi-organ models. These advancements address the urgent need for more efficient drug development processes, surpassing the limitations of traditional 2D culture systems and animal testing in terms of throughput and resource utilization. Moreover, modern organ-on-chip systems hold immense potential for uncovering the mechanisms underlying currently incurable diseases, with significant milestones already attained in this area. However, future challenges will revolve around balancing standardization, reproducibility, and reliability in replicating disease microenvironments and drug responses (Brancato et al., 2020), as well as advancing towards patient-derived testing platforms for personalized precision medicine.

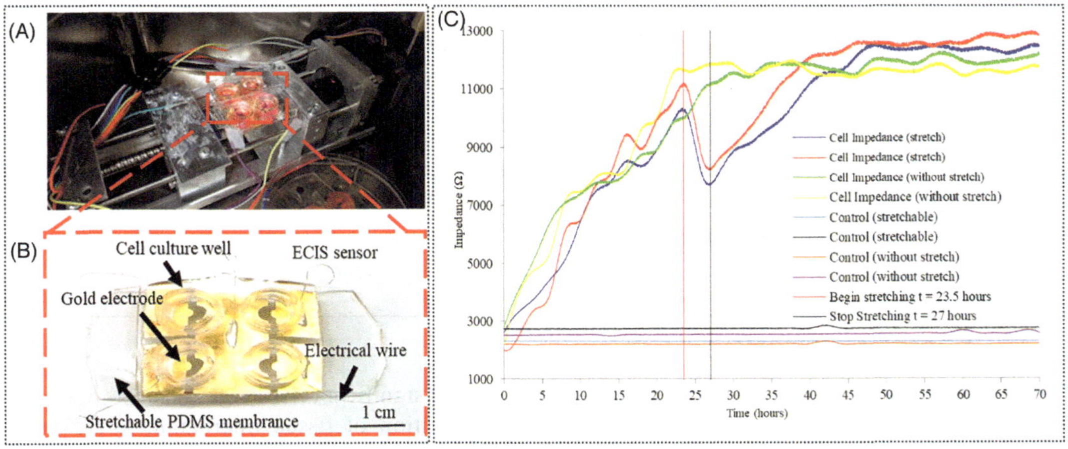

FIGURE 10.12

(A) Linear motor was used to cyclically elongate the stretchable ECIS sensor. The linear motor with the stretchable ECIS sensor clamped on it was placed inside the incubator. (B) The stretchable ECIS sensor. (C) The impedance responses from 4 stretchable sensors with and without 8% stretch.

(C) Zhang, X., Wang, W., Li, F., & Voiculescu, I. (2017). Stretchable impedance sensor for mammalian cell proliferation measurements. Lab on a Chip, 17*(12), 2054–2066. https://doi.org/10.1039/c7lc00375g.*

10.4.3(b) Stretchable LoC devices

The stretchability of the LoC could become an important component of organ-on-chip technology, that involves creating miniature versions of human organs on tiny chips. The stretchable LoC devices replicate the functions and structures of organs, allowing scientists to study their behaviour and test the effects of drugs or diseases in a more accurate and efficient manner closed to the behaviour of human body. The organ-on-chip technology allows researchers to mimic the complexity of human organs in lab conditions and adding the stretchability condition creates devices that could mimic the movements of the human body, for example the heartbeat.

A stretchable LoC with the possibility to perform ECIS measurements was presented in the literature (Zhang et al., 2017). The ECIS sensor, fabricated on a stretchable membrane is shown in Fig. 10.12A–B. The LoC device was used to apply mechanical pulses on the mammalian cells and to study the cell proliferation as a function of mechanical elongation. Innovative ECIS electrodes were fabricated on a stretchable substrate and ECIS measurements on mammalian cells exposed to cyclic strain of 8% were successfully demonstrated. Bovine aortic endothelial cells (BAEC) were used to evaluate the cell proliferation influenced by mechanical stimulus, because they undergo in vivo cyclic physiologic elongation produced by the blood circulation in the arteries. The BAEC were cultured on the ECIS sensor mounted on a linear motor and kept inside a cell culture incubator. The impedance data was acquired at the end of each stretch/release cycle. Microelectrodes fabricated from gold (Au) on the stretchable membrane were used to record the impedance values of endothelial cell membranes during the proliferation process. The device had miniature dimensions. The impedance responses from the sensors with and without 8% stretch during the first 70 h after BAECs cell seeding with low density is shown in Fig. 10.12C. The impedance response from the 4 ECIS sensors with BAECs cell was related to the cell number and densities. As reported in the literature, endothelial cells exhibit an increase in proliferation in response to stretch andthis behavior was also demonstrated by experiments performed with stretchable ECIS electrodes, see Fig. Fig. 10.12C. The ECIS technique is a powerful technique to monitor in real-time cell viability, attachment, proliferation and apoptosis. The stretchable membrane integrated with the ECIS sensor can simulate and replicate the dynamic environment of organism and enable the analysis of cell's activity such as cells proliferation.

10.5 Chapter summary

The integration of microengineering with cell biology presents significant challenges in the development of LoC devices. One of the main hurdles is the reconstitution of tissue microarchitecture within these devices. This requires precise control over the biochemical milieu and dynamic mechanical microenvironments to accurately mimic the complex interactions within human tissues. Additionally, ensuring that these microdevices can predict human responses with high accuracy remains a challenge, particularly for applications in drug discovery and environmental toxicology testing.

Looking ahead, the field aims to develop interconnected organ-on-a-chip systems, often referred to as 'human-on-a-chip'. These systems would model multiple organs linked by a microfluidic circulatory system, simulating whole-body responses to drugs and toxins. Such advancements could revolutionize the way we approach drug efficacy and safety testing, moving away from animal models towards more reliable human-based systems. Moreover, scaling up from lab-based prototypes to commercially viable products is a critical step. The future of LoC technology lies in creating chips that are user-friendly, reproducible, and suitable for high-throughput screening applications, making them accessible to a broader range of researchers and industries (Huh et al., 2011).

AI disclosure

During the preparation of this work the author(s) used ChatGPT in order to Language proofreading. After using this tool/service, the author(s) reviewed and edited the content as needed and take(s) full responsibility for the content of the publication.

References

Achyuthan, K. E., & Whitten, D. G. (2007). Design considerations for high throughput screening and in vitro diagnostic assays. *Combinatorial Chemistry and High Throughput Screening, 10*(6), 399–412. https://doi.org/10.2174/138620707781996475.

Ahadian, S., Civitarese, R., Bannerman, D., Mohammadi, M. H, Lu, R., Wang, E., Davenport-Huyer, L., Lai, B., Zhang, B., Zhao, Y., Mandla, S., Korolj, A., & Radisic, M. (2018). Organ-on-a-chip platforms: A convergence of advanced materials, cells, and microscale technologies. *Advanced Healthcare Materials, 7*(2), 2192–2640. https://doi.org/10.1002/adhm.201700506.

Azimzadeh, M., Khashayar, P., Amereh, M., Tasnim, N., Hoorfar, M., & Akbari, M. (2022). Microfluidic-based oxygen (O2) sensors for on-chip monitoring of cell, tissue and organ metabolism. *Biosensors, 12*(1), 6. https://doi.org/10.3390/bios12010006.

Beißner, N., Reichl, S., & Lorenz, T. (2016). *Organ on chip microsystems for pharmatechnology: Manipulation of fluids, particles, droplets, and cells* (pp. 299–339). Springer International Publishing. https://doi.org/10.1007/978-3-319-26920-7_11.

Boot, R. C., Roscani, A., van Buren, L., Maity, S., Koenderink, G. H., & Boukany, P. E. (2023). High-throughput mechanophenotyping of multicellular spheroids using a microfluidic micropipette aspiration chip. *Royal Society of Chemistry, Netherlands Lab on a Chip, 23*(7), 1768–1778. https://doi.org/10.1039/d2lc01060g.

Bovard, D., Sandoz, A., Luettich, K., Frentzel, S., Iskandar, A., Marescotti, D., Trivedi, K., Guedj, E., Dutertre, Q., Peitsch, M. C., & Hoeng, J. (2018). A lung/liver-on-a-chip platform for acute and chronic toxicity studies. *Lab on a Chip, 18*(24), 3814–3829. https://doi.org/10.1039/c8lc01029c.

Brancato, V., Oliveira, J. M., Correlo, V. M., Reis, R. L., & Kundu, S. C. (2020). Could 3D models of cancer enhance drug screening? *Elsevier Ltd, Portugal Biomaterials, 232*. https://doi.org/10.1016/j.biomaterials.2019.119744.

Burger, R., Kurzbuch, D., Gorkin, R., Kijanka, G., Glynn, M., McDonagh, C., & Ducrée, J. (2015). An integrated centrifugo-opto-microfluidic platform for arraying, analysis, identification and manipulation of individual cells. *Lab on a Chip, 15*(2), 378–381. https://doi.org/10.1039/C4LC01002G.

Byung, H., Park, J., Lee, H., Jung, J. H, Oh, S. J, Lee, D. C., & Seok Seo, T. (2015). A centrifuge-based stepwise chemical loading disc for the production of multiplex anisotropic metallic nanoparticles. *RSC Advances, 5*(3), 1846–1851. https://doi.org/10.1039/C4RA13778G.

Cliffe, F. E., Madden, C., Costello, P., Devitt, S., Mukkunda, S. R., Keshava, B. B., Fearnhead, H. O., Vitkauskaite, A., Dehkordi, M. H., Chingwaru, W., Przyjalgowski, M., Rebrova, N., & Lyons, M. (2023). Mera: A scalable high throughput automated micro-physiological system. *SLAS Technology, 28*(4), 230–242. https://doi.org/10.1016/j.slast.2023.01.004.

Cunha, M. L., Da Silva, S. S., Stracke, M. C., Zanette, D. L., Aoki, M. N., & Blanes, L. (2022). Sample preparation for lab-on-a-chip systems in molecular diagnosis: A review. *Analytical Chemistry, 94*(1), 41–58. https://doi.org/10.1021/acs.analchem.1c04460.

Damiati, S., Kompella, U. B., Damiati, S. A., & Kodzius, R. (2018). Microfluidic devices for drug delivery systems and drug screening. *Genes, 9*(2). https://doi.org/10.3390/genes9020103.

Davenport Huyer, L., Montgomery, M., Zhao, Y., Xiao, Y., Conant, G., Korolj, A., & Radisic, M. (2015). Biomaterial based cardiac tissue engineering and its applications. *Biomedical Materials, 10*(3), 034004. https://doi.org/10.1088/1748-6041/10/3/034004.

De Haan, P., Ianovska, M. A., Mathwig, K., Van Lieshout, G. A. A., Triantis, V., Bouwmeester, H., & Verpoorte, E. (2019). Digestion-on-a-chip: A continuous-flow modular microsystem recreating enzymatic digestion in the gastrointestinal tract. *Royal Society of Chemistry, Netherlands Lab on a Chip, 19*(9), 1599–1609. https://doi.org/10.1039/c8lc01080c.

Ding, C., Chen, X., Kang, Q., & Yan, X. (2020). Biomedical application of functional materials in organ-on-a-chip. *Frontiers in Bioengineering and Biotechnology, 8*, 2296–4185. https://doi.org/10.3389/fbioe.2020.00823.

Ducrée, J., Haeberle, S., Lutz, S., Pausch, S., Von Stetten, F., & Zengerle, R. (2007). The centrifugal microfluidic bio-disk platform. *Journal of Micromechanics and Microengineering, 17*(7), S103. https://doi.org/10.1088/0960-1317/17/7/S07.

Faley, S. L., Boghdeh, N. A., Schaffer, D. K., Spivey, E. C., Alem, F., Narayanan, A., Wikswo, J. P., & Brown, J. A. (2025). Gravity-perfused airway-on-a-chip optimized for quantitative BSL-3 studies of SARS-CoV-2 infection: Barrier permeability, cytokine production, immunohistochemistry, and viral load assays. *Lab on a chip, 25*(23), 6231–6247. https://doi.org/10.1039/D5LC00510H.

Focke, M., Kosse, D., Müller, C., Reinecke, H., Zengerle, R., & Von Stetten, F. (2010). Lab-on-a-foil: Microfluidics on thin and flexible films. *Royal Society of Chemistry, Germany Lab on a Chip, 10*(11), 1365–1386. https://doi.org/10.1039/c001195a.

Gabbin, B., Meraviglia, V., Angenent, M. L., Oostwaard, D. W-, Sol, W., Mummery, C. L., Rabelink, T. J., Meer, B. J., Berg, C. W., & Bellin, M. (2023). Heart and kidney organoids maintain organ-specific function in a microfluidic system. *Materials Today Bio, 23*, 100818. https://doi.org/10.1016/j.mtbio.2023.100818.

Gorkin, R., Park, J., Siegrist, J., Amasia, M., Lee, B. S., Park, J. M., Kim, J., Kim, H., Madou, M., & Cho, Y. K. (2010). Centrifugal microfluidics for biomedical applications. *Lab on a Chip, 10*(14), 1758–1773. https://doi.org/10.1039/b924109d.

Huh, D., Hamilton, G. A., & Ingber, D. E. (2011). From 3D cell culture to organs-on-chips. *Trends in Cell Biology, 21*(12), 745–754. https://doi.org/10.1016/j.tcb.2011.09.005.

Ingber, D. E. (2022). Human organs-on-chips for disease modelling, drug development and personalized medicine. *Nature Reviews Genetics, 23*(8), 467–491. https://doi.org/10.1038/s41576-022-00466-9.

Jung, J. H., Choi, S. J., Park, B. H., Choi, Y. K., & Seo, T. S. (2012). Ultrafast rotary PCR system for multiple influenza viral RNA detection. *Lab on a Chip, 12*(9), 1598–1600. https://doi.org/10.1039/c2lc21269b.

Kadimisetty, K., Song, J., Doto, A. M., Hwang, Y., Peng, J., Mauk, M. G., Bushman, F. D., Gross, R., Jarvis, J. N., & Liu, C. (2018). Fully 3D printed integrated reactor array for point-of-care molecular diagnostics. *Biosensors and Bioelectronics, 109*, 156–163. https://doi.org/10.1016/j.bios.2018.03.009.

Kamei, K., Kato, Y., Hirai, Y., Ito, S., Satoh, J., Oka, A., Tsuchiya, T., Chen, Y., & Tabata, O. (2017). Integrated heart/cancer on a chip to reproduce the side effects of anti-cancer drugs in vitro. *RSC Advances, 7*(58), 36777–36786. https://doi.org/10.1039/c7ra07716e.

Keller, M., Wadle, S., Paust, N., Dreesen, L., Nuese, C., Strohmeier, O., Zengerle, R., & von Stetten, F. (2015). Centrifugo-thermopneumatic fluid control for valving and aliquoting applied to multiplex real-time PCR on off-the-shelf centrifugal thermocycler. *RSC Advances, 5*(109), 89603–89611. https://doi.org/10.1039/C5RA16095B.

Kutluk, H., Bastounis, E. E., & Constantinou, I. (2023). Integration of extracellular matrices into organ-on-chip systems. *Advanced Healthcare Materials, 12*(20). https://doi.org/10.1002/adhm.202203256.

Lee, S., Khoo, V. S. L., Medriano, C. A. D., Lee, T., Park, S. Y., & Bae, S. (2019). Rapid and in-situ detection of fecal indicator bacteria in water using simple DNA extraction and portable loop-mediated isothermal amplification (LAMP)PCR methods. *Water Research, 160*, 371–379. https://doi.org/10.1016/j.watres.2019.05.049.

Leung, C. M., de Haan, P., Ronaldson-Bouchard, K., Kim, G. A., Ko, J., Rho, H. S., Chen, Z., Habibovic, P., Jeon, N. L., Takayama, S., Shuler, M. L., Vunjak-Novakovic, G., Frey, O., Verpoorte, E., & Toh, Y. C. (2022). A guide to the organ-on-a-chip. *Nature Reviews Methods Primers, 2*(1). https://doi.org/10.1038/s43586-022-00118-6.

Liu, D., Liang, G., Zhang, Q., & Chen, B. (2013a). Detection of Mycobacterium tuberculosis Using a capillary-array microsystem with integrated DNA extraction, loop-mediated isothermal amplification, and fluorescence detection. *Analytical Chemistry, 85*(9), 4698–4704. https://doi.org/10.1021/ac400412m.

Liu, F., Li, F., Nordin, A., & Voiculescu, I. (2013b). A novel cell-based hybrid acoustic wave biosensor with impedimetric sensing capabilities. *Sensors, 13*(3), 3039–3055. https://doi.org/10.3390/s130303039.

Liu, F., Nordin, A. N., Li, F., & Voiculescu, I. (2014). A lab-on-chip cell-based biosensor for label-free sensing of water toxicants. *Lab on a Chip, 14*(7), 1270–1280. https://doi.org/10.1039/c3lc51085a.

Madou, M., Zoval, J., Jia, G., Kido, H., Kim, J., & Kim, N. (2006). Lab on a CD. *Annual Review of Biomedical Engineering, 8*(1), 601–628. https://doi.org/10.1146/annurev.bioeng.8.061505.095758.

Mansor, A. F. M., Ibrahim, I., Zainuddin, A. A., Voiculescu, I., & Nordin, A. N. (2018). Modeling and development of screen-printed impedance biosensor for cytotoxicity studies of lung carcinoma cells. *Medical and Biological Engineering and Computing, 56*(1), 173–181. https://doi.org/10.1007/s11517-017-1756-1.

Mansor, A. FM., Sedek, S., Nordin, A. N, Mohammad, M., Hashim, Y. ZH.-Y., Nordin, A. FW., Goh, K. L., Azih, M. N. A., Abdul Razak, M. R., Ibrahim, A., & Hin How, S. (2023). Application of real-time cell analysis biosensor technology for drug cytotoxicity studies in primary lung cancer cells. *Biomedical Materials & Devices, 1*(2), 825–835. https://doi.org/10.1007/s44174-022-00045-w.

Mark, D., Haeberle, S., Roth, G., Stetten, F. V., & Zengerle, R. (2010a). Microfluidic lab-on-a-chip platforms: Requirements, characteristics and applications. *Chemical Society Reviews, 39*(3), 1153–1182. https://doi.org/10.1039/b820557b.

Mark, D., Haeberle, S., Roth, G., Von Stetten, F., & Zengerle, R. (2010b). Microfluidic lab-on-a-chip platforms: Requirements, characteristics and applications. In *NATO science for peace and security series A: Chemistry and biology* (pp. 305–376). Springer. https://doi.org/10.1007/978-90-481-9029-4_17.

Maschmeyer, I., Lorenz, A. K., Schimek, K., Hasenberg, T., Ramme, A. P., Hübner, J., Lindner, M., Drewell, C., Bauer, S., Thomas, A., Sambo, N. S., Sonntag, F., Lauster, R., & Marx, U. (2015). A four-organ-chip for interconnected long-term co-culture of human intestine, liver, skin and kidney equivalents. *Lab on a Chip, 15*(12), 2688–2699. https://doi.org/10.1039/c5lc00392j.

Monteduro, A. G., Rizzato, S., Caragnano, G., Trapani, A., Giannelli, G., & Maruccio, G. (2023). Organs-on-chips technologies – A guide from disease models to opportunities for drug development. *Biosensors and Bioelectronics, 231*. https://doi.org/10.1016/j.bios.2023.115271.

Oh, S. J., & Seo, T. S. (2019). Combination of a centrifugal microfluidic device with a solution-loading cartridge for fully automatic molecular diagnostics. *Analyst, 144*(19), 5766–5774. https://doi.org/10.1039/c9an00900k.

Oliva-Cárdenas, A., Fernández-Zamora, F., Santana-Rodríguez, E., Sordo-Puga, Y., de la Vargas-Hernández, M. C., Rodríguez-Moltó, M. P., Pérez-Pérez, D., Sardina-González, T., Duarte, C. A., León-Goñi, A., Blanco-Gámez, D., Contreras-Pérez, F., Valdés-Faure, O., Hernández-Prado, R., Acosta-Lago, E., Sosa-Testé, I., & Suárez-Pedroso, M. F. (2021). Safety and immunogenicity in piglets of two immunization schedules initiated at two or three weeks of age with Porvac®, a classical swine fever subunit marker vaccine. *Centro de Biotecnologia y Biomedicina, Clinical Biotec. Universidad Católica del Oriente (UCO), 6*(3), 1896–1901. https://doi.org/10.21931/RB/2021.06.03.3.

Park, B. H., Jung, J. H., Zhang, H., Lee, N. Y., & Seo, T. S. (2012a). A rotary microsystem for simple, rapid and automatic RNA purification. *Lab on a chip, 12*(20), 3875–3881. https://doi.org/10.1039/c2lc40487g.

Park, B. H., Kim, D., Jung, J. H., Oh, S. J., Choi, G., Lee, D. C., & Seo, T. S. (2015). An advanced centrifugal microsystem toward high-throughput multiplex colloidal nanocrystal synthesis. *Sensors and Actuators, B: Chemical, 209*, 927–933. https://doi.org/10.1016/j.snb.2014.12.067.

Park, J., Sunkara, V., Kim, T. H., Hwang, H., & Cho, Y. K. (2012b). Lab-on-a-disc for fully integrated multiplex immunoassays. *Analytical Chemistry, 84*(5), 2133–2140. https://doi.org/10.1021/ac203163u.

Paul, R., Ostermann, E., Chen, Y., Saville, A. C., Yang, Y., Gu, Z., Whitfield, A. E., Ristaino, J. B., & Wei, Q. (2021). Integrated microneedle-smartphone nucleic acid amplification platform for in-field diagnosis of plant diseases. *Biosensors and Bioelectronics, 187*, 113312. https://doi.org/10.1016/j.bios.2021.113312.

Payne, E. M., Holland-Moritz, D. A., Sun, S., & Kennedy, R. T. (2020). High-throughput screening by droplet microfluidics: Perspective into key challenges and future prospects. *Lab on a Chip, 20*(13), 2247–2262. https://doi.org/10.1039/d0lc00347f.

Poddar, S., Fink, G., Haselmayr, W., & Wille, R. (2022). A generic sample preparation approach for different microfluidic labs-on-chips. *IEEE Transactions on Computer-Aided Design of Integrated Circuits and Systems, 41*(11), 4612–4625. https://doi.org/10.1109/tcad.2021.3135323.

Podwin, L., Przystupski, D., Kubicki, W., Śniadek, J. K, Kulbacka, J., Wymysłowski, A., … Dziuban, J. A. (2020). Lab-on-Chip Platform for Culturing and Dynamic Evaluation of Cells Development. *Micromachines, 11*(2), 196. https://doi.org/10.3390/MI11020196. doi:10.3390/mi11020196.

Sanka, R., Lippai, J., Samarasekera, D., Nemsick, S., & Densmore, D. (2019). 3DμF - Interactive design environment for continuous flow microfluidic devices. *Scientific Reports, 9*(1). https://doi.org/10.1038/s41598-019-45623-z.

Seidel, S., Winkler, K. F., Kurreck, A., Cruz-Bournazou, M. N., Paulick, K., Groß, S., & Neubauer, P. (2024). Thermal segment microwell plate control for automated liquid handling setups. *Lab on a Chip, 24*(8), 2224–2236. https://doi.org/10.1039/d3lc00714f.

Shu, B., Zhang, C., & Xing, D. (2017). A sample-to-answer, real-time convective polymerase chain reaction system for point-of-care diagnostics. *Biosensors and Bioelectronics, 97*, 360–368. https://doi.org/10.1016/j.bios.2017.06.014.

Strohmeier, O., Keller, M., Schwemmer, F., Zehnle, S., Mark, D., von Stetten, F., Zengerle, R., & Paust, N. (2015). Centrifugal microfluidic platforms: Advanced unit operations and applications. *Chemical Society Reviews, 44*(17), 6187–6229. https://doi.org/10.1039/c4cs00371c.

Strohmeier, O., Marquart, N., Mark, D., Roth, G., Zengerle, R., & von Stetten, F. (2014). Real-time PCR based detection of a panel of food-borne pathogens on a centrifugal microfluidic "LabDisk" with on-disk quality controls and standards for quantification. *Analytical Methods, 6*(7), 2038. https://doi.org/10.1039/c3ay41822g.

Subudhi, S. K., & Das, S. (2023). Reliability of lab-on-a-chip technologies for wearable electronics: A perspective. *Frontiers in Sensors, 4*. https://doi.org/10.3389/fsens.2023.1283402.

Szymański, P. ł, Markowicz, M., & Mikiciuk-Olasik, E. (2012). Adaptation of high-throughput screening in drug discovery—Toxicological screening tests. *International Journal of Molecular Sciences, 13*(1), 427–452. https://doi.org/10.3390/ijms13010427.

Trieu, P. T., & Lee, N. Y. (2019). Paper-based all-in-one origami microdevice for nucleic acid amplification testing for rapid colorimetric identification of live cells for point-of-care testing. *Analytical Chemistry, 91*(17), 11013–11022. https://doi.org/10.1021/acs.analchem.9b01263.

Vecchio, G., Fenech, M., Pompa, P. P., & Voelcker, N. H. (2014). Lab-on-a-chip-based high-throughput screening of the genotoxicity of engineered nanomaterials. *Small, 10*(13), 2721–2734. https://doi.org/10.1002/smll.201303359.

Voiculescu, I., Li, F., Liu, F., Zhang, X., Cancel, L. M., Tarbell, J. M., & Khademhosseini, A. (2013). Study of long-term viability of endothelial cells for lab-on-a-chip devices. *Sensors and Actuators, B: Chemical, 182*, 696–705. https://doi.org/10.1016/j.snb.2013.03.030.

Voiculescu, I., Toda, M., Inomata, N., Ono, T., & Li, F. (2018). Nano and microsensors for mammalian cell studies. *Micromachines., 9*(9), 439. https://doi.org/10.3390/mi9090439.

Wang, C. H., Lien, K. Y., Wu, J. J., & Lee, G. B. (2011). A magnetic bead-based assay for the rapid detection of methicillin- resistant Staphylococcus aureus by using a microfluidic system with integrated loop-mediated isothermal amplification. *Lab on a Chip, 11*(8), 1521–1531. https://doi.org/10.1039/c0lc00430h.

Wang, M. (2023). High-throughput screening technologies for drug discovery. *Theoretical and Natural Science, 15*(1), 18–23. https://doi.org/10.54254/2753-8818/15/20240387.

Wang, S., Liu, N., Zheng, L., Cai, G., & Lin, J. (2020). A lab-on-chip device for the sample-in-result-out detection of viable Salmonella using loop-mediated isothermal amplification and real-time turbidity monitoring. *Lab on a Chip, 20*(13), 2296–2305. https://doi.org/10.1039/d0lc00290a.

Wang, Y. I., Carmona, C., Hickman, J. J., & Shuler, M. L. (2018). Multiorgan microphysiological systems for drug development: Strategies, advances, and challenges. *Advanced Healthcare Materials, 7*(2). https://doi.org/10.1002/adhm.201701000.

Wei, L., Li, W., Entcheva, E., & Li, Z. (2020). Microfluidics-enabled 96-well perfusion system for high-throughput tissue engineering and long-term all-optical electrophysiology. *Lab on a Chip, 20*(21), 4031–4042. https://doi.org/10.1039/d0lc00615g.

Wu, J., Gao, Y., Xi, J., You, X., Zhang, X., Zhang, X., Cao, Y., Liu, P., Chen, X., & Luan, Y. (2022). A high-throughput microplate toxicity screening platform based on Caenorhabditis elegans. *Ecotoxicology and Environmental Safety, 245*, 114089. https://doi.org/10.1016/j.ecoenv.2022.114089.

Yang, L., Pijuan-Galito, S., Rho, H. S., Vasilevich, A. S., Eren, A. D., Ge, L., Habibović, P., Alexander, M. R., De Boer, J., Carlier, A., Van Rijn, P., & Zhou, Q. (2021). High-throughput methods in the discovery and study of biomaterials and materiobiology. *Chemical Reviews, 121*(8), 4561–4677. https://doi.org/10.1021/acs.chemrev.0c00752.

Zaidon, N., Nordin, A. N., Ismail, A. F., & Cheung, M. W. (2016). Malaysia Serpentine microfluidic structures for concentration gradient generators. *Symposium on Design, Test, Integration and Packaging of MEMS/MOEMS*. Institute of Electrical and Electronics Engineers Inc https://doi.org/10.1109/DTIP.2016.7514870.

Zhang, X., Wang, W., Li, F., & Voiculescu, I. (2017). Stretchable impedance sensor for mammalian cell proliferation measurements. *Lab on a Chip, 17*(12), 2054–2066. https://doi.org/10.1039/c7lc00375g.

CHAPTER

11 Data transfer, analysis, and management technology for biomedical devices (IoT/IoE)

Sawal Hamid Md Ali

Faculty of Engineering and Built Environment, University Kebangsaan Malaysia, Bangi, Selangor, Malaysia

11.1 Introduction

The convergence of biomedical devices and the Internet of Things (IoT) has created unprecedented opportunities for transforming healthcare delivery (Aledhari et al., 2022; Chatterjee & Armentano, 2020; Istepanian et al., 2016; Sethi & Sarangi, 2017). Traditionally, biomedical monitoring systems were constrained by their reliance on stand-alone instruments, limited storage, and offline analysis. Patients were often required to attend clinics for periodic measurements, while clinicians relied on retrospective data to make treatment decisions. This model provided only snapshots of health status and was insufficient for continuous or preventive care. With the emergence of IoT, biomedical devices have become connected, intelligent, and capable of real-time communication, enabling a paradigm shift toward proactive and personalized healthcare.

At the core of this transformation lies the ability to **transfer, analyze, and manage biomedical data** efficiently. Miniaturized sensors embedded in wearable, implantable, or portable devices capture physiological, biochemical, and behavioral signals. These data are transmitted via wired or wireless communication protocols to local gateways, cloud platforms, or edge servers, where they are stored, processed, and analyzed. Advanced machine learning (ML) and artificial intelligence (AI) models transform raw data into meaningful clinical insights, while dashboards, mobile applications, and web interfaces provide intuitive access for both patients and healthcare providers.

The importance of such systems extends beyond individual health monitoring. On a larger scale, connected biomedical devices form the foundation of **digital health ecosystems**, integrating hospital infrastructures, telemedicine platforms, and population health databases. This integration enables not only remote monitoring and early detection of disease but also large-scale analytics for epidemiological studies and resource optimization. For example, IoT-based tracking systems using RFID and near-field communication (NFC) improve drug traceability in hospitals, while wearable sensors combined with AI support long-term management of chronic conditions such as diabetes and cardiovascular disease.

Despite the rapid growth of biomedical IoT applications, several challenges remain. Issues of interoperability, energy efficiency, data privacy, and regulatory compliance must be addressed to ensure reliable adoption. Moreover, while cloud computing provides powerful analytics and storage, the emergence of **edge and fog computing** offers solutions for latency-sensitive biomedical applications. Ethical considerations are equally important, particularly as AI increasingly participates in diagnostic and therapeutic decision-making.

BioMEMS Devices. DOI: https://doi.org/10.1016/B978-0-443-24015-7.00015-2

FIGURE 11.1

Conceptual overview of a biomedical IoT system: Sensors capture physiological data, processors, and communication modules transmit information to cloud or edge platforms, analytics generate insights, and visualization interfaces provide actionable feedback to clinicians and patients.

This chapter explores the enabling technologies that make biomedical IoT possible, focusing on **data transfer, analysis, and management**. It begins by examining the fundamental building blocks of communication protocols, sensors, processing units, storage, and analytics. The discussion then advances toward visualization methods and concludes with integrated system-level architectures that combine these components into cohesive biomedical IoT ecosystems. Case studies of wearables, implantable devices, smart hospitals, and telemedicine illustrate real-world implementations. Finally, the chapter discusses ongoing challenges and future directions, offering insights into how biomedical IoT will shape the next generation of healthcare systems (Fig. 11.1).

11.2 Communication technologies

The success of biomedical IoT systems depends on robust communication technologies. Without reliable channels for data transfer, even the most advanced sensors and processing units cannot deliver clinical value. Communication technologies in biomedical IoT must meet strict requirements: They must be energy efficient to accommodate battery-constrained wearables and implants, secure to protect sensitive patient information, and sufficiently flexible to operate across a range of clinical and home-care environments. This section examines the spectrum of communication technologies available for biomedical devices, beginning with traditional wired solutions and moving through short-range, medium-range, and wide-area wireless approaches. The discussion concludes with a synthesis of trade-offs, illustrating how protocol selection depends on specific healthcare applications.

11.2.1 Wired communication

Although wireless technologies dominate the biomedical IoT landscape, wired communication remains foundational in certain scenarios (Tavera et al., 2021; Alam et al., 2021; Sethi & Sarangi, 2017). Wired connections such as **USB and Ethernet** provide unparalleled reliability, low latency, and high data throughput, making them indispensable for hospital-based monitoring systems and diagnostic equipment. Electroencephalography (EEG) systems, for instance, often rely on wired connections to transmit multi-channel neural signals in real time. Similarly, Ethernet links are widely used in imaging modalities such as magnetic resonance imaging (MRI) and computed tomography (CT), where large data volumes must be transferred rapidly to processing servers.

In addition to reliability, wired connections offer strong resistance to electromagnetic interference, an important consideration in clinical environments crowded with electronic devices. However, the limitations of wired systems are equally apparent. Cables restrict patient mobility and comfort, making them unsuitable for wearable or long-term monitoring applications. Consequently, wired communication is often relegated to **device calibration, firmware updates, or charging**, while wireless solutions are preferred for continuous operation.

11.2.2 Wireless short-range communication

The advent of **wireless short-range communication** has been transformative for biomedical IoT, particularly in wearable and implantable devices. These protocols emphasize low energy consumption and compact hardware integration, making them suitable for miniaturized, body-worn systems (Alam et al., 2021; Tavera et al., 2021).

One of the most widely adopted technologies is **Bluetooth Low Energy (BLE)**. BLE was specifically designed for low-power applications, offering data rates of up to 2 Mbps while consuming significantly less energy than classic Bluetooth. Its presence in virtually all modern smartphones makes it a natural gateway for biomedical wearables. Devices such as smartwatches, adhesive ECG patches, and continuous glucose monitors routinely use BLE to transmit physiological data to mobile applications. From there, data can be uploaded to cloud platforms for storage and analysis. For example, the **Dexcom G6 glucose monitoring system** employs BLE to continuously stream glucose levels to a smartphone, providing users with real-time alerts and trend analysis.

Another important protocol is **near field communication (NFC)**, which enables wireless data exchange over distances of only a few centimeters. NFC is particularly advantageous for implantable devices, where power availability is severely constrained. Patients can simply place a smartphone over the implant site to retrieve sensor data. A notable example is the **Abbott FreeStyle Libre**, a flash glucose monitoring system that uses NFC to allow patients to scan their interstitial glucose sensor with a smartphone. Beyond implants, NFC is also used in smart packaging to verify medication authenticity and monitor adherence.

Radio-frequency identification (RFID), though similar in principle to NFC, has distinct applications in biomedical IoT. Passive RFID tags, which do not require an onboard power source, are widely used for patient identification through RFID-enabled wristbands. In hospital settings, RFID-tagged pharmaceuticals and equipment improve workflow efficiency by ensuring accurate tracking and reducing errors. Active RFID systems, which contain a power source, extend the range of communication and enable real-time location monitoring of assets such as infusion pumps or wheelchairs. These applications illustrate how short-range communication protocols can enhance both patient safety and hospital operations.

11.2.3 Wireless medium-range communication

While short-range protocols are sufficient for personal monitoring devices, broader healthcare ecosystems often require medium-range communication. **Wi-Fi** is one of the most prominent protocols in this category, providing high data throughput and reliable connectivity within homes and hospitals. Wi-Fi's capability to handle large data volumes makes it suitable for bandwidth-intensive applications, such as transmitting medical images, continuous multi-channel vital sign monitoring, or integrating multiple bedside devices in intensive care units (ICUs).

However, Wi-Fi is relatively energy-intensive compared to short-range protocols. For this reason, its use is often restricted to stationary devices with stable power supplies, such as hospital monitoring equipment or home-based telehealth hubs. In contrast, battery-powered wearables and implants typically avoid Wi-Fi for continuous communication, though they may use it intermittently for data synchronization when docked in charging stations.

In addition to Wi-Fi, protocols such as **ZigBee and Z-Wave** have also been explored in healthcare environments, particularly in smart home and assisted-living applications. Their ability to form mesh networks allows sensors distributed throughout a facility to communicate efficiently, enabling continuous monitoring of elderly patients for falls, activity levels, or environmental conditions.

11.2.4 Wireless wide-area communication

For applications that extend beyond local environments, **wide-area wireless communication** is essential. Cellular technologies, particularly **4G LTE and 5G**, are increasingly integrated into biomedical IoT systems. Cellular connectivity allows wearable or implantable devices to transmit data directly to cloud platforms without requiring intermediary gateways. This is particularly valuable in emergency applications: For example, a wearable cardiac monitor may automatically send alerts to a physician if abnormal rhythms are detected, regardless of the patient's location.

The introduction of **5G** represents a milestone for biomedical IoT. With its ultralow latency (as low as 1 millisecond) and support for massive device connectivity, 5G enables advanced applications such as real-time telemedicine consultations, remote robotic surgery, and large-scale epidemic surveillance. Moreover, its network slicing capabilities allow healthcare applications to be prioritized, ensuring reliable performance even during high traffic conditions. Looking ahead, **6G** is expected to integrate AI directly into communication infrastructures, enabling predictive network management and autonomous data handling for biomedical applications.

Another class of wide-area technologies is **low-power wide area networks (LPWANs)**, which includes protocols such as LoRa and NB-IoT. These networks are designed to transmit small amounts of data over long distances while consuming minimal energy (Sethi & Sarangi, 2017). They are particularly suited to distributed health monitoring systems, such as environmental sensors tracking air quality for respiratory health studies or large-scale vaccination cold chain monitoring. While LPWAN technologies cannot support high-bandwidth biomedical data, their ability to connect vast numbers of low-cost devices makes them indispensable for public health monitoring.

11.2.5 Trade-offs in biomedical IoT communications

Each communication protocol offers unique advantages, but none is universally optimal for all biomedical applications. The selection of a technology involves balancing trade-offs among range, power consumption, bandwidth, latency, and security. Short-range protocols such as BLE and NFC excel in energy efficiency and integration with consumer devices but are unsuitable for large-scale health infrastructures. Wi-Fi and cellular networks provide high throughput and wide coverage but consume more energy, limiting their use in miniaturized wearables and implants. LPWAN solutions

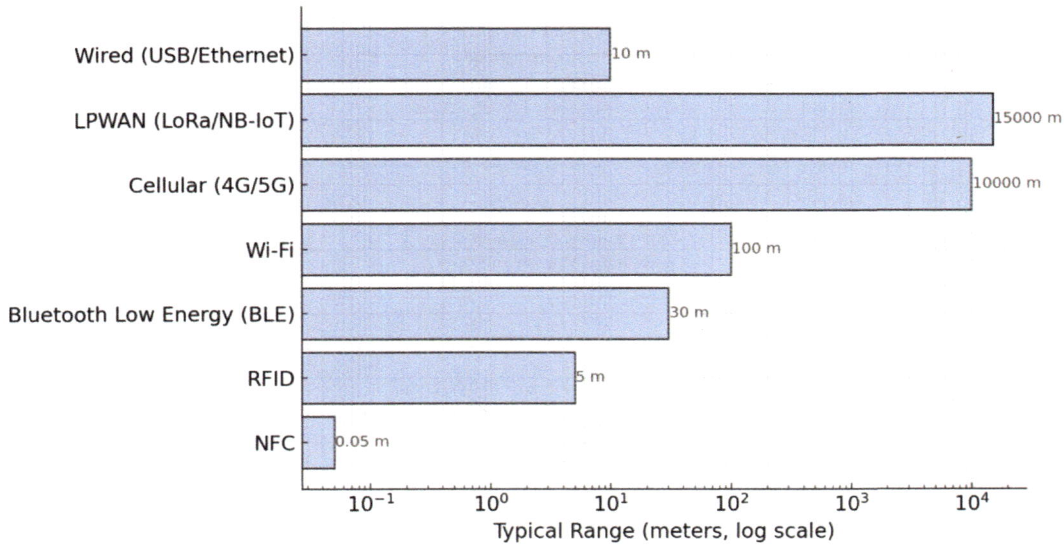

FIGURE 11.2

A comparative overview of communication protocols in terms of range, power consumption, bandwidth, and latency.

FIGURE 11.3

A typical workflow in a wearable system, showing how physiological data are transferred from the sensor to the cloud and ultimately to a clinician's dashboard.

offer long-range, low-power operation but sacrifice bandwidth and latency, making them best suited for population-level monitoring rather than individual patient care (Fig. 11.2).

In addition to technical considerations, security and privacy remain paramount. Biomedical data are among the most sensitive categories of personal information, and breaches can have severe ethical and legal consequences. Thus, all communication protocols must incorporate encryption, authentication, and compliance with regulations such as the Health Insurance Portability and Accountability Act (HIPAA) in the United States or the General Data Protection Regulation (GDPR) in the European Union.

In practice, hybrid systems often provide the most effective solution. A wearable glucose monitor, for instance, may use BLE to transmit data to a smartphone, which then employs cellular or Wi-Fi to forward the data to a cloud platform. Similarly, hospital systems may combine RFID for patient identification with Wi-Fi for centralized monitoring dashboards. The orchestration of these diverse protocols within a coherent system-level architecture is therefore central to the success of biomedical IoT (Fig. 11.3).

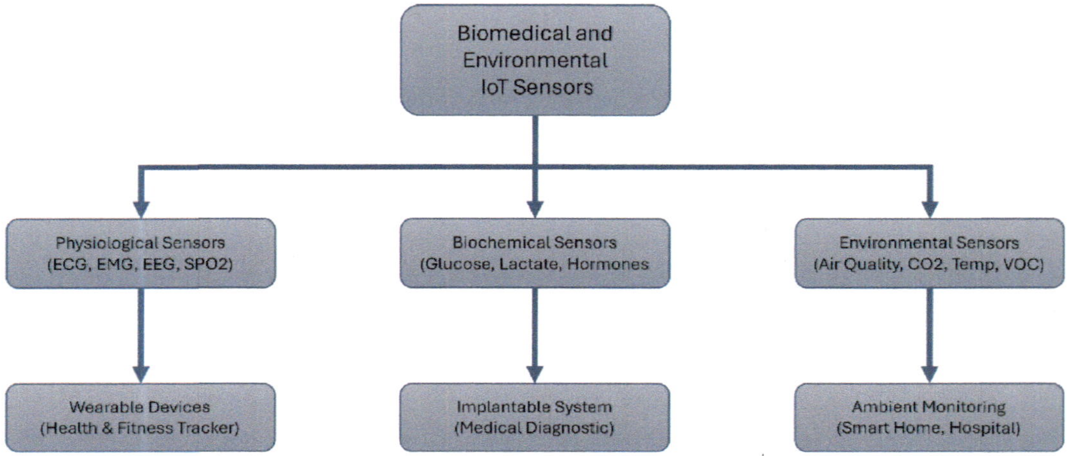

FIGURE 11.4

Category of biomedical and environmental IoT Sensors.

11.3 Sensors and data acquisition

Sensors are at the heart of biomedical IoT systems. They serve as the interface between the biological world and the digital domain, capturing physiological, biochemical, and environmental parameters that describe a patient's health status. The ability to sense accurately, continuously, and noninvasively determines the reliability of the entire biomedical IoT ecosystem. Data acquisition involves not only the sensing mechanism itself but also signal conditioning, digitization, calibration, and preprocessing to ensure data integrity before transmission to processing units or cloud servers.

11.3.1 Overview of biomedical sensors

Biomedical sensors convert physiological or biochemical quantities into measurable electrical signals. Depending on their function, they can be broadly categorized into **physiological sensors, biochemical sensors**, and **environmental sensors** as shown in Fig. 11.4 . Each class is designed with specific transduction principles and materials to ensure sensitivity, selectivity, and biocompatibility.

Physiological sensors detect electrical or mechanical signals generated by the human body. Examples include electrocardiogram (ECG) electrodes for heart activity, electromyography (EMG) sensors for muscle activity, and electroencephalography (EEG) electrodes for brain activity (Alam et al., 2021; Tavera et al., 2021; Yang et al., 2014). Nonelectrical physiological parameters such as body temperature, respiration rate, and blood pressure are also measurable using thermistors, piezoresistive membranes, and optical sensors, respectively.

Biochemical sensors, on the other hand, rely on molecular recognition. These sensors detect chemical substances or biomarkers in biological fluids such as blood, saliva, or sweat. Glucose sensors, lactate sensors, and cholesterol sensors are prominent examples that employ enzymatic or electrochemical transduction mechanisms. Recent advances in **BioMEMS technology** have enabled miniaturized biochemical sensors capable of integrating multiple sensing modalities on a single chip, allowing simultaneous detection of several biomarkers in real time.

Environmental sensors complement physiological measurements by providing contextual information about the patient's surroundings. Parameters such as air quality, humidity, and temperature can significantly influence physiological responses, particularly for individuals with respiratory or cardiovascular conditions. Incorporating environmental data into biomedical IoT systems allows for holistic health assessment and contextualized interpretation of physiological readings.

11.3.2 Principles of data acquisition

The process of data acquisition begins with **signal transduction**, where the sensor converts a physical or chemical stimulus into an electrical output. This analog signal is often weak and susceptible to noise; hence, **signal conditioning circuits** are employed to amplify, filter, and stabilize the signal. For example, in ECG acquisition, low-noise instrumentation amplifiers are used to extract microvolt-level cardiac signals while suppressing motion artifacts and power-line interference.

Following conditioning, the analog signal undergoes **analog-to-digital conversion (ADC)**, transforming it into a digital representation suitable for processing by microcontrollers or digital signal processors (DSPs). The resolution and sampling rate of the ADC must match the characteristics of the measured signal. High-resolution ADCs (12–16 bits) are typically used in precision biomedical sensors to preserve subtle signal variations, while lower-resolution converters may suffice for slower-changing physiological parameters such as temperature.

The **choice of ADC architecture** plays a significant role in determining both the performance and cost of a biomedical IoT system. Among the most commonly used ADC types are **Successive Approximation Register (SAR)** and **Sigma-Delta (Σ-Δ)** converters. SAR ADCs provide a good balance between speed and accuracy, with resolutions typically ranging from 8 to 16 bits and sampling rates up to several mega samples per second (MSPS). They are well suited for multi-channel physiological monitoring applications such as ECG and EEG, where moderate speed and high precision are required. In contrast, Sigma-Delta ADCs achieve extremely high resolution—up to 24 bits—by oversampling and noise shaping, making them ideal for low-frequency, high-accuracy measurements such as biopotential or chemical sensing. However, Sigma-Delta ADCs tend to have lower sampling rates and higher power consumption, which can limit their use in portable or real-time systems. For applications that demand very high-speed acquisition, such as ultrasound imaging or neural recording arrays, **Pipeline ADCs** may be employed, offering sampling rates of tens to hundreds of MSPS but at the expense of higher power and cost.

Selecting the appropriate ADC architecture therefore involves careful consideration of the target signal bandwidth, required resolution, power budget, and cost constraints. For instance, a low-cost wearable pulse oximeter may employ a 12-bit SAR ADC integrated into a microcontroller, whereas a laboratory-grade electrophysiology system might rely on an external 24-bit Sigma-Delta ADC to achieve superior precision. These trade-offs illustrate that the ADC is not merely a supporting component but a determining factor in the overall performance and affordability of biomedical IoT systems.

Once digitized, data often undergo **preprocessing** before transmission. Preprocessing can include baseline correction, noise suppression, normalization, or feature extraction to reduce data size and

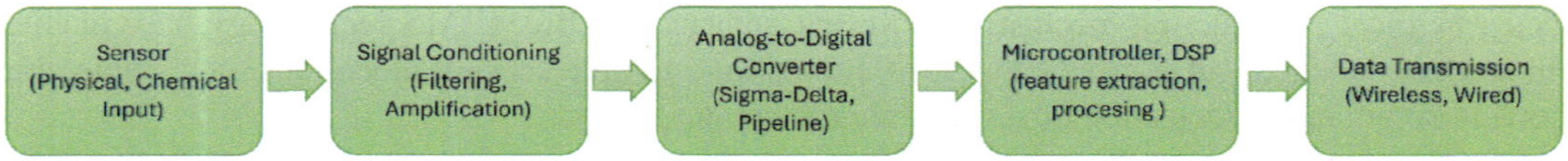

FIGURE 11.5

Typical data flow in IoT system.

enhance the reliability of downstream analytics. For example, a wearable EMG system might perform local feature extraction—such as root mean square (RMS) calculation—on the microcontroller before sending data to the cloud, conserving power and bandwidth. Fig. 11.5 shows the typical data flow in an IoT system.

11.3.3 Miniaturization and integration

A major trend in biomedical sensing is the **miniaturization and integration** of multiple sensors into compact, wearable, or implantable platforms. Advances in **microelectromechanical systems (MEMS)** and **nanotechnology** have enabled sensors that are lightweight, flexible, and biocompatible (Tang et al., 2018; Yang et al., 2014). MEMS pressure sensors, for instance, are used in catheters for continuous blood pressure monitoring, while accelerometer-based motion sensors track patient activity and posture.

Integration of multiple sensors on a single platform allows **multimodal sensing**, where physiological, biochemical, and mechanical signals are captured concurrently. This approach provides a comprehensive picture of a patient's health, enabling the correlation of different signals. For instance, a single wearable patch can measure heart rate, skin temperature, and sweat electrolyte concentration simultaneously, providing multidimensional insight into hydration and cardiovascular status.

Furthermore, integration extends beyond hardware. The co-design of sensors with low-power microcontrollers, wireless communication modules, and energy-harvesting systems enables truly autonomous devices. These integrated platforms represent the foundation of smart healthcare wearables that can operate continuously without frequent maintenance.

11.3.4 Calibration, reliability, and biocompatibility

Biomedical sensors must meet stringent requirements for accuracy, repeatability, and long-term stability. **Calibration** ensures that sensor outputs remain consistent over time and across varying environmental conditions. In clinical devices, calibration procedures are often automated, using internal references or software compensation algorithms.

Reliability is equally important, particularly in implantable systems where replacement is invasive. Materials must withstand physiological environments without degradation or drift. Encapsulation and surface treatments are employed to prevent biofouling and corrosion.

Biocompatibility is a critical factor in sensor design. Materials that contact skin or tissue must be nontoxic, hypoallergenic, and capable of long-term use. Silicone elastomers, parylene coatings, and medical-grade stainless steel are commonly used to ensure compatibility with biological systems. These design considerations not only ensure patient safety but also improve data quality by minimizing interference and inflammation responses.

11.3.5 Challenges and future directions in sensing

Despite significant advances, several challenges persist in biomedical sensing and data acquisition. The **power consumption** of continuous monitoring systems remains a key limitation, motivating research into ultralow-power sensors and energy harvesting from body heat or motion. **Data accuracy** can be compromised by motion artifacts, temperature fluctuations, or individual variability, emphasizing the need for adaptive calibration and sensor fusion techniques.

Future developments are likely to focus on **soft and flexible sensors** that conform to body surfaces, **implantable biosensors** capable of long-term operation, and **self-powered systems** that integrate micro-energy harvesters. Another emerging direction is **AI-assisted sensing**, where local ML algorithms help optimize sampling strategies, detect anomalies, and reduce redundant data transmission. Together, these trends are pushing biomedical sensors toward becoming intelligent, autonomous agents within the IoT ecosystem.

11.4 Processing units: Microcontrollers and microprocessors

In biomedical IoT systems, the processing unit acts as the *central intelligence* that bridges the sensing and communication layers. Once a signal has been captured and converted into digital form, the processing unit executes critical tasks such as signal analysis, data compression, feature extraction, and device control. It determines how efficiently the system can manage real-time operations while meeting constraints in energy, size, and cost. Depending on the complexity of the application, biomedical IoT devices may employ either **microcontroller-based systems (MCUs)** or **microprocessors (MPUs)**, each with distinct advantages in terms of performance and architecture.

11.4.1 Role of the processing unit in biomedical IoT

The processing unit serves multiple essential functions beyond mere computation. It coordinates the **data flow** between the sensors, memory, and communication modules; performs **real-time signal processing** for noise filtering and feature extraction; and manages **device-level operations** such as power control, timing synchronization, and sensor calibration. In some advanced systems, the processor also executes **embedded AI** models to classify health conditions or predict physiological trends directly at the device level.

For instance, a wearable ECG patch may use its processor to detect abnormal cardiac patterns such as arrhythmias locally before transmitting only the relevant events to the cloud. This **edge processing** capability reduces bandwidth usage and enhances privacy, as sensitive raw data never leave the device. Similarly, implantable neural recording systems often rely on embedded digital signal processors (DSPs) to perform real-time spike sorting, enabling high-speed analysis with minimal power consumption.

11.4.2 Microcontrollers

Microcontrollers are compact, integrated computing systems that include a central processing unit (CPU), memory, and input/output peripherals on a single chip. They are the workhorses of most biomedical IoT devices, particularly those that must be small, low cost, and energy efficient.

Typical microcontrollers used in biomedical applications include the **ARM Cortex-M series, Microchip PIC**, and **Atmel AVR** families. Among these, ARM Cortex-M processors dominate modern medical-grade wearables due to their favorable balance between performance and power efficiency. For example, the Cortex-M4 and M7 series support single-cycle multiply-accumulate operations, enabling efficient digital signal processing for ECG or EEG signal filtering. The **Reduced Instruction Set Computing-V (RISC-V)** architecture, an emerging open-source alternative, is also gaining popularity in biomedical device development because of its flexibility and cost-effectiveness in custom chip design.

Microcontrollers are particularly advantageous for **real-time data acquisition and control** tasks. They typically operate at clock frequencies between 10 and 200 MHz and provide 32-bit precision, sufficient for most physiological signal processing applications. Their integrated peripherals—such as analog-to-digital converters (ADCs), timers, and serial communication interfaces—simplify hardware design, making them ideal for compact wearable systems. Many MCUs also include built-in wireless transceivers for BLE or Wi-Fi, eliminating the need for external modules.

An example is the **Nordic nRF52840** MCU, which combines a 32-bit ARM Cortex-M4 processor with integrated BLE and NFC connectivity. This type of system-on-chip (SoC) architecture enables complete biomedical sensor nodes that are small, lightweight, and capable of continuous wireless operation with minimal power consumption (Amin and Hossain, 2021; Li et al., 2022; García-Ramírez et al., 2020).

11.4.3 Microprocessors

Microprocessors, in contrast, are more powerful and are typically used in devices that require extensive computation or multitasking capabilities. Unlike microcontrollers, microprocessors lack integrated memory and peripherals; instead, they interface with external memory, storage, and input/output controllers. This architecture allows for higher clock speeds—often exceeding 1 GHz—and support for complex operating systems such as **Linux** or **Android**.

In biomedical IoT, microprocessors are used in systems where **data analysis and visualization** occur locally, or where multiple sensors and communication channels must be managed simultaneously. Examples include portable diagnostic devices, advanced wearable health monitors, and medical imaging workstations. The **ARM Cortex-A** series and **Intel Atom** processors are commonly found in such systems, while low-power options such as the **Raspberry Pi** or **BeagleBone Black** boards have become popular for research and prototyping purposes.

Microprocessors can handle **high-dimensional data** and execute sophisticated algorithms, including ML models, for real-time decision support. For example, an ARM Cortex-A53 processor embedded in a handheld ultrasound system can perform on-board image reconstruction, reducing dependence on external computing resources. However, this computational power comes at the cost of increased energy consumption, larger form factor, and higher system cost, which limit their use in miniaturized or battery-powered biomedical wearables.

11.4.4 System-on-chip integration

Recent advances in semiconductor technology have blurred the traditional boundaries between micro-controllers and microprocessors through the emergence of **SoC** architectures. SoCs integrate multiple

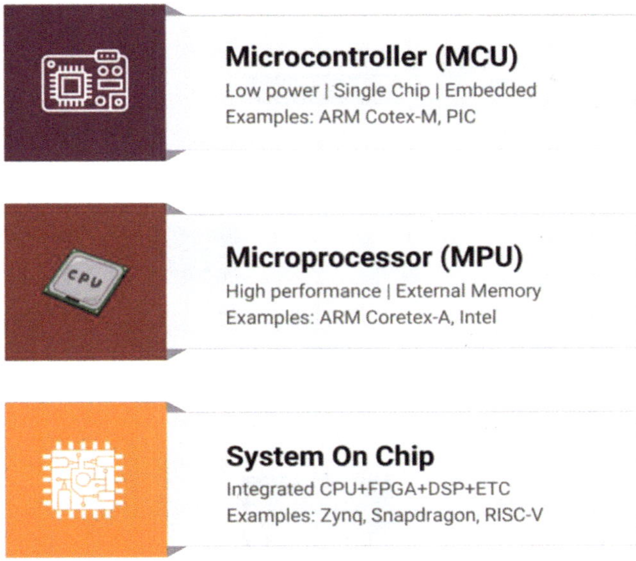

FIGURE 11.6

Three types of central processing unit in IoT system.

subsystems—including CPUs, DSPs, GPUs, wireless transceivers, and hardware accelerators—on a single silicon die.

SoC platforms such as **Texas Instruments' CC2650, Qualcomm Snapdragon Wear**, and **NVIDIA Jetson Nano** demonstrate how processing, communication, and AI capabilities can coexist within a unified architecture. In biomedical IoT, SoCs are particularly valuable for **edge computing** applications, where data must be processed locally to minimize latency and preserve privacy. For instance, AI-enabled SoCs can perform real-time heart rate variability (HRV) analysis or respiratory anomaly detection directly within the wearable device, sending only summary metrics to the cloud.

The integration of processing and communication modules on a single chip also simplifies design and reduces power consumption, two crucial factors for long-term wearable and implantable use.

Fig. 11.6 shows three different types of central processing unit typically used in the IoT system with some examples on each of them, and Fig. 11.7 shows an overview of an IoT system with all the hardware or components used.

11.4.5 Operating systems and real-time control

The selection of a suitable **operating system (OS)** is a crucial design decision in biomedical IoT systems, as it dictates how hardware resources, timing, and data management are handled. Whether an OS is required—or feasible—depends largely on the processing hardware, application complexity, and power constraints.

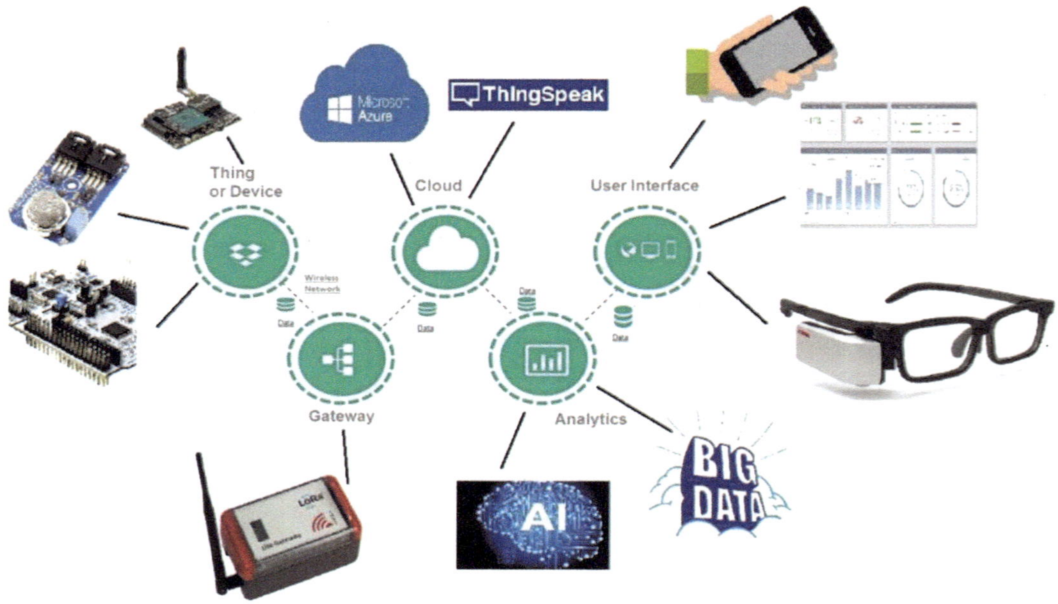

FIGURE 11.7

The overview of an IoT system.

MCUs typically do **not require a conventional OS**. They execute firmware stored in embedded memory using a single control loop or event-driven scheduling. This design allows for minimal overhead, low latency, and extremely low power consumption—qualities ideal for continuous biomedical monitoring such as heart-rate sensors, glucose monitors, or wearable motion detectors. In such systems, deterministic behavior is often achieved through **bare-metal programming**, where each hardware interrupt and peripheral is explicitly managed by the developer.

However, as biomedical IoT applications grow in complexity—handling multiple sensors, wireless communication, and data logging simultaneously—developers often turn to **real-time operating systems (RTOS)** (Li et al., 2022; Sethi & Sarangi 2017; Amin and Hossain, 2021). An RTOS provides lightweight multitasking while ensuring that critical functions (e.g., ECG sampling or arrhythmia detection) are executed within strict timing constraints. Popular RTOS options include **FreeRTOS, Zephyr**, and **ARM Mbed OS**, which are optimized for embedded platforms like ARM Cortex-M and RISC-V microcontrollers. These OSs provide deterministic task scheduling, inter-process communication, and built-in support for network stacks such as Bluetooth and Wi-Fi, simplifying the development of real-time medical applications.

In contrast, MPUs almost always rely on a full-featured OS, given their higher computational power and memory capacity. Such systems frequently employ **Linux-based distributions**—for instance, **Debian, Ubuntu**, or **Yocto Linux**—which offer advanced file systems, user management, and networking capabilities. In portable diagnostic instruments, handheld ultrasound devices, and hospital gateways, a Linux environment allows integration with complex middleware, databases, and graphical user interfaces (UIs). For consumer health wearables or telemedicine tablets, **Android** is often preferred because of its mature ecosystem and compatibility with cloud-based healthcare platforms.

Between these two extremes, some embedded systems adopt a **hybrid approach**, using lightweight real-time kernels layered on top of simplified Linux environments to achieve both responsiveness and user-interface flexibility. An example is **PREEMPT-RT Linux**, which extends standard Linux with deterministic scheduling suitable for medical devices requiring precise timing.

The choice of OS thus directly impacts not only system performance but also regulatory certification and security compliance. Systems running RTOS or bare-metal firmware are often easier to validate for medical certification due to their predictability and reduced attack surface. Conversely, Linux-based systems offer greater versatility and connectivity but require rigorous security management and patching strategies.

Ultimately, the decision to use or omit an OS—and the type chosen—depends on the **application's timing requirements, computational load, safety standards, and power budget**. A single-lead ECG patch may function flawlessly on a bare-metal microcontroller, while a multi-sensor telehealth gateway with data visualization requires a full-fledged Linux or Android system. The ability to tailor the software environment to the biomedical application's needs remains one of the most important aspects of system-level design. Fig. 11.8 presents a decision tree for the selection of microprocessors or microcontrollers in IoT systems.

11.4.6 Comparative analysis and design considerations

Selecting between a microcontroller, microprocessor, or SoC depends on the target application, computational demand, and energy constraints. For low-power devices such as fitness trackers or glucose sensors, microcontrollers provide sufficient performance with minimal battery usage. In contrast, telemedicine hubs, portable imaging systems, and hospital gateways benefit from microprocessors' higher processing power and operating system flexibility.

Designers must also consider factors such as **cost, form factor, thermal management**, and **regulatory compliance**. For instance, implantable devices require processors with ultralow leakage currents and strict electromagnetic compatibility (EMC) performance, while hospital-based systems prioritize computational throughput and software scalability.

The growing adoption of **AI accelerators** and **edge AI microcontrollers** (e.g., ARM Ethos-U series or RISC-V-based neural processing units) suggests a convergence toward hybrid platforms capable of performing both real-time control and intelligent inference. These emerging architectures will enable the next generation of biomedical IoT systems to deliver personalized analytics at the point of care without relying entirely on cloud computing.

11.4.7 RISC-V architecture and its potential in biomedical IoT

The evolution of biomedical IoT systems has traditionally been driven by proprietary processor architectures such as ARM and x86, which dominate the embedded and general-purpose computing markets. However, in recent years, the emergence of **RISC-V** (pronounced *"risk-five"*) has reshaped

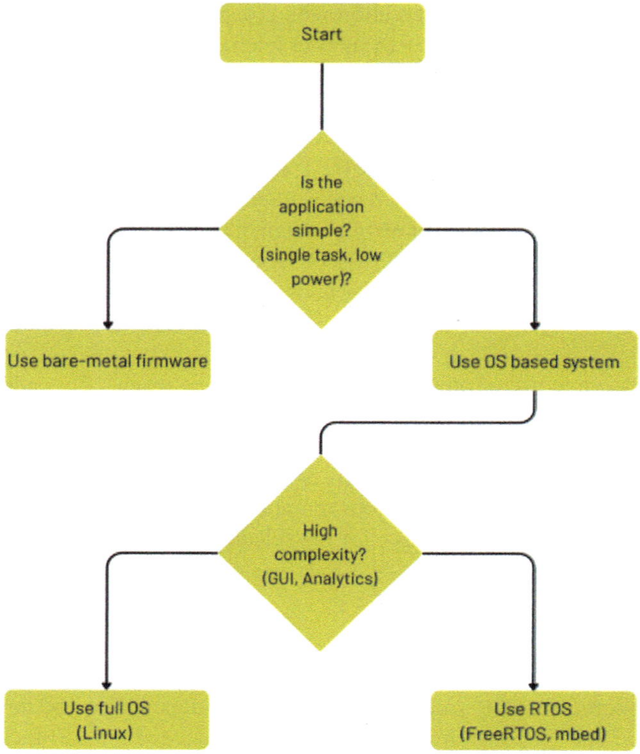

FIGURE 11.8

A decision tree for the selection of microprocessors or microcontrollers in IoT system.

the landscape of processor design by offering an **open and modular Instruction Set Architecture (ISA)**. Originating from the University of California, Berkeley in 2010, RISC-V has rapidly gained international attention as a transformative platform that promotes innovation, cost reduction, and customization across both academic and industrial domains (García-Ramírez et al., 2020).

At its core, RISC-V is founded on the principles of **RISC**—a design philosophy that simplifies processor instruction sets to improve performance, scalability, and energy efficiency. What distinguishes RISC-V from earlier RISC architectures is its **open-source licensing model**. Unlike proprietary ISAs such as ARM, RISC-V is not bound by licensing fees or restrictive intellectual property (IP) agreements, enabling designers, universities, and startups to develop processors tailored to specific needs without incurring high royalty costs. This openness makes RISC-V particularly attractive for **biomedical and academic research**, where affordability and flexibility are often major constraints.

RISC-V is designed as a **modular and extensible architecture**, comprising a small base instruction set with optional standard extensions for multiplication, atomic operations, floating-point arithmetic, and vector processing. This modularity allows system designers to create processors optimized for their target applications—ranging from ultra-low-power wearable sensors to AI-enabled medical imaging devices. For example, a wearable heart monitor might use a minimalist RISC-V core implementing only integer and multiplication extensions to minimize power consumption, while a diagnostic imaging device could integrate vector extensions to accelerate real-time image reconstruction.

A defining feature of RISC-V is its capacity for **custom instruction extensions**. Designers can introduce domain-specific instructions or hardware accelerators that enhance performance in targeted applications. In biomedical IoT, such customization could include dedicated signal processing instructions for ECG or EEG analysis, or low-latency encryption engines for secure patient data transmission. This flexibility empowers research institutions and companies to experiment with **application-specific integrated circuits (ASICs)** or **SoC** designs without depending on closed vendor ecosystems.

From a performance perspective, RISC-V processors can scale from small embedded cores (RV32E and RV32IM) used in microcontrollers to 64-bit and 128-bit implementations (RV64 and RV128) for high-performance computing. The **RISC-V Vector (RVV)** extension, in particular, offers a compelling advantage for biomedical AI applications, as it supports efficient parallel computation of matrix and vector operations used in neural network inference and biosignal analytics. This capability aligns with the growing demand for **edge AI** in healthcare, where localized inference reduces latency and preserves patient privacy by minimizing cloud data dependency.

The global ecosystem surrounding RISC-V is expanding rapidly. Organizations such as **RISC-V International, SiFive, Andes Technology**, and **Esperanto Technologies** are actively developing processors, compilers, and design tools that enhance its accessibility. Major semiconductor manufacturers, including Western Digital, Intel, and NVIDIA, have also incorporated RISC-V cores into their products for control and management tasks. In academia, RISC-V has become a platform for teaching computer architecture and developing experimental biomedical processors, enabling students and researchers to understand and modify every layer of the hardware design.

In the context of **biomedical IoT**, the adoption of RISC-V brings several strategic advantages. First, it supports **local innovation and sovereignty** in processor development, allowing research institutions and startups to create custom chips without relying on foreign IP restrictions. Second, its open ecosystem accelerates **collaborative research** between universities and industries in domains such as low-power design, neural computing, and secure data management. Finally, RISC-V aligns well with the healthcare sector's emphasis on **trust, transparency, and verifiability**, as open hardware architectures allow full inspection of security mechanisms—an increasingly critical feature in medical applications handling sensitive patient data.

The future potential of RISC-V in biomedical IoT lies in its synergy with **AI, edge computing, and custom chip fabrication**. As healthcare devices become smarter and more interconnected, RISC-V offers an adaptable foundation for developing processors that are not only cost-efficient but also optimized for biomedical workloads. From wearable health trackers and implantable biosensors to hospital automation systems, RISC-V provides the flexibility to design processors that balance precision, power efficiency, and scalability.

In summary, RISC-V represents more than a technical innovation—it signifies a shift toward **open, inclusive, and application-specific hardware design** in the biomedical IoT era. By empowering developers to customize, optimize, and democratize processor technology, RISC-V is poised to play a central role in the next generation of intelligent and sustainable healthcare systems. Fig. 11.9 shows the RISC-V ecosystem.

Architecture	Description	Typical Performance	Power Consumption	Operating System	Cost	Advantages	Limitations	Typical Biomedical Applications
Microcontroller (MCU)	Single-chip processor integrating CPU, memory, ADC, and peripherals. Optimized for control and low-power applications.	10–200 MHz; 8–32 bit	Very Low (<100 mW)	Bare-metal or RTOS (e.g., FreeRTOS and Zephyr)	Low	Compact, inexpensive, low power, deterministic timing	Limited computational power, small memory	Wearables, ECG/EMG patches, glucose monitors, implantable sensors
Microprocessor (MPU)	High-performance CPU requiring external memory and peripherals. Suitable for multitasking and OS-based systems.	500 MHz – 3 GHz; 32–64 bit	Moderate to High (0.5–5 W)	Full OS (Linux, Android)	Moderate to High	High computing capacity, supports GUI and analytics, flexible software ecosystem	Larger form factor, higher cost, and power	Telemedicine hubs, portable imaging devices, diagnostic workstations
System-on-Chip (SoC)	Integrated platform combining CPU, GPU, DSP, wireless, and AI accelerator on a single chip.	100 MHz – 2 GHz (varies by subsystem)	Low to Moderate (0.2–2 W)	RTOS or Full OS	Moderate	Compact integration, efficient for AI/edge computing, reduced latency	Complex design, limited customization, higher development cost	Smart wearables, edge AI systems, autonomous diagnostic devices
RISC-V Platform	Open-source ISA enabling custom MCUs, MPUs, or SoCs with tailored instruction sets for biomedical workloads.	Scalable (from <50 MHz to multi-GHz)	Configurable (ultralow to high)	Bare-metal, RTOS, or Linux	Variable (often low)	Open architecture, customizable, license-free, suitable for research and innovation	Toolchain maturity still evolving, smaller ecosystem than ARM	Custom biomedical processors, research platforms, low-cost edge AI systems

This table highlights the progressive scaling of processing architectures from **low-power embedded microcontrollers** to **AI-enabled SoC** and **open RISC-V platforms**. The choice depends on application requirements—for instance, **continuous biosignal monitoring** demands low power and deterministic control (favoring MCUs), while **telemedicine and diagnostic imaging** benefit from the computational

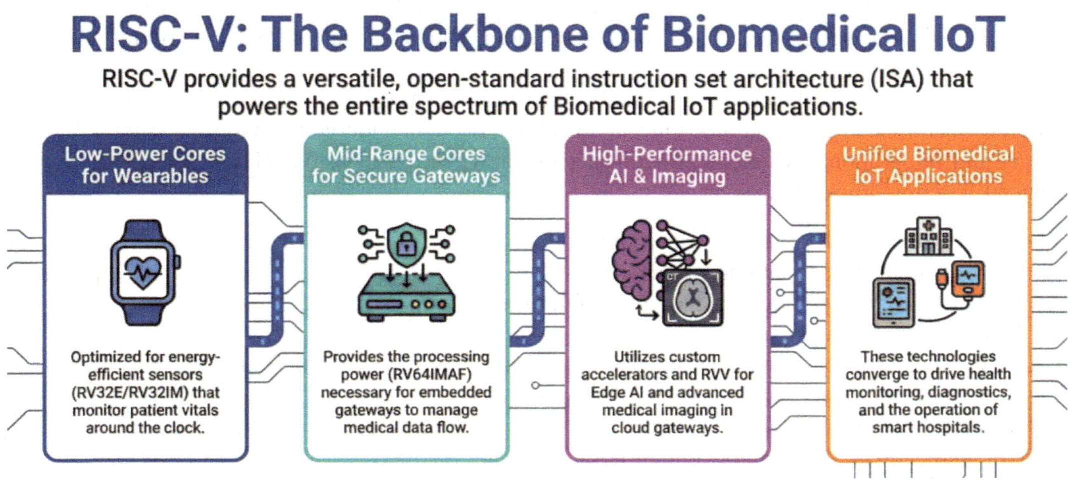

FIGURE 11.9

RISC-V ecosystem.

power of MPUs or SoCs. RISC-V uniquely bridges both ends of the spectrum by offering scalable, customizable designs that balance flexibility, cost, and performance.

11.5 Data storage and management

In biomedical IoT systems, the continuous flow of physiological data—from sensors and embedded processors to remote servers—necessitates robust mechanisms for **data storage, management, and security**. Unlike traditional medical devices, which store data locally or transmit intermittently, IoT-based biomedical systems generate continuous streams of time-sensitive information. Efficiently capturing, transmitting, and managing this data is essential for enabling real-time diagnostics, personalized medicine, and long-term healthcare analytics.

The architecture for data storage and management has evolved from centralized cloud-based platforms to more distributed and intelligent systems involving **edge and fog computing**. These technologies aim to balance latency, energy efficiency, scalability, and privacy while complying with stringent healthcare regulations (Hossain et al., 2019; Li et al., 2022; Hossain and Muhammad, 2016; Amin and Hossain, 2021).

11.5.1 Cloud storage and computing in biomedical IoT

Cloud computing serves as the backbone of modern biomedical IoT infrastructures. It provides virtually unlimited computational power, storage capacity, and accessibility, allowing data from millions of biomedical devices to be aggregated and analyzed at scale. Popular commercial platforms such as

Amazon Web Services (AWS), Microsoft Azure, and **Google Cloud Platform (GCP)** offer specialized IoT and healthcare services designed to meet both performance and compliance requirements (Li et al., 2022; Amin and Hossain, 2021).

In a typical cloud-based workflow, biomedical devices transmit processed or raw data through secure internet connections to cloud servers. Once in the cloud, the data undergo further processing, long-term storage, and integration with medical databases or analytical dashboards. This architecture supports advanced functionalities such as:

- **Scalable storage:** Cloud servers can dynamically adjust capacity to accommodate varying data volumes from different patients or devices.
- **Big data analytics:** Centralized storage enables the use of ML and AI algorithms to identify health trends, detect anomalies, or predict diseases.
- **Remote accessibility:** Clinicians and caregivers can access patient data from anywhere, supporting telemedicine and distributed healthcare services.
- **Redundancy and reliability:** Cloud providers maintain multiple data replicas across geographically distributed centers, ensuring data integrity and disaster recovery.

However, cloud-based systems also introduce **challenges**. Transmitting raw biomedical data continuously to remote servers consumes significant bandwidth and power. Latency can hinder real-time decision-making in applications such as cardiac monitoring or insulin regulation. Additionally, concerns over **data privacy, compliance, and ownership** remain central—especially when data cross international boundaries or are processed by third-party vendors.

11.5.2 Edge and fog computing for biomedical systems

To overcome the limitations of fully cloud-dependent architectures, **edge** and **fog computing** paradigms have emerged. These approaches shift part of the data processing and storage closer to the data source—on local gateways, routers, or even the biomedical devices themselves.

Edge computing refers to the processing of data directly at or near the sensing node. For example, a wearable ECG monitor may compute HRV or detect arrhythmias locally before sending only summarized results to the cloud. This reduces latency, minimizes bandwidth usage, and enhances data privacy by keeping sensitive raw signals within the local network.

Fog computing, often seen as an intermediary layer between the edge and the cloud, extends this idea to include nearby gateways or base stations capable of aggregating and filtering data from multiple devices. Fog nodes perform tasks such as temporal data caching, local analytics, and real-time alerts. For instance, in a hospital environment, patient monitoring systems may transmit continuous data streams to a local fog server that issues alerts for abnormal vital signs while asynchronously uploading noncritical data to the cloud for record-keeping.

The advantages of edge and fog computing include the following:

- **Low latency:** Immediate processing enables real-time response to critical biomedical events.
- **Reduced bandwidth:** Only processed or compressed data are sent to the cloud, lowering transmission costs.
- **Improved privacy:** Sensitive raw data remain within local networks, reducing exposure to external breaches.

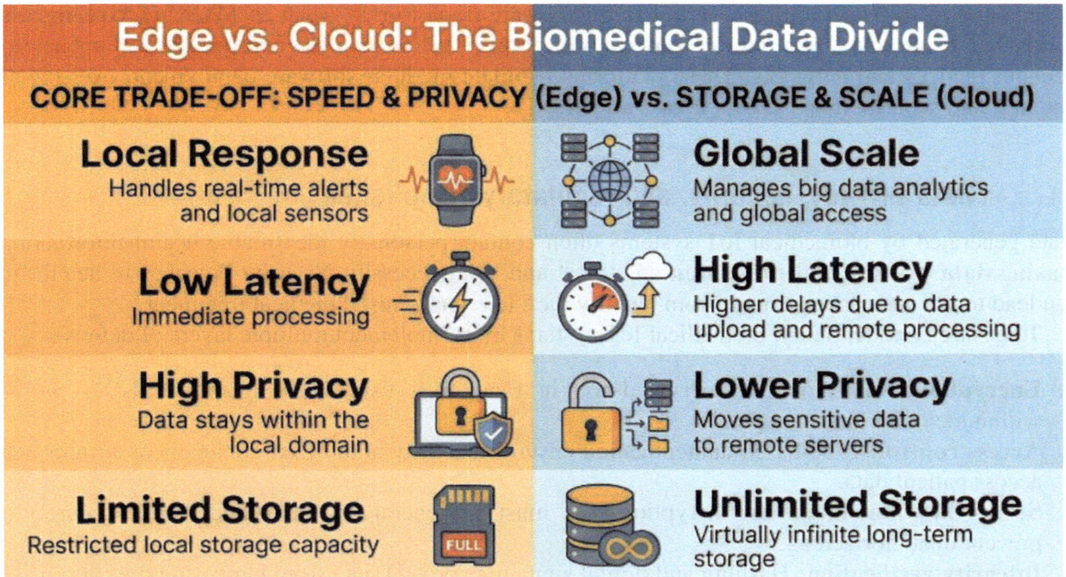

FIGURE 11.10

The division of edge and cloud computing in biomedical IoT devices.

- **Enhanced reliability:** Local analytics continue functioning even when cloud connectivity is lost.

However, these benefits come with trade-offs in computational capacity, storage, and system complexity. Edge nodes must be designed to handle intermittent connectivity, dynamic workloads, and security vulnerabilities similar to those in larger cloud systems. Fig. 11.10 shows the division between edge and cloud computing especially in biomedical IoT devices.

11.5.3 Hybrid architectures: Integrating edge, fog, and cloud

A growing trend in biomedical IoT is the adoption of **hybrid architectures** that combine the strengths of cloud, edge, and fog computing into a unified framework. In these systems, real-time or latency-sensitive operations are performed locally at the edge, while long-term analytics and storage are handled by the cloud.

For example, an **AI-based fall detection system** might analyze motion data on a smartwatch (edge), send critical alerts to a home gateway (fog), and store long-term activity logs in a cloud database for clinical review. This division of labor optimizes the system's responsiveness, scalability, and power consumption.

Hybrid architectures also enable **context-aware intelligence**. By distributing workloads intelligently, systems can dynamically decide where computation should occur based on network conditions, data priority, and device energy state. Edge nodes may perform lightweight anomaly detection, fog nodes handle correlation across patients or devices, and the cloud executes global predictive labeled.

To support interoperability, several IoT middleware frameworks—such as **AWS IoT Greengrass, Azure IoT Edge**, and **OpenFog Reference Architecture**—allow seamless coordination among heterogeneous devices and networks. These platforms facilitate modular software deployment, over-the-air updates, and security management across all tiers.

11.5.4 Data Privacy, security, and regulatory compliance

Data generated by biomedical IoT systems often contain personally identifiable health information, making **data privacy and protection** an ethical and legal necessity. Security breaches in healthcare can lead to serious consequences, from identity theft to compromised medical treatment.

To ensure confidentiality, biomedical IoT systems must implement multiple layers of defense:

1. **Encryption:** Data should be encrypted both in transit (e.g., using TLS/SSL protocols) and at rest within local or cloud storage.
2. **Access control:** Role-based authentication ensures that only authorized clinicians or systems can access patient data.
3. **Secure key management:** Encryption keys must be generated, stored, and rotated securely to prevent unauthorized use.
4. **Integrity verification:** Hashing and digital signatures protect data from tampering during transfer or storage.

From a regulatory perspective, compliance with **international data protection standards** is essential.

- In the **United States**, the **HIPAA** mandates strict policies for the storage, transmission, and disclosure of medical data.
- In **Europe**, the **GDPR** governs data handling with an emphasis on user consent and transparency.
- Many countries in Asia, including Malaysia, are developing national frameworks aligned with these principles to ensure safe digital healthcare transformation.

In addition, **blockchain-based storage systems** are emerging as a complementary approach to enhance data integrity and auditability (Rahman et al., 2020; Sun et al., 2019; Tariq et al., 2025). By maintaining immutable transaction logs across distributed nodes, blockchain ensures that patient data cannot be altered without detection, while enabling traceability and consent management.

11.5.5 Future directions in biomedical data management

As biomedical IoT systems expand in scale and complexity, future data management strategies will likely focus on **intelligent, decentralized, and privacy-preserving models**. Federated learning frameworks—where AI models are trained locally on edge devices and only share model parameters instead of raw data—will become increasingly important in medical AI research.

Moreover, **energy-aware data management** will play a vital role in prolonging device lifetime. Adaptive data compression, selective synchronization, and context-aware data prioritization will optimize both network and energy efficiency.

The convergence of **quantum-safe encryption, homomorphic computation**, and **zero-trust architectures** will redefine the standards of secure biomedical data handling, ensuring resilience against future cyberthreats.

11.6 Data analytics and artificial intelligence

Biomedical IoT systems generate continuous streams of physiological, biochemical, and behavioral data from wearable, implantable, and ambient sensors. To transform this raw data into clinically meaningful insights, advanced analytics and AI techniques are employed. These methods encompass a spectrum of computational strategies—ranging from classical signal processing to deep learning (DL)—each serving distinct roles in **diagnostics, monitoring, and personalized medicine**.

The integration of AI into biomedical IoT represents a paradigm shift: devices no longer merely record data; they interpret, predict, and even intervene autonomously. This convergence of sensor technology, cloud computing, and intelligent algorithms forms the foundation of *smart healthcare ecosystems*, capable of continuous assessment and adaptive decision-making.

11.6.1 Signal processing in biomedical IoT

Signal processing constitutes the first layer of data analytics in biomedical IoT systems. Raw sensor data—whether electrical, optical, chemical, or mechanical—must be conditioned and transformed into usable digital signals.

Typical signal processing steps include **filtering, denoising, feature extraction**, and **time–frequency analysis**. For example, an ECG sensor may acquire analog voltage fluctuations that represent cardiac activity. These signals are first amplified and filtered to remove noise such as baseline wander or power-line interference. Subsequent digital filters, such as **finite impulse response (FIR)** or **wavelet transforms**, isolate key morphological features such as the P-wave, QRS complex, and T-wave.

Similarly, for **EEG** and **EMG** signals, **Fourier transform** or **short-time Fourier transform (STFT)** methods are used to analyze frequency-domain characteristics associated with cognitive load or muscular activation. Modern embedded processors even enable **on-device feature extraction**, reducing the need for transmitting high-volume raw data to the cloud.

Beyond classical filtering, **adaptive signal processing** techniques such as the **Kalman filter** and **empirical mode decomposition (EMD)** are increasingly applied in wearable devices to handle nonstationary signals typical of real-world monitoring environments.

These preprocessed features—peak intervals, frequency bands, entropy, or statistical measures—serve as the essential input for ML models, linking the physical domain of sensing with the computational domain of AI.

11.6.2 Machine learning for biomedical diagnostics

ML algorithms play a pivotal role in automating diagnostic processes within biomedical IoT systems. Unlike rule-based systems, ML models learn patterns directly from data, enabling the detection of subtle physiological changes that may not be evident through conventional analysis (Al-Turjman & Deebak, 2020; Chen et al., 2017; Hayyolalam et al., 2021).

In **ECG monitoring**, for example, ML classifiers such as **support vector machines (SVM), random forests**, and **K-nearest neighbors (KNN)** can differentiate between normal sinus rhythm and arrhythmic patterns (e.g., atrial fibrillation or premature ventricular contractions). Similarly, **EEG**-based ML systems can identify epileptic seizures or sleep stages with high accuracy.

For **continuous glucose monitoring (CGM)**, regression-based algorithms—such as **linear regression, decision trees**, or **gradient boosting machines (GBMs)**—are used to predict blood glucose levels from multi-sensor data (skin impedance, temperature, and heart rate).

The ML workflow typically involves several stages:

1. **Data preprocessing:** Cleaning, normalization, and outlier removal.
2. **Feature engineering:** Extraction of domain-specific attributes (e.g., HRV and spectral entropy).
3. **Model training and validation:** Using labeled datasets to learn mapping functions.
4. **Deployment and inference:** Integration of trained models into edge or cloud environments for real-time decision support.

Edge-based ML is especially significant for **latency-sensitive healthcare applications**. Lightweight models such as **TinyML** or **TensorFlow Lite** allow predictive algorithms to run directly on microcontrollers, enabling offline diagnostics even in resource-constrained settings (Chen et al., 2017; Soumma et al., 2025).

11.6.3 Deep learning in imaging and wearable analytics

While traditional ML relies on engineered features, **DL** models learn hierarchical representations directly from raw data. In biomedical IoT, DL techniques have revolutionized imaging diagnostics, wearable analytics, and multimodal data fusion.

Convolutional neural networks (CNNs) are the cornerstone of medical image analysis, enabling the detection of abnormalities in X-rays, MRI, and ultrasound scans. For example, CNN-based architectures such as **ResNet, U-Net**, and **DenseNet** have achieved human-level performance in identifying tumors or lesions from medical images.

In the context of wearable analytics, DL models extend beyond static images to **time-series data. Recurrent neural networks (RNNs)** and **long short-term memory (LSTM)** networks capture temporal dependencies, making them ideal for ECG, EEG, and accelerometer data. For instance, an LSTM network trained on wearable ECG data can recognize arrhythmias in real time with adaptive confidence levels.

Recent advances integrate **multimodal** DL, combining physiological signals, environmental data, and user behavior to infer comprehensive health states. **Attention mechanisms** and **transformer architectures** have further enhanced model interpretability, allowing systems to focus on the most relevant features in complex datasets.

The deployment of DL in IoT environments, however, requires addressing computational and energy constraints. Techniques such as **model pruning, quantization**, and **knowledge distillation** enable compact models suitable for real-time execution on embedded hardware.

11.6.4 Predictive analytics for personalized medicine

The ultimate goal of biomedical IoT and AI integration is **personalized healthcare**—where predictive models anticipate health risks and tailor interventions to individual profiles. Predictive analytics leverage large-scale longitudinal datasets, combining physiological measurements, lifestyle parameters, and historical medical records to model individual health trajectories.

For example, AI-driven **early warning systems** can detect deviations from baseline physiological patterns, signaling potential cardiac or metabolic events before they manifest clinically. Predictive models trained on population data can be dynamically adjusted using each user's personal data, creating **adaptive healthcare loops** that evolve over time.

An emerging paradigm in this domain is **federated learning**, where models are trained collaboratively across multiple devices without transferring raw data to a central server. This approach preserves patient privacy while harnessing distributed intelligence.

Predictive AI systems are also instrumental in **therapy optimization**—for example, adjusting medication dosage or physical activity recommendations based on real-time monitoring. In future healthcare ecosystems, predictive analytics will serve as the core decision engine linking patients, clinicians, and digital platforms.

11.6.5 Ethical, legal, and clinical integration

The adoption of AI in healthcare raises important ethical and regulatory considerations. AI-driven biomedical systems must comply with **medical device regulations (MDR, FDA)** and demonstrate transparency, fairness, and explainability. Clinicians must be able to interpret algorithmic outputs and understand their limitations to ensure patient safety.

Moreover, ensuring **data fairness** and avoiding algorithmic bias is critical. Datasets used for model training must represent diverse populations to prevent inequities in diagnosis or treatment.

Ethical frameworks such as **Responsible AI** and **Human-in-the-Loop (HITL)** systems are increasingly emphasized to maintain clinical oversight while benefiting from automation.

11.7 Visualization and user interfaces

In biomedical Internet-of-Things (IoT) systems, data collection and analytics are only meaningful if the resulting information is effectively communicated to end users. Visualization and UI technologies serve as the final layer in the biomedical IoT architecture, transforming complex physiological and environmental data into interpretable, actionable insights (Istepanian et al., 2016; Sethi & Sarangi, 2017). Through appropriate interface design, continuous streams of sensor data can be simplified into clinically useful representations that support decision-making, promote user engagement, and ensure timely intervention.

UIs in biomedical IoT can be broadly divided into **clinician-oriented dashboards, patient-facing mobile applications**, and **web platforms for telemedicine**. Each category emphasizes different priorities—clinical accuracy, usability, accessibility, and privacy—yet they share the same underlying goal: To bridge the gap between machine intelligence and human interpretation.

11.7.1 Dashboards for clinicians

Dashboards designed for clinical use integrate heterogeneous data sources into a single visual workspace, allowing healthcare professionals to rapidly assess patient status. In hospital networks or remote monitoring centers, such dashboards provide aggregated views of vital signs, ECG morphology, blood glucose trends, oxygen saturation, and other physiological metrics collected from distributed IoT sensors (Soumma et al., 2025).

Effective clinical dashboards emphasize information hierarchy. Key indicators—such as heart rate, blood pressure, and respiration rate—are presented prominently, often accompanied by color-coded thresholds (e.g., green representing normal range, amber indicating warning level, and red signifying critical conditions). Secondary panels display temporal trends, comparative baselines, and annotations generated by AI-assisted diagnostic algorithms. Interactive filters enable clinicians to drill down into specific patients, sensors, or time windows when anomalies are detected.

From an implementation perspective, hospital dashboards are increasingly web-based, employing responsive front-end frameworks such as React or Angular that connect to secure API gateways for retrieving data from cloud databases or hospital information systems (HIS). Integration with electronic health record (EHR) systems ensures continuity of patient data, while compliance with data protection regulations such as HIPAA and GDPR maintains patient confidentiality (European Commission, 2023). The interface thus functions not merely as a visualization tool but as an intelligent **clinical decision-support layer**, uniting physiological monitoring with electronic documentation and predictive analytics.

11.7.2 Mobile health applications for patients

While dashboards cater primarily to clinicians, mobile health (m-Health) applications provide a personalized platform for patients to view and manage their physiological data. These applications translate sensor measurements—such as heart rate, step count, sleep quality, or glucose concentration—into daily summaries, progress indicators, and motivational feedback (Fig. 11.11).

A well-designed mobile interface promotes self-awareness and adherence to medical advice. Typically, such applications incorporate a home dashboard that summarizes daily activity and vital signs alongside personalized health scores. Notification features provide timely alerts for abnormal readings or reminders for medication intake, while graphical visualization tools allow users to monitor their progress over time. Increasingly, AI-generated recommendations are integrated to offer adaptive feedback based on user behavior and physiological trends.

Recent developments have extended mobile applications into more natural and inclusive forms of interaction. Conversational and emotion-aware interfaces, for example, enable users to communicate with their applications through voice commands or empathetic feedback mechanisms. This is particularly beneficial for older adults and individuals with limited digital literacy, ensuring accessibility and engagement. Secure communication channels allow patients to share data with clinicians in real time, closing the feedback loop between remote monitoring and medical consultation.

11.7.3 Web applications for telemedicine and remote care

Telemedicine platforms extend the visualization capabilities of biomedical IoT systems beyond individual users, supporting large-scale remote monitoring and care delivery. These web-based systems consolidate data from multiple patients and present it within a multi-user dashboard, allowing healthcare providers to oversee patient populations in real time.

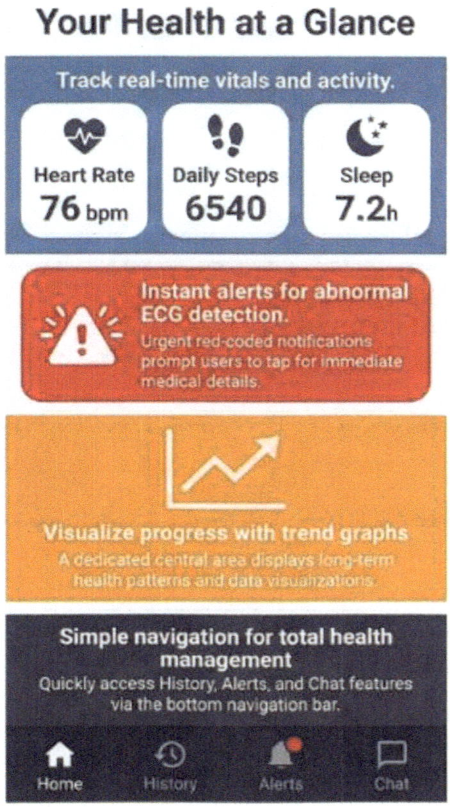

FIGURE 11.11

A mobile health application interface that presents a summary panel with key physiological indicators such as heart rate, step count, and sleep duration. It also includes an alert banner for abnormal ECG activity and a bottom navigation bar that links to history, alert, and chat modules, enabling integrated patient–clinician interaction.

A typical telemedicine interface integrates several functional layers. The patient management panel organizes individuals under care, with searchable filters by name, risk category, or location. The main visualization area displays current vital parameters alongside historical trends, often enhanced by AI algorithms that highlight abnormal patterns or predict deteriorating conditions. An alert center summarizes system-generated notifications, while built-in communication modules support chat, video consultation, or document exchange between clinicians and patients.

From a system architecture perspective, telemedicine dashboards are commonly developed using modular designs, where biomedical data from IoT gateways are transmitted to a cloud repository, analyzed by AI services, and visualized using frameworks such as Grafana, Plotly Dash, or Power BI

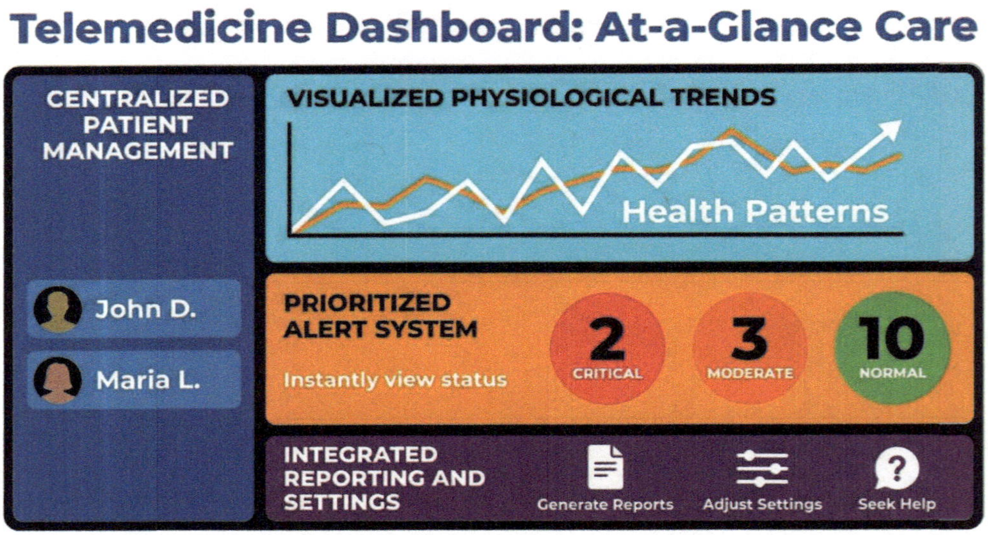

FIGURE 11.12

A conceptual layout of a telemedicine web dashboard consisting of a left-hand navigation panel for patient selection, a central chart area for physiological trends, and a series of color-coded indicators that provide at-a-glance summaries of patient risk levels. The resulting system enables efficient triaging and clinical decision-making, even across geographically distributed healthcare networks.

Embedded. Real-time updates rely on protocols such as WebSocket or MQTT to minimize latency and maintain continuous connectivity between devices and servers (Fig. 11.12).

11.7.4 Design principles for biomedical IoT interfaces

The design of biomedical interfaces must adhere to both technical rigor and human-centered design principles. Interfaces must accommodate diverse users—from clinicians interpreting diagnostic signals to patients tracking their personal health—while ensuring reliability, accessibility, and security.

The first design consideration is clarity. Interfaces should highlight essential metrics prominently while minimizing visual clutter. Context sensitivity is equally important: The same dataset may be visualized differently depending on whether the end user is a doctor, patient, or caregiver. Consistency in iconography, color coding, and terminology helps users develop familiarity and trust with the system, while accessibility features such as scalable fonts and high-contrast color palettes make interfaces inclusive for users with visual impairments. Responsiveness ensures that the visualization layout adapts seamlessly across different devices, from smartphones to large clinical monitors. Finally, privacy-by-design principles—such as user authentication, role-based access, and anonymized data visualization—are critical for maintaining compliance with regulatory frameworks and building user confidence.

Recent research trends are shifting toward immersive and multimodal visualization paradigms. Augmented and virtual reality environments are being explored to display biomedical data in three-dimensional space, enabling novel perspectives for surgical planning, rehabilitation monitoring, and anatomical education. These innovations point toward a future in which biomedical IoT interfaces evolve from static dashboards into adaptive, interactive environments that deepen human understanding of physiological processes.

11.8 System-level architecture of biomedical IoT systems

Biomedical IoT systems rely on a complex yet structured integration of sensing, processing, communication, storage, analytics, and visualization components (Chatterjee & Armentano, 2020; Li et al., 2022; Xu et al., 2014; Aledhari et al., 2022; Amin and Hossain, 2021). The system-level architecture represents the holistic design that coordinates these subsystems to enable seamless acquisition, transmission, interpretation, and presentation of biomedical data. By establishing an end-to-end digital healthcare pipeline, this architecture transforms raw physiological signals into meaningful, actionable insights for patients, clinicians, and researchers.

A typical biomedical IoT system follows a **multilayered structure** comprising: (i) The **sensor layer**, which acquires physiological and environmental data; (ii) the **processing and communication layer**, which manages local computation and wireless data transfer; (iii) the **data storage and management layer**, which maintains data repositories at the edge or in the cloud; (iv) the **data analytics layer**, which employs AI for decision support; and (v) the **visualization and feedback layer**, which translates information into UIs for clinical or personal health applications. These layers collectively form a continuous feedback loop between the physical and digital worlds.

11.8.1 Overview of layered biomedical IoT architecture

The architecture of a biomedical IoT system begins with the **sensor and acquisition layer**, which serves as the interface between the biological world and the electronic domain. This layer comprises wearable, implantable, or ambient sensors capable of measuring vital parameters such as heart rate, blood pressure, body temperature, oxygen saturation (SpO_2), or ECG signals. The analog outputs from these sensors are converted into digital data through integrated **ADCs** before transmission to the next stage.

The **processing and edge layer** forms the computational core of the system. Built around microcontrollers, microprocessors, or SoC architectures, this layer performs local signal conditioning, filtering, and preliminary analysis. Edge processing reduces communication overhead and latency while preserving privacy by preventing raw physiological data from being transmitted directly over the network. Embedded communication modules such as **BLE, Wi-Fi**, or **Zigbee** facilitate wireless data exchange between wearable devices and gateway nodes (e.g., smartphones or edge servers).

The next layer, **communication and networking**, enables reliable data transfer from edge devices to centralized or distributed storage systems. Protocols such as **MQTT, HTTPs, CoAP**, or **LoRaWAN** are employed depending on data rate requirements and deployment scale. In hospital environments, data may traverse secured local networks (LAN), whereas community healthcare systems rely on cellular or satellite connectivity.

The **data management layer** provides long-term storage and retrieval functionality. Cloud infrastructures—such as AWS IoT Core, Google Cloud Healthcare, or Microsoft Azure—offer scalable storage, structured databases, and real-time query support. In latency-sensitive or privacy-critical scenarios, **edge and fog computing** paradigms allow processing to occur near the data source, ensuring responsiveness and data sovereignty.

The **data analytics and intelligence layer** integrates ML and DL models that transform raw data into diagnostic or predictive insights. These algorithms can perform classification (e.g., arrhythmia detection), regression (e.g., glucose prediction), or clustering (e.g., activity pattern recognition). Model outputs are relayed to the visualization layer, forming the basis of clinical or self-monitoring applications.

Finally, the **visualization and feedback layer** completes the system architecture by converting processed data into user-friendly interfaces. Dashboards for clinicians, mobile applications for patients, and telemedicine portals for remote monitoring provide real-time updates and alerts. Through this interface, users receive personalized feedback and recommendations, closing the loop between sensing, computation, and clinical decision-making.

11.8.2 Data flow and functional integration

The biomedical IoT workflow is inherently cyclical, characterized by continuous interaction between sensing, computation, and response. The process begins with **data acquisition** from biosensors, followed by **local processing and filtering** at the edge. The processed data are transmitted through the network to a **cloud or local server**, where they undergo further analysis, classification, and visualization.

Each subsystem is functionally integrated to maintain seamless operation. For instance, signal conditioning in the microcontroller directly affects the performance of downstream analytics, while cloud-based model updates influence on-device parameter tuning. This closed-loop structure enables adaptive optimization—improving detection accuracy, reducing false alarms, and supporting personalized healthcare models (Soumma et al., 2025; Yang et al., 2014).

11.8.3 Hardware–software co-design and integration challenges

The development of biomedical IoT systems requires an integrated approach to hardware and software co-design. On the hardware side, compact and low-power components must be selected to ensure wearability and long battery life. Signal integrity, electromagnetic compatibility, and sensor calibration represent key design constraints. On the software side, RTOS coordinate data acquisition and communication tasks while maintaining synchronization with cloud services.

Power efficiency remains one of the dominant challenges. Energy harvesting and adaptive power management are often implemented to extend the operational lifespan of wearable and implantable devices. **Reliability** and **latency** are equally critical, especially for continuous monitoring and emergency applications.

Interoperability poses another significant barrier. Biomedical IoT systems often combine devices from multiple vendors using different standards. Adherence to interoperability frameworks such as **IEEE 11073** (Health Informatics), **HL7 FHIR** (Fast Healthcare Interoperability Resources), and **ISO/IEEE 802.15.6** (Body Area Networks) is therefore crucial to ensure cross-platform compatibility.

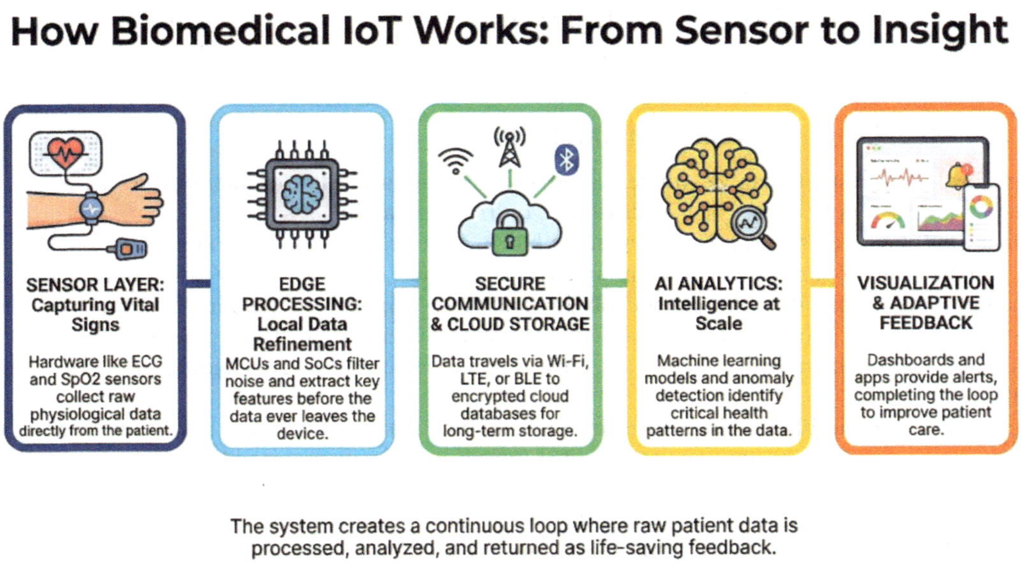

FIGURE 11.13

Typical implementation of an end-to-end biomedical IoT framework.

Finally, **security and privacy** underpin the trustworthiness of the entire system. Encrypted communication, device authentication, and anonymized data storage must be enforced to prevent unauthorized access and data leakage. The increasing incorporation of AI models necessitates transparent and explainable algorithms to align with ethical and regulatory standards for medical applications.

11.8.4 Example: End-to-end biomedical IoT system

A representative implementation of an end-to-end biomedical IoT framework is illustrated in Fig. 11.13. The system begins with a network of wearable sensors (ECG, SpO_2, temperature, and motion) connected to a portable microcontroller or embedded SoC (Soumma et al., 2025). The local device aggregates and preprocesses the data before transmitting it wirelessly to a smartphone or gateway. The gateway serves as an intermediary node, performing additional computation and uploading the information to a secure cloud server.

Within the cloud, data from multiple patients are stored, processed, and analyzed by AI algorithms. The outcomes are transmitted to a web dashboard or mobile interface for visualization, enabling clinicians and patients to monitor physiological trends and receive alerts in real time. The architecture can support multiple operational modes, including continuous remote monitoring, emergency detection, and preventive analytics.

The modular design also allows scalability—additional sensors or analytics modules can be incorporated without disrupting the existing structure. This flexibility makes the architecture suitable for diverse applications ranging from hospital telemetry systems to community-based chronic disease management programs.

11.8.5 Modular vs. monolithic architectures

Biomedical IoT systems can be implemented using either **modular** or **monolithic** architectures, depending on design priorities such as flexibility, cost, and reliability.

In a **modular architecture**, each functional block—sensing, processing, communication, storage, and visualization—is implemented as an independent module with standardized interfaces. This approach enhances scalability and maintainability, allowing designers to upgrade or replace individual subsystems without altering the entire system. Modular designs are particularly suited for research, prototyping, and commercial systems that must accommodate evolving technologies or new sensor modalities.

By contrast, **monolithic architectures** integrate all functional units within a single device or tightly coupled system. Such configurations offer higher integration density, reduced latency, and often better power efficiency. However, they are less flexible to modify and may require complete redesigns when upgrading individual components. Monolithic systems are commonly employed in regulated medical devices or implants where stability, robustness, and certification are paramount.

The selection between modular and monolithic architectures is determined by application context. While modularity favors innovation and scalability, monolithic integration remains the preferred approach for compact, high-reliability medical instrumentation. The future of biomedical IoT is expected to merge both paradigms through **hybrid architectures**, where modular interfaces coexist with tightly integrated subsystems on shared hardware platforms.

11.9 Case studies and applications

Biomedical IoT technologies have rapidly evolved from conceptual frameworks into tangible systems that are actively transforming healthcare. The integration of BioMEMS, miniaturized sensors, communication modules, data storage infrastructures, and advanced analytics has made it possible to continuously monitor physiological, biochemical, and lifestyle parameters (Chatterjee & Armentano, 2020; Xu et al., 2014; Aledhari et al., 2022). This section highlights representative applications across wearables, implantable devices, smart hospital systems, and telemedicine. Each case study demonstrates how data transfer, analysis, and management technologies converge to deliver practical healthcare benefits.

11.9.1 Wearable health monitors

Wearables are among the most widespread and familiar applications of biomedical IoT. These devices typically consist of body-worn sensors, low-power microcontrollers, wireless communication modules, and cloud-based or mobile applications for data analysis and visualization. They provide continuous streams of physiological and lifestyle data, empowering both patients and clinicians with actionable insights (Alam et al., 2021; Soumma et al., 2025; Tavera et al., 2021; Yang et al., 2014).

One of the most established examples is **CGM**. CGM devices employ minimally invasive electrochemical sensors that measure glucose concentrations in interstitial fluid. Data are transmitted wirelessly, often via BLE, to a mobile application where patients can track glucose trends. Cloud-based analytics further enhance the system by predicting glycemic excursions, allowing diabetic patients to anticipate hyper- or hypoglycemic events and adjust insulin dosage accordingly. This reduces the reliance on episodic finger-prick tests and supports more precise disease management.

Cardiovascular health monitoring has also benefited from IoT-enabled wearables. Devices such as smart patches and smartwatches incorporate electrodes for continuous **ECG acquisition**. Unlike conventional Holter monitors that are bulky and limited to short-term recordings, these IoT-based devices can capture cardiac activity over extended periods. Data are transmitted to mobile devices and subsequently to cloud platforms, where ML algorithms can detect arrhythmias such as atrial fibrillation in near real time. Clinicians are alerted when anomalies occur, enabling timely intervention.

Beyond vital signs, IoT technologies are extending into lifestyle-related health tracking. **Diet monitoring systems** represent an emerging frontier in biomedical IoT. Traditional self-reporting of dietary intake is often inaccurate and burdensome. To overcome this, researchers have developed wearable or device-assisted solutions. Examples include smart utensils that detect food type and portion size, wearable bite counters that monitor chewing patterns, and mobile applications that use camera-based food recognition to estimate calorie intake. When integrated with cloud-based AI, these systems can provide personalized dietary recommendations, track long-term nutrition trends, and support obesity prevention and management. Such systems exemplify how biomedical IoT can integrate physiological and behavioral data to create holistic health insights.

Wearable health monitors also include **activity and sleep trackers**, which combine accelerometers, gyroscopes, and photoplethysmography (PPG) sensors to assess movement, heart rate, and sleep quality. By correlating activity and rest cycles with physiological markers, these devices help detect sleep disorders, monitor rehabilitation progress, and guide personalized fitness regimens. Collectively, wearable monitors highlight the power of IoT systems to bridge clinical and consumer health domains.

11.9.2 Implantable IoT devices

Implantable devices represent another critical application area where IoT technologies enhance traditional medical implants. By embedding wireless communication capabilities into devices such as pacemakers, neurostimulators, and BioMEMS-based implants, clinicians can remotely monitor patient status, adjust therapy, and detect early signs of malfunction.

Connected pacemakers and defibrillators exemplify this integration. Traditionally, cardiac implants required in-person follow-up visits to evaluate device performance. Modern IoT-enabled implants, however, can transmit data about heart rhythms, device battery life, and lead function directly to hospital servers or physician dashboards. These data streams allow for continuous oversight and the automatic generation of alerts in the event of arrhythmias or technical issues, significantly improving patient safety and reducing hospital readmissions.

Similarly, **neurostimulators** used in conditions such as epilepsy or Parkinson's disease benefit from IoT integration. These devices deliver targeted electrical stimulation to neural circuits while simultaneously transmitting usage and patient response data. Remote monitoring enables clinicians to fine-tune stimulation parameters, optimizing therapy outcomes on an individualized basis.

A particularly promising development is the use of **BioMEMS-based implants**. These miniature systems can incorporate pressure, chemical, or flow sensors to track physiological changes inside the body. For example, implantable pressure sensors can continuously monitor intraocular pressure in glaucoma patients, transmitting data externally for early detection of disease progression. By linking BioMEMS sensors with IoT infrastructures, implants evolve from passive therapeutic tools to active participants in a dynamic healthcare ecosystem.

11.9.3 Smart hospital systems

At the institutional level, hospitals are embracing IoT technologies to improve patient safety, streamline workflows, and enhance operational efficiency. One of the most widely adopted technologies is **RFID**. Patients wearing RFID-enabled wristbands can be uniquely identified at every point of care, ensuring that they receive the correct medications and procedures. Similarly, RFID-tagged pharmaceuticals enable automated verification processes that reduce dispensing errors. These systems generate real-time logs that are integrated into HIS, ensuring traceability and accountability throughout the care pathway.

In addition, **connected ICUs** illustrate the transformative potential of biomedical IoT in critical care. Traditionally, ICU monitoring relies on isolated devices such as ventilators, infusion pumps, and vital sign monitors. IoT integration enables these devices to communicate with centralized dashboards, providing clinicians with a consolidated view of patient status. Data analytics can then be applied to detect early signs of deterioration, trigger alerts, and assist in clinical decision-making. AI-driven predictive models further enhance these systems by identifying patients at risk of sepsis, respiratory failure, or cardiac arrest hours before such events occur.

Smart hospitals also deploy IoT for **asset management and logistics** (Hossain et al., 2019; Li et al., 2022; Hossain and Muhammad, 2016; Amin and Hossain, 2021). Sensors track the location and availability of critical equipment such as infusion pumps and wheelchairs, reducing delays and optimizing resource allocation. Beyond improving efficiency, these applications demonstrate how IoT-driven data transfer and management technologies contribute directly to patient safety and hospital performance.

11.9.4 Telemedicine and remote patient monitoring

The COVID-19 pandemic accelerated the adoption of telemedicine, with IoT serving as a cornerstone of remote healthcare delivery. Patients at home can use IoT-enabled devices such as wireless blood pressure cuffs, pulse oximeters, or wearable thermometers to collect vital signs. These data are transmitted via smartphones or dedicated hubs to cloud platforms, where healthcare providers can access them in real time.

A typical remote monitoring system involves three layers: The **sensing layer**, where physiological data are collected; the **communication layer**, where data are transmitted securely via wireless protocols; and the **application layer**, where physicians review data trends, conduct virtual consultations, and make clinical decisions. Advanced systems integrate AI-driven triage modules that prioritize patients based on severity. For example, a patient showing a sudden drop in oxygen saturation could be flagged for immediate medical attention, while others with stable readings are monitored routinely.

Telemedicine systems also integrate lifestyle and behavioral data, enabling holistic remote care. For example, integrating diet monitoring data with physiological signals allows clinicians to provide more personalized advice for managing chronic diseases such as diabetes and cardiovascular disorders. By reducing the need for frequent hospital visits, telemedicine improves accessibility in underserved areas and alleviates the burden on healthcare facilities.

11.9.5 Lessons learned from applications

From wearables to telemedicine, these case studies highlight recurring themes in biomedical IoT deployment. **Interoperability** between heterogeneous devices and data standards remains a major challenge, often requiring middleware or standardized frameworks such as HL7 and FHIR. **Power management** is critical for both wearable and implantable devices, driving innovation in low-power electronics and energy harvesting techniques. Furthermore, **data privacy and security** are paramount, as sensitive biomedical information must be safeguarded through encryption, authentication, and regulatory compliance (e.g., HIPAA, GDPR). Finally, the integration of **AI** has proven essential for transforming raw data into actionable clinical insights, but issues such as explainability, bias, and ethical responsibility remain pressing concerns.

These lessons underscore the dual nature of biomedical IoT: While it holds immense potential for transforming healthcare delivery, its success hinges on robust system-level design, secure data management, and careful attention to ethical and regulatory frameworks.

11.10 Future directions and research challenges

The convergence of biomedical engineering, microelectronics, and IoT technologies has ushered in a transformative era in personalized and connected healthcare. The systems described throughout this chapter represent the foundation of a rapidly evolving ecosystem where physiological sensing, intelligent processing, and real-time decision support are becoming increasingly autonomous and integrated. Yet, the field of biomedical IoT continues to face complex research challenges that demand innovation across multiple domains—ranging from hardware miniaturization and energy efficiency to AI, data governance, and regulatory alignment.

The next decade will see the emergence of **intelligent, context-aware, and self-adaptive biomedical systems** that move beyond simple monitoring toward proactive intervention. These systems will not only collect and analyze data but will also learn from individual- and population-level trends to make predictive and preventive decisions. Such advancements will redefine healthcare delivery, transforming it from reactive treatment into **continuous, adaptive wellness management**.

11.10.1 Evolution toward intelligent and autonomous biomedical IoT

The early generation of biomedical IoT devices focused primarily on data acquisition and remote transmission. However, modern systems increasingly emphasize **distributed intelligence**, where data interpretation occurs at multiple layers—from the sensor node to the edge gateway and cloud analytics platform. The future trajectory points toward **autonomous IoT ecosystems** that integrate sensing, decision-making, and actuation in a seamless feedback loop.

In this context, **BioMEMS technologies** serve as the physical interface between biological signals and digital intelligence. When combined with miniaturized electronics, microfluidics, and nanoscale transducers, BioMEMS devices enable continuous, noninvasive monitoring of physiological parameters. Their integration with AI-based analytics allows not only for real-time anomaly detection but also for adaptive response mechanisms—for instance, automated insulin delivery in response to glucose fluctuations or electrostimulation for cardiac rhythm correction.

The evolution toward autonomy will require advances in **cross-domain integration**: Sensors that are self-calibrating, communication modules that dynamically adapt to environmental conditions, and embedded AI that optimizes power and performance on demand. These trends signify the transition from "connected" devices to **cognitive biomedical systems** capable of understanding, predicting, and responding to complex human health states.

11.10.2 Emerging research themes

11.10.2.1 AI-on-Chip and edge intelligence

As AI algorithms become increasingly integral to biomedical signal analysis, their deployment at the edge—directly on microcontrollers or SoC platforms—represents a key frontier. The development of **AI-on-Chip architectures** enables real-time inference close to the data source, reducing latency and dependency on cloud resources. Low-power neural accelerators and quantized AI models are being optimized for biosignal classification tasks such as ECG arrhythmia detection, EMG gesture recognition, and sleep stage analysis.

RISC-V-based and neuromorphic processors are particularly promising in this domain, allowing reconfigurable, domain-specific computation with minimal power consumption.

11.10.2.2 6G and next-generation connectivity

The forthcoming **6G communication paradigm** will expand biomedical IoT capabilities through ultralow latency (<1 ms), terahertz-bandwidth communication, and centimeter-level localization. Such performance will be critical for time-sensitive medical applications such as telesurgery, robotic-assisted diagnostics, and emergency response systems. The integration of 6G with **edge computing and AI orchestration** will support seamless device interconnectivity, allowing real-time monitoring of thousands of patients within a shared healthcare network (Hassan et al., 2022; Khan et al., 2023; Tang et al., 2018; García-Ramírez et al., 2020; Hassija et al., 2024; Parihar et al., 2023).

11.10.2.3 Digital twins in healthcare

The concept of a **digital twin**—a dynamic virtual model of a physical entity—is gaining traction in personalized medicine (Hassan et al., 2022; Hassija et al., 2024). A patient's digital twin continuously updates itself based on real-world sensor data, providing clinicians with a predictive model of the patient's physiology. This allows simulation of potential disease trajectories, treatment outcomes, and early risk detection. Integrating biomedical IoT with digital twin frameworks could revolutionize clinical decision-making by enabling individualized health simulations and proactive care strategies (Tang et al., 2018).

11.10.2.4 *Energy harvesting and self-powered systems*

Power supply remains a persistent limitation for wearable and implantable medical devices. Future biomedical IoT systems aim to achieve **energy autonomy** through ambient energy harvesting mechanisms such as thermoelectric, piezoelectric, and biofuel cell technologies. Self-powered BioMEMS platforms would eliminate the need for frequent recharging or battery replacement, particularly for implanted sensors, thereby improving patient comfort and system longevity.

11.10.2.5 *Quantum and nano-biosensing*

At the frontier of biomedical detection, **quantum and nanoscale sensors** offer unprecedented sensitivity for molecular and biochemical analysis. Quantum tunneling, nitrogen-vacancy (NV) centers in diamond, and nanoelectromechanical systems (NEMS) are being investigated for detecting ultralow concentrations of biomarkers and pathogens. Integration of these sensors within IoT architectures will enable **real-time molecular diagnostics** at the point of care, pushing the boundaries of early disease detection and personalized treatment monitoring (Khan et al., 2023; Parihar et al., 2023).

11.10.3 Technical challenges

Despite the rapid progress, biomedical IoT systems face several engineering challenges that constrain large-scale deployment. **Data interoperability** remains a major barrier, as diverse devices and platforms often use proprietary formats and incompatible communication protocols. The absence of unified standards complicates data fusion and cross-platform integration. **Latency and reliability** are also critical, particularly for medical-grade systems where delayed or lost data could endanger patient safety.

Another persistent issue lies in the **power–performance trade-off**. As the computational complexity of AI models increases, energy consumption rises, especially in continuous monitoring scenarios. Designing ultralow-power circuits, memory-efficient neural networks, and adaptive task scheduling mechanisms will be vital.

Hardware miniaturization introduces its own set of challenges, such as managing heat dissipation, ensuring biocompatibility, and preventing signal interference. Meanwhile, **security vulnerabilities** grow proportionally with system complexity, demanding multilayer encryption, secure boot mechanisms, and hardware root-of-trust implementations.

11.10.4 Ethical, privacy, and regulatory considerations

Biomedical IoT systems process highly sensitive personal data, necessitating strict adherence to ethical and regulatory principles. The rise of AI-driven analytics introduces new concerns regarding **algorithmic transparency, fairness, and accountability**. In medical contexts, decision-support systems must be interpretable and auditable to ensure trust among clinicians and patients.

From a governance perspective, frameworks such as the **HIPAA**, the **European GDPR**, and the **MDR** set essential guidelines for data handling and system certification (European Commission, 2023). However, global disparities in standards continue to hinder interoperability across borders. In the near future, harmonization efforts led by ISO, IEEE, and HL7 are expected to define unified frameworks for secure medical data exchange and AI validation.

Ethically, there is also a growing discourse on the **ownership and monetization of health data**. The transition from institutional to patient-controlled data paradigms—where individuals govern their own medical records via secure digital identities—will redefine the balance between privacy and accessibility.

11.10.5 Vision for the future

The future of biomedical IoT lies in **fully integrated, intelligent healthcare ecosystems** that merge physiological sensing, AI-driven analytics, and responsive actuation into a single adaptive framework. These systems will operate autonomously, predicting health risks, initiating interventions, and updating their internal models based on user behavior and environmental context.

Within this paradigm, **BioMEMS technologies** will continue to evolve as the foundational enablers of human–machine interfacing, offering biocompatible, miniaturized platforms for seamless physiological monitoring. The incorporation of AI-on-Chip computing and real-time cloud coordination will enable scalable deployment across diverse healthcare environments—from hospitals and smart homes to rural clinics and disaster zones.

Ultimately, the vision aligns with the **United Nations Sustainable Development Goal (SDG) 3 – Good Health and Well-being**, emphasizing accessible, affordable, and proactive healthcare for all (United Nations Development Programme, 2024). Biomedical IoT systems, through continuous innovation and ethical integration, will not only extend the human lifespan but also improve its quality—transforming global healthcare from a reactive system into an intelligent, preventive, and inclusive network of well-being.

11.11 Chapter summary

The integration of biomedical devices with IoT technologies has fundamentally reshaped the landscape of modern healthcare. Through the combined evolution of sensors, communication networks, embedded processors, cloud infrastructures, and intelligent data analytics, the biomedical IoT ecosystem has become a powerful enabler of continuous, real-time, and patient-centered health management (Aledhari et al., 2022; Chatterjee & Armentano, 2020; Soumma et al., 2025; Chen et al., 2017; Li et al., 2022; Amin and Hossain, 2021). This chapter has explored the complete technological chain—from data acquisition to storage, analysis, and visualization—illustrating how each component contributes to the formation of intelligent, connected biomedical systems.

At the foundation of this ecosystem lie **BioMEMS and miniaturized sensors**, which serve as the physiological interface between humans and digital intelligence. Advances in microfabrication and nanotechnology have allowed the creation of highly sensitive and biocompatible sensing platforms capable of monitoring an extensive range of biological, physical, and chemical parameters. When coupled with efficient **microcontrollers, microprocessors, and SoC architectures**, these devices can perform local data processing, feature extraction, and secure communication, all while maintaining ultralow power operation—an essential requirement for wearables and implantable devices.

The communication layer, consisting of **wired and wireless technologies** such as BLE, NFC, Wi-Fi, and long-range IoT protocols, forms the nervous system of biomedical networks. These technologies enable seamless data transmission between devices, gateways, and cloud servers, ensuring that physiological signals are reliably captured and interpreted in near real time. Complementing these are **cloud and edge computing infrastructures**, which provide scalable environments for data management, advanced analytics, and secure storage. Together, they constitute the digital backbone that allows raw physiological data to be transformed into actionable health insights.

As the chapter has emphasized, **data analytics and AI** represent the brain of the biomedical IoT framework. Through ML, DL, and predictive modeling, raw sensor signals can be transformed into diagnostic and prognostic information. Applications such as ECG anomaly detection, emotion recognition, and dietary pattern analysis illustrate how AI-driven analytics are moving biomedical systems from mere monitoring toward decision-making and adaptive intervention. This analytical intelligence, when visualized through intuitive dashboards, mobile applications, or telemedicine platforms, bridges the gap between patients and clinicians—creating a closed-loop system of monitoring, diagnosis, and care.

The transition from individual technologies to a **system-level architecture** marks a critical evolution in biomedical IoT design. The convergence of sensors, processors, communication modules, storage, and visualization into unified architectures has given rise to scalable, modular, and interoperable healthcare solutions. Whether in wearable devices, smart hospital systems, or remote patient monitoring platforms, the system-level approach ensures that data integrity, latency, and usability are optimized for clinical reliability. Furthermore, as interoperability standards mature and cross-domain data fusion becomes more sophisticated, these integrated systems will enable holistic understanding of human health—merging physiological, behavioral, and environmental information into a single analytical framework.

Looking ahead, biomedical IoT will continue to expand toward **personalized and preventive healthcare paradigms**. The emerging generation of systems will integrate **AI-on-Chip, 6G connectivity, digital twins**, and **quantum biosensing**, paving the way for autonomous, context-aware, and predictive medical solutions (Khan et al., 2023; Tang et al., 2018; García-Ramírez et al., 2020; Parihar et al., 2023). These advances will enable healthcare systems that not only diagnose disease but also anticipate risk, adapt to patient behavior, and recommend interventions before symptoms appear. At the same time, developments in energy harvesting and self-powered BioMEMS will enhance device sustainability, while advances in ethical AI and secure data governance will preserve patient trust and privacy.

In summary, the technologies and frameworks discussed throughout this chapter collectively illustrate the transformation of biomedical systems from isolated devices into **intelligent, interconnected, and adaptive ecosystems**. The future of healthcare will be characterized by systems that continuously learn from data, personalize treatment to the individual, and operate seamlessly across devices and platforms. Biomedical IoT thus stands at the nexus of engineering, data science, and medicine—empowering a new era of preventive, predictive, and participatory healthcare that places the individual at the center of an intelligent, data-driven health continuum.

References

Aledhari, M., Razzak, R., Qolomany, B., Al-Fuqaha, A., & Saeed, F. (2022). Biomedical IoT: enabling technologies, architectural elements, challenges, and future directions. *IEEE Access: Practical Innovations, Open Solutions, 10*, 31306–31339.

Alam, M., Reaz, M. B. I., Ali, M. A. M., & Samad, S. A. (2021). Wireless body area network technologies for healthcare: A review. *IEEE Sensors Journal, 21*(3), 2456–2471.

Amin, S. U., & Hossain, M. S. (2021). Edge intelligence and internet of things in healthcare: A survey. *IEEE Access, 9*, 45–59. https://doi.org/10.1109/ACCESS.2020.3045115.

Al-Turjman, F., & Deebak, B. D. (2020). Edge intelligence for IoT-enabled smart healthcare systems. *Future Generation Computer System, 109*, 475–486.

Chatterjee, S., & Armentano, R. L. (2020). Internet of Things for biomedical engineering: A review of the state-of-the-art, challenges, and future directions. *IEEE Reviews Biomedical Engineering, 13*, 202–211.

Chen, M., Hao, Y., Hwang, K., Wang, L., & Wang, L. (2017). Disease prediction by machine learning over big healthcare data. *IEEE Access, 5*, 8869–8879.

Commission, European (2023). *Medical device regulation (MDR) and in vitro diagnostic regulation (IVDR).* https://health.ec.europa.eu/medical-devices-vitro-diagnostics_en.

García-Ramírez, R., Chacon-Rodriguez, A., MolinaRobles, R., Castro-Gonzalez, R., Solera-Bolanos, E., Madrigal Boza, G., Oviedo, M., Salazar-Sibaja, D., Sanchez-Jimenez, D., Fonseca-Rodriguez, M., Arrieta-Solorzano, J., Rimolo-Donadio, R., Arnaud, A., Miguez, M., & Gak, J. (2020). Siwa: A custom RISC-V based system on chip (SoC) for low power medical applications. *Microelectronics Journal, 98*, 104753. https://doi.org/10.1016/j.mejo.2020.104753.

Hassija, V., Chamola, V., De, R., Das, S., Chakrabarti, A., Sangwan, K. S., & Pandey, A. (2024). A survey on digital twins: Enabling technologies, use cases, application, open issues, and more. *IEEE Journal of Selected Areas in Sensors, 2*, 84–107. https://doi.org/10.1109/JSAS.2024.3523856.

Hossain, M. S., & Muhammad, G. (2016). Cloud-assisted Industrial Internet of Things (IIoT) – Enabled framework for health monitoring. *Computer Networks, 101*(C), 192–202. https://doi.org/10.1016/j.comnet.2016.01.009.

Hossain, M. S., Muhammad, G., & Rahman, S. M. (2019). Cloud-assisted industrial Internet of Things (IIoT) enabled framework for health monitoring. *Future Generation Computer System, 95*, 444–453.

Hayyolalam, V., Aloqaily, M., Ozkasap, O., & Guizani, M. (2021). Edge intelligence for empowering IoT-based healthcare systems. *IEEE Wireless Communications, 28*, 6–14. https://doi.org/10.1109/MWC.001.2000345.

Istepanian, R. S. H. & Woodward B. (2016) m-Health: Fundamentals and applications. Wiley-IEEE Press.

Parihar, A., Yadav, S., Sadique, M. A., Ranjan, P., Kumar, N., Singhal, A., Khare, V., Khan, R., Natarajan, S., & Srivastava, A. K. (2023). Internet-of-medical-things integrated point-of-care biosensing devices for infectious diseases: Toward better preparedness for futuristic pandemics. *Bioengineering & Translational Medicine, 8*(3), e10481. https://doi.org/10.1002/btm2.1048.

Sethi, P., & Sarangi, S. R. (2017). Internet of things: Architectures, protocols, and applications. *Journal of Electrical and Computer Engineering, 2017*, 9324035. https://doi.org/10.1155/2017/9324035.

Soumma, S. B., Mamun, A., & Ghasemzadeh, H. (2025). AI-powered wearable sensors for health monitoring and clinical decision making. *Current Opinion in Biomedical Engineering, 36*, 100628.

Sun, Y., Lo, F. P.-W., & Lo, B. (2019). Security and privacy for the Internet of Medical Things enabled healthcare systems: A survey. *IEEE Access, 7*, 183339–183355. https://doi.org/10.1109/ACCESS.2019.2960617.

Tang, X., Wang, X., Cattley, R., Gu, F., & Ball, A. D. (2018). Energy harvesting technologies for achieving self powered wireless sensor networks in machine condition monitoring: A review. *Sensors, 18*(12), 4113. https://doi.org/10.3390/s18124113.

Tariq, U. U., Sabrina, F., Rashid, M. M., Gordon, S., Lin, Y., Wang, Z., & Azad, S. (2025). Blockchain-based secured data sharing in healthcare: A systematic literature review. *IEEE Access, 13*, 45415–45435. doi:10.1109/ACCESS.2025.3547953.

Tavera, C. A., Ortiz, J. H., Khalaf, O. I., Saavedra, D. F., & Aldhyani, T. H. H. (2021). Wearable wireless body area networks for medical applications. *Computational and Mathematical Methods in Medicine, 2021*(1), 5574376.

United Nations Development Programme. (2024). *Sustainable development goal 3: Ensure healthy lives and promote well-being for all at all ages*. UNDP.

Xu, L. D., He, W., & Li, S. (2014). Internet of things in industries: A survey. *IEEE Transactions on Industrial Informatics, 10*(4), 2233–2243. https://doi.org/10.1109/TII.2014.2300753.

Yang, G., Xie, L., Mäntysalo, M., Zhou, X., Pang, Z., Da Xu, L., Kao-Walter, S., Chen, Q., & Zheng, L.-R. (2014). A health-IoT platform based on intelligent packaging, unobtrusive bio-sensor, and intelligent medicine box. *IEEE Transactions on Industrial Informatics, 10*(4), 2180–2191. https://doi.org/10.1109/TII.2014.2307795.

CHAPTER

12 Bringing new BioMEMS devices to the market including case studies

Shama Kazmi
Monash Institute of Medical Engineering (MIME), Monash University, Melbourne, VC, Australia

12.1 Introduction: A growing market and demand for new BioMEMS devices

Biological micro-electro-mechanical systems or BioMEMS is an exciting class of biomedical products, which has been gaining steady commercial traction through their practical and important applications across a range of fields, particularly biomedicine and healthcare. BioMEMS devices offer a number of useful benefits, and it is considered that there is substantial opportunity to maximize these benefits by developing, commercializing, and deploying new, innovative, and practical BioMEMS products for the market.

BioMEMS have applications in fields such as tissue engineering, diagnostics, therapeutics, sensing, and others, which means they are potentially a powerful tool in the biomedical innovator's toolkit. However, as with the commercialization of other medtech devices, BioMEMS products must navigate and cross the commercialization "valley of death," which is encountered in bringing new products to the market. It is often said that this valley is where good ideas go to die, in other words, many translational innovations fail to be productized because they encounter hurdles, such as funding, IP strategy, regulatory or reimbursement challenges to name a few, along the "valley". In reality, there can be multiple such valleys along the translational pathway of bringing a new BioMEMS product to life. A few perspectives are given here on how BioMEMS can cross the broad commercialization valley of death and go on to become successful products and services. Understanding the user and the market for the product are vital.

12.2 A significant market opportunity

The global BioMEMS market has been estimated to reach a value of over USD $15.5 billion by 2027, with strong growth rate of approximately 12% CAGR (IndustryARC). The rapidly growing international biomedical industry and steady technical advances in platform technologies such as methods of fabrication and microfluidics have created a strong demand for BioMEMS devices. Additionally, technologies such as personalized genomics, proteomics, and pharmaco-omics are increasingly gaining acceptance as users become more familiar with them. Many of these proteomics and genomics approaches rely on the utilization of BioMEMS devices, whether on their own, or in combination with other technologies. While genomics and proteomics do not traditionally rely on micro-electro-mechanical systems, increasingly these tools are being classified as BioMEMS devices.

BioMEMS Devices. DOI: https://doi.org/10.1016/B978-0-443-24015-7.00006-1

Technologies such as personalized genomics and proteomics are putting power back in the hands of the patient, or the medical consumer, as well as their treating clinicians. The evolving trend of an empowered medical consumer indicates that demand for BioMEMS devices is likely to continue, as people become more aware about new biomedical products launched on the market, how they work and their impact, and thus start demanding certain forms of therapeutic products to better manage their health and alleviate their ailments. Today, with the rapid dissemination of information and a trend toward influencer marketing, medical consumers are more informed and sophisticated than ever, and actively engage with their biomedical devices. This means that an effective strategy for bringing a new BioMEMS device to market must factor in the needs of these key stakeholders, the technology users, and seek to meet them “where they are at,” i.e., a whole person-centered view should be taken when aiming for new BioMEMS devices to be adopted. Bringing along medical consumers as key participants in the journey to health and well-being will see medical device manufacturers find better acceptance and uptake of their products. This will uplift and facilitate the entire translational and product development pathway. This consumer-centric product-market fit effect can be seen at every stage of a device’s development, from the use of a new medtech product to its application and the intended outcome. Therefore, the role of the consumer is key.

Importantly, and as with any medtech product, a rigorous intellectual property (IP) and commercialization strategy is vital in bringing new BioMEMS products to users. This chapter will explore various IP and commercialization aspects of BioMEMS devices, with a particular application to devices being developed out of higher education settings. It is recognized that many universities have dynamic and rapidly expanding research programs focusing on BioMEMS devices. Often, despite no shortage of diverse and interesting research ideas, it can be a challenge to translate and commercialize these ideas into products and services in formats suitable for the market. These challenges in translation of research are not new, since the commercialization of new technologies has been frequently said to encounter a valley of death problem, mentioned above, where new ideas struggle to make it across the hurdles of commercialization (the "valley") to reach the other side, that is for translation into viable products and services. Nevertheless, there are specific steps that can be taken to overcome the commercialization valley of death for BioMEMS devices, and these points are considered further in this chapter. Importantly, these steps should be considered at the very outset of the translational process, and then throughout the journey. Beginning with a translational and commercialization mindset and a "commercial filter" at the outset can set up a BioMEMs project up for success.

12.3 The importance of a sound intellectual property and commercialization strategy

As with the commercialization of all medtech products and services, sound intellectual property strategy and commercialization strategies are required to advance BioMEMS projects. BioMEMS devices are unique in that they are multifaceted in nature, in terms of the underlying technology, effect, placement in patients, and the intended clinical outcomes, meaning that a range of underlying technologies and scientific paradigms typically inform these devices and guide their use. A number of different technologies and technical components can be incorporated into a single BioMEMS device. Importantly, the underlying technologies of BioMEMS devices are critical in informing their function as inventions and ultimately as innovations for the market. This is because the function, or how the device operates

Table 12.1 Features of BioMEMS devices and their technology development considerations.

Features of BioMEMS devices	Technology development considerations
Miniaturization	• What is the performance to size ratio? • Is the device efficient? • What is the manufacturing complexity?
Biocompatibility	• Does the device have long-term stability? • Is the device compatible for sterilization? • What is its allergenicity?
Sensing and detection	• Does the device have wireless conductivity? • Does the device offer multiplexed detection? • Does the device offer biomolecule detection?
Drug delivery	• Is there a controlled release mechanism? • Does the device provide targeted delivery? • Does the device provide personalized medicine?
Lab-on-a-Chip systems	• Does the device perform analytical assays? • Does the device provide high-throughput screening? • Does the device integrate cell culture systems?
Implantable devices	• Can the device carry out sensing and monitoring? • Does the device deliver therapeutics? • Is the device stable in the long term?
Point-of-care diagnostics	• Does the device detect and monitor diseases? • Does the device have a user-friendly interface? • Is the device portable and accessible?

according to its purpose, is often tied to the technical parameters of the device. These technical parameters are a key part of the BioMEMS invention.

Consider the technical features and parameters of BioMEMS devices outlined in Table 12.1 and several of their technology development considerations, such as the parameters which inform the device's function and performance. The technology development considerations will be different for different devices, and this is what shapes the effectiveness and purpose of the device.

These technological considerations will ultimately inform how the device functions and its usability.

12.4 TRIZ principles

A useful paradigm to consider when developing the technical features and parameters of a new BioMEMS device is the paradigm of TRIZ principles. TRIZ, which stands for the Russian acronym of theory of inventive problem-solving, is a consolidated system of creative and inventive problem-solving

(Gadd, 2011). TRIZ contains 40 core principles that can guide a line of enquiry into the development of a new invention.

Just like the technological considerations outlined in Table 12.1, TRIZ principles inform an innovator about the various technical parameters relating to their new BioMEMS invention. TRIZ invites the inventor to consider different technical dimensions and specifications, thus supporting the inventor to optimize each aspect of their technology. TRIZ also invites the inventor to consider which aspects of their technological invention will be practical and implementable. Optimizing the various TRIZ parameters can lead to new insights and sometimes these result in new inventions, particularly when the device performs better or achieves a new or suprising outcome, that is unlike anything that has been done before. These surprising advantages can often form the basis of new inventive concepts. TRIZ is a useful tool in the BioMEMS innovator's toolkit and emerging inventors are advised to familiarize themselves with the TRIZ system and rules. The "TRIZ 40—Your TRIZ tool" website is a good place to begin an exploration of TRIZ principles (TRIZ40, 2016).

12.5 Commercialization of new BioMEMS devices

Technology transfer, or commercialisation, is the formal process of transferring new scientific discoveries and innovations to another party for commercial development. Traditionally, universities managed this process by patenting new inventions and licensing the rights in the IP asset to industry partners. The rights may also be licensed to the researchers themselves, in the case of university spinouts.

The process of commercialisation can occur in concurrence with publishing the research, however the timeline of patenting and publication needs to be managed carefully. Additionally, along with patenting, other forms of IP asset management can be considered as well, such as trademarks, copyright and management of know-how. Today, the role of technology transfer has expanded significantly. Modern technology transfer professionals now manage a broader mandate, including corporate engagement, internal invention development, startup creation, and regional economic growth.

12.6 Principles for commercialization of IP

Briefly, the fundamental principles underlying IP commercialization are considered below. Among other key benefits, commercialization of IP is said to serve as a:

- **Form of innovation incentive**—returns on research investment, in the form of proceeds from IP commercialization, can incentivize and galvanize inventors to create new inventions, and turn the wheel of innovation, leading to a more prosperous society in the long term.
- **Vehicle for economic growth**—IP commercialization encourages new venture formation through entrepreneurship and the creation of jobs as well as the upskilling of new talent, thus improving economic well-being and societal progress.
- **Mode of attraction of investment**—by developing a sound landscape for IP commercialization, with excellent legal, business, strategy and design firms, new companies will be attracted to become established in a region, drawing in venture capital, angel investment, and amplifying grant funding activities. This can often lead to the establishment of innovation hubs or tech precincts where the collocation of organizations can turbocharge technology development. These strategic partnerships

can be leveraged to unlock further investment, capability, and capacity of research and industry organizations that will further drive the economic success of the region.
- **Method of enhancing strategic partnerships**—commercializing IP relies on the formation of strong partnerships between public–private–academic–Government partners, and thus helps to embed an ecosystem of strategic partnerships within an innovation ecosystem.

In the section below, a typical translational research and innovation pathway for a BioMEMS device developed in a university or higher education setting is considered. The new innovation begins its translational journey as a research output and then is then shaped into an IP asset, in conjunction with the work of the technology transfer office (TTO) at the university. Subsequently, this IP asset can be translated to industry (or to a university spin-out company), with the practicable goal of being developed into a product or service. Legal mechanisms such as licensing or assignments of IP may be used to enable the translation of research IP. This translational process requires adopting a "valuecreation lens" at every stage and acting swiftly to pivot the project to new applications if something has been shown along the way not to be effective. The characteristics of being agile, responsive, and open to customer feedback are key descriptors of success in research commercialization.

12.7 Translation of research into research outputs

The translation of research into tangible outputs has a number of steps. The various technical considerations are important, but very often, so are the commercial considerations.

The process of generating a sound technology begins with the act of inventing. The word 'invent' is derived from the Latin word in venire which means "to come upon" or "to discover". Therefore, new discoveries are often the basis of an invention.

Scoping these new discoveries into intellectual property assets is central to the commercialization process. Below, we explore how a new invention might be originated, and the steps in framing the invention as an intellectual property asset, working closely with a particular research organization's technology transfer office. It is important to note that each organization's technology transfer office has its nuances to their practice and what is outlined below is a high-level general overview of what the practice could look like.

12.7.1 Research and development

Each of the technology development considerations listed in Table 12.1 has particular implications on research. For example, the performance, stability, conductivity, or monitoring capability of a BioMEMS device could be new—better and different—than anything that has been done before.

By asking the right questions, the researcher is able to hone in on valuable technical insights. If there is potentially new and inventive subject matter in the BioMEMS device, it could be valuable subject matter as a research output for licensing to industry. Thus, the technical basis of a project often informs its commercial potential. A sound technology, which is validated with solid technical evidence and backed by data, is more likely to overcome the initial technical hurdles in the journey of commercialization. Commercial potential of a new BioMEMS device relies on ensuring that the device can substantially differentiate itself from other things on the market, and that there will be a quantifiable value-add in developing it further to a productized format. Therefore, the technology development considerations underlying new BioMEMS research projects are fundamentally important in an assessment of patentability and commercial potential.

12.7.2 How do technology development considerations originate?

The translational research and innovation process is itself a ripe source of technological considerations. The research innovator could have begun their research program with a foundational questions as to whether things could be done differently or better than before. They may have chosen to benchmark it against established products or services. Alternatively, it could be a fresh line of research enquiry.

The insight that it may be possible to do things better or differently is often the first step in the innovation process, where a research innovator realizes that the new and inventive aspects of their research can be applied into something useful, for the benefit and improvement of health and social well-being and the alleviation of ill health and suffering. While these considerations may begin as high-level questions, they can be refined with additional research and inquiry. Adding new parameters can help to scope new aspects of the emerging innovation. The new parameters should be defined as a set of acceptance criteria that can evolve as the research continues. Continuously setting and adapting the acceptance criteria will help to shape the research enquiry to the point where it best begins to address the user's unmet needs. As the sophistication of the research enquiry develops, the researcher unravels more new and unique insights about their project that can progress it toward a more commercially ready format. The goal of asking good questions throughout the research and innovation process is to continually clarify the technical aspects of the project, ensuring that the researcher does not become simply wedded to the solutions they are developing. It helps to add an element of stringency and a "self-checking mechanism" to the design of new technologies. The TRIZ system mentioned above is a good method of checking one's research and development framework and adds a level of stringency to product development. It helps to ensure that the technical parameters are feasible and more likely to lead to a practical outcome in the form of a product or service.

Ensuring that the technology is safe and effective is important. In asking the right questions, a research and innovation team additionally comes also closer to addressing the question of what the market wants or needs. The new BioMEMS product or service that the team is working toward, is then more likely to meet a "market fit," i.e., to meet the needs of the target end-users who will use the new solutions.

The journey toward adequately addressing the product-market fit relies on conducting detailed assessments, including an evaluation of core topics such as business case analysis, market research, regulatory assessment, clinical and manufacturing strategy, product development, and IP strategy to name a few. These topics are important in the overall assessment of commercialization of a new product or service, and they run in parallel, meaning that the research innovator has to delicately balance their attention to a range of complex matters, while prioritizing what is best for the technology.

Importantly, the analysis of a product-market fit often begins with healthcare consumers themselves, such as patients, clinicians, policymakers, peak body representatives, and others who have a clear stake in the healthcare innovation development process. A number of techniques can be used to work with these stakeholders to unlock their insights, such as focus groups, co-design and consumer-led sessions, surveys and polls, observational studies, and others. These techniques can bring about key insights from healthcare consumer groups and inform a research project team where the unmet needs lie and which are the right problems to solve. Projects that are foundationally based on the right unmet needs are more likely to be successful, as arguably they are designed to address valid gaps in the market. These gaps in the market are also potential commercial opportunities, which are open for exploration in the form of useful new solutions. It has frequently been said in the research sector that it is important to solve real and right problems rather than be a "set of solutions looking for a problem." Incorporating the voices of consumers early and into the design of BioMEMS projects will provide projects with strategic focus and help the research innovator map out their projects.

12.7.3 Practical aspects of commercialization

Having understood that commercialization of BioMEMS devices is about moving research outputs into new BioMEMS products and services that address a healthcare reality, where they will benefit the people who need them the most, it is imperative to understand the mechanism for commercialization in research-based organizations such as universities and academic medical centers.

In organizations such as universities and hospital-based academic medical centers, commercialization is usually within the remit of technology transfer offices (TTOs). These are internal functions with organizations that liaise with inventors, file IP assets such as patent applications, manage and maintain these IP assets and license them to licensee companies, or to university spin-outs. In many cases, TTOs are also responsible for innovation and entrepreneurship within the organization and play a key role in supporting the development of the translational research ecosystem.

In the case of spin-outs, some equity in the company is usually retained by the university or academic medical center. In the case of traditional out-licensing of an IP asset, there is usually a milestone payment at the time of licensing, followed by royalty payments on products or services sold. In the case of both spin-outs and traditional licenses, there are potentially returns for the inventors according to the respective organization's IP policy. These returns are usually in the form of a percentage of the commercialization revenue. Usually, in accordance with the IP policy, if there is more than one inventor, these returns are distributed on terms agreed or set into place by virtue of the policy.

12.8 Invention disclosure process

The first step in commercialization in the university setting typically begins with submission of an invention disclosure form (IDF) to the university's TTO for evaluation and a determination of the filing of a new patent application.

In some cases, other forms of IP protection may be recommended as well, including for instance to keep the IP as confidential information or know-how within the team. This is not explored here in detail, but the strategy adopted by the TTO will usually be based on a combination of factors such as, including whether there are imminent publications, the type of technology and a range of risk factors from publication. Managing IP in the form of confidential information or know-how is not without its challenges, particularly when inventors move across organizations or choose to work with multiple collaborators. Nondisclosure agreements (NDAs) may be used.

In Table 12.2, the key elements of a typical university IDF form are presented along with guidance notes to its completion. BioMEMS research innovators are advised to consider these guidance notes in

Table 12.2 Key points to consider in an IDF form.

Themes and description	Key points to consider
Theme 1: Describe the technology	• Describe the details of the specific inventive concept—at a high level what is your new technology and what does it do? How might it be developed into a product or a service? • Consider listing all the known technical features/components that are part of the inventive concept (e.g., if the inventive concept is a "chair," technical features would be "legs, base, joining component, backing, seat padding, etc.") • What are the alternative manifestations of the inventive concept, e.g., what are some different ways in which the technology/product could work? • Consider the "must-have" features and the "good-to-have features"—what is absolutely essential for the inventive concept to work? What are the features that go over-and-above the essential features? • If certain features can be swapped for other features, what are they? Why would it be possible to swap to other technical features and how?
Theme 2: What is inventive?	• What is the problem to be solved by the inventive concept? • Why is that problem not being solved by other methods or technologies? • What kind of commercial opportunity is presented by solving the problem? • How can the proposed inventive concept provide a value-add over and above what is currently in existence? In other words, what are the unique selling points of the current inventive concept? • Would this inventive concept have been obvious to the scientist in your field of activity? (e.g., another bioengineer, or another paramedic, or another confocal specialist…) • If it would have been obvious to this sort of skilled expert, is there a barrier, due to which it has not been done before? What are those barriers?
Theme 3: How might the invention be used?	• List the key uses or applications of the inventive concept. • How else might the invention be used or applied? • Is use in a different field or area possible, and practical? • How might the average users use this product/service? In other words, how could the product/service be intergrated into their daily workflow?

(continued on next page)

Table 12.2 Key points to consider in an IDF form—cont'd

Themes and description	Key points to consider
Theme 4: How is the invention better than existing technologies/products?	• Consider all the technical features and the specific advantages they bring; • Is there a benefit achieved by virtue of the invention as a whole, that separate parts could not have achieved on their own? • What is the closest competing technology? • What are some distinguishing features to the closest competing technology? • Is there an advantage in terms of efficiency, effectiveness, speed, or scale? • Does the inventive concept do something differently than the current approaches? • Is the benefit achieved as a result of a broader systems level effect? What is the systems effect?
Theme 5: What key data is there to support your invention?	• Along with your IDF submission, attach all pre-print or published manuscripts relevant to the technology and marking all the documents clearly • State that are strictly confidential and that are not • If something has been pre-published, please describe the prior publication, listing all relevant details • Attach all poster presentations, slides, exhibits, magazine extracts, book chapters, etc. • Attach details of all data and experiments that are relevant to the evaluation of this invention disclosure form. Attaching the data itself is not necessary, unless it helps to explain the concept. It is useful to provide an explanation of the data package and what the results show. • Attach details of any press releases, feature articles, newspaper, or magazine clippings that would be relevant to this inventive concept • Attach details of all relevant grant applications • Attach details of any collaboration or other contractual agreements relating to this technology
Theme 6: What's next in terms of your technical development?	• Explain the key planned experiments • Describe any work planned with collaborators and what is hoped to be achieved • Include details of any funding mechanisms that will support the next phase of this project and the steps taken to secure this funding • Explain what is missing in order to progress to the next level of prototype development and your early thoughts on how to overcome these hurdles.

detail before approaching their university's TTO for the consideration of a new patent application. The subject matter presented here will help to inform how the TTO initially assesses and ranks the project. Relevant and insightful information in the IDF form will enable the team to get the right support, as shown in Table 12.2.

Notably, there is a delicate balance of timing between a public disclosure (such as a publication, but also other disclosures) and a new patent application filing. If a publication (of any nature) is imminent, the research innovator should contact their TTO right away, since generally speaking, TTOs will not support patent filings after a public disclosure of the inventive concept. Patent applications filed after a public disclosure can run into trouble during patent prosecution on the grounds of novelty. Some territories, such as Singapore, the UK, and others, have limited grace period provisions, which are patent law exceptions as to disclosures preceding a patent filing. These grace periods are often limited in scope, and advice should be sought from the TTO whether availing this form of limited exception is possible in a given territory. As far as possible, the BioMEMS research innovator is advised to contact their TTO before a public disclosure of any nature.

Once the above information has been provided in an IDF to a TTO, they would typically begin an internal evaluation process, which is an assessment of the technology's IP situation and the commercial potential of the project and invention disclosure. Often, TTOs have internal patent review committees to advise on whether or not a patent application should be filed, based on an evaluation of the scientific and market aspects. These patent review committees have been proven to be quite helpful in supporting TTOs to manage their patent filing decisions and budgets and to support projects which are promising, with high IP and commercial potential, impact opportunity and strong teams. A patent review committee's decision does not need to be binding on the TTO, but can be influential in shaping the patent portfolio and IP strategy, as shown in Fig. 12.1.

12.8.1 Patent filing

Following submission of an invention disclosure and the technology transfer office's assessment process, based on an internal decision, a provisional patent application may be filed. This is typically a patent application which is kept secret at the patent office with none of its contents, other than key applicant and application details, made available publicly. The provisional patent application locks in a priority date for the invention, i.e., the date of filing.

Under the Paris Convention, a patent applicant typically has 12 months to file a corresponding application in member states claiming priority from the first filed patent application. The Paris Convention therefore establishes a 12-month right of priority for patent applications, while an applicant can rely on the filing date of the priority (initial) patent application. Alternatively, within the same 12-month period, an applicant can file an international application under the Patent Cooperation Treaty (PCT), preserving the priority date while postponing entry into the national phase. The PCT is a treaty which provides for international patent applications (PCT applications), which streamlines the process for seeking patent protection in multiple PCT territories. PCT applications are not international patents, and ultimately, the patents would need to be granted in respective member territories, where they undergo patent prosecution. New provisional applications can be filed, but the subject matter should not have been disclosed publicly before the new provisional's filing date. The research innovator must take care that their own disclosures do not impede upon the filing of their own patent applications.

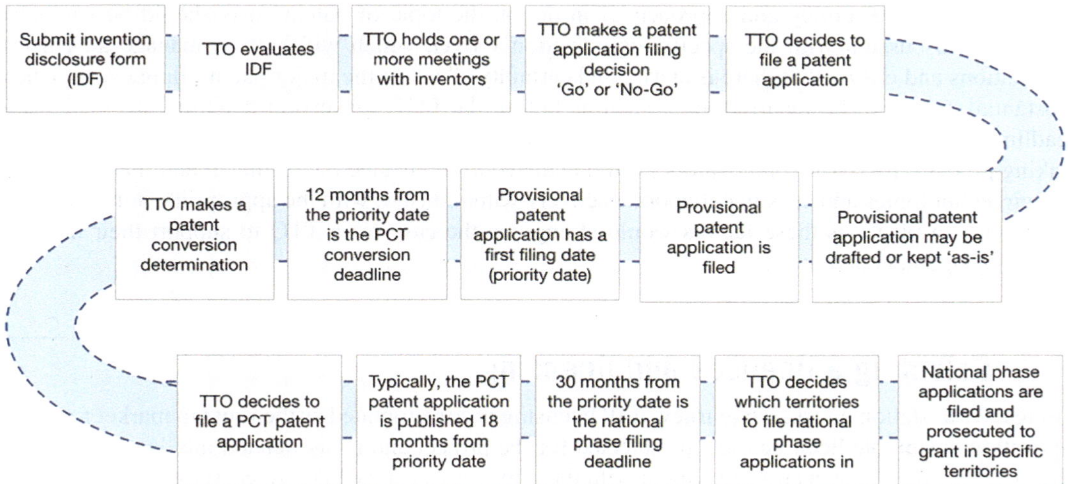

FIGURE 12.1

Journey of a new patent application from initial assessment to filing . The figure describes the journey of an invention disclosure, to a first filed provisional patent application, to a PCT patent application and ultimately a suite of national phase patent applications. The process has many steps and associated deadlines. Each phase of the patent application filing is associated with different costs and the costs may add up over the lifetime of the patent.

A PCT patent application is one filed under the Patent Cooperation Treaty (PCT). An international application, it gives patent applicants the ability to "enter" the national phase in one or more territories at the national phase deadline. While a PCT patent application is not an international patent, it is an international application, and the distinction is important.

If a PCT patent application is filed before or at the 12 month mark from the filing date, it will go on to be published at the 18 month mark, i.e., made available for the general public to access, such as on the World Intellectual Property Organisation's Patentscope database. As mentioned above, notably, a PCT application is not a patent. In order for patents to be granted in respective territories, as mentioned above, the PCT application will need to enter the national phase, meaning national phase patent applications will need to be filed in specific territories where a patent grant is being sought. Only once they have been filed in these specific territories, can the patents be prosecuted and may eventually be granted in those territories, and provide applicants with the right to exclude others from practising their invention. Sometimes, depending on the patent prosecution process, a patent might not be granted in one or more territories, due to objections issues by the respective patent offices. It is therefore important to frame a strong PCT patent application (on which the national phase filings will be based), to maximize the potential for patent grants. Working with a qualified patent attorney is a key aspect of this process, and will typically be guided by the respective TTOs.

The process of filing patent applications can involve a substantial fee, which is often initially borne by the TTO. Sometimes, TTOs pass on these costs by requiring licensee companies to bear the costs of the patent filing, before or at the time of licensing. Patent cost management is typically in line with an organization's IP policy and approach. If in doubt, the topic of patent costs should be navigated via a clear discussion with the specific organization's TTO, which will help to manage all parties' expectations and ensure that there is clarity and certainty ahead of the major patent filing events where substantial costs can be incurred, e.g., such as before the PCT conversion deadline, national phase deadline, and others. Again, a patent review committee may be called upon to support in the decision-making process. BioMEMS innovators should endeavour to advance the technical and market aspects of their technologies and seek to collaborate actively with industry, with the appropriate non-disclosure agreements in place, as these factors would strengthen the case for a TTO to support their onwards patent management activities.

12.9 Selecting a licensee and licensing

Along the translational research journey, if IP licensing is contemplated as the route to market for a new innovation, a suitable licensee must be selected for the project and its associated intellectual property. The licensee enters a licensing agreement with the university (via its TTO) to exercise the invention on agreed terms in exchange for the payment of milestone and/or royalty fees.

Selecting the right licensee can take a significant amount of time and sometimes no suitable licensees can be found. The university's business development teams can be very helpful in identifying the best prospective licensee companies and assessing their suitability for a licensing deal. One other way of maximizing the chances of finding the right licensee is to consider partnering with companies or industry partners that a researcher is already working with in their daily research. For instance, if a researcher is hosting a suite of BioMEMS projects with a specific assay provider, e.g., company ABC, it may be feasible to deploy a new innovation in the assay space to company ABC. This is particularly feasible if the new innovation uses company ABC's platform technologies and if the validation experiments are based on these platform devices. In such circumstances, integration of the licensed IP into an existing company's portfolio may be easier.

In this situation, the researcher and their TTO can begin a process of relationship-building and negotiations with company ABC to understand their appetite for a new innovation project and their interest in licensing university-developed IP. Sometimes, the IP might be jointly developed, with a time-limited option to the company to license the technology (option-to-license).

A similar process would apply to new prospective licensee companies which are not previously known to the researcher or their TTO, but which could be viable means of bringing new products and services to the market. In this case, there would be additional levels of due diligence by the TTO and business development teams to assess the suitability of the new licensee partner company, the viability and strength of their business and their long-term ability to form an enduring partnership with the university.

Once a suitable licensee company has been identified, the process of deal-making typically begins with completion of a term sheet document, where there is a "meeting of the minds" on the key terms of the licensing agreement. The term sheet is exchanged a few times between the university and the licensee company's representatives, each time adding in elements from their negotiation, and making the relevant edits to ensure both party's needs are adequately covered in these draft terms. Once the draft terms have been confirmed and a version of the term sheet has been agreed upon, the document is sent to a legal team for drafting into a full license agreement. Upon signing, this agreement is binding upon the parties on the terms they agree upon. The process is quite similar for licensing IP to a university spin-out, however in the case of the latter, the university may hold some equity in the company. For details of this process, reference should always be made to the respective university's IP policy.

Following a licensing event, the licensee company typically develops the new BioMEMS product or service to a format where it can be used by end-users. This process would also involve navigating the right regulatory pathways and seeking to have the BioMEMS device adopted by clinical champions who would advocate for the new product and support its widest uptake in the real-world. As with the beginning of the process where unmet needs and user insights were harnessed, and a new solution has been developed, testing the product to seek the inputs of consumers is key, as this will shape how the product is developed and performs once on the market.

Now that the process of commercialization has been outlined in detail, from the submission of an IDF to its evaluation, filing, selecting of a licensee and licensing, it is clear that it is an exciting process where the BioMEMS research innovator has a significant opportunity to make a difference.

At this juncture, it is opportune to consider the perspectives of several researchers in the space of commercialization of BioMEMS research.

Below, two leading researchers from Monash University in Clayton, Australia, have been interviewed for their perspectives. Dr Tuncay Alan and Dr Reza Nosrati are working on exciting BioMEMS projects with translational impact.

12.10 Case study: Dr Tuncay Alan's "PALM" technology

Dr Tuncay Alan is the lead inventor of the PALM technology and has authored over 80 publications, holds eight patents, and has secured approximately $3 million in research funding. PALM offers a smart inhaler that reimagines respiratory drug delivery through an innovative, palm-sized delivery platform. Dr Alan's research interests sit at the intersection of experimental physics and engineering. Over the past decade, his work has focused on the complex physics arising from interactions between vibrating micro- and nanoscale structures and surrounding fluid media.

He leads the Dynamic Micro Devices Laboratory at Monash University, Australia, where research integrates nanofabrication, acoustics, and advanced experimental techniques. The laboratory investigates the ejection and interaction of micrometre-scale aqueous droplets, nanoparticle synthesis, and the development of novel sensing technologies. In addition to his research leadership, Dr Alan serves as Director of Industry Engagement within the Department of Mechanical and Aerospace Engineering at Monash University.

Dr Alan was awarded a PhD in Engineering (Theoretical and Applied Mechanics) from Cornell University in the United States in 2007. Prior to joining Monash University, he undertook postdoctoral research at Delft University of Technology in the Netherlands and University College London in the United Kingdom. In 2015, he spent six months as a visiting scientist at the Paul Scherrer Institute in Switzerland.

As an Associate Professor in the Department of Mechanical Engineering at Monash University, Dr Alan's research centers on microsystems and microfluidics, with a particular emphasis on the physics emerging from interactions between fluids and nonlinear micro-mechanical oscillators. He has contributed to the development of MEMS-based technologies aimed at improving drug synthesis, inorganic nanomaterial production, and diagnostic tools. A key area of impact has been the design of microfluidic systems that enable more effective drug delivery for chronic respiratory conditions, advancing targeted therapies and patient care.

Early in his BioMEMS research journey, Dr Alan identified interdisciplinary collaboration as a critical success factor. Working alongside experts in materials science, chemistry, pharmacy, clinical medicine, molecular biology, and industry enabled the integration of diverse perspectives and methodologies. These collaborations proved essential for addressing complex challenges in drug synthesis, delivery, and diagnostics.

He also recognized the importance of continuous innovation in micro- and nanofabrication techniques. Advancing MEMS devices requires ongoing improvements in fabrication precision, functionality, and scalability to ensure translation into real-world applications. Experimental validation, iterative design processes, and early incorporation of user needs were further insights that shaped his approach to BioMEMS research and development.

Dr Alan's interest in research commercialization emerged from the broad applicability of his work across both fundamental and applied domains. Collaboration with the Monash Institute of Medical Engineering (MIME), a joint initiative bridging medicine, IT, design, and engineering, provided a platform for clinician-driven, applied research projects focused on addressing clearly defined clinical problems. These collaborations led to the development of new medical technologies, resulting in patents and design registrations and enabling the translation of research outcomes into commercially viable solutions.

The commercialization of BioMEMS devices is particularly compelling due to the rapid pace of innovation and the potential to transform healthcare. Advances in this field enable precise diagnostics, personalized and targeted drug delivery, and minimally invasive medical interventions. Commercialization is the mechanism through which these laboratory innovations reach patients and healthcare providers, delivering real-world impact.

One of the most significant commercialization projects to emerge from Dr Alan's laboratory is PALM, an award-winning aerosol drug delivery platform. PALM delivers medications to the lungs more effectively than existing solutions while enhancing both patient experience and clinical insight. Designed to fit comfortably in the palm of a user's hand and cover the mouth and nostrils, the device incorporates built-in sensors to measure a patient's breathing cycle and deliver medication at the optimal moment, eliminating coordination challenges. The particle size can also be adjusted on demand to meet individual patient needs, maximizing lung deposition, capabilities not currently available in any commercial device.

The PALM platform has already achieved key milestones, including the establishment of a reliable and scalable fabrication process for its microfluidic components, validation of particle size control through pharmacopeia-standard testing, and comprehensive IP protection.

Dr Alan emphasizes several important principles in research commercialization. Securing intellectual property protection prior to publication is essential to establishing competitive advantage, and close collaboration with university IP and patent management teams is critical. Early engagement with clinicians, patients, and industry stakeholders ensures that feedback informs development from the outset. Building strong teams to conduct market research and develop robust commercialization strategies is equally important, despite these activities often being outside a scientist's traditional skillset.

Conversely, he considers that overcomplicating technology and ignoring stakeholder feedback can significantly undermine commercialization efforts. Failure to establish strategic partnerships with industry and end users is also likely to impede successful translation. On that note, forming the right project teams is fundamental to success.

Effective teams bring together diverse expertise and ensure access to key stakeholders, including patients and clinicians. While the core PALM team remains compact, it collaborates closely with industrial designers, clinicians across multiple disease areas, biochemists, pharmacists, and engineers to support product development, device characterization, manufacturing, and optimization.

For emerging research innovators pursuing commercialization and entrepreneurship, Dr Alan advises maintaining flexibility beyond a purely academic mindset and prioritizing simplicity in technology design by incorporating industry feedback early.

Looking ahead, PALM is progressing toward commercialization, with current efforts focused on product development for clinical testing environments. Animal studies are planned to commence in the short term, alongside exploration of new partnerships with contract development and manufacturing organizations and pharmaceutical companies.

12.11 Case study: Dr Reza Nosrati's BioMEMS sperm selection device

Dr Reza Nosrati is an Associate Professor in Mechanical Engineering and the Director of the Applied Microfluidics & Bioengineering (AMB) Lab at Monash University. He is widely regarded as a pioneer in the application of microfluidics to fertility and assisted reproduction, with his research playing a defining role in shaping the field of microfluidics for fertility. His work has opened new opportunities for sperm analysis and selection approaches that more closely replicate in vivo processes.

Dr Nosrati began his academic journey by completing his bachelor's and master's degrees in mechanical engineering in Iran before moving to Canada to undertake his PhD at the University of Toronto under the supervision of Professor David Sinton. From early on, he was drawn to engineering challenges that extended beyond traditional disciplinary boundaries, particularly those at the intersection of engineering, cell biology, and medicine. His doctoral research focused on developing novel engineering tools for sperm analysis and selection, an endeavor that initially required him to rapidly acquire biological and medical knowledge. This interdisciplinary immersion ultimately positioned him to bridge engineering and biomedical science, enabling the development of innovative and clinically relevant solutions.

His research interests center on microfluidic and BioMEMS technologies for biomedical applications, with a particular focus on infertility treatment. His work uses microfluidics to recreate physiologically relevant in vitro environments, leading to new insights into reproduction processes in vivo and the development of novel technologies for sperm analysis and selection. Dr Nosrati believes the field is only just beginning to make a meaningful impact, not only through high-quality academic publications but also through the translation of technologies that improve clinical outcomes and directly benefit patients.

Reflecting on his early research journey in BioMEMS, Dr Nosrati highlights several key insights. These include recognizing the transformative potential of MEMS and microfluidic devices in biomedical applications, developing a deep understanding of biological and medical challenges, and dedicating substantial time to reviewing existing literature. He emphasizes that these steps are critical for identifying knowledge gaps, support needs, and collaborative opportunities. He also underscores the importance of mentorship, noting that Professor David Sinton played a pivotal role in guiding his development and continues to be a source of inspiration.

Dr Nosrati's interest in research commercialization emerged during his PhD, when he developed his first patented technology, a radial microfluidic device for sperm selection. The potential for this technology to significantly improve fertility treatment practices sparked his curiosity about how academic innovations could be translated into real-world clinical solutions.

What excites him most about the commercialization of BioMEMS devices is their potential to transform patient journeys and improve quality of life. At the same time, he acknowledges that commercialization within a university environment is challenging, requiring sustained effort and resilience, often without immediate reward. During difficult periods, his motivation remains grounded in the long-term impact these technologies could have if successfully brought to market.

Among his commercialization efforts, Dr Nosrati and his team are advancing a portfolio of fertility-focused technologies, including next-generation sperm selection devices designed to outperform existing methods. These devices aim to deliver higher-quality cells more efficiently while automating processes that are currently manual and subjective. In addition, his group has developed the world's first noninvasive technology to enhance sperm motility using high-frequency ultrasound. This approach has the potential to revive sperm motility, expand infertility treatment options, and assist in detecting testicular sperm in surgical samples, a process that remains clinically challenging. Collectively, these technologies aim to standardize and automate assisted reproduction practices, improving outcomes while reducing the socioeconomic burden on patients.

Drawing on his experience, Dr Nosrati offers several practical insights into research commercialization. He stresses the importance of building a strong, multidisciplinary team and aligning research activities early with proof-of-concept data required for translation. He notes that academics often design projects based on scientific passion rather than development pathways or business planning and that addressing this mismatch early can significantly accelerate progress. He also emphasizes careful commercialization planning, cultivating trusted collaborators, and clearly articulating the value proposition for both investors and end-users. Conversely, he cautions against underestimating regulatory requirements, delaying proactive engagement in commercialization, or neglecting continuous stakeholder feedback throughout the process.

According to Dr Nosrati, forming the right project team is one of the most critical determinants of success in biomedical device commercialization. Beyond access to equipment and infrastructure, the team must collectively possess technical expertise, regulatory knowledge, market insight, and the ability to execute effectively. Strong teams are characterized by trust, clear communication, defined accountability, adaptability, and a shared focus on delivering outcomes.

For emerging research innovators, Dr Nosrati encourages engagement with entrepreneurship and business training, which is increasingly embedded within modern biomedical and engineering programs. He views these skills as essential tools for navigating the complexities of commercialization and translating research into impact.

Looking ahead, Dr Nosrati is enthusiastic about the next phase of his commercialization journey. His focus is on scaling existing technologies, expanding applications across broader areas of reproductive health, and forming new partnerships to accelerate market entry. Ultimately, he hopes to see his BioMEMS innovations delivering tangible benefits, technologies that improve human health and quality of life, and contribute to a better future.

12.12 Chapter summary

In summary, the development and commercialization of BioMEMS devices can be immensely beneficial for research innovators and pave the start of an exciting journey in bringing new products and services to life. Intellectual property protection, management, and commercialization are fundamentally important in this process, as is the formation of strong teams and partnerships. The right partnerships can be vital in unlocking success in the translational journey. BioMEMS devices are bringing benefits across a range of applications, particularly healthcare, and it will be interesting to track this industry over time, as research innovators develop and scale their exciting projects in this space.

The author is a project manager at the Monash Institute of Medical Engineering (MIME) at Monash University in Clayton, Melbourne, Australia. In previous roles, she has worked at a technology transfer office in the healthcare space and as a trainee patent attorney in legal services. At MIME, she is leading a commercialization acceleration initiative and a range of other associated externally facing programs.

AI disclosure

During the preparation of this work, the author(s) used ChatGPT in order to prepare Table 12.1— to prepare content in line with the technical requirements of the chapter. The author also used ChatGPT to prepare the case study summaries. After using this tool/service, the author(s) reviewed and edited the content as needed and take(s) full responsibility for the content of the publication.

References

Gadd, K. (2011). *TRIZ for Engineers: Enabling Inventive Problem Solving*. John Wiley & Sons, Ltd https://onlinelibrary.wiley.com/doi/book/10.1002/9780470684320.

TRIZ40, Solve a Technical Problem, 2016. https://www.triz40.com/TRIZ_GB.php.

CHAPTER

Emerging technologies and future outlooks 13

Azrul Azlan Hamzah[a], Anis Nurashikin Nordin[b] and Jumril Yunas[a]
[a]*Institute of Microengineering and Nanoelectronics (IMEN), Universiti Kebangsaan Malaysia, Bangi, Selangor, Malaysia,* [b]*Department of Electrical and Computer Engineering, International Islamic University Malaysia, Kuala Lumpur, Malaysia*

Fine-tuning cellular machinery using biological microelectromechanical systems (BioMEMS) constitute an emerging set of technologies at the crossroads between miniaturization and microsystems integration, which strives to minimize bodily intervention. As the world evolves, so to do cutting-edge developments transforming healthcare methods as well as diagnostics and medicines. BioMEMS technology is leading to new approaches in bioanalysis and medical care from lab-on-a-chip platforms, implantable sensors for patients suffering from chronic diseases or cancer, e.g. glucose monitors used by diabetic people on a daily basis; drug delivery systems with programmed release rates (i.e., jet injection).

Recent innovations in nanotechnology, microfabrication and materials science have opened the door for development of more precise, reliable devices with greater BioMEMS functionality These improvements facilitate the development of rapid response diagnostics, bedside testing and even targeted therapies making a great leap forward in healthcare resources from an operational (time) point-of-view to the perspective centered on patient needs. In addition, artificial intelligence and machine learning coupled with data analytics of BioMEMS are leading to systems that can make smarter decisions in less time, less sample usage as well accurate results for medical diagnostics.

All of this pushes the frontier and opens up a wide range of possibilities for what BioMEMS can do as new technologies such as nanofabrication, 3D bioprinting or wearable biosensors begin to expand. The aim of these innovations are to enhance sensitivity and functionality of existing systems, as well expand the applicability into regenerative medicine, tissue engineering) or remote healthcare monitoring.

In this chapter, we will look at cutting-edge innovations in BioMEMS technology and how they might be used to address contemporary biological concerns. We will also address the future of BioMEMS, focusing on major themes that are expected to drive innovation in healthcare, diagnostics, and therapeutic devices. Understanding these growing patterns allows academics, engineers, and healthcare professionals to better predict and contribute to the next generation of BioMEMS devices, which will influence the future of medical science.

13.1 Nanotechnology and its integration in BioMEMS

Nanotechnology has arisen as a revolutionary force in a variety of sectors, most notably in biomedicine, where it has been integrated with microelectromechanical systems (MEMS) to create BioMEMS. These

BioMEMS Devices. DOI: https://doi.org/10.1016/B978-0-443-24015-7.00010-3

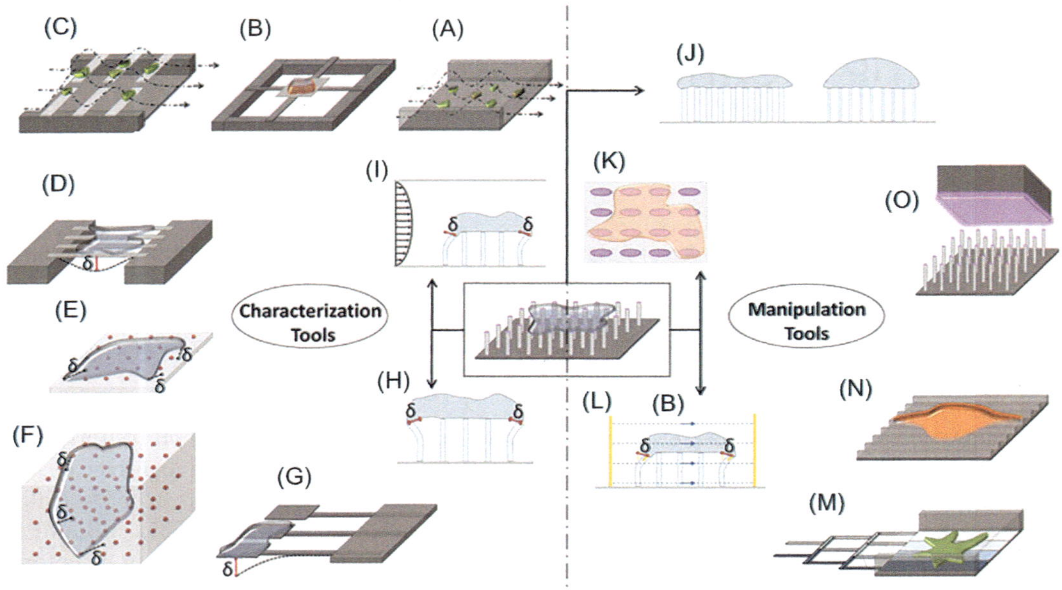

FIGURE 13.1

BioMEMS devices in cell mechanics. The tools for cells are categorized into two types: characterization tools, measuring physical properties, and manipulation tools, applying external effects. Techniques include microfluidics, cantilevers, micropillars, nanoribbons, and magnetic nanowires.

From Unal, M., Alapan, Y., Jia, H., Varga, A. G., Angelino, K., Aslan, M., Sayin, I., Han, C., Jiang, Y., Zhang, Z., & Gurkan, U. A. (2014). Micro and nano-scale technologies for cell mechanics. Nanobiomedicine, 1. *https://doi.org/10.5772/59379.*

systems take advantage of nanoparticles' unique features to improve the functioning and efficiency of biomedical devices, allowing for advances in diagnostics, drug administration, and therapeutic applications.

The junction of nanotechnology and BioMEMS is defined by the miniaturization of devices capable of performing complicated biological activities. Polysilicon–chromium–gold intracellular chips have potential for multifunctional biomedical applications. Understanding the interaction between nanostructures and biological systems is crucial for improving their safety and effectiveness (Patiño et al., 2016). This is critical because the physicochemical characteristics of nanomaterials may considerably impact biological responses, which are an important concern in the design of BioMEMS (Dahlan et al., 2022). BioMEMS technologies are very useful for investigating cell mechanics and dynamics. Recent advances have enabled researchers to develop platforms for analyzing cellular behavior at new sizes. This includes cancer cell mechanics and organ-on-chip systems as shown in Fig. 13.1 (Unal et al., 2014). The capacity to modify and view cells in real time with BioMEMS has far-reaching implications for understanding disease causes and designing targeted therapeutics.

Furthermore, the importance of nanomaterials in the production of BioMEMS cannot be denied. The use of graphene, carbon nanotubes, and magnetic nanoparticles has resulted in the creation of highly sensitive biosensors and drug delivery devices (Dahlan et al., 2022). These materials not only improve the performance of BioMEMS, but they also aid in the miniaturization and integration of various capabilities into one device. Magnetic films and nanoparticles, for example, have been used in micropump systems to regulate drug delivery, demonstrating nanomaterials' potential to transform treatment techniques.

BioMEMS' adaptability extends to a wide range of biological applications, including implanted devices and smart stents. These devices are intended to interact smoothly with biological surroundings, and their biocompatibility is frequently improved using sophisticated surface engineering approaches (Barkam et al., 2013). The use of nanotechnology into these systems improves performance and patient outcomes since they may be adjusted to specific therapeutic requirements. In addition to medication delivery, BioMEMS are being investigated for their potential use in diagnostics and monitoring. The development of microfluidic devices using nanomaterials offers fine control over fluid dynamics at the microscale, allowing for speedy and reliable diagnostic testing (Kim et al., 2015). This feature is especially useful in point-of-care situations, where prompt diagnosis can have a substantial influence on patient treatment.

The integration of nanotechnology into BioMEMS represents a significant advancement in biomedical engineering. Nano materialization is an effective way of developing solutions to modern problems in healthcare, exploiting the exceptional properties that arise at atomic scales. Continuous investigation into the applications of these technologies expect to improve diagnostics, as well as drug delivery and therapeutic adventures that would subsequently translate into better patient results.

13.2 Innovations in microfabrication and 3D bioprinting for BioMEMS

Microfabrication coupled with 3D bioprinting has revolutionized the development of BioMEMS devices, which have unrivaled capability in fabricating biostructures at the level as small as several micrometers for biomedical applications. These advances allow for the fabrication of intricate, biocompatible architectures that are essential for applications as diverse as drug delivery and tissue engineering. These new advances continue to expand on the potential for next-generation BioMEMS healthcare in highly individual patient specific medical solutions that test what we believe these technologies are capable of beyond simple therapies.

Microfabrication and 3D bioprinting innovations are critical in increasing the capabilities of BioMEMS, allowing the development of complex devices that can interact with biological systems on a micro scale. The combination of these technologies has resulted in the development of devices that are not only functional but also customized to fulfil specific biological requirements. Traditional microfabrication techniques, including as soft lithography and photolithography, have been supplemented by 3D printing, which enables quick prototyping of complicated geometries and structures that are frequently unfeasible using traditional methods. For example, a new production method called 3D Printing, Ink Casting, and Micromachined Lamination (3D PICLμM) combines numerous processes to make biological microdevices with improved functionality and precision (Kundu et al., 2018). This method exemplifies how the convergence of different fabrication strategies can yield devices that are more versatile and accessible to researchers and clinicians alike.

The flexibility of 3D printing is most obvious in its use to manufacture patient-specific devices. For example, demonstrated the use of fugitive glue to construct extremely flexible and biocompatible substrates, which may be employed in many BioMEMS applications, including biosensors and drug delivery systems (Li et al., 2015). This capacity to customize devices based on specific patient anatomy is a big step forward in personalized medicine, enabling individualized treatment methods that enhance patient outcomes. Furthermore, the use of 3D printing methods to create magneto-responsive devices has opened up new opportunities for developing smart BioMEMS that can respond dynamically to external stimuli. investigated the creation of silicone soft structures via controlled magneto-capillary reconfiguration, demonstrating how such devices can imitate the behavior of natural biomaterials (Roh et al., 2019).

Furthermore, the materials employed in 3D bioprinting affect the performance and biocompatibility of BioMEMS devices. Advances in polymer science have resulted in the development of bio-based polymers with desirable characteristics for biomedical applications. For example, (Garcia-Ramirez et al., 2021) emphasized the importance of polymer-based bio microelectromechanical systems in animal cell investigations, citing their flexibility, transparency, and simplicity of construction. These traits are required for developing devices that may successfully interact with biological systems while retaining structural integrity and functioning. Furthermore, the use of sophisticated materials such as shape memory polymers in 3D printing enables the construction of devices that can alter shape or function in response to environmental stimuli, increasing the flexibility of BioMEMS (Wei et al., 2017).

13.3 Wearable and implantable BioMEMS: Next-generation health monitoring

Wearable and implantable BioMEMS are the next step in health monitoring, providing continuous, real-time tracking of vital signs and physiological data with incredible precision. These gadgets, which include microsensors and actuators, integrate seamlessly with the human body, offering vital data for personalized therapy and early illness diagnosis. Materials, wireless communication, and power efficiency advancements have accelerated the development of these BioMEMS devices, which are revolutionizing healthcare by enabling non-invasive, long-term monitoring. Wearable and implantable BioMEMS have the potential to transform how medical data is collected, analyzed, and utilized in patient care as technology advances.

The development of wearable and implanted BioMEMS technology represents a substantial shift in health monitoring paradigms, allowing for continuous and real-time surveillance of physiological data. These gadgets use modern sensor technologies to deliver non-invasive, personalized healthcare solutions that are especially useful for controlling chronic conditions and improving overall health outcomes. BioMEMS integration into wearable and implantable devices has the potential to transform health monitoring by enabling early detection, rapid intervention, and increased patient participation.

Wearable gadgets have gained popularity due to their ability to monitor vital signs continuously and discreetly. Recent studies have directed their attention to the evolution of these aspects, underlining that they play an essential role in personalized healthcare. Reflecting the importance of wearable sensors to identify key markers for non-invasive health monitoring, in particular with an ageing demographic (Guk et al., 2019). Speed might be good, but the researchers encourage you to take your time and delve into exponential changes in wearable health devices, now including smart textiles and electronic clothes that extends monitoring of health. These improvements point toward a future filled with health monitoring symbiotically paired devices being far more user-friendly as well integrated in our daily lives.

The Internet of Things (IoT) has expanded the capabilities of wearable devices, allowing for seamless data collecting and processing. highlights that IoT-enabled wearable devices can provide real-time health monitoring, which is critical for handling public health crises like infectious disease epidemics (Sivaraman, 2019). This interconnection enables not only the collection of health data but also to control the delivery of drug dosage, which may be used for predictive analytics and personalized health interventions. Remote monitoring improves patient adherence to treatment regimens and overall health management (Loncar-Turukalo et al., 2019).

In addition to wearables, implantable BioMEMS devices are becoming important instruments in health monitoring. These sensors can continuously monitor physiological data, such as heart rate and glucose levels, from within the body. For example, highlight the relevance of flexible and wearable biosensors in real-time health monitoring, which can greatly lessen the pressure on centralized healthcare services (Song et al., 2023). The development of such technologies is especially important in managing chronic illnesses, because timely data delivery and management of the patients may lead to improved health outcomes and fewer hospital visits.

The use of BioMEMS into health monitoring devices answers the need for personalized healthcare solutions. Emphasizes that personalized health monitoring is becoming more reliant on wearable sensors that can adapt to particular patient demands, improving the efficacy of health interventions (Du, 2020). This personalization is critical for increasing patient engagement and adherence, since users are more inclined to use gadgets that address their individual health needs.

The future of wearable and implantable BioMEMS technology seems promising, with continuous research aimed at improving functionality and user experience. Materials science and microfabrication advancements are opening the way for the creation of increasingly complex and adaptable gadgets that can easily integrate into daily life. As these technologies advance, they have the ability to shift health monitoring from reactive to proactive, allowing individuals to take greater control of their health than ever before.

13.4 Future trends in BioMEMS: Artificial intelligence, remote monitoring, and beyond

The integration of Artificial Intelligence (AI) and remote monitoring into BioMEMS is poised to transform healthcare delivery. AI-driven BioMEMS devices offer enhanced data analysis, enabling real-time, precise diagnostics, and personalized treatment options. Additionally, remote monitoring through wearable and implantable BioMEMS technologies allows for continuous, non-invasive tracking of physiological parameters, improving patient outcomes and reducing healthcare costs. As AI algorithms evolve, coupled with advancements in biosensing and communication technologies, BioMEMS will become smarter and more autonomous, ushering in a new era of predictive medicine and proactive healthcare management. This chapter explores these future trends and their potential impact.

AI is emerging as a critical component in the advancement of BioMEMS technology. BioMEMS devices may use machine learning algorithms to analyze massive volumes of sensor data and deliver actionable insights and predictive analytics. For example, as shown in Fig. 13.2 (e.g., DNA microarrays,

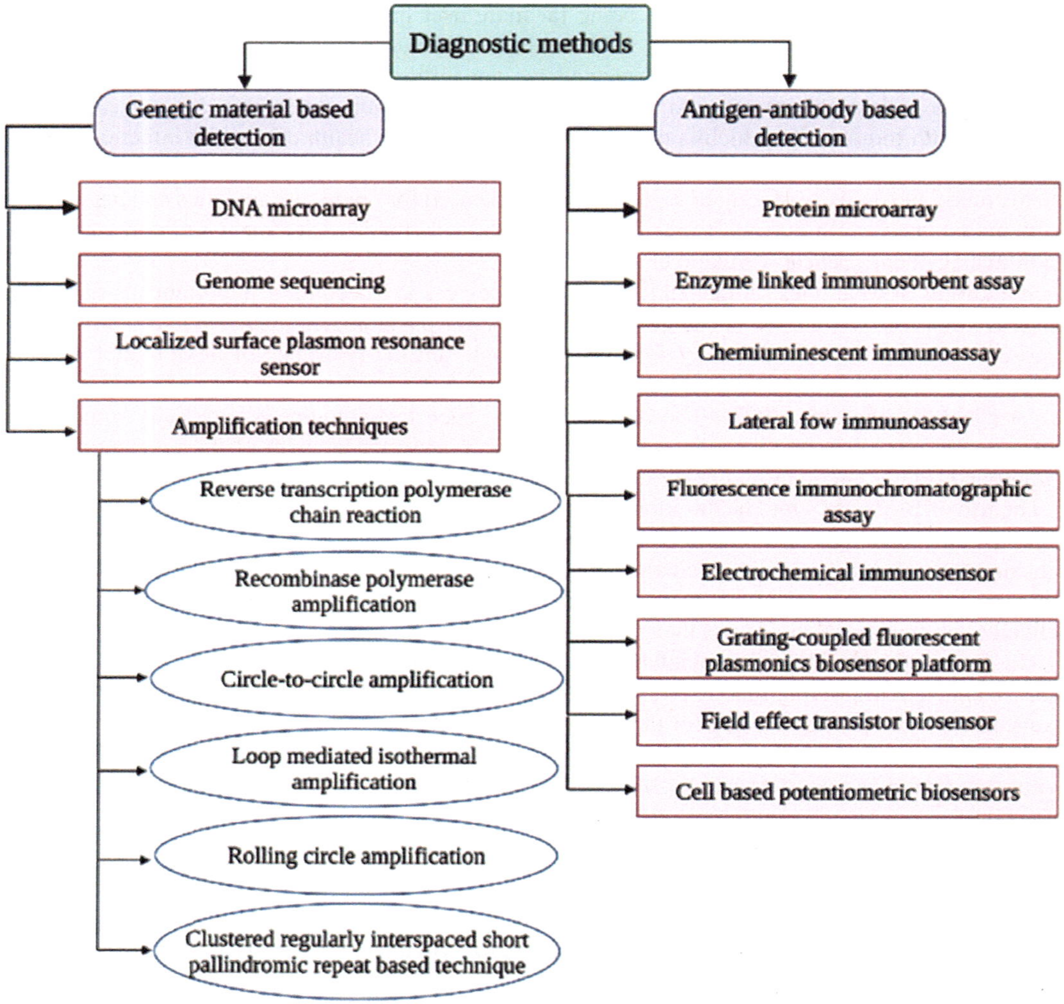

FIGURE 13.2

A schematic illustration of the numerous COVID-19 diagnostic procedures.

From Preethi, M., Roy, L., Lahkar, S., & Borse, V. (2022). Outlook of various diagnostics and nanodiagnostic techniques for COVID-19. Biosensors and Bioelectronics: X, 12. *https://doi.org/10.1016/j.biosx.2022.100276.*

genome sequencing) and antigen-antibody-based detection (e.g., protein microarrays, immunosensors) generate significant data, which can be combined with AI to create intelligent healthcare solutions. AI's ability to recognize patterns in these large datasets enables early diagnosis and timely interventions, enhancing diagnostic accuracy and patient outcomes (Preethi et al., 2022).

Remote monitoring is another important development influencing the future of BioMEMS. The capacity to continually monitor patients' health parameters from the comfort of their own homes is becoming more practical because to improvements in wireless communication technology and IoT integration. show an innovative strategy that uses direct electromagnetic connection to monitor physiological changes without the need of intrusive sensors (Labus et al., 2019). This approach demonstrates how BioMEMS may evolve to give real-time health monitoring while maintaining patient comfort.

The growing popularity of remote monitoring technology is especially significant in the context of chronic illness management. BioMEMS devices may be developed to continually monitor vital indicators such as heart rate, glucose levels, and blood pressure, giving healthcare doctors useful information to adapt treatment approaches. emphasize that the multidisciplinary character of BioMEMS enables the creation of devices that may be incorporated into daily life, supporting proactive health management. This change from reactive to proactive healthcare is critical for improving patient outcomes while lowering healthcare expenses.

As BioMEMS technology progress, ethical concerns about data security and patient privacy will become more essential. addresses the importance of strong security measures in IoT connection, especially when BioMEMS devices generate sensitive health data (Jaime et al., 2023). Ensuring the security and integrity of patient information is critical as these technologies become increasingly common in healthcare settings. Future advances in BioMEMS must prioritize ethical frameworks that safeguard patient autonomy while encouraging innovation.

13.5 Innovation in microfluidics for the future health care system

Microfluidics that refers to the manipulation and control of fluids within channels that typically range inn size from tens to hundreds of micrometers enables the precise handling of extremely small fluid volumes, making it highly suitable for applications in drug delivery, artificial organs, and implantable therapeutic devices. Microfluidic drug delivery systems often incorporate components such as micropumps, microvalves, reservoirs, and sensors, which together form closed-loop therapeutic platforms. These integrated systems represent a core aspect of modern BioMEMS technologies, allowing the combination of sensing, actuation, and control functions in compact and biocompatible devices. In future healthcare systems, microfluidic drug delivery platforms are expected to provide highly controlled and targeted administration of therapeutic agents. By regulating flow rates at the microliter or nanoliter scale, these systems can deliver drugs with high accuracy, reduce side effects, and improve the overall treatment efficiency. These capabilities are particularly important for long-term therapies that require precise dosing. Typical applications include continuous insulin delivery systems for artificial pancreas devices, implantable chemotherapy pumps, and hormone or pain management infusion systems. Microfluidics also plays a critical role in the development of artificial organs that aim to replicate the functions of biological organs using engineered systems. Many organs, such as the kidneys, pancreas, and liver, rely on fluid transport, filtration, and biochemical exchange at the microscale. Microfluidic technologies can mimic these physiological processes by enabling controlled fluid flows, selective filtration, and biochemical interactions. For example, microfluidic channels and membranes can simulate natural blood circulation, support nutrient and drug transport, and facilitate waste removal from the body. These capabilities are essential for artificial kidney and liver systems, as well as organ-on-chip platforms that replicate cellular microenvironments. Furthermore, the integration of sensors and actuators within microfluidic devices allows real-time monitoring and adjustment of organ functions, thereby supporting the development of closed-loop therapeutic systems. Overall, future microfluidic technologies for drug delivery and artificial organs are expected to enable fully implantable closed-loop therapeutic devices. These systems will likely incorporate advanced control algorithms, AI-based decision-making, and biocompatible materials to support personalized and precision medicine.

13.6 Chapter summary and future outlooks

MEMS (BioMEMS) technology is rapidly advancing in the biomedical field, representing one of the most dynamic areas in medical innovation. In the future, wearable MEMS biosensors are expected to enable ongoing, individualized health monitoring. This will be facilitated by advancements in more adaptable and stretchable sensors designed to monitor sweat, interstitial fluid, movement, ECG, and other parameters. Implantable MEMS are expected to feature smaller, biocompatible sensors with on-device actuation capabilities for drug delivery or stimulation, paving the way for closed-loop therapies to treat conditions such as diabetes, Parkinson's disease, kidney issues, and heart failure. Additionally, microfabricated chips that mimic organ micro-environments, known as organ-on-a-chip, along with microfluidic MEMS, are set to reduce animal testing, accelerate candidate screening, and facilitate patient-specific models for tumors, liver, and heart conditions. This trend is likely to spur the development of multi-organ "body-on-chip" platforms and their integration with high-throughput assays. The incorporation of AI and IoT in sensor signal conditioning and data processing will enable BioMEMS devices to provide clinically relevant metrics rather than raw data, enhancing diagnostics, anomaly detection, and minimizing bandwidth and privacy concerns. Regulatory and data security considerations will become increasingly significant. Furthermore, advanced materials such as conductive polymers, nanocomposites, hydrogels, and additive manufacturing techniques such as 3D printing are expected to broaden the range of useful form factors, particularly for point-of-care testing, wearable patches, and custom implants. Despite the vast potential for BioMEMS technology development, challenges such as regulatory hurdles (clinical validation and approval timelines), reproducibility, scalability, and data security remain significant obstacles to be addressed.

References

Barkam, S., Saraf, S., & Seal, S. (2013). Fabricated micro-nano devices for in vivo and in vitro biomedical applications. *Wiley Interdisciplinary Reviews: Nanomedicine and Nanobiotechnology, 5*(6), 544–568. https://doi.org/10.1002/wnan.1236.

Dahlan, N. A., Thiha, A., Ibrahim, F., Milić, L., Muniandy, S., Jamaluddin, N. F., Petrović, B., Kojić, S., & Stojanović, G. M. (2022). Role of nanomaterials in the fabrication of bioNEMS/MEMS for biomedical applications and towards pioneering food waste utilisation. *Nanomaterials, 12*(22). https://doi.org/10.3390/nano12224025.

Du, X. (2020). Roll-to-roll printing of wearable sensors and devices for personalized health monitoring. *American Journal of Biomedical Science & Research, 7*, 495–498. https://doi.org/10.34297/AJBSR.2020.07.001208.

Garcia-Ramirez, R., Cerda-Kipper, A. S., Alvarez, D., Reveles-Huizar, S., Garza-Abdala, J. A., & Hosseini, S. (2021). Latest updates on the advancement of polymer-based biomicroelectromechanical systems for animal cell studies. *Advances in Polymer Technology, 2021*. https://doi.org/10.1155/2021/8816564.

Guk, K., Han, G., Lim, J., Jeong, K., Kang, T., Lim, E.-K., & Jung, J. (2019). Evolution of wearable devices with real-time disease monitoring for personalized healthcare. *Nanomaterials, 9*(6). https://doi.org/10.3390/nano9060813.

Jaime, F. J., Muñoz, A., Rodríguez-Gómez, F., & Jerez-Calero, A. (2023). Strengthening privacy and data security in biomedical microelectromechanical systems by IoT communication security and protection in smart healthcare. *Sensors (Basel, Switzerland), 23*(21). https://doi.org/10.3390/s23218944.

Kim, J., Kang, K.-N., Jin, Y., Goettert, J., & Ajmera, P. K. (2015). Hydrodynamic focusing micropump module with PDMS/nickel-particle composite diaphragms for microfluidic systems. *Microsystem Technologies, 21*(1), 65–73. https://doi.org/10.1007/s00542-014-2089-8.

Kundu, A., Ausaf, T., & Rajaraman, S. (2018). 3D printing, ink casting and micromachined lamination (3D PICLμM): A makerspace approach to the fabrication of biological microdevices. *Micromachines, 9*(2). https://doi.org/10.3390/mi9020085.

Labus, K. M., Sutherland, C., Notaros, B. M., Ilić, M. M., Chaus, G., Keiser, D., & Puttlitz, C. M. (2019). Direct electromagnetic coupling for non-invasive measurements of stability in simulated fracture healing. *Journal of Orthopaedic Research, 37*(5), 1164–1171. https://doi.org/10.1002/jor.24275.

Li, W., Lucioni, T., Guo, X., Smelser, A., & Guthold, M. (2015). Highly stretchable, biocompatible, striated substrate made from fugitive glue. *Materials, 8*(6), 3508–3518. https://doi.org/10.3390/ma8063508.

Loncar-Turukalo, T., Zdravevski, E., Machado da Silva, J. é, Chouvarda, I., & Trajkovik, V. (2019). Literature on wearable technology for connected health: Scoping review of research trends, advances, and barriers. *Journal of Medical Internet Research, 21*(9). https://doi.org/10.2196/14017.

Patiño, T., Soriano, J., Amirthalingam, E., Durán, S., González-Campo, A., Duch, M., Ibáñez, E., Barrios, L., Antonio Plaza, J., Pérez-García, L., & Nogués, C. (2016). Polysilicon-chromium-gold intracellular chips for multifunctional biomedical applications. *Nanoscale, 8*(16), 8773–8783. https://doi.org/10.1039/C5NR09022A.

Preethi, M., Roy, L., Lahkar, S., & Borse, V. (2022). Outlook of various diagnostics and nanodiagnostic techniques for COVID-19. *Biosensors and Bioelectronics: X, 12*. https://doi.org/10.1016/j.biosx.2022.100276.

Roh, S., Okello, L. B., Golbasi, N., Hankwitz, J. P., Liu, J. A.-C., Tracy, J. B., & Velev, O. D. (2019). 3D-printed silicone soft architectures with programmed magneto-capillary reconfiguration. *Advanced Materials Technologies, 4*(4). https://doi.org/10.1002/admt.201800528.

Sivaraman, H. (2019). IoT-enabled healthcare monitoring: A systematic review of wearable devices. *Information Technology in Industry, 7*, 78–86. https://doi.org/10.17762/itii.v7i3.815.

Song, Z., Zhou, S., Qin, Y., Xia, X., Sun, Y., Han, G., Shu, T., Hu, L., & Zhang, Q. (2023). Flexible and wearable biosensors for monitoring health conditions. *Biosensors, 13*(6). https://doi.org/10.3390/bios13060630.

Unal, M., Alapan, Y., Jia, H., Varga, A. G., Angelino, K., Aslan, M., Sayin, I., Han, C., Jiang, Y., Zhang, Z., & Gurkan, U. A. (2014). *Micro and Nano-Scale Technologies for Cell Mechanics. Nanobiomedicine., 1*. https://doi.org/10.5772/59379.

Wei, H., Zhang, Q., Yao, Y., Liu, L., Liu, Y., & Leng, J. (2017). Direct-write fabrication of 4D active shape-changing structures based on a shape memory polymer and its nanocomposite. *ACS Applied Materials and Interfaces, 9*(1), 876–883. https://doi.org/10.1021/acsami.6b12824.

Index

Page numbers followed by "*f*" and "*t*" indicate, figures and tables respectively.

N

O

9780443240157